6th Edition Revised & Enlarged

Topics & Problems added to Appendix

ENGINEERING MATHEMATICS

Volume - III

for the students of 3rd and 4th Semesters,

B.Tech Course as per WBUT

M 302 [*ECE, EE, EEE, AEIE, ICE, BME*]

M 402 [*CE, ME / PE / AUE, CHE*]

By

B.K. Pal & K. Das

By the same authors

➲ **ENGINEERING MATHEMATICS**
Volume I - IV

➲ **BBA MATHEMATICS**
Volume I & II

➲ **BCA MATHEMATICS**
Volume I - IV

By K. Das

➲ **NUMERICAL METHODS**

By B. K. Pal

➲ **DIPLOMA ENGINEERING MATHEMATICS**
Volume I & II

6th Edition Revised & Enlarged

Topics & Problems added to Appendix

ENGINEERING MATHEMATICS

Volume - III

***for* 3rd and 4th Semesters of B.Tech (WBUT)**

M 302 [*ECE, EE, EEE, AEIE, ICE, BME*]

M 402 [*CE, ME / PE / AUE, CHE*]

B. K. PAL, *M.Sc., Ph.D.*

(*P. N. Banerjee Gold Medalist & Recipient of University Gold Medal*)

Head of the Department of Mathematics

Maulana Azad College, Kolkata

Formerly, Reader in Mathematics

Kalyani Government Engineering College, Nadia, West Bengal

K. DAS, *M.Sc., B.Ed., Ph.D.*

Assistant Professor and Head of the Department of Mathematics

Kalyani Government Engineering College, Nadia, West Bengal

U. N. DHUR & SONS PRIVATE LIMITED

KOLKATA - 700 073

Published by :

Dr. Purnendu Dhar, M.Sc. (Chem.), Ph.D. *on behalf of*

M/s. U. N. DHUR & SONS PRIVATE LIMITED

2A, Bhawani Dutta Lane, Kolkata - 700 073

Phone : (033) 3292 3854 / 2241-9573

Mobile : 92300 83459 / 94330 17104

ISBN-978-93-80673-79-0

First	Edition	2005
Second	Edition	2006
Reprint		2007
Third	Edition	2008
Fourth	Edition	2009
Fifth	Edition	2011
Reprint		2012
Sixth	Edition	2013

***Price :* Rupees Three hundred only**

Printed by :

Shreedurga Press

73A, Amhrst Row, Kolkata - 700 009

Preface to the 6th edition

This edition is thoroughly revised, and re-arranged according the ***latest syllabus of W.B.U.T, 2013*** *with the addition of various topics and problems in the* ***APPENDIX divided into four separate Modules*** *as per the*

Subject Code : M 302 & M 402.

The ***complete Solution of up-to-date W.B.U.Tech Question Papers*** *are also included at the end of the book.*

We do hope that, this edition will be indispensible for the for the students of

Subject Code : **M 302 (ECE, EE, AEIE, ICE, BME)**
M 402 (ME, PE, CHE)

to cover the precribed syllabus in time and will ***secure high marks in the examination.***

We express our heartiest thanks to all the Professors and Teachers for referring our books to the students.

All the suggestions for improvement of the book will be accecpted with thanks.

We are thankful to our publisher for bringing out this edition in time.

Kolkata — **B. K. Pal**
June, 2013 — **K. Das**

PREFACE

Engineering Mathematics being one of the most essential subjects of the B.Tech., course offered by the Engineering or Technological Institutes.

This text book is written according to the syllabus of WBU Tech covering the entire course and also as reference book to the all India Engineering Courses in four separate volumes.

The present volume (Vol IIIA) has been designed for B. Tech 3rd semester, i.e. Second year 1st semester **for the sudents of Computer Science & Information Technology under WBU Technology syllabus .**

Sincere attempts have been made to present the topics in a simple and lucid manner to create interest into the subject along with various types of worked out problems for securing high marks in the examination .

Valumes I, II and IV, published previously, have achived acceceptation to the Professors and have become popular to the students considering the utility of the same.

We take this opportunity to express our gratitude to all the Teachers & Professors who recomended our books to the Students and sent their valuable comments.

We are thankful to Dr. Purnendu Dhar, M. Sc.(Chem)., Ph.D.,Director, M/s. U. N. Dhur & Sons Pvt. Ltd. for publishing the book in time.

Any positive suggestion shall be acknowledged with thanks.

This work would not also be possible without the constant encouragement and enthusiasm of our family members .

Kalyani Govt. Engg. College
Nadia, West Bengal.
Dated : July, 2005.

Bidyut Kumar Pal
Kalidas Das

Revised Syllabus of B.Tech course According to West Bengal University of Technology MATHEMATICS

Code: M 302 / M 402

Contacts: 3L +1T = 4

Credits: 4

Note 1: The entire syllabus has been divided into four modules.

Note 2: Structure of Question Paper (There will be two groups in the paper:)

Group A: Ten questions, each of 2 marks, are to be answered out of a total of 15 questions, covering the entire syllabus.

Group B: Five questions, each carrying 10 marks, are to be answered out of (at least) 8 questions.

Students should answer at least one question from each module.

[At least 2 questions should be set from each of **Modules II & IV. At least 1 question** should be set from each of **Modules I & III.** Sufficient questions should be set covering the whole syllabus for alternatives.]

Module I: Fourier Series & Fourier Transform [8L]

Topic: Fourier Series:

Sub-Topics: Introduction, Periodic functions: Properties, Even & Odd functions: Properties, Special wave forms: Square wave, Half wave Rectifier, Full wave Rectifier, Saw-toothed wave, Triangular wave. (1)

Euler's Formulae for Fourier Series, Fourier Series for functions of period 2I, Fourier Series for functions of period 2I, Dirichlet's conditions, Sum of Fourier series. Examples. (1)

Theorem for the convergence of Fourier Series (statement only). Fourier Series of a function with its periodic extension. Half Range Fourier Series: Construction of Half range Sine Series, Construction of Half range Cosine Series. Parseval's identity (statement only). Examples. (2)

Topic: Fourier Transform:

Sub-Topics: Fourier Integral Theorem (statement only), Fourier Transform of a function, Fourier Sine and Cosine Integral Theorem (statement only), Fourier Cosine & Sine Transforms.

Fourier, Fourier Cosine & Sine Transforms of elementary functions. (1)

Properties of Fourier Transform: Linearity, Shifting, Change of scale, Modulation. Examples.

Fourier Transform of Derivatives. Examples. (1)

Convolution Theorem (statement only), Inverse of Fourier Transform, Examples. (2)

Module II : Calculus of Complex Variable [13L]

Topic: Introduction to Functions of a Complex Variable.

Sub-Topics: Complex functions, Concept of Limit, Continuity and Differentiability. (1)

Analytic functions, Cauchy-Riemann Equations (statement only). Sufficient condition for a function to be analytic. Harmonic function and Conjugate Harmonic function, related problems. (1)

Construction of Analytic functions: Milne Thomson method, related probs. (1)

Topic: Complex Integration.

Sub-Topics: Concept of simple curve, closed curve, smooth curve & contour. Some elementary properties of complex Integrals. Line integrals along a piecewise smooth curve. Examples. (2)

Cauchy's theorem (statement only). Cauchy-Goursat theorem (statement only). Examples. (1)

Cauchy's integral formula, Cauchy's integral formula for the derivative of an analytic function, Cauchy's integral formula for the successive derivatives of an analytic function. Examples. (2)

Taylor's series, Laurent's series. Examples (1)

Topic: Zeros and Singularities of an Analytic Function & Residue Theorem.

Sub-Topics: Zero of an Analytic function, order of zero, Singularities of an analytic function. Isolated and non-isolated singularity, essential singularities. Poles: simple pole, pole of order m.

Examples on determination of singularities and their nature. (1)

Residue, Cauchy's Residue theorem (statement only), problems on finding the residue of a given function, evaluation of definite integrals:

$$\int_0^{\infty} \frac{\sin x}{x} dx, \int_0^{2\pi} \frac{d\theta}{a + b\cos\theta + c\sin\theta}, \int_c \frac{P(z)}{Q(z)} dz$$

(elementary cases, $P(z)$ & $Q(z)$ are polynomials of 2nd order or less). (2)

Topic: Introduction to Conformal Mapping.

Sub-Topics: Concept of transformation from z-plane to w-plane. Concept of Conformal Mapping. Idea of some standard transformations. Bilinear Transformation and determination of its fixed point. (1)

Module III: Probability [8L]

Topic: Basic Probability Theory

Sub-Topics: Classical definition and its limitations. Axiomatic definition.

Some elementary deduction: (i) $P(O) = 0$, (ii) $0 \le P(A) \le 1$, (iii) $P(A') = 1 - P(A)$

etc. where the symbols have their usual meanings. Frequency interpretation of probability. (1)

Addition rule for 2 events (proof) & its extension to more than 2 events (statement only). Related problems.Conditional probability & Independent events. Extension to more than 2 events (pairwise & mutual independence). Multiplication Rule. Examples. Baye□fs theorem (statement only) and related problems. (3)

Topic: Random Variable & Probability Distributions. Expectation.

Sub-Topics: Definition of random variable. Continuous and discrete random variables. Probability density function & probability mass function for single variable only. Distribution function and its properties (without proof). Examples. Definitions of Expectation & Variance, properties & Examples. (2)

Some important discrete distributions: Binomial & Poisson distributions and related problems.

Some important continuous distributions: Uniform, Exponential, Normal distributions and related problems. Determination of Mean & Variance for Binomial, Poisson & Uniform distributions only. (2)

Module IV: Partial Differential Equation (PDE) and Series solution of Ordinary Differential Equation (ODE) [13L]

Topic: Basic concepts of PDE.

Sub-Topics: Origin of PDE, its order and degree, concept of solution in PDE. Introduction to different

methods of solution: Separation of variables, Laplace & Fourier transform methods. (1)

Topic: Solution of Initial Value & Boundary Value PDE's by Separation of variables, Laplace & Fourier transform methods.

Sub-Topics: PDE I: One dimensional Wave equation. (2)

PDE II: One dimensional Heat equation. (2)

PDE III: Two dimensional Laplace equation. (2)

Topic: Introduction to series solution of ODE.

Sub-Topics: Validity of the series solution of an ordinary differential equation. General method to solve $P_0y'' + P_1y' + P_2y = 0$ and related problems. (2)

Topic: Bessel's equation.

Sub-Topics: Series solution, Bessel function, recurrence relations of Bessel's Function of first kind. (2)

Topic: Legendre's equation.

Sub-Topics: Series solution, Legendre function, recurrence relations and orthogonality relation. (2)

CONTENTS

MODULE-I

Fourier Series & Fourier Transform

1.1 FOURIER SERIES

1.2 FOURIER TRANSFORM

MODULE-II

Calculus of Complex Variable

2.1 FUNCTION, LIMIT, CONTINUITY AND ANALYTIC FUNCTIONS

MODULE-III

Probability

MODULE - I
FOURIER SERIES
AND
FOURIER TRANSFORM

1.1 FOURIER SERIES

1.1.1. Introduction :

In many engineering problems, especially in the study of periodic phenomenae e.g in conduction of heat, electro-dynamics etc. the periodic functions occur frequently. These functions can be expressed as a series of sines and cosines which is known as Fourier Series. Expression into this series is used to solve several difficult problems in technology since 1700 A.D. Though the rigorous theory of Fourier Series is complicated its application is so simple. Its field of application is more wide than that of Taylor's/ Macluirin's series because many discontinuous function can be expressed into a Fourier Series.

In this chapter we first introduce perodic functions and then develope its Fourier series.

1.1.2. Some Special Functions.

In this article we introduce the following functions which are very well known to the readers.

Periodic Functions and its Properties : A function $f(x)$ is said to be periodic if there exists a positive number T such that $f(x+T)=f(x)$ for all values of x. For example $f(x)=\sin x$ is a periodic function of period 2π because $f(x+2\pi)=\sin(x+2\pi)=\sin x=f(x)$.

Note that a periodic function is defined on $(-\infty,\infty)$.

(i) Sum or difference of two periodic functions with common period T is also periodic with period T.

(ii) If $f(x)$ is periodic with period T then $f(ax)$ is periodic with period $\frac{T}{a}$ e.g $\sin 3x$ is periodic with period $\frac{2\pi}{3}$.

(iii) Any function $f(x)$ defined on a finite interval $[a,b]$ can be extended to a 'periodic function $f(x)$ of period $T=b-a$' directly by defining $f(x+T)=f(x)$

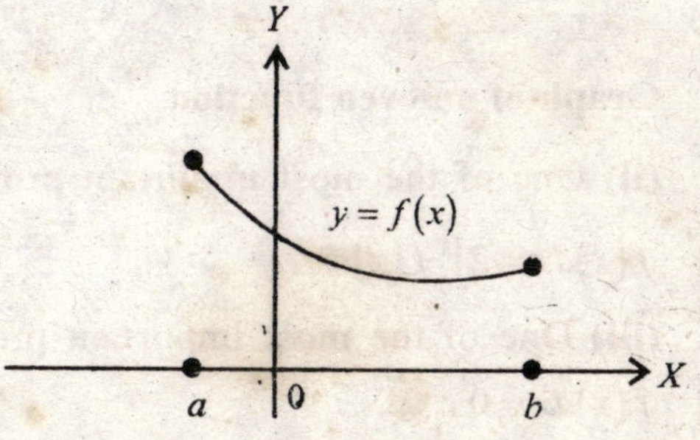

Fig. 1

for all x. This can be illustrated graphically. Let the adjacent Fig. 1 shows the graph of $f(x)$.

After extension to a periodic function of period $T = b - a$ we have the following graph (Fig. 2).

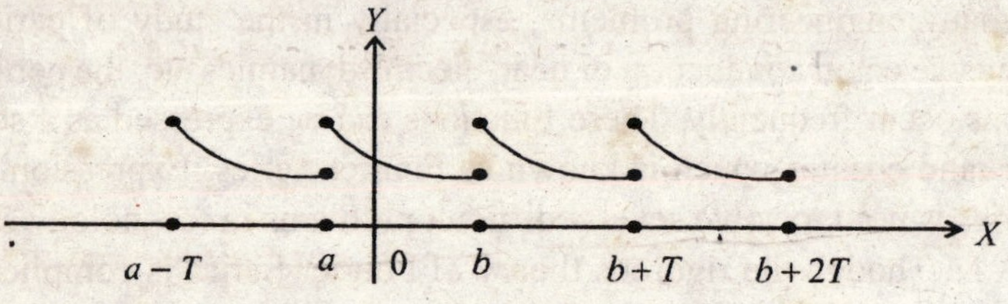

Fig. 2

This extended function is defined everywhere.

Even and Odd Function; Their Properties

A funtion $f(x)$ is said to be even funtion if $f(-x) = f(x)$ for all values of x ; e.g the functions $\cos x$, x^2 all are even functions .

A function $f(x)$ is said to be odd function if $f(-x) = -f(x)$ for all values of x. e.g. the functions $\sin x$, x^3 etc are all odd functions.

(i) The graph of these functions shown in the following figures :

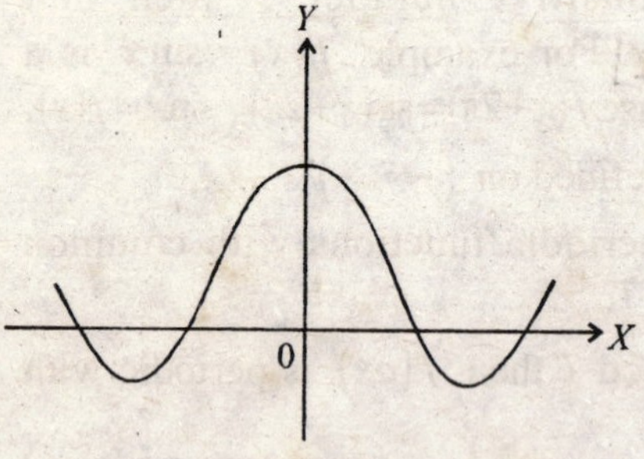

Graph of an even function

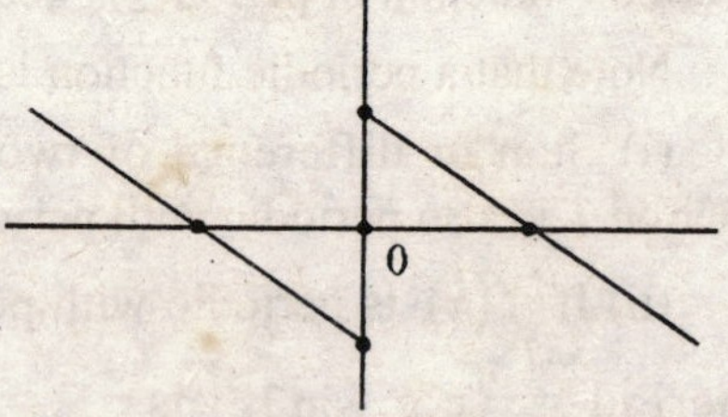

Graph of an odd function

(ii) One of the most important properties of even function is $\int_{-a}^{a} f(x)dx = 2\int_{0}^{a} f(x)dx$.

(iii) One of the most important properties of odd function is $\int_{-a}^{a} f(x)dx = 0$.

1.1.3. Typical Waveform

The graph of every periodic function runs like a wave — this is wave-form. Below we show some typical wave–form which are usually met in communication engineering :

(i) *Square Waveform* : Consider the periodic function $f(x)$ defined by

$$f(x) = -k, \ -a < x < 0$$
$$= k, \ 0 < x \le a$$

and $f(x+2a) = f(x)$ for all x.

Graph of this periodic function is shown below.

This kind of graph is known as Square Waveform.

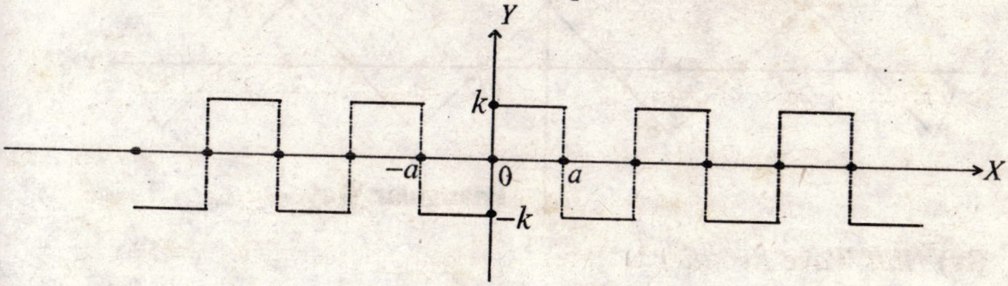

Square Waveform

(ii) *Saw – toothed Waveform* :

Consider the periodic function $f(x)$ defined by $f(x) = x$, $-a < x \le a$ and $f(x+2a) = f(x)$ for all x.

Graph of this periodic function is shown below. This kind of graph is known as Saw -toothed Waveform.

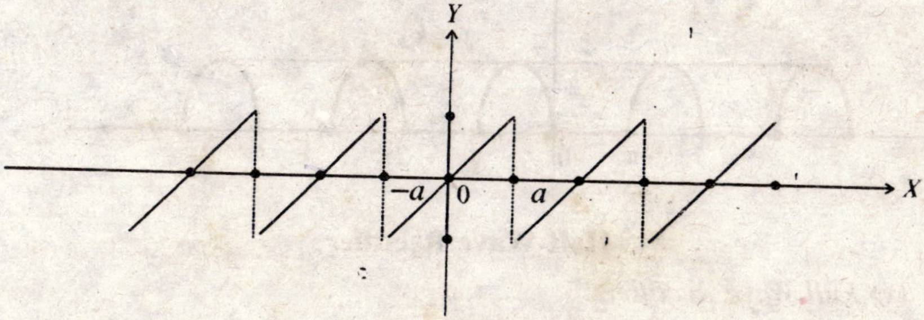

Saw-toothed Waveform

(iii) *Triangular Waveform* :

Consider the periodic functions $f(x)$ defined by

$$f(x) = 1 + \frac{2x}{a}, \; -a < x \le 0$$

$$= 1 - \frac{2x}{a}, \; 0 \le x \le a$$

and $f(x+2a) = f(x)$ for all x.

Graph of this periodic function is shown below. This kind of graph is known as Triangular Waveform.

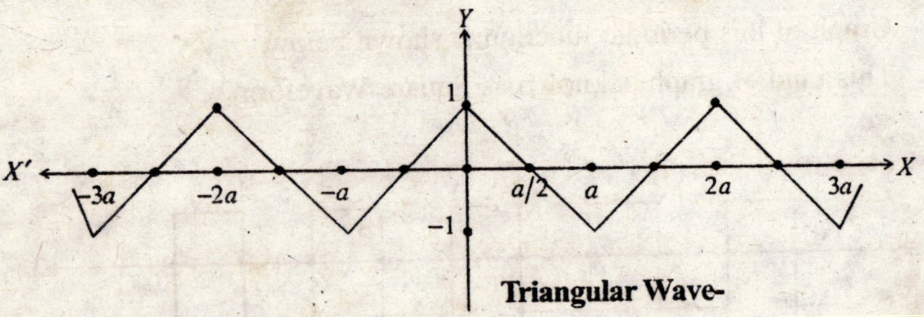

Triangular Wave-

(iv) *Half Wave Rectifier* :

Consider the periodic function $f(x)$ defined by

$$f(x) = k \sin x, \; 0 \le x \le \pi$$

$$= 0, \; \pi \le x \le 2\pi$$

and $f(x+2\pi) = f(x)$ for all x.

Graph of this periodic function is shown below. This kind of graph is known as Half-Wave Rectifier.

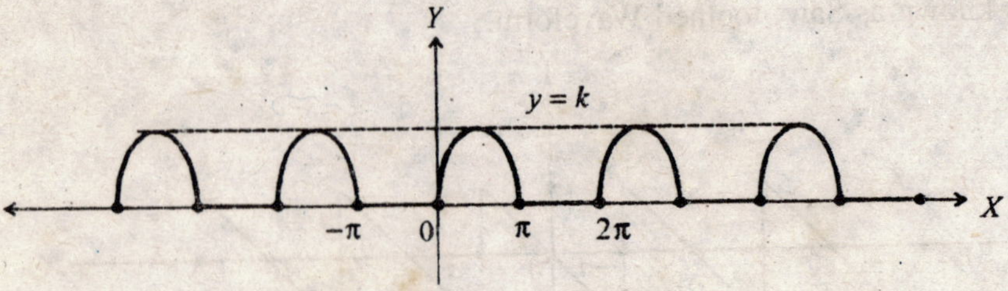

Half-Wave Rectifier

(v) *Full Wave Rectifier* :

Consider the periodic function $f(x) = k \sin x, \; 0 \le x < \pi$ and $f(x+\pi) = f(x)$ for all values of x.

Graph of this periodic function is shown below. This kind of graph is known as Full Wave Rectifier.

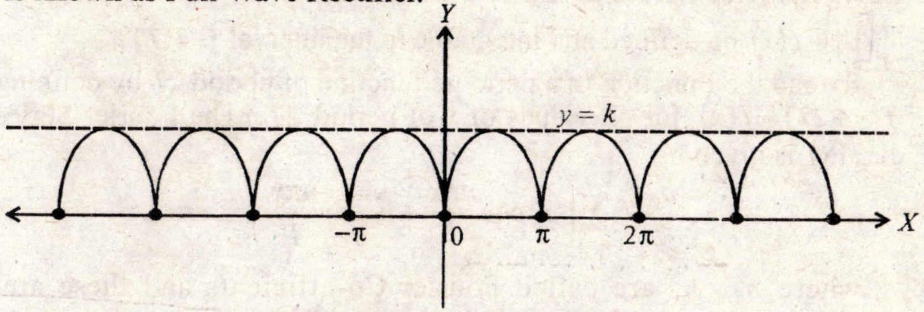

Full-Wave Rectifier

Illustrative Examples

Ex. 1. *Extend the function* $f(x)=0, -3<x<0$

$$=x^2, 0<x<3$$

to a periodic function. Then draw its graph.

This function is defined on the interval $(-3,3)$ only. To extend in periodic form just define the function on $(-\infty,\infty)$ by the rule $f(x+6)=f(x)$ for all values of x.

The graph is

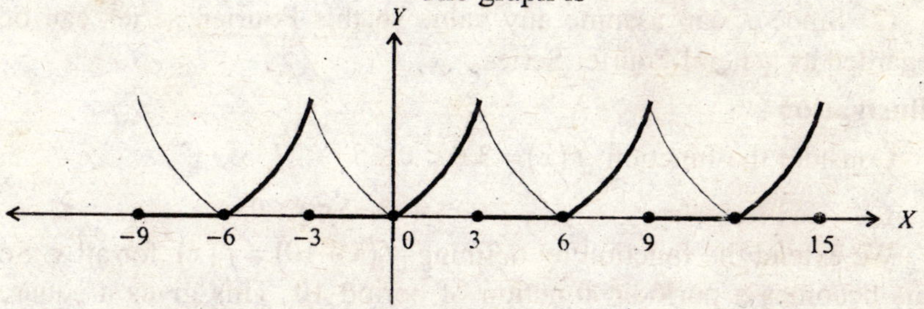

Ex. 2. *Find the period of the function* $f(x)=2|\cos^2 x|$. *Draw its graph. Find which type of Waveform is this.*

Since period of $\cos x$ is 2π so period of $\cos 2x$ is $\frac{2\pi}{2}=\pi$ $\therefore |\cos^2 x|$ has period π $\therefore f(x)$ is periodic of period π.

The graph is shown below :

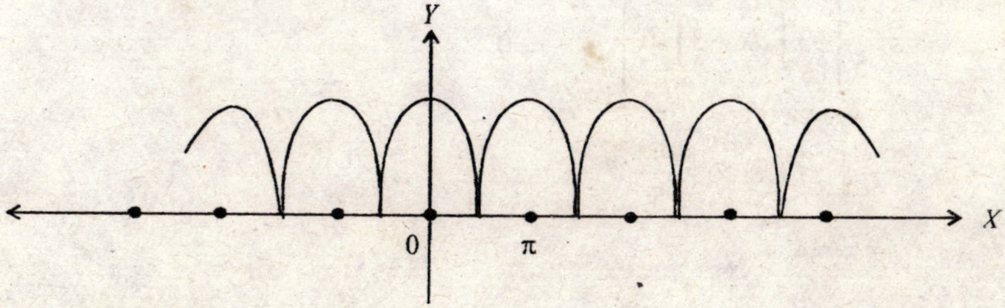

From the graph we see it is a Full Wave Rectifier.

1.1.4. Fourier Series as Euler's Formula

Let $f(x)$ be defined and integrable in the interval $(-T,T)$.

Extend the Function to a periodic function of period $2T$ by defining $f(x+2T)=f(x)$ for all values of x of period $2T$. The Fourier Series of $f(x)$ is given by

$$\frac{a_0}{2}+\sum_{n=1}^{\infty}\left(a_n\cos\frac{n\pi x}{T}+b_n\sin\frac{n\pi x}{T}\right)$$

.where a_n, b_n are called Fourier Co-efficients and these are,

according to Euler, $a_0=\frac{1}{T}\int_{-T}^{T}f(x)dx$

$$a_n=\frac{1}{T}\int_{-T}^{T}f(x)\cos\frac{n\pi x}{T}dx$$

$$b_n=\frac{1}{T}\int_{-T}^{T}f(x)\sin\frac{n\pi x}{T}dx,\ n=1,2,3\cdots\cdots s$$

Note : (1) There has no gurantee that every function $f(x)$ equals its Fourier Series. In the subsequent article we shall discuss the conditions under which this equality would hold good.

(2) Since T can assume any value so this Fourier Series can be regarded as general Fourier Series.

Illustration

Consider the function $f(x)=3, 0<x\leq 5$

$$=-3, -5<x\leq 0$$

We extend the function by defining $f(x+10)=f(x)$ for all x. So this becomes a periodic function of period 10. This gives a square waveform.

Its Fourier co-efficients, according to Euler Formula, are

$$a_0=\frac{1}{5}\int_{-5}^{5}f(x)dx$$

$$=\frac{1}{5}\left\{-3\int_{-5}^{0}dx+3\int_{0}^{5}dx\right\}\qquad=0$$

$$a_n=\frac{1}{5}\int_{-5}^{5}f(x)\cos\frac{n\pi x}{5}dx$$

$$= \frac{1}{5}\left\{-3\int_{-5}^{0}\cos\frac{n\pi x}{5}\,dx + 3\int_{0}^{5}\cos\frac{n\pi x}{5}\,dx\right\}$$

$$= \frac{1}{5}\left\{-3\int_{0}^{5}\cos\frac{n\pi x}{5}\,dx + 3\int_{0}^{5}\cos\frac{n\pi x}{5}\,dx\right\} = 0$$

and $b_n = \frac{1}{5}\int_{-5}^{5} f(x)\sin\frac{n\pi x}{5}\,dx$

$$= \frac{2}{5}\int_{0}^{5} f(x)\sin\frac{n\pi x}{5}\,dx \quad \left[\because f(x)\sin\frac{n\pi x}{5} \text{ is even function}\right]$$

$$= \frac{6}{5}\int_{0}^{5}\sin\left(\frac{n\pi x}{5}\right)dx = \frac{-6}{5}\left[\frac{\cos\frac{n\pi x}{5}}{\frac{n\pi}{5}}\right]_0^5$$

$$= -\frac{6}{n\pi}(\cos n\pi - 1) = \frac{6(1-\cos n\pi)}{n\pi}$$

Therefore the Fourier series of $f(x)$ is

$$\frac{0}{2} + \sum_{n=1}^{\infty}\left(0\cdot\cos\frac{n\pi x}{5} + \frac{6(1-\cos n\pi)}{n\pi}\sin\frac{n\pi x}{5}\right)$$

i.e. $\sum_{n=1}^{\infty}\frac{6}{\pi}\cdot\frac{1-\cos n\pi}{n}\sin\frac{n\pi x}{5}$

i.e. $\frac{6}{\pi}\left\{(1-\cos\pi)\sin\frac{\pi x}{5} + \frac{(1-\cos 2\pi)}{2}\sin\frac{2\pi x}{5} + \frac{(1-\cos 3\pi)}{3}\sin\frac{3\pi x}{5} + \cdots\cdots\right\}$

We see $f(0) = -3$ but the values of the Fourier Series at $x = 0$

is $\frac{6}{\pi}\{0+0+0+\cdots\cdots\} = 0$.

Fourier Series of a Function of Period 2π

The above Fourier Series for $T = \pi$ i.e. the Fourier series for the function $f(x)$ defined and integrable on $(-\pi,\pi)$ and $f(x+2\pi) = f(x)$ for all values of x, is

$$\frac{a_0}{2} + \sum_{n=1}^{\infty}(a_n\cos nx + b_n\sin nx)$$

where the Fourier Co-efficients are

$$a_0 = \frac{1}{\pi}\int_{-\pi}^{\pi} f(x)dx$$

$$a_n = \frac{1}{\pi}\int_{-\pi}^{\pi} f(x)\cos nx\,dx$$

and $b_n = \frac{1}{\pi}\int_{-\pi}^{\pi} f(x)\sin nx\,dx$, for $n = 1,2,3\cdots\cdots$

Illustration

Consider the function, $f(x) = x^2, -\pi < x \le \pi$

This function is defined on the interval $(-\pi, \pi]$.

We extend this by defining $f(x+2\pi) = f(x)$ for all values of x. This is a periodic function of period 2π. Its Fourier co-efficients are

$$a_0 = \frac{1}{\pi}\int_{-\pi}^{\pi} x^2 dx = \frac{2}{\pi}\int_0^{\pi} x^2 dx = \frac{2\pi^2}{3}$$

$$a_n = \frac{1}{\pi}\int_{-\pi}^{\pi} f(x)\cos nx\,dx$$

$$= \frac{1}{\pi}\int_{-\pi}^{\pi} x^2 \cos nx\,dx = \frac{2}{\pi}\int_0^{\pi} x^2 \cos nx\,dx \quad [\because x^2 \cos nx \text{ is even function}]$$

$$= \frac{2}{\pi}\left\{\left[x^2\frac{\sin nx}{n}\right]_0^{\pi} - \int_0^{\pi} 2x\frac{\sin nx}{n}dx\right\}$$

$$= -\frac{4}{\pi n}\left\{\left[-x\frac{\cos nx}{n}\right]_0^{\pi} + \int_0^{\pi}\frac{\cos nx}{n}dx\right\}$$

$$= -\frac{4}{\pi n}\left\{-\frac{\pi\cos n\pi}{n} + \frac{1}{n}\left[\frac{\sin nx}{n}\right]_0^{\pi}\right\} = \frac{4\cos n\pi}{n^2}$$

and $b_n = \frac{1}{\pi}\int_{-\pi}^{\pi} x^2 \sin nx\,dx = 0$ $[\because x^2 \sin nx$ is odd function$]$

So, the Fourier series of $f(x)$ is

$$\frac{1}{2}\cdot\frac{2\pi^2}{3} + \sum_{n=1}^{\infty}\left(\frac{4\cos n\pi}{n^2}\cos nx + 0\cdot\sin nx\right)$$

i.e. $\frac{\pi^2}{3} + 4\sum_{n=1}^{\infty}\frac{\cos n\pi\cos nx}{n^2}$.

1.1.5. Fourier Expansion and its Conditions

We have earlier remarked that the Fourier Series of a function $f(x)$ may not be equal to $f(x)$. We also gave an example in this regard.

However when the Fourier series of $f(x)$ becomes equal to $f(x)$ we say the series as a Fourier Expansion of $f(x)$.

Next we are going to state the condition under which a function will have Fourier Expansion.

Dirichlet's Conditions.

A function $f(x)$ defined on $[-T,T]$ is said to satisfy Dirichlet's Conditions if it satisfies **any one** of the following two conditions :

(1) $f(x)$ is bounded in $[-T,T]$ and the interval $[-T,T]$ can be decomposed into a finite number of sub-intervals such that $f(x)$ is monotonic (increasing or decreasing) on each of the sub-intervals.

(2) $f(x)$ has a finite number of points of infinite discontinuity in $[-T,T]$.

When arbitrary small neighbourhood of these points are excluded from $[-T,T]$ $f(x)$ becomes bounded in the remaining part and this remaining part can be decomposed into a finite number of sub-intervals such that $f(x)$ is monotonic in each of the sub-intervals. Moreover the improper integral $\int_{-\pi}^{\pi} f(x)dx$ is absolutely convergent.

Illustrations.

(i) Let $f(x)=x-3,\ -3\le x\le 0$

$$=3-x,\ 0<x\le 3.$$

Let us draw the graph of $f(x)$ below :

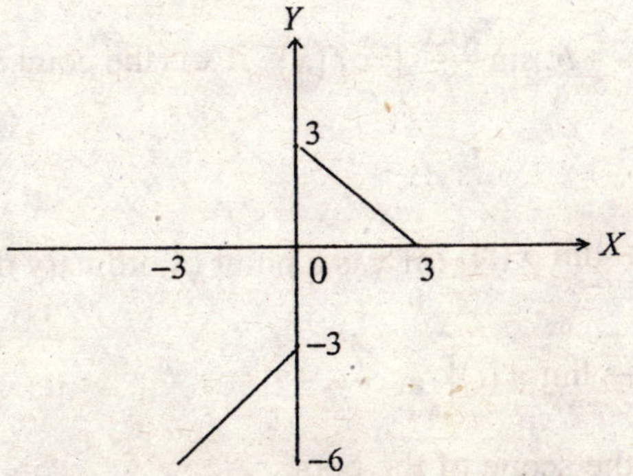

From the graph we see $-6 \le f(x) \le 3$ i.e. $f(x)$ is bounded in $[-3,3]$.

The interval $[-3,3]$ is decomposed as $[-3,3]=[-3,0]\cup[0,3]$ such that $f(x)$ is increasing in $[-3,0]$ and decreasing in $[0,3]$. So we conclude this function $f(x)$ satisfies Dirichlet's Condition.

(ii) Let $f(x)=\dfrac{1}{\sqrt{1-x}}$, $-1\le x\le 1$.

Its graph is shown below :

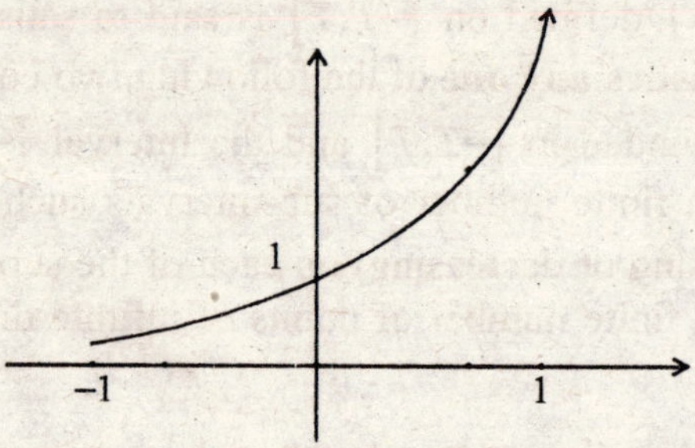

Though this function is not bounded in $[-1,1]$ it has one (finite) point of infinite discontinuity (at $x=1$). If a neighbourhood of $x=1$ is excluded the functions becomes bounded and $f(x)$ is monotonic increasing in the remaining part.

Moreover the improper integral $\int_{-1}^{1}\dfrac{dx}{\sqrt{1-x}}$ is absolutely convergent.

So this function satisfies Dirichlet's Condition.

Theorem

If a function $f(x)$ defined on $[-T,T]$ satisfies Dirichlet's Condition then its Fourier series is convergent and the Fourier series

$$\frac{a_0}{2}+\sum_{n=1}^{\infty}\left(a_n\cos\frac{n\pi x}{T}+b_n\sin\frac{n\pi x}{T}\right)=f(x)$$, if x is the point of continuity of $f(x)$

$$=\frac{1}{2}\left\{\lim_{t\to x+}f(t)+\lim_{t\to x-}f(t)\right\}$$, if x is a point of ordinary discontinuity

$$=\frac{1}{2}\left\{\lim_{t\to -T+}f(t)+\lim_{t\to T-}f(t)\right\}$$ at $x=\pm T$.

Proof. Beyond the scope of the book.

Illustration

Again consider the function

$f(x)=x^2,\ -\pi<x\leq\pi$

Extending this to periodic by defining $f(x+2\pi)=f(x)$ for all values of x we got its Fourier Series as (see a previous illustration)

$$\frac{\pi^2}{3}+4\sum_{n=1}^{\infty}\frac{\cos n\pi\cos nx}{n^2}$$

The graph of $f(x)$ in $[-\pi,\pi]$ is

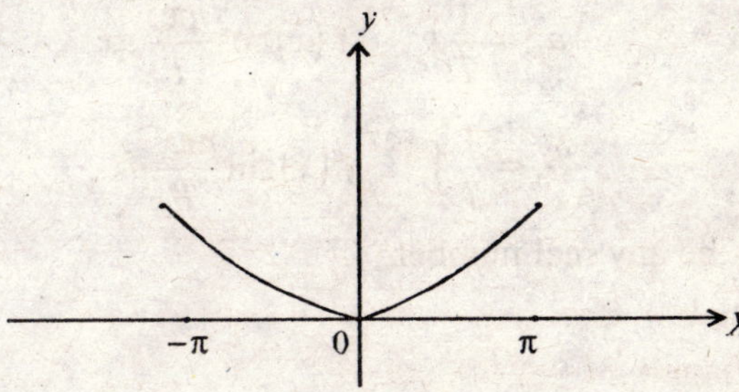

From the graph it is clear that $f(x)$ is bounded and monotonic in two sub intervals $[-\pi,0]$ and $[0,\pi]$, So $f(x)$ satisfies Dirichlet's Condition. $f(x)$ is continuous every where in $(-\pi,\pi)$.

Now, $\frac{1}{2}\left\{\lim_{t\to-\pi+}f(t)+\lim_{t\to\pi-}f(t)\right\}$

$$=\frac{1}{2}\left\{\lim_{t\to-\pi}t^2+\lim_{t\to\pi}t^2\right\}=\frac{1}{2}\left(\pi^2+\pi^2\right)=\pi^2=f(\pi)$$

So, the Fourier Expansion of the given function is

$$f(x)=\frac{\pi^2}{3}+4\sum_{n=1}^{\infty}\frac{\cos n\pi\cos nx}{n^2}.$$

1.1.6. Change of Interval

We introduce Fourier Series of a function $f(x)$ which is primarily defined on the interval $[-T,T]$ and then extending it to a periodic wave.

But in many engineering problem the function may appear as defined primarily on an interval $[c,c+2T]$ where c may be any real number. In that case also we have no trouble of getting its Fourier Series. In fact the following theorem assures so.

Theorem

If $f(x)$ be defined and integrable in $[c, c+2T]$ and $f(x+2T) = f(x)$ for all values of x (i.e. $f(x)$ periodic of period $2T$) then the Fourier Series of $f(x)$ is also

$$\frac{a_0}{2} + \sum_{n=1}^{\infty}\left(a_n \cos\frac{n\pi x}{T} + b_n \sin\frac{n\pi x}{T}\right)$$

where the Fourier co-efficients are

$$a_0 = \frac{1}{T}\int_c^{c+2T} f(x)dx$$

$$a_n = \frac{1}{T}\int_c^{c+2T} f(x)\cos\frac{n\pi x}{T}dx$$

$$b_n = \frac{1}{T}\int_c^{c+2T} f(x)\sin\frac{n\pi x}{T}dx$$

where c may be any real number.

Proof. Omitted.

Illustration.

Consider the function $f(x) = x^2$, $0 < x \le 2\pi$.

Here the function is defined in $[0, 2\pi]$. Here $c = 0$, $c + 2T = 2\pi$ $\therefore T = \pi$.

$\therefore$ the Fourier co-efficients are

$$a_0 = \frac{1}{\pi}\int_0^{0+2\pi} x^2 dx = \frac{8\pi^2}{3}$$

$$a_n = \frac{1}{\pi}\int_0^{0+2\pi} x^2 \cos nx\, dx = \frac{1}{\pi}\int_0^{2\pi} x^2 \cos nx\, dx = \frac{4}{n^2}.$$

(can be had by integ by parts)

$$b_n = \frac{1}{\pi}\int_0^{0+2\pi} x^2 \sin nx\, dx = \frac{1}{\pi}\int_0^{2\pi} x^2 \sin nx\, dx = -\frac{4\pi}{n}$$

(evaluations are not shown).

So the Fourier Series of the given function (after extending to a periodic function of period 2π) is

$$\frac{4\pi^2}{3} + \sum_{n=1}^{\infty}\left(\frac{4}{n^2}\cos nx - \frac{4\pi}{n}\sin nx\right).$$

Clearly $f(x)$ satisfies Dirichlet's Condition. So the Fourier expansion is

$$\frac{4\pi^2}{3}+\sum_{n=1}^{\infty}\left(\frac{4}{n^2}\cos nx-\frac{4\pi}{n}\sin nx\right)=x^2,\ 0<x\leq 2\pi .$$

1.1.7. Half Range Series : Sine or Cosine

A trigonometric series like the Fourier Series is called a Half Range Series if only sine terms or only cosine terms are present.

When only sine terms are present the series is called *Half Range Sine series*; when only cosine terms are present the series is called *Half Range Cosine Series*.

When a half range series corresponding to a function is desired, the function is generally defined in the interval $(0,T)$ which is half of the interval $(-T,T)$.

Construction of Half Range Sine Series.

Let $f(x)$ be a function defined and integrable on the interval $(0,T)$. We extend the domain of definition to $[-T,0]$ defining by $f(-x)=-f(x)$. This extension is shown in the adjacent figure. Then this extended $f(x)$ becomes odd in the interval $[-T,T]$.

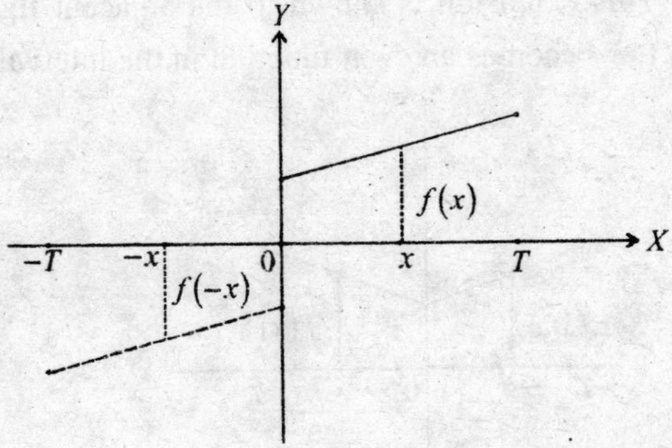

Therefore, $a_0=\frac{1}{T}\int_{-T}^{T}f(x)dx$

$=0 \qquad \because f(x)$ is odd.

$$a_n = \frac{1}{T}\int_{-T}^{T} f(x)\cos\frac{n\pi x}{T}dx$$

$$= 0 \quad [\because f(x)\cos\frac{n\pi x}{T} \text{ is odd function}]$$

and $$b_n = \frac{1}{T}\int_{-T}^{T} f(x)\sin\frac{n\pi x}{T}dx$$

$$= \frac{2}{T}\int_{0}^{T} f(x)\sin\frac{n\pi x}{T}dx \quad [\because f(x)\sin\frac{n\pi x}{T} \text{ is even function}]$$

The Fourier series of $f(x)$ becomes

$$\frac{0}{2}+\sum_{n=1}^{\infty}\left(0\cos\frac{n\pi x}{T}+b_n\sin\frac{n\pi x}{T}\right) \quad \text{i.e. } \sum_{n=1}^{\infty} b_n\sin\frac{n\pi x}{T}$$

which is the required Half Range Sine Series. Obviously if $f(x)$ satisfies Dirichlets condition in $[0,T]$ then this series is convergent and the value is as for (Full Range) Fourier Series.

Construction of Half Range Cosine Series.

Let $f(x)$ be a function defined and integrable on the interval $(0,T)$. We extend the domain of definition to $[-T,0]$ defining by $f(-x)=f(x)$. This extension is shown in the adjacent figure. Then this extended $f(x)$ becomes an even function in the interval $[-T,T]$.

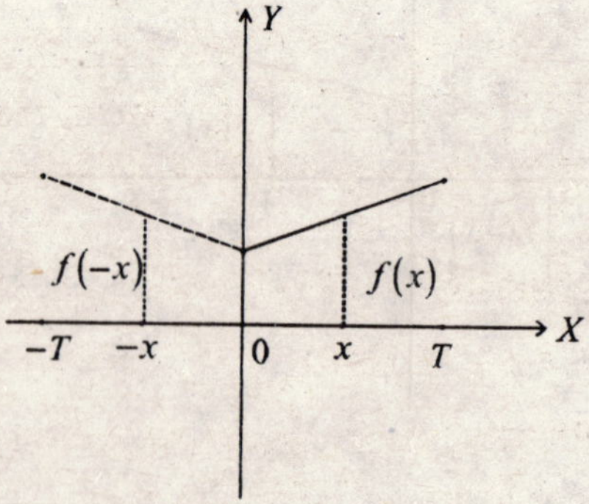

Therefore, $$a_0 = \frac{1}{T}\int_{-T}^{T} f(x)dx = \frac{2}{T}\int_{0}^{T} f(x)dx \qquad \because f(x) \text{ is even}$$

$$a_n = \frac{1}{T}\int_{-T}^{T} f(x)\cos\frac{n\pi x}{T}dx$$

$$=\frac{2}{T}\int_0^T f(x)\cos\frac{n\pi x}{T}dx\ \left[\because f(x)\cos\frac{n\pi x}{T}\text{ is even function}\right]$$

and $b_n=\frac{1}{T}\int_{-T}^{T}f(x)\sin\frac{n\pi x}{T}dx=0\ \left[\because f(x)\sin\frac{n\pi x}{T}\text{ is odd function}\right]$

Consequently the Fourier Series of $f(x)$ becomes

$$\frac{a_0}{2}+\sum_{n=1}^{\infty}\left(a_n\cos\frac{n\pi x}{T}+0\cdot\sin\frac{n\pi x}{T}\right)\text{ i.e. }\frac{a_0}{2}+\sum_{n=1}^{\infty}a_n\cos\frac{n\pi x}{T}.$$

which is the required Half Range Cosine Series.

Here also this series converges according as $f(x)$ satisfies Dirichlet's Condition.

Note. In particular if the interval of definition becomes $[0,\pi]$ then

(1) the Half Range Sine series becomes $\sum_{n=1}^{\infty}b_n\sin nx$ where

$$b_n=\frac{2}{\pi}\int_0^{\pi}f(x)\sin nx\,dx$$

(2) the Half Range Cosine series becomes $\frac{a_0}{2}+\sum_{n=1}^{\infty}a_n\cos nx$

where $a_0=\frac{2}{\pi}\int_0^{\pi}f(x)dx$ and $a_n=\frac{2}{\pi}\int_0^{\pi}f(x)\cos nx\,dx$.

Illustration.

(i) Consider the function $f(x)=x,\ 0\le x\le 2$.

To get its Half-Range Sine Series we extend the function to the interval $(-2,0)$ defining by $f(-x)=-f(x)$. This extended function becomes odd. Therefore $a_0=\frac{1}{2}\int_{-2}^{2}f(x)dx=0\ \because f(x)$ is odd function

$$a_n=\frac{1}{2}\int_{-2}^{2}f(x)\cos\frac{n\pi x}{2}dx=0\quad\left[\because f(x)\cos\frac{n\pi x}{2}\text{ is odd function}\right]$$

and $b_n=\frac{1}{2}\int_{-2}^{2}f(x)\sin\frac{n\pi x}{2}dx=\int_0^2 x\sin\frac{n\pi x}{2}dx$

$$=\left[\frac{-2}{n\pi}x\cos\frac{n\pi x}{2}\right]_0^2-\int_0^2(\cdot\cdot)\frac{2}{n\pi}\cos\frac{n\pi x}{2}dx=\frac{4(-1)^{n+1}}{n\pi}.$$

Hence the Half Range Sine Series of the function is

$$\frac{0}{2}+\sum_{n=1}^{\infty}\left(0\cdot\cos\frac{n\pi x}{2}+\frac{4(-1)^{n+1}}{n\pi}\sin\frac{n\pi x}{2}\right)=\frac{4}{\pi}\sum_{n=1}^{\infty}\frac{(-1)^{n+1}}{n}\sin\frac{n\pi x}{2}$$

$$=\frac{4}{\pi}\left[\sin\frac{\pi x}{2}-\frac{1}{2}\sin\frac{2\pi x}{2}+\frac{1}{3}\sin\frac{3\pi x}{2}-\ldots\ldots \text{ up to } \infty\right]$$

(ii) Consider the function $f(x)=\sin x$, $0<x\le\pi$. To get its Half Range Cosine series we extend the function to the interval $(-\pi,0)$ defining by $f(-x)=f(x)$. With this extension $f(x)$ becomes even on the interval $(-\pi,\pi)$. Then $a_0=\frac{2}{\pi}\int_0^{\pi}\sin x\,dx=\frac{4}{\pi}$

$$a_n=\frac{2}{\pi}\int_0^{\pi}\sin x\cos nx\,dx \tag{1}$$

$$=\frac{1}{\pi}\int_0^{\pi}\{\sin(x+nx)+\sin(x-nx)\}dx$$

$$=\frac{1}{\pi}\left[-\frac{\cos(n+1)x}{n+1}+\frac{\cos(n-1)x}{n-1}\right]_0^{\pi} \quad \text{for } n\ne 1$$

$$=\frac{-2(1+\cos n\pi)}{\pi(n^2-1)}=\frac{2(1+\cos n\pi)}{\pi(1-n^2)} \quad \text{for } n\ne 1$$

Now $a_1=0$ (from (1)) and $b_n=0$.

$\therefore$ The Half-Range Cosine Series becomes

$$\frac{1}{2}\cdot\frac{4}{\pi}+\sum_{n=2}^{\infty}\left(\frac{2(1+\cos n\pi)}{\pi(1-n^2)}\cos nx+0\cdot\sin nx\right)$$

$$=\frac{2}{\pi}-\frac{2}{\pi}\sum_{n=2}^{\infty}\frac{1+\cos n\pi}{n^2-1}\cos nx$$

$$=\frac{2}{\pi}-\frac{4}{\pi}\left(\frac{\cos 2x}{2^2-1}+\frac{\cos 4x}{4^2-1}+\frac{\cos 6x}{6^2-1}+\ldots\ldots\right).$$

1.1.8. Parseval's Identity

If the Fourier Series of a function $f(x)$ converges uniformly to $f(x)$ in the interval $(-T,T)$ then

$$\frac{1}{T}\int_{-T}^{T}\{f(x)\}^2 dx = \frac{a_0^{\ 2}}{2} + \sum_{n=1}^{\infty}\left(a_n^{\ 2} + b_n^{\ 2}\right) \tag{1}$$

where a_n, b_n are Fourier Co-efficients of $f(x)$.

Proof. Beyond the scope of the book.

Note. 1. The relation (1) is Parseval's identity. This is valid under less restrictive conditions than that imposed in the above theorem.

2. Corresponding to Half Range sine series $\sum_{n=1}^{\infty} p_n \sin nx$ the Persaval's indentity would be $\frac{2}{T}\int_0^T \{f(x)\}^2 dx = \sum_{n=1}^{\infty} b_n^2$ (under the above condition) since here $f(x)$ is extended to an even function.

3. Corresponding to Half Range cosine series $\frac{a_0}{2} + \sum_{n=1}^{\infty} a_n \cos nx$ the Persavals indentity would be $\frac{2}{T}\int_0^T \{f(x)\}^2 dx = \frac{a_0^2}{2} + \sum_{n=1}^{\infty} a_n^2$ since here $f(x)$ is extended to an odd function, i.e $\{f(x)\}^2$ is extended to even.

Illustration.

Consider the function $f(x) = -x$, $-2 < x \le 0$

$= x$, $0 \le x \le 2$.

We see $f(x)$ is an even function. Extending this to a periodic function defining by $f(x+4) = f(x)$ we get the graph as follow :

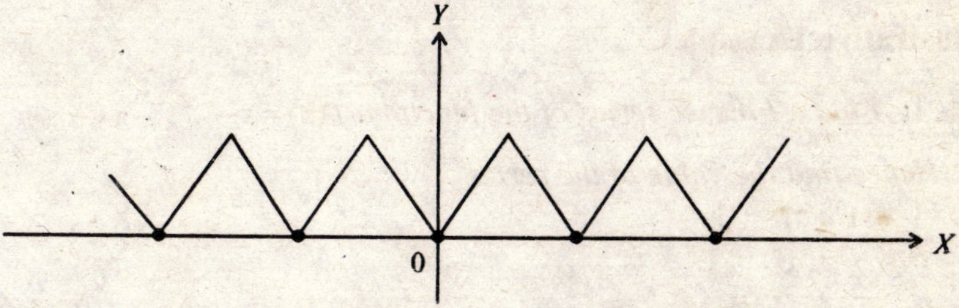

Here $a_0 = \frac{1}{2}\int_{-2}^{2} f(x)dx = \frac{1}{2}\int_{-2}^{0} -x dx + \frac{1}{2}\int_0^2 x dx = 2$

$$a_n = \frac{1}{2}\int_{-2}^{2} f(x)\cos\frac{n\pi x}{2}dx = \frac{2}{2}\int_0^2 x\cos\frac{n\pi x}{2}dx$$

$$\left[\because f(x)\cos\frac{n\pi x}{2} \text{ is an even function}\right]$$

$$= \left[x\left(\frac{2}{n\pi}\sin\frac{n\pi x}{2}\right) - 1\left(\frac{-4}{n^2\pi^2}\cos\frac{n\pi x}{2}\right)\right]_0^2$$

$$= \frac{4}{n^2\pi^2}(\cos n\pi - 1) \text{ for } n \neq 0.$$

$$b_n = \frac{1}{2}\int_{-2}^{2} f(x)\sin\frac{n\pi x}{2}dx = 0 \left[\because f(x)\sin\frac{n\pi x}{2} \text{ is an odd function}\right]$$

Again the function $f(x)$ satisfies Dirichlet's Condition and it is ontinuous everywhere (which is seen from its graph).

So its Parseval's Identity is

$$\frac{1}{2}\int_{-2}^{2}\{f(x)\}^2 dx = \frac{2^2}{2} + \sum_{n=1}^{\infty}\left[\left\{\frac{4}{n^2\pi^2}(\cos n\pi - 1)\right\}^2 + 0^2\right]$$

$$\text{or, } \frac{1}{2}\left[\int_{-2}^{0}(-x)^2 dx + \int_0^2 x^2 dx\right] = 2 + \sum_{n=1}^{\infty}\frac{16}{n^4\pi^4}(\cos n\pi - 1)^2$$

$$\text{or, } \frac{1}{2}\left[\frac{x^3}{3}\right]_{-2}^{2} = 2 + \frac{64}{\pi^4}\left(\frac{1}{1^4} + \frac{1}{3^4} + \frac{1}{5^4} + \cdots\cdots\right)$$

$$\text{or, } \frac{8}{3} = 2 + \frac{64}{\pi^4}\left(\frac{1}{1^4} + \frac{1}{3^4} + \frac{1}{5^4} + \cdots\cdots\right)$$

$$\text{or, } \frac{1}{1^4} + \frac{1}{3^4} + \frac{1}{5^4} + \cdots\cdots = \frac{\pi^4}{96},$$ an interesting result is obtained.

Illustrative Examples.

Ex. 1. *Find a Fourier series of the function* $f(x) = x - x^2$, $-\pi < x \leq \pi$.

Hence find the value of the series

$$\frac{1}{1^2} - \frac{1}{2^2} + \frac{1}{3^2} - \frac{1}{4^2} + \cdots\cdots$$

The function is defined on $(-\pi, \pi]$ primarily. We extend by defining outside as $f(x+2\pi) = f(x)$. It becomes a periodic function of period 2π.

The Fourier Co-efficients are

$$a_0 = \frac{1}{\pi}\int_{-\pi}^{\pi}\left(x - x^2\right)dx = -\frac{2\pi^2}{3}$$

$$a_n = \frac{1}{\pi}\int_{-\pi}^{\pi}\left(x - x^2\right)\cos nx\,dx$$

$$= \frac{1}{\pi}\left\{\left[\left(x - x^2\right)\frac{\sin nx}{n}\right]_{-\pi}^{\pi} - \int_{-\pi}^{\pi}(1-2x)\frac{\sin nx}{n}dx\right\}$$

$$= \frac{-1}{n\pi}\int_{-\pi}^{\pi}(1-2x)\sin nx\,dx$$ {equating 2nd part}

$$= \frac{-1}{n\pi}\left\{\left[-(1-2x)\frac{\cos nx}{n}\right]_{-\pi}^{\pi} - \int_{-\pi}^{\pi}(-2)\frac{(-)\cos nx}{n}dx\right\}$$

$$= \frac{1}{n\pi}\left\{\frac{(1-2\pi)\cos n\pi}{n} - \frac{(1+2\pi)\cos n\pi}{n} + \frac{2}{n}\int_{-\pi}^{\pi}\cos nx\,dx\right\}$$

$$= \frac{4(-1)^{n+1}}{n^2} \quad \text{for } n \neq 0.$$

$$b_n = \frac{1}{\pi}\int_{-\pi}^{\pi}\left(x - x^2\right)\sin nx\,dx$$

$$= \frac{1}{\pi}\left\{\left[\frac{-\left(x - x^2\right)\cos nx}{n}\right]_{-\pi}^{\pi} - \int_{-\pi}^{\pi}(1-2x)\frac{-\cos nx}{n}dx\right\}$$

$$= \frac{-1}{\pi}\left\{\frac{\left(\pi - \pi^2\right)\cos n\pi}{n} - \frac{\left(-\pi - \pi^2\right)\cos n\pi}{n} - \frac{1}{n}\int_{-\pi}^{\pi}(1-2x)\cos nx\,dx\right\}$$

$$= \frac{2(-1)^{n+1}}{n}$$

So the Fourier series of the given function is

$$\frac{1}{2}\left(-\frac{2\pi^2}{3}\right) + \sum_{n=1}^{\infty}\left\{\frac{4(-1)^{n+1}}{n^2}\cos nx + \frac{2\cdot(-1)^{n+1}}{n}\sin nx\right\}$$

$$= -\frac{\pi^2}{3} + 4\sum_{n=1}^{\infty}\frac{(-1)^{n+1}}{n^2}\cos nx + 2\sum_{n=1}^{\infty}\frac{(-1)^{n+1}}{n}\sin nx$$

Now, $y=x-x^2$ is a parabola shown in the following figure. From the figure

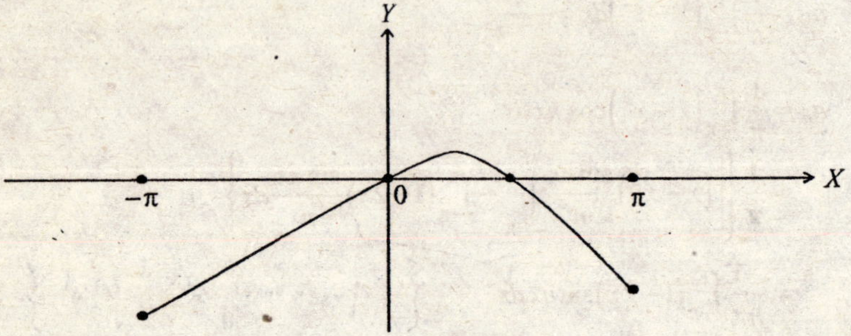

we see it is bounded and monotonic in two subintervals. So $f(x)$ satisfies Dirichlet's condition. Since the function is continuous in $(-\pi,\pi)$ so

$$x-x^2=-\frac{\pi^2}{3}+4\sum_{n=1}^{\infty}\frac{(-1)^{n+1}}{n^2}\cos nx+2\sum_{n=1}^{\infty}\frac{(-1)^{n+1}}{n}\sin nx$$

Putting $x=0$ we get

$$0=-\frac{\pi^2}{3}+4\sum_{n=1}^{\infty}\frac{(-1)^{n+1}}{n^2}+2\sum_{n=1}^{\infty}\frac{(-1)^{n+1}}{n}\times 0$$

or, $\frac{\pi^2}{3}=4\left(\frac{1}{1^2}-\frac{1}{2^2}+\frac{1}{3^2}-\frac{1}{4^2}+\cdots\cdots\right)$

or, $\frac{1}{1^2}-\frac{1}{2^2}+\frac{1}{3^2}-\frac{1}{4^2}+\cdots\cdots=\frac{\pi^2}{12}$.

Ex. 2. *Write the Fourier Series of the function*

$f(x)=0,\ -5<x<0$

$\quad =3,\ 0<x<5$

and $f(x)$ is periodic function of period 10. How should $f(x)$ be defined at $x=-5$, $x=0$ and $x=5$ so that its Fourier series converges to $f(x)$ for $-5\le x\le 5$. [W.B.U.Tech 2003]

The function is defined primarily on $(-5,5)$. Therefore the Fourier Co-efficients,

$$a_0=\frac{1}{5}\int_{-5}^{5}f(x)dx \qquad =\frac{1}{5}\left\{\int_{-5}^{0}o\,dx+\int_{0}^{5}3dx\right\}=3$$

For $n \neq 0$ $a_n = \frac{1}{5}\int_{-5}^{5} f(x)\cos\frac{n\pi x}{5}dx$

$$= \frac{1}{5}\int_0^5 3\cos\frac{n\pi x}{5}dx \text{ [by def. of } f(x)]$$

$$= \frac{3}{5}\left[\frac{5}{n\pi}\sin\frac{n\pi x}{5}\right]_0^5 = \frac{3}{n\pi}\{\sin n\pi - 0\} = 0$$

and $b_n = \frac{1}{5}\int_{-5}^{5} f(x)\sin\frac{n\pi x}{5}dx = \frac{1}{5}\int_0^5 3\sin\frac{n\pi x}{5}dx$

$$= \frac{3}{5}\left[\frac{-5}{n\pi}\cos\frac{n\pi x}{5}\right]_0^5 = -\frac{3}{n\pi}\left[\cos\frac{n\pi x}{5}\right]_0^5$$

$$= -\frac{3}{n\pi}\{\cos n\pi - \cos 0\} = \frac{3}{n\pi}(1-\cos n\pi) \quad = \frac{3}{n\pi}\left(1-(-1)^n\right).$$

So the Fourier Series is

$$\frac{3}{2} + \sum_{n=1}^{\infty}\left\{0\cdot\cos\frac{n\pi x}{5} + \frac{3}{n\pi}\left(1-(-1)^n\right)\sin\frac{n\pi x}{5}\right\}$$

$$= \frac{3}{2} + \frac{3}{\pi}\sum_{n=1}^{\infty}\frac{1-(-1)^n}{n}\sin\frac{n\pi x}{5}$$

$$= \frac{3}{2} + \frac{3}{\pi}\left\{2\sin\frac{\pi x}{5} + \frac{2}{3}\sin\frac{3\pi x}{5} + \frac{2}{5}\sin\frac{5\pi x}{5} + \cdots\cdots\right\}$$

$$= \frac{3}{2} + \frac{6}{\pi}\left\{\sin\frac{\pi x}{5} + \frac{1}{3}\sin\frac{3\pi x}{5} + \frac{1}{5}\sin\frac{5\pi x}{5} + \cdots\cdots\right\}$$

We draw the graph of the function below :

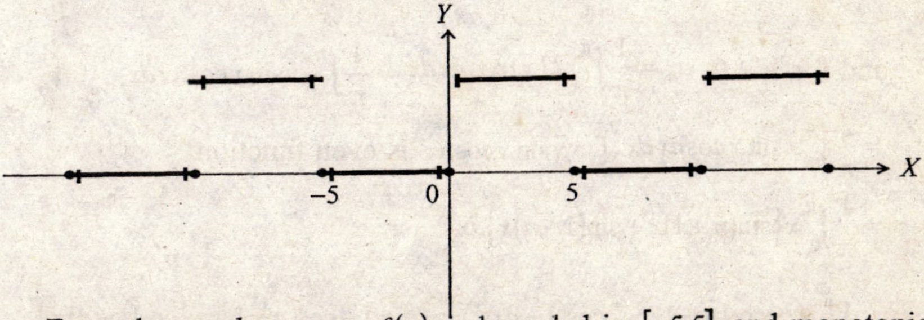

From the graph we see $f(x)$ is bounded in $[-5,5]$ and monotonic in the two sub-intervals $(-5,0)$ and $(0,5)$.

So $f(x)$ satisfies Dirichlet's Condition in $(-5,5)$. It is discontinuous at $x=0$. Therefore at $x=0$ the Fourier series converges to

$$\frac{1}{2}\left\{\lim_{t\to 0+} f(t)+\lim_{t\to 0-} f(t)\right\}=\frac{1}{2}\left\{\lim_{t\to 0+} 3+\lim_{t\to 0-} 0\right\}=\frac{3}{2}.$$

So for convergence of $f(x)$ at $x=0$ we should define $f(0)=\frac{3}{2}$.

At the end points $x=-5$ and $x=5$ the Fourier Series converges to

$$\frac{1}{2}\left\{\lim_{t\to -5+} f(t)+\lim_{t\to 5-} f(t)\right\}$$

$$=\frac{1}{2}\left\{\lim_{t\to -5+} 0+\lim_{t\to 5-} 3\right\}=\frac{3}{2}.$$

So for convergence of $f(x)$ at $x=-5$ and at $x=5$ we should define $f(-5)=\frac{3}{2}$, $f(5)=\frac{3}{2}$.

Ex. 3. *Obtain the Fourier expansion of* $x\sin x$ *is* $-\pi\le x\le\pi$ *and deduce that*

$$\frac{\pi}{4}=\frac{1}{2}+\frac{1}{1.3}-\frac{1}{3.5}+\frac{1}{5.7}-\cdots\cdots$$

$f(x)=x\sin x$ satisfies Dirichlet's condition in $(-\pi,\pi)$ and it is continuous everywhere so we can say it can be expanded into its Fourier series, i. e.

$$f(x)=\frac{a_0}{2}+\sum_{n=1}^{\infty}(a_n\cos nx+b_n\sin nx).$$

$$a_0=\frac{1}{\pi}\int_{-\pi}^{\pi} f(x)dx=\frac{2}{\pi}\int_{0}^{\pi} x\sin x\,dx=2$$

and for $n\neq 0$ $a_n=\frac{1}{\pi}\int_{-\pi}^{\pi} f(x)\cos nx\,dx=\frac{1}{\pi}\int_{-\pi}^{\pi} x\sin x\cos nx\,dx$

$$=\frac{2}{\pi}\int_{0}^{\pi} x\sin x\cos nx\,dx \quad [\because x\sin x\cos nx \text{ is even function}]$$

$$=\frac{1}{\pi}\int_{0}^{\pi} x\{\sin(n+1)x+\sin(1-n)x\}dx$$

$$=\frac{1}{\pi}\left[x\left(\frac{-\cos(n+1)x}{n+1}-\frac{\cos(1-n)x}{1-n}\right)\right]_0^{\pi}-\int_0^{\pi}\left\{\frac{-\cos(n+1)x}{n+1}-\frac{\cos(1-n)x}{1-n}\right\}dx$$

$$=\frac{1}{\pi}\left[-\pi\left\{\frac{\cos(n+1)\pi}{n+1}+\frac{\cos(1-n)\pi}{1-n}\right\}\right]+\frac{1}{\pi}\int_0^{\pi}\left(\frac{\cos(n+1)x}{n+1}+\frac{\cos(1-n)x}{1-n}\right)dx$$

$$=-\left\{\frac{\cos(1-n)\pi}{1-n}+\frac{\cos(n+1)\pi}{n+1}\right\}+\frac{1}{\pi}\left[\frac{\sin(n+1)x}{(n+1)^2}+\frac{\sin(1-n)x}{(1-n)^2}\right]_0^{\pi}$$

$$=-\left\{\frac{(-1)^{1-n}}{1-n}+\frac{(-1)^{n+1}}{1+n}\right\}=-\left\{\frac{(-1)^{n+1}}{1-n}+\frac{(-1)^{n+1}}{1+n}\right\}\quad\left[\because(-1)^{1-n}=(-1)^{n+1}\right]$$

$$=-(-1)^{n+1}\left\{\frac{1}{1-n}+\frac{1}{1+n}\right\}=(-1)^n\frac{1+n+1-n}{1-n^2}$$

$$=\frac{2\cdot(-1)^n}{\left(1-n^2\right)}\quad\text{for } n\neq 1.$$

Now, $a_1=\frac{1}{\pi}\int_{-\pi}^{\pi}f(x)\cos x\,dx=\frac{1}{\pi}\int_{-\pi}^{\pi}x\sin x\cos x\,dx=\frac{2}{\pi}\int_0^{\pi}x\sin x\cos x\,dx$

$$=\frac{1}{\pi}\int_0^{\pi}x\sin 2x\,dx=\frac{1}{\pi}\left\{\left[\frac{-x\cos 2x}{2}\right]_0^{\pi}+\int_0^{\pi}\frac{\cos 2x}{2}dx\right\}$$

$$=\frac{1}{\pi}\left\{-\frac{\pi}{2}+\frac{1}{2}\int_0^{\pi}\cos 2x\,dx\right\}=\frac{1}{\pi}\left\{-\frac{\pi}{2}+\frac{1}{2}\left[\frac{\sin 2x}{2}\right]_0^{\pi}\right\}=\frac{1}{\pi}\left\{-\frac{\pi}{2}+\frac{1}{2}\times 0\right\}=-\frac{1}{2}$$

$$b_n=\frac{1}{\pi}\int_{-\pi}^{\pi}f(x)\sin nx\,dx=\frac{1}{\pi}\int_{-\pi}^{\pi}x\sin x\,\sin nx\,dx$$

$=0$ [$\because x\sin x\,\sin nx$ is odd function]

$$\therefore x\sin x=\frac{1}{2}\cdot 2+(a_1\cos x+b_1\sin x)+\sum_{n=2}^{\infty}\left\{\frac{2(-1)^n}{\left(1-n^2\right)}\cos nx+0\cdot\sin nx\right\}$$

or, $x\sin x=1-\frac{1}{2}\cos x+\sum_{n=2}^{\infty}\frac{2(-1)^n}{(1+n)(1-n)}\cos nx$

$$=1-\frac{1}{2}\cos x+2\left\{\frac{1}{3\cdot(-1)}\cos 2x+\frac{-1}{4(-2)}\cos 3x+\frac{1}{5(-3)}\cos 4x+\cdots\cdots\right\}$$

or, $x\sin x = 1 - \frac{1}{2}\cos x - 2\left\{\frac{\cos 2x}{1.3} - \frac{\cos 3x}{2.4} + \frac{\cos 4x}{3.5} - \cdots\cdots\right\}.$

Putting $x = \frac{\pi}{2}$ we get

$$\frac{\pi}{2}\cdot\sin\frac{\pi}{2} = 1 - \frac{1}{2}\cos\frac{\pi}{2} - 2\left\{\frac{\cos\pi}{1.3} - \frac{\cos 3\cdot\frac{\pi}{2}}{2.4} + \frac{\cos 2\pi}{3.5} - \cdots\cdots\right\}$$

or, $\frac{\pi}{2} = 1 - 2\left\{-\frac{1}{1.3} + \frac{1}{3.5} - \frac{1}{5.7} + \cdots\cdots\right\}$

or, $\frac{\pi}{4} = \frac{1}{2} - \left\{-\frac{1}{1.3} + \frac{1}{3.5} - \frac{1}{5.7} + \cdots\cdots\right\}$ or, $\frac{\pi}{4} = \frac{1}{2} + \frac{1}{1.3} - \frac{1}{3.5} + \frac{1}{5.7} - \cdots\cdots$

Ex. 4. *Draw the graph of the function and state the Wave-form :*

$f(x) = \pi + 2x,\ -\pi < x < 0$

$\quad = \pi - 2x,\ 0 \le x \le \pi.$

Deduce that $\frac{1}{1^2} + \frac{1}{3^2} + \frac{1}{5^2} + \cdots\cdots = \frac{\pi^2}{8}$

Graph of $f(x)$ (after extending by $f(x+2\pi) = f(x)$) is

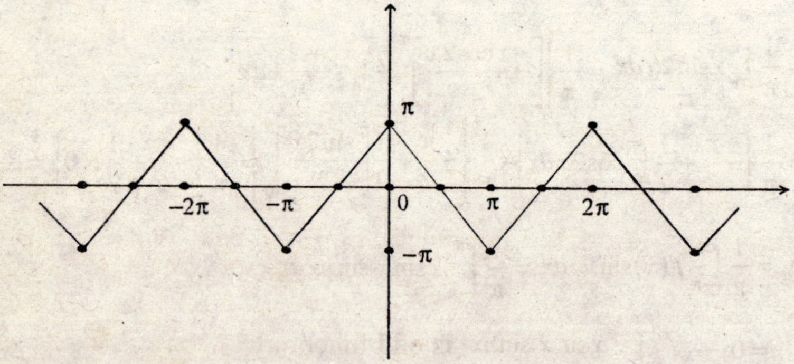

It gives Triangular Wave form.

From the graph we see $f(x)$ is even function.

Now, $a_0 = \frac{1}{\pi}\int_{-\pi}^{\pi} f(x)dx = \frac{2}{\pi}\int_0^{\pi} f(x)dx \quad \because f(x)$ is even

$= \frac{2}{\pi}\int_0^{\pi}(\pi - 2x)dx = \frac{2}{\pi}\left[\pi x - x^2\right]_0^{\pi} = \frac{2}{\pi}\left(\pi^2 - \pi^2\right) = 0.$

Now, $a_n = \frac{1}{\pi}\int_{-\pi}^{\pi} f(x)\cos nx\,dx = \frac{2}{\pi}\int_0^{\pi} f(x)\cos nx\,dx$ [$\because f(x)\cos nx$ is even]

$$=\frac{2}{\pi}\int_0^{\pi}(\pi-2x)\cos nx\,dx=\frac{2}{\pi}\left\{\pi\int_0^{\pi}\cos nx\,dx-2\int_0^{\pi}x\cos nx\,dx\right\}$$

$$=\frac{2}{\pi}\left[\pi\left[\frac{\sin nx}{n}\right]_0^{\pi}-2\left\{\left[x\frac{\sin nx}{n}\right]_0^{\pi}-\int_0^{\pi}\frac{\sin nx}{n}dx\right\}\right]$$

$$=\frac{2}{\pi}\left[\frac{2}{n}\int_0^{\pi}\sin nx\,dx\right]=\frac{-4}{\pi n}\left[\frac{\cos nx}{n}\right]_0^{\pi}=-\frac{4}{\pi n^2}(\cos n\pi-1)$$

$$=-\frac{4}{\pi n^2}\left((-1)^n-1\right)=\frac{4}{\pi n^2}\left(1-(-1)^n\right)\text{ for } n\neq 0$$

For $n>0$, $b_n=\frac{1}{\pi}\int_{-\pi}^{\pi}f(x)\sin nx\,dx=0$

[$\because f(x)\sin nx$ is odd function]

$\therefore$ the Fourier series of $f(x)$ is

$$\frac{0}{2}+\sum_{n=1}^{\infty}\frac{4}{\pi n^2}\left(1-(-1)^n\right)\cos nx$$

$$=\frac{4}{\pi}\left(\frac{2}{1^2}\cos x+\frac{2}{3^2}\cos 3x+\frac{2}{5^2}\cos 5x+\cdots\cdots\right)$$

$$=\frac{8}{\pi}\left(\frac{1}{1^2}\cos x+\frac{1}{3^2}\cos 3x+\frac{1}{5^2}\cos 5x+\cdots\cdots\right)$$

From the graph we see $f(x)$ is bounded in the interval $[-\pi,\pi]$ and it is monotonic in the two sub-intervals $(-\pi,0)$ and $(0,\pi)$. So $f(x)$ satisfies Dirichlet's Condition. Moreover we see, from the graph, $f(x)$ is continuous everywhere. So

$$f(x)=\frac{8}{\pi}\left(\frac{1}{1^2}\cos x+\frac{1}{3^2}\cos 3x+\frac{1}{5^2}\cos 5x+\cdots\cdots\right)$$

Putting $x=0$ we get, $f(0)=\frac{8}{\pi}\left(\frac{1}{1^2}+\frac{1}{3^2}+\frac{1}{5^2}+\cdots\cdots\right)$

or, $\pi=\frac{8}{\pi}\left(\frac{1}{1^2}+\frac{1}{3^2}+\frac{1}{5^2}+\cdots\cdots\right)$ or, $\frac{1}{1^2}+\frac{1}{3^2}+\frac{1}{5^2}+\cdots\cdots=\frac{\pi^2}{8}$.

Ex. 5. *Find a series of sines and cosines of multiples of x which represents* $f(x)$ *in the interval* $-\pi < x < \pi$ *when*

$$f(x)=0,\ -\pi < x \le 0$$
$$=\frac{\pi x}{4},\ 0 < x < \pi .$$

Hence deduce that $1+\frac{1}{3^2}+\frac{1}{5^2}+\cdots\cdots=\frac{\pi^2}{8}$.

The Fourier co-efficients,

$$a_0=\frac{1}{\pi}\int_0^{\pi}\frac{\pi x}{4}dx=\frac{\pi^2}{8}$$

and $a_n=\frac{1}{\pi}\int_{-\pi}^{\pi}f(x)\cos nx\,dx=\frac{1}{\pi}\int_0^{\pi}\frac{\pi x}{4}\cos nx\,dx=\frac{1}{4n^2}(\cos n\pi-1)$ for $n\neq 0$.

$$b_n=\frac{1}{\pi}\int_{-\pi}^{\pi}f(x)\sin nx\,dx=\frac{1}{\pi}\int_0^{\pi}\frac{\pi x}{4}\sin nx\,dx=-\frac{\pi}{4n}\cos n\pi .$$

Obviously $f(x)$ satisfies Dirichlet's condition. So, as $f(x)$ is continuous in

$$-\pi < x < \pi,\ f(x)=\frac{1}{2}\cdot\frac{\pi^2}{8}+\sum_{n=1}^{\infty}\left\{\frac{1}{4n^2}(\cos n\pi-1)\cos nx-\frac{\pi}{4n}\cos n\pi\sin nx\right\}$$

or, $f(x)=\frac{\pi^2}{16}+\frac{1}{2}\left(-\cos x+\frac{\pi}{2}\sin x\right)-\frac{\pi}{8}\sin 2x+\cdots\cdots$

when $-\pi < x < \pi$.

At $x=\pi$, the Fourier series converges to

$$\frac{1}{2}\left\{\lim_{t\to-\pi+}f(t)+\lim_{t\to-\pi-}f(t)\right\}=\frac{1}{2}\left(0+\lim_{t\to\pi}\frac{\pi t}{4}\right)=\frac{\pi^2}{8}.$$

$\therefore$ Putting $x=\pi$ in the F-series we get

$$\frac{\pi^2}{8}=\frac{\pi^2}{16}+\frac{1}{2}\left(1+\frac{1}{3^2}+\frac{1}{5^2}+\cdots\cdots\right)\text{ or, }1+\frac{1}{3^2}+\frac{1}{5^2}+\cdots\cdots=\frac{\pi^2}{8}.$$

Ex. 6. *Expand* $\cos px$ *in* $[-\pi,\pi]$ *(p not being an integer) in Fourier series. Deduce the value of* $\sum_{n=0}^{\infty}(-1)^n\left(\frac{1}{n+p}+\frac{1}{n+1-p}\right)$.

Here $f(x) = \cos px$ is an even function. Its Fourier Co-efficients are

$$a_0 = \frac{1}{\pi}\int_{-\pi}^{\pi} \cos px\, dx = \frac{1}{p\pi}[\sin px]_{-\pi}^{\pi} \quad = \frac{2}{p\pi}\sin p\pi$$

$$a_n = \frac{1}{\pi}\int_{-\pi}^{\pi} \cos px \cos nx\, dx$$

$$= \frac{2}{\pi}\int_0^{\pi} \cos px \cos nx\, dx \quad [\because \cos px \cos nx \text{ is even function}]$$

$$= \frac{1}{\pi}\int_0^{\pi} \{\cos(p+n)x + \cos(p-n)x\} dx$$

$$= \frac{\sin(p+n)\pi}{(p+n)\pi} + \frac{\sin(p-n)\pi}{(p-n)\pi}.$$

Now $\sin(p+n)\pi = \sin(p-n)\pi = (-1)^n \sin p\pi$.

$$\therefore a_n = \frac{\sin(p+n)\pi}{(p+n)\pi} + \frac{\sin(p-n)\pi}{(p-n)\pi} = \frac{(-1)^n}{\pi}\frac{2p\sin p\pi}{p^2 - n^2}.$$

Now, $b_n = 0$ since $\cos px \sin nx$ is odd function.

Again $\cos px$ satisfies Dirichlet's Condition and since $\cos px$ is continuous everywhere therefore its Fourier Expansion is

$$\cos px = \frac{\sin p\pi}{p\pi} + \sum_{n=1}^{\infty}\left\{\frac{(-1)^n 2p\sin p\pi}{\pi(p^2 - n^2)}\cos nx + 0\cdot \sin nx\right\}$$

$$\text{or, } \cos px = \frac{\sin p\pi}{p\pi} + \sum_{n=1}^{\infty}\frac{(-1)^n 2p\sin p\pi}{\pi(p^2 - n^2)}\cos nx$$

Putting $x = 0$ we get

$$1 = \frac{\sin p\pi}{p\pi} + \sum_{n=1}^{\infty}\frac{(-1)^n 2p\sin p\pi}{\pi(p^2 - n^2)} = \frac{\sin p\pi}{p\pi} + \frac{2p\sin p\pi}{\pi}\sum_{n=1}^{\infty}\frac{(-1)^n}{p^2 - n^2}$$

$$= \frac{\sin p\pi}{\pi}\left\{\frac{1}{p} - \frac{1}{p+1} - \frac{1}{p-1} + \frac{1}{p+2} + \frac{1}{p-2} - \frac{1}{p+3} - \frac{1}{p-3} + \cdots\cdots\right\}$$

$$\text{or, } \frac{\pi}{\sin p\pi} = \sum_{n=0}^{\infty}(-1)^n\left(\frac{1}{n+p} + \frac{1}{n+1-p}\right).$$

Ex. 7. *Find a Fourier series to represent* x^2 *in the interval* $(-l,l)$.

The Fourier Co-efficients.

$$a_0 = \frac{1}{l}\int_{-l}^{l} x^2 dx = \frac{2l^2}{3}$$

$$a_n = \frac{1}{l}\int_{-l}^{l} x^2 \cos nx\, dx = \frac{2}{l}\int_{0}^{l} x^2 \cos nx\, dx \quad (\because x^2 \cos nx \text{ is even}) \text{ for } n \neq 0$$

$$= \frac{4l^2 \cos n\pi}{n^2\pi^2} \quad \text{[doing integ by part twice]}$$

$$= \frac{4l^2(-1)^n}{n^2\pi^2},\ n \neq 0.$$

and $b_n = \frac{1}{l}\int_{-l}^{l} x^2 \sin nx dx = 0$ $\qquad \because x^2 \sin nx$ is odd function.

Again the graph of $y = x^2$ is shown in the following figure (it is a parabola).

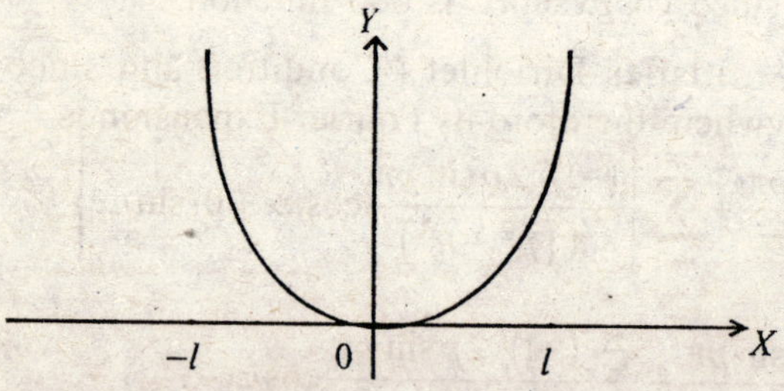

This is bounded in $[-l,l]$ and monotonic in the two sub-interval $(-l,0)$ and $(0,l)$. So it satisfies Dirichlet's Condition. Since it is continuous everywhere so

$$x^2 = \frac{1}{2}\left(\frac{2l^2}{3}\right) + \sum_{n=1}^{\infty}\left(\frac{4l^2(-1)^n}{n^2\pi^2}\cos nx + 0\cdot \sin nx\right)$$

or, $x^2 = \frac{l^2}{3} + \frac{4l^2}{\pi^2}\sum_{n=1}^{\infty}\frac{(-1)^n \cos nx}{n^2}$ which is the required Fourier Expansion.

Ex. 8. *Find the Fourier series of the function* e^{-x} *in the inerval* $0 < x < 2\pi$.

The interval is changed.

Here $c = 0$, $c + 2T = 2\pi$ $\therefore T = \pi$. $\therefore$ the Fourier co-efficients are

$$a_0 = \frac{1}{\pi}\int_0^{0+2\pi} e^{-x}\,dx = \frac{1-e^{-2\pi}}{\pi}$$

and $a_n = \frac{1}{\pi}\int_0^{0+2\pi} e^{-x}\cos nx\,dx = \frac{1}{\pi}\int_0^{2\pi} e^{-x}\cos nx\,dx$

$$= \frac{1}{\pi}\left\{\left[e^{-x}\frac{\sin nx}{n}\right]_0^{2\pi} + \int_0^{2\pi} e^{-x}\frac{\sin nx}{n}\,dx\right\}$$

$$= \frac{1}{\pi}\cdot\frac{1}{n}\int_0^{2\pi} e^{-x}\sin nx\,dx$$

$$= \frac{1}{n\pi}\left\{-\left[e^{-x}\frac{\cos nx}{n}\right]_0^{2\pi} - \int_0^{2\pi} e^{-x}\frac{\cos nx}{n}\,dx\right\}$$

$$= \frac{1}{n\pi}\left\{-\left(e^{-2\pi}\frac{\cos 2n\pi}{n} - \frac{1}{n}\right) - \frac{1}{n}\int_0^{2\pi} e^{-x}\cos nx\,dx\right\}$$

$$a_n = \frac{1}{n\pi}\left\{-\left(\frac{e^{-2\pi}}{n} - \frac{1}{n}\right) - \frac{1}{n}\pi a_n\right\}$$

or, $\pi n a_n = \frac{1}{n} - \frac{e^{-2\pi}}{n} - \frac{\pi a_n}{n}$

or, $\pi a_n\left(n + \frac{1}{n}\right) = \frac{1-e^{-2\pi}}{n}$

$$\therefore a_n = \frac{1-e^{-2\pi}}{\pi\left(n^2+1\right)}$$

$$b_n = \frac{1}{\pi}\int_0^{0+2\pi} e^{-x}\sin nx\,dx = \frac{1}{\pi}\int_0^{2\pi} e^{-x}\sin nx\,dx$$

or, $b_n = \frac{n\left(1-e^{-2\pi}\right)}{\pi\left(n^2+1\right)}$ (As above).

The graph of e^{-x} is shown below in $[0, 2\pi]$

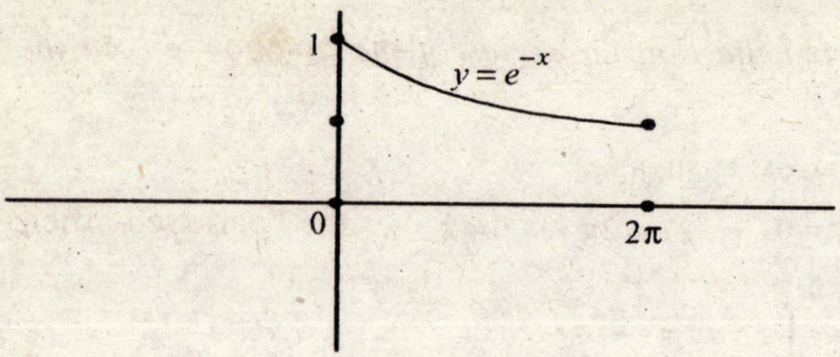

From the graph we see e^{-x} is monotonic in $[0,2\pi]$ and continuous. So the Fourier expansion, after extending it to a period ic function of period 2π,

$$e^{-x}=\frac{1}{2}\cdot\frac{1-e^{-2\pi}}{\pi}+\sum_{n=1}^{\infty}\left\{\frac{1-e^{-2\pi}}{\pi\left(n^2+1\right)}\cos nx+\frac{n\left(1-e^{-2\pi}\right)}{\pi\left(n^2+1\right)}\sin nx\right\}$$

$$\text{or, } e^{-x}=\frac{1-e^{-2\pi}}{\pi}\left\{\frac{1}{2}+\sum_{n=1}^{\infty}\left(\frac{\cos nx}{n^2+1}+\frac{n\sin nx}{n^2+1}\right)\right\}$$

$$\text{or, } e^{-x}=\frac{1-e^{-2\pi}}{\pi}\left\{\frac{1}{2}+\sum_{n=1}^{\infty}\frac{1}{n^2+1}(\cos nx+n\sin nx)\right\}.$$

Ex. 9. $f(x)=\frac{1}{4}-x,\ 0<x\le\frac{1}{2}$

$=x-\frac{3}{4},\ \frac{1}{2}<x<1.$ [W.B.U.Tech 2005]

Find the half range sine series of this function. State whether $f(x)$ equals this series. How $f(x)$ would be defined at $x=0$ so that the series converges to $f(x)$.

The function is defined on $(0,1)$. To get its Half-Range sine series we extend the function to the interval $(-1,0)$ by defining $f(-x)=-f(x)$. This extended function becomes odd. Therefore,

$$a_0=\frac{1}{1}\int_{-1}^{1}f(x)dx=0$$

,

$$\text{and } a_n=\frac{1}{1}\int_{-1}^{1}f(x)\cos\frac{n\pi x}{1}dx=\int_{-1}^{1}f(x)\cos n\pi x\,dx=0$$

$$b_n=\frac{1}{1}\int_{-1}^{1}f(x)\sin\frac{n\pi x}{1}dx=2\int_{0}^{1}f(x)\sin n\pi x\,dx$$

$$= 2\left\{\int_0^{1/2}\left(\frac{1}{4}-x\right)\sin n\pi x\,dx + \int_{1/2}^{1}\left(x-\frac{3}{4}\right)\sin n\pi x\,dx\right\} \tag{1}$$

Now, $\int_0^{1/2}\left(\frac{1}{4}-x\right)\sin n\pi x\,dx = \left[-\left(\frac{1}{4}-x\right)\frac{\cos n\pi x}{n\pi}\right]_0^{1/2} - \int_0^{1/2}(-1)\frac{-\cos n\pi x}{n\pi}\,dx$

$$= \left[\left(x-\frac{1}{4}\right)\frac{\cos n\pi x}{n\pi}\right]_0^{1/2} - \frac{1}{n\pi}\int_0^{1/2}\cos n\pi x\,dx$$

$$= \left(\frac{1}{2}-\frac{1}{4}\right)\frac{\cos\frac{n\pi}{2}}{n\pi} - \left(-\frac{1}{4}\right)\frac{1}{n\pi} - \frac{1}{n\pi}\left[\frac{\sin n\pi x}{n\pi}\right]_0^{1/2}$$

$$= \frac{\cos\frac{n\pi}{2}}{4n\pi} + \frac{1}{4n\pi} - \frac{1}{(n\pi)^2}\sin\frac{n\pi}{2}.$$

Similarly we can find the second part

$$\int_{1/2}^{1}\left(x-\frac{3}{4}\right)\sin n\pi x\,dx = -\frac{\cos\frac{n\pi}{2}}{4n\pi} - \frac{\cos n\pi}{4n\pi} - \frac{\sin\frac{n\pi}{2}}{n^2\pi^2}.$$

Then from (1), after simplification, we get

$$b_n = \frac{1}{2n\pi}\left\{1-(-1)^n\right\} - \frac{4\sin\frac{n\pi}{2}}{n^2\pi^2}.$$

The graph of $f(x)$ on $(-1,1)$ is shown below :

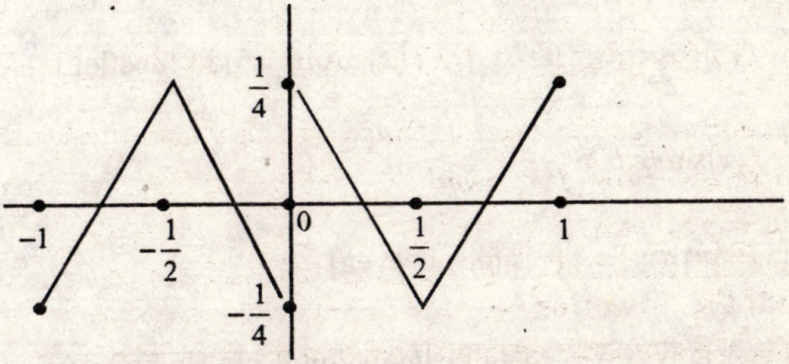

Obviously it satisfies Dirichlet's Condition but discontinuous at $x=0$. The series converges at $x=0$ to the value

$$\frac{1}{2}\left\{\lim_{t\to 0-} f(t) + \lim_{t\to 0+} f(t)\right\}$$

$$= \frac{1}{2}\left\{-\frac{1}{4}+\frac{1}{4}\right\} = 0.$$

$\therefore f(x)=0$ at $x=0$ so that the series converges to $f(x)$.

The series is

$$\frac{0}{2}+\sum_{n=1}^{\infty}\left[\frac{1}{2n\pi}\left\{1-(-1)^n\right\}-\frac{4\sin\frac{n\pi}{2}}{n^2\pi^2}\right]\sin\frac{n\pi x}{1}$$

$$=\sum_{n=1}^{\infty}\left[\frac{1}{2n\pi}\left\{1-(-1)^n\right\}-\frac{4\sin\frac{n\pi}{2}}{n^2\pi^2}\right]\sin n\pi x$$

$$=\left(\frac{1}{2\pi}\cdot 2-\frac{4\cdot 1}{\pi^2}\right)\sin\pi x+\left(\frac{1}{4\pi}\times 0-\frac{0}{2^2\pi^2}\right)\sin 2\pi x+\left(\frac{1}{6\pi}\cdot 2-\frac{4(-1)}{3^2\cdot\pi^2}\right)\sin 3\pi x+$$

$$=\left(\frac{1}{\pi}-\frac{4}{\pi^2}\right)\sin\pi x+\left(\frac{1}{3\pi}+\frac{4}{3^2\pi^2}\right)\sin 3\pi x+\left(\frac{1}{5\pi}-\frac{4}{5^2\pi^2}\right)\sin 5\pi x+\ldots\ldots$$

Ex. 10. *Find the sine series which represents the function* $f(x)=\pi-x$ *in* $0<x<\pi$.

To get its Half range sine series we extend the function to the interval $(-\pi,0)$ defining by $f(-x)=-f(x)$. This extended function becomes odd. Therefore,

$$a_0=\frac{1}{\pi}\int_{-\pi}^{\pi}f(x)dx=0$$

$$a_n=\frac{1}{\pi}\int_{-\pi}^{\pi}f(x)\cos nx\,dx=0 \qquad [\because f(x)\cos nx \text{ is odd function}]$$

$$b_n=\frac{1}{\pi}\int_{-\pi}^{\pi}f(x)\sin nx\,dx$$

$$=\frac{2}{\pi}\int_0^{\pi}f(x)\sin nx\,dx \quad [\because f(x)\sin nx \text{ is even}]$$

$$=\frac{2}{\pi}\int_0^{\pi}(\pi-x)\sin nx\,dx=\frac{2}{n} \quad \text{(detail calculation is not shown)}$$

$\therefore$ the Fourier series is

$$\frac{0}{2}+\sum_{n=1}^{\infty}\left(0\cdot\cos nx+\frac{2}{n}\sin nx\right)=2\sum_{n=1}^{\infty}\frac{\sin nx}{n}$$

$$=2\left(\frac{\sin x}{1}+\frac{\sin 2x}{2}+\frac{\sin 3x}{3}+\ldots\ldots\right)$$

which is the required sine series.

Ex. 11. *Expand the following function $f(x)$ into a Fourier cosine series :*

$$f(x)=1,\quad 0\le x<\frac{1}{3}\pi$$
$$=0,\quad \frac{1}{3}\pi\le x\le\frac{2}{3}\pi$$
$$=-1,\quad \frac{2}{3}\pi<x\le\pi.$$

Is the series convergent at all values of x.

To get the Fourier cosine series we extend the function in the interval $[-\pi,0]$ defining by $f(-x)=f(x)$. This extension is shown in the following figure. The extended function becomes an even function in the interval $[-\pi,\pi]$.

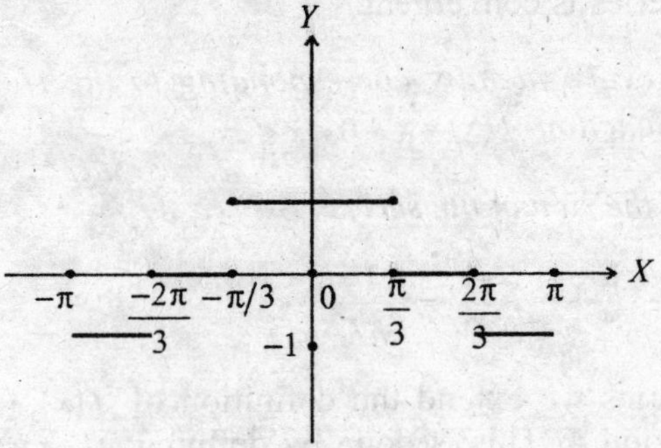

Now, $a_0=\frac{1}{\pi}\int_{-\pi}^{\pi}f(x)dx=\frac{2}{\pi}\left\{\int_0^{\pi/3}dx+\int_{\pi/3}^{2\pi/3}0\cdot 1dx+\int_{2\pi/3}^{\pi}-1\cdot dx\right\}$

$$=\frac{2}{\pi}\left\{[x]_0^{\pi/3}-[x]_{2\pi/3}^{\pi}\right\}=0.$$

and for $n\ne 0$ $a_n=\frac{1}{\pi}\int_{-\pi}^{\pi}f(x)\cos\frac{n\pi x}{\pi}dx$

$$=\frac{2}{\pi}\int_0^{\pi}f(x)\cos nx\,dx\quad [\because f(x)\cos nx \text{ is even function}]$$

$$=\frac{2}{\pi}\left\{\int_0^{\pi/3}1\cdot\cos nx\,dx+\int_{\pi/3}^{\frac{2\pi}{3}}0\cdot\cos nx\,dx+\int_{\frac{2\pi}{3}}^{\pi}-1\cdot\cos nx\,dx\right\}$$

$$= \frac{2}{\pi}\left\{\left[\frac{\sin nx}{n}\right]_0^{\pi/3} - \left[\frac{\sin nx}{n}\right]_{2\pi/3}^{\pi}\right\} = \frac{2}{\pi}\left\{\frac{1}{n}\left(\sin\frac{n\pi}{3}\right) + \frac{1}{n}\left(\sin\frac{2n\pi}{3}\right)\right\}$$

$$= \frac{2}{n\pi}\left(\sin\frac{n\pi}{3} + \sin\frac{2n\pi}{3}\right), \text{ for } n \neq 0.$$

$b_n = 0$, since $f(x)\sin nx$ is odd function.

So the Fourier series is

$$\frac{0}{2} + \sum_{n=1}^{\infty}\frac{2}{n\pi}\left(\sin\frac{n\pi}{3} + \sin\frac{2n\pi}{3}\right)\cos nx = \sum_{n=1}^{\infty}\frac{2}{n\pi}\left(\sin\frac{n\pi}{3} + \sin\frac{2n\pi}{3}\right)\cos nx.$$

Obviously the function satisfies Dirichlet's Condition on the interval $[-\pi, \pi]$.

So this Fourier series is convergent.

Ex. 12. *Write Parseval's identity corresponding to the Half Range cosine series of the function* $f(x) = x$, $0 < x < 2$.

Hence determine the sum of the series

$$\frac{1}{1^4} + \frac{1}{2^4} + \frac{1}{3^4} + \cdots\cdots + \frac{1}{n^4} + \cdots\cdots$$

To get cosine series we extend the definition of $f(x)$ to that of even function of period 4. This is done by defining $f(-x) = f(x)$ in $-2 < x \leq 0$.

Then, $a_0 = \frac{1}{2}\int_{-2}^{2} f(x)dx = \int_0^2 x\,dx = 2$

and $a_n = \frac{1}{2}\int_{-2}^{2} f(x)\cos\frac{n\pi x}{2}dx = \frac{2}{2}\int_0^2 f(x)\cos\frac{n\pi x}{2}dx$ $\left[\because f(x)\cos\frac{n\pi x}{2} \text{ is even}\right]$

$= \int_0^2 x\cos\frac{n\pi x}{2}dx = \frac{4}{n^2\pi^2}(\cos n\pi - 1)$ for $n \neq 0$ (obtained by integrating by parts)

$b_0 = \frac{1}{\pi}\int_{-2}^{2} f(x)\sin\frac{n\pi x}{2}dx = 0$ $\left[\because f(x)\sin\frac{n\pi x}{2} \text{ is odd function}\right]$.

Obviously $f(x)$ satisfies Dirichlet's Condition and continuous everywhere. So it can be expanded as

$$f(x)=\frac{2}{2}+\sum_{n=1}^{\infty}\left\{\frac{4}{n^2\pi^2}(\cos n\pi-1)\cos\frac{n\pi x}{2}+0\cdot\sin\frac{n\pi x}{2}\right\}$$

$$=1+\sum_{n=1}^{\infty}\frac{4}{n^2\pi^2}(\cos n\pi-1)\cos\frac{n\pi x}{2}.$$

Then the Parseval's identity becomes

$$\frac{1}{2}\int_{-2}^{2}\{f(x)\}^2=\frac{2^2}{2}+\sum_{n=1}^{\infty}\left[\left\{\frac{4}{n^2\pi^2}(\cos n\pi-1)\right\}^2+0^2\right]$$

or, $\frac{1}{2}\int_{-2}^{2}x^2dx=2+\sum_{n=1}^{\infty}\frac{16}{n^4\pi^4}(\cos n\pi-1)^2$

or, $\frac{8}{3}=2+\sum_{n=1}^{\infty}\frac{16}{n^4\pi^4}\left\{(-1)^n-1\right\}^2$

or, $\frac{8}{3}-2=\frac{16}{\pi^4}\left\{\frac{4}{1^4}+\frac{4}{3^4}+\frac{4}{5^4}+\cdots\cdots\infty\right\}$

or, $\frac{2}{3}=\frac{16\times4}{\pi^4}\left(\frac{1}{1^4}+\frac{1}{3^4}+\frac{1}{5^4}+\cdots\cdots\right)$

or, $\frac{\pi^4}{96}=\frac{1}{1^4}+\frac{1}{3^4}+\frac{1}{5^4}+\cdots\cdots$ (1)

Second part : Let $S=\frac{1}{1^4}+\frac{1}{2^4}+\frac{1}{3^4}+\cdots\ \cdots+\frac{1}{n^4}+\cdots\cdots\infty$.

Then, $S=\left(\frac{1}{1^4}+\frac{1}{3^4}+\frac{1}{5^4}+\cdots\cdots\right)+\left(\frac{1}{2^4}+\frac{1}{4^4}+\frac{1}{6^4}+\cdots\cdots\right)$

or, $S=\frac{\pi^4}{96}+\frac{1}{2^4}S$ from (1)

or, $S-\frac{1}{16}S=\frac{\pi^4}{96}$ or, $S=\frac{\pi^4}{90}$

$\therefore \frac{1}{1^4}+\frac{1}{2^4}+\frac{1}{3^4}+\cdots\cdots=\frac{\pi^4}{90}.$

EXERCISES

1. Find the period of the function $\cos 2x$, $\sin 2\pi x$, $4+\sin 2\pi x$.

2. Drawing the graph of the periodic function $f(x)$ (of period 2π) given by

   $f(x)=0$, $-\pi<x<0$

   $=\sin x$, $0<x<\pi$

   show that it is half waveform.

3. State the Waveform given by the periodic function

   $f(x)=\pi+x$, $-\pi<x<0$

   $=\pi-x$, $0<x<\pi$

   and $f(x+2\pi)=f(x)$.

4. Draw the graph of the function $f(x)$ given by

   $f(x)=-\frac{\pi}{4}$, $-\pi<x<0$

   $=\pi/4$, $0\le x\le\pi$ and $f(x+2\pi)=f(x)$.

   Show that it is square waveform.

5. Draw the graph of the function $f(x)=\frac{x}{2}$, $-\pi<x\le\pi$ and $f(x+2\pi)=f(x)$. Show that it is Saw toothed Waveform.

6. Draw the graph of the function $f(x)$ given by $f(x)=x^2$, $-\pi<x\le\pi$ and $f(x+2\pi)=f(x)$.

7. Draw the graph of the periodic function (of period k) given by $f(x)=x^2$, $-k<x\le 0$.

8. Draw the graph of the periodic function $f(x)=|x|$, $-1<x\le 1$ and $f(x+2)=f(x)$ for all x.

9. Find the period of the function $f(x)=|\sin x|$. Draw its graph. Indicate which type of waveform is this.

10. Show that the function $f(x)=x$, $0<x<2$ gives a saw-toothed wave form.

11. Find whether the following function is even or odd :

    (i) $f(x)=2$, $0<x<3$

    $=-2$, $-3<x<0$.

    (ii) $f(x)=\cos x$, $0<x<\pi$

    $=0$, $\pi<x<2\pi$.

(iii) $f(x)=x(10-x)$, $0<x<10$.

12. Show that every function is a sum of a even function and an odd function. [*W.B.U.Tech 2005*]

[Hint : Express $f(x)=\frac{1}{2}\{f(x)+f(-x)\}+\frac{1}{2}\{f(x)-f(-x)\}$

$=\phi(x)+\psi(x)$.]

13. Show that for $-\pi<x<\pi$,

$$x\cos x=-\frac{1}{2}\sin x+2\left(\frac{2}{1.3}\sin 2x-\frac{3}{2.4}\sin 3x+\frac{4}{3.5}\sin 4x-\cdots\cdots\right)$$

14. Expand the Fourier series for $f(x)=x$ over $-\pi\le x\le\pi$. Hence deduce that $1-\frac{1}{3}+\frac{1}{5}-\cdots\cdots=\frac{\pi}{4}$.

15. Find the Fourier series corresponding to the function

$f(x)=\pi-x$, $0<x<\pi$

$=x-\pi$, $-\pi<x<0$.

What is the value of the series at $x=0$ and at $x=\pi$.

16. Obtain the Fourier series of the function

$f(x)=0$, $-\pi<x\le 0$

$=\pi$, $0<x\le\pi$.

Find whether the series is convergent.

17. Find the Fourier series of the function

$f(x)=0$, $-2<x<0$

$=1$, $0<x<2$ and $f(x+4)=f(x)$.

Find $f(0)$, $f(-2)$ and $f(2)$ so that the Fourier series converges to $f(x)$ for $-2\le x\le 2$.

18. Find the Fourier expansion of x^2 on $[-\pi,\pi]$.

Hence find the value of

(i) $1+\frac{1}{2^2}+\frac{1}{3^2}+\cdots\cdots$ $\pi^2/6$ Putting $x=\pi$

(ii) $1-\frac{1}{2^2}+\frac{1}{3^2}-\frac{1}{4^2}+\cdots\cdots$ $\pi^2/12$ [*W.B.U.Tech 2005*]

Putting $x=0$

(iii) $1+\frac{1}{3^2}+\frac{1}{5^2}+\cdots\cdots$ $\pi^2/8$

Putting $x=$

19. Prove that $x+x^2=\frac{\pi^2}{3}-4\left\{\frac{\cos x}{1^2}-\frac{\cos 2x}{2^2}+\frac{\cos 3x}{3^2}-\ldots\ldots\right\}$

$$+2\left\{\frac{\sin x}{1}-\frac{\sin 2x}{2}+\frac{\sin 3x}{3}-\ldots\ldots\right\}$$

Hence find the value of $\frac{1}{1^2}+\frac{1}{2^2}+\frac{1}{3^2}+\ldots\ldots$

20. Find the Fourier series for the function defined by

$f(x)=1,\ 0<x<\pi$

$=0,\ x=0$

$=-1,\ -\pi<x<0.$

Tell whether the series is an expansion of $f(x)$.

Hence deduce the value of the series, $1-\frac{1}{3}+\frac{1}{5}-\frac{1}{7}+\ldots\ldots$

21. Find the Fourier series of the function

$f(x)=-k,\ -\pi<x<0$

$=k,\ 0<x<\pi$ and $f(x+2\pi)=f(x)$.

22. Find the Fourier series of the function

$f(x)=0,\ -2<x<-1$

$=k,\ -1<x<1$

$=0,\ 1<x<2.$

Draw the graph of the extended periodic function $f(x)$ of period 4. Mention the Waveform given.

23. A sinusoidal voltage $E\sin wt$ is passed through a half- wave rectifier which clips the negative portion of the wave. Find the Fourier series of the resulting periodic function.

$f(t)=0,\quad -\frac{T}{2}<t<0$

$=E\sin wt,\ 0<t<\frac{T}{2},\quad$ where $T=\frac{2\pi}{w}$.

24. Find the waveform given by the function $f(x)=x+\pi,\ -\pi<x<\pi$ and $f(x+2\pi)=f(x)$. Obtain its Fourier series.

25. Find the Fourier expansion for $f(x)$,

if $f(x) = -\pi,\ -\pi < x < 0$

$= x,\ 0 < x < \pi$.

Hence deduce that $\frac{1}{1^2} + \frac{1}{3^2} + \frac{1}{5^2} + \cdots\cdots = \frac{\pi^2}{8}$.

[Hint : at $x = 0$ the Four. series $= \frac{1}{2}\left\{\lim_{t \to 0-} f(t) + \lim_{t \to 0+} f(t)\right\} = -\frac{\pi}{2}$]

26. Expand $f(x) = e^{-x}$ as a Fourier series in the interval $(-4,4)$.

27. Find the Fourier series for the function

$f(x) = 1 + \frac{2x}{\pi},\ -\pi \le x \le 0$

$= 1 - \frac{2x}{\pi},\ 0 \le x \le \pi$.

Hence deduce that $\frac{1}{1^2} + \frac{1}{3^2} + \frac{1}{5^2} + \cdots\cdots = \frac{\pi^2}{8}$.

28. Find the Fourier series corresponding to the function

$f(x) = -x,\ -\pi < x < 0$

$= 0;\ 0 < x < \pi$.

29. Find the Fourier series of the function $|\sin x|,\ -\pi < x \le \pi$. Find whether this Fourier series is a Fourier expansion of $|\sin x|$.

30. Let $f(x) = x,\ 0 \le x \le \frac{\pi}{2}$

$= \pi - x,\ \frac{\pi}{2} \le x \le \pi$

$= -f(-x),\ -\pi \le x \le 0$.

Show that $f(x)$ satisfies Dirichlet's condition on $[-\pi,\pi]$. Obtain the Fourier series for f in $[-\pi,\pi]$.

31. Find the Fourier series of the function

$f(x) = -\frac{\pi}{4},\ -\pi \le x \le 0$

$= \frac{\pi}{4},\ 0 < x \le \pi$.

32. Find the Fourier series of the function

$$f(x)=0,\ -\pi<x<0$$
$$=\frac{1}{2},\ x=0$$
$$=1,\ 0<x<\pi.$$

Does it converge at $x=0$. Find the value at this point.

33. Find the Fourier series corresponding to the function

$$f(x)=x,\ -\pi\le x\le 0$$
$$=2x,\ 0\le x\le \pi.$$

Does the series equals $f(x)$ everywhere ?

34. Draw the graph of the following periodic function of period 2π. Identity the waveform given by it. Expand the function into its Fourier expansion :

$$f(x)=0,\ -\pi\le x\le 0$$
$$=\sin x,\ 0<x\le \pi.$$

Hence show that $\frac{1}{1.3}-\frac{1}{3.5}+\frac{1}{5.7}-\cdots,\cdots\infty=\frac{1}{4}(\pi-2)$.

35. Prove that $f(x)=\frac{\pi^2}{16}+\sum_{n=1}^{\infty}\frac{1}{4\pi^2}(\cos n\pi-1)\cos nx-\sum_{n=1}^{\infty}\frac{\pi}{4n}\cos n\pi\sin nx$

when

$$f(x)=0,\ -\pi<x\le 0$$
$$=\frac{\pi x}{4},\ 0<x<\pi.$$

Hence deduce that $1+\frac{1}{3^2}+\frac{1}{5^2}+\frac{1}{7^2}+\cdots\cdots=\frac{\pi^2}{8}$.

36. Find the Fourier expansion of the following function f of period 2π defined as $f(x)=0,\ -\pi\le x\le 0$

$$=x,\ 0<x\le\pi.$$

Find the sum of the series at $x=-5\pi$. Hence deduce that $1+\frac{1}{3^2}+\frac{1}{5^2}+\cdots\cdots=\frac{\pi^2}{8}$.

37. Show that on $-\pi \le x \le \pi$,

$$\frac{1}{2}\pi - x \sim \frac{1}{2}\pi - 2\left(\sin x - \frac{1}{2}\sin 2x + \frac{1}{3}\sin 3x - \cdots\cdots\right).$$

38. Draw the graph of the function $f(x)=0,\ -\pi<x<0$

$$=x^2,\ 0<x<\pi.$$

If $f(2\pi+x)=f(x)$, obtain Fourier series of $f(x)$.

39. Find the Fourier series of the function

$$f(x)=x^2,\ 0\le x\le \pi$$
$$=-x^2,\ -\pi\le x\le 0.$$

40. Find a Fourier expansion for $f(t)=1-t^2,\ -1\le t\le 1$.

41. Expand the function $\sin ax$ in Fourier series in $(-\pi,\pi)$.

42. Find the Fourier series of $|x|,\ -\pi<x<\pi$.

Hence deduce that $\frac{1}{1^2}+\frac{1}{3^2}+\frac{1}{5^2}+\frac{1}{7^2}+\cdots\cdots=\frac{\pi^2}{8}$.

43. Expand $f(x)=|\cos x|$ as a Fourier series in $(-\pi,\pi)$. Does it converge to $f(x)$ for all values of x. Give reasons.

44. If $f(x)=x+x^2,\ -\pi<x<\pi$, be repersented in Fourier series as $\frac{a_0}{2}+\sum_{n=1}^{\infty}(a_n\cos nx+b_n\sin nx)$ then find a_0. [*W.B.U Tech, 2004*]

45. A function $f(x)=x^2, -\pi<x\le\pi$ is represented by a Fourier Series as $f(x)=\frac{a_0}{2}+\sum_{n=1}^{\infty}(a_n\cos nx+b_n\sin nx)$. Find a_0 and show that $b_n=0$ for all n. [*W.B.U.Tech 2005*]

46. Expand $x\sin x,\ 0<x<2\pi$ in a Fourier series.

47. Expand $f(x)=x^2,\ 0<x<2\pi$ in a Fourier series. Hence find the value of $\frac{1}{1^2}+\frac{1}{2^2}+\frac{1}{3^2}+\cdots\cdots$

48. $f(x)$ with period 2π is defined by the relation

$$f(x)=\frac{\pi-x}{2},\ 0<x<2\pi$$
$$=0;\ x=0.$$

Find its Fourier series .

49. Express $f(x)=\pi-x$ in a cosine series in $0<x<\pi$. Prove that the sum of the series for $x=\pm\pi$ is 0. Hence deduce the sum $\frac{1}{1^2}+\frac{1}{3^2}+\frac{1}{5^2}+\cdots\cdots$

50. Given $f(x)=\frac{1}{4}\pi x, \quad 0\le x\le\frac{\pi}{2}$

$$=\frac{1}{4}\pi(\pi-x);\ \frac{\pi}{2}\le x\le\pi.$$

Expand $f(x)$ in an infinite series of

(i) Cosines of multiples of x in $[0,\pi]$.

(ii) Sines of multiples of x.

51. Expand $f(x)=x$, $0<x<2$ in a half range cosine series.

52. Find the half Range sine expansion of the function

$$f(x)=\frac{2k}{l}x, \quad 0<x<\frac{l}{2}$$
$$=\frac{2k}{l}(l-x),\ \frac{l}{2}<x<l.$$

Mention its waveform also.

53. Find the Half Range Cosine series of the function

$$f(x)=\frac{\pi}{3},\ 0\le x<\frac{\pi}{3}$$
$$=0,\ \frac{\pi}{3}<x<\frac{2\pi}{3}$$
$$=-\frac{\pi}{3},\ \frac{2\pi}{3}<x\le\pi$$
$$=\frac{\pi}{6},\ x=\frac{\pi}{3}$$
$$=-\frac{\pi}{6},\ x=\frac{2\pi}{3}.$$

Does the series equal $f(x)$ everywhere ?

54. Prove that if c be a constant then

$$c=\frac{4}{\pi}c\left(\sin x+\frac{1}{3}\sin 3x+\frac{1}{5}\sin 5x+\cdots\cdots\right),\ 0<x<\pi.$$

55. Show that on $0<x<\pi$,

$$x^2=\frac{2}{\pi}\left\{\left(\frac{\pi^2}{1}-\frac{4}{1^3}\right)\sin x-\frac{\pi^2}{2}\sin 2x+\left(\frac{\pi^2}{3}-\frac{4}{3^3}\right)\sin 3x-\cdots\cdots\right\}.$$

56. Expand $f(x)=\cos x$, $0\le x<\pi$ in Fourier sine series in $0\le x\le\pi$.

57. Find the Half Range sine series of the function

$$f(x)=0,\ 0\le x<\frac{\pi}{2}$$
$$=\frac{\pi}{2},\ \frac{\pi}{2}<x<\pi.$$

Define $f(\pi/2)$ and $f(\pi)$ so that the obtained series equals $f(x)$ everywhere.

58. Find the Half Range sine series of the function defined by

$$f(x)=\frac{1}{3}\pi,\ 0<x<\frac{\pi}{3}$$
$$=0,\ \frac{\pi}{3}<x<\frac{2\pi}{3}$$
$$=-\frac{\pi}{3},\ \frac{2\pi}{3}<x<\pi.$$

Find $f(0)$, $f(\pi)$, $f\left(\frac{\pi}{3}\right)$, $f\left(\frac{2\pi}{3}\right)$ so that the series converges to $f(x)$ for all values of x.

59. Prove that $\frac{\pi}{2}-x=\frac{4}{\pi}\left(\frac{\cos x}{1^2}+\frac{\cos 3x}{3^2}+\frac{\cos 5x}{5^2}+\cdots\cdots\right)$, $0\le x\le\pi$.

60. Prove that the function $f(x)=\frac{1}{2}k-x$, $0<x<k$ is represented by the series $\frac{k}{\pi}\sum_{n=1}^{\infty}\frac{1}{n}\sin\frac{2n\pi x}{k}$.

61. If $f(x)=\frac{(\pi-x)^2}{4}$ in $0\le x\le 2\pi$, show that $f(x)=\frac{\pi^2}{12}+\sum_{n=1}^{\infty}\frac{\cos nx}{n^2}$.

62. Find the Fourier series to represent the function $f(x)$ given by

$$f(x)=x,\ 0\le x\le\pi$$
$$=2\pi-x,\ \pi\le x\le 2\pi.$$

Hence find the value of $\frac{1}{1^2}+\frac{1}{3^2}+\frac{1}{5^2}+\cdots\cdots$

63. Obtain the Fourier series for $f(x)=\pi x$ in $[0,2]$.

64. Find the Half-Range cosine series for the function $f(x)=(x-1)^2$ in the interval $0<x<1$.

Hence find the value of $8\left(\frac{1}{1^2}+\frac{1}{3^2}+\frac{1}{5^2}+\cdots\cdots\right)$.

65. Find the half-range sine series for the function $f(t)=t-t^2$, $0<t<1$.

66. Find the half range sine series of e^x in $0<x<1$.

67. Find a half-range cosine series for

$$f(x)=ax, \quad 0\le x\le l/2$$
$$=a(l-x), \quad \frac{l}{2}\le x\le l.$$

Hence find the value of $\frac{1}{1^2}+\frac{1}{3^2}+\frac{1}{5^2}+\cdots\cdots$.

68. Expand the followng function $f(x)$ in a series of sines, where

$$f(x)=\sin x, \quad 0\le x\le\frac{\pi}{4}$$
$$=\cos x, \quad \frac{\pi}{4}\le x\le\frac{\pi}{2}.$$

69. Expand $f(x)=2-x, \quad 0<x<4$

$$=x-6, \quad 4<x<8$$

in a Fourier series of period 8.

70. Find a series of cosines of multiples of x representing the function $f(x)=x$ on the closed interval $0\le x\le\pi$.

Hence deduce $\frac{1}{1^2}+\frac{1}{3^2}+\frac{1}{5^2}+\cdots\cdots=\frac{\pi^2}{8}$.

71. Find a series of sines corresponding to the function x on $0\le x\le\pi$. Find whether this series can be assumed as an expansion of x.

72. Show that

$$\cos mx=\frac{2}{\pi}\sin mx\left(\frac{1}{2m}+\frac{m\cos x}{1^2-m^2}-\frac{m\cos 2x}{2^2-m^2}+\frac{m\cos 3x}{3^2-m^2}-\cdots\cdots\right)$$

73. State whether the function (i) $\sin\frac{1}{x}$ (ii) $\operatorname{cosec} x$ can be expanded into Fourier series.

74. Prove that for $0 \le x \le \pi$

$$x(\pi - x) = \frac{\pi^2}{6} - \left(\frac{\cos 2x}{1^2} + \frac{\cos 4x}{2^2} + \frac{\cos 6x}{3^2} + \cdots\cdots\right).$$

Using Parseval's Identity show that $\sum_{n=1}^{\infty}\frac{1}{n^4} = \frac{\pi^4}{90}$.

75. Prove that for $0 \le x \le \pi$

$$x(\pi - x) = \frac{8}{\pi}\left(\frac{\sin x}{1^3} + \frac{\sin 3x}{3^3} + \frac{\sin 5x}{5^3} + \cdots\cdots\right).$$

Using Parseval's identity show that $\sum_{n=1}^{\infty}\frac{1}{n^6} = \frac{\pi^6}{945}$.

76. Using Parseval's identity show that

(i) $\sum_{n=1}^{\infty}\frac{1}{(2n-1)^4} = \frac{\pi^4}{96}$ (ii) $\sum_{n=1}^{\infty}\frac{1}{(2n-1)^6} = \frac{\pi^6}{960}$.

ANSWERS

1. π, 1, 1; **3.** Triangular; **9.** π; Full wave Rectifier

11. (i) odd; (ii) neither even nor odd; (iii) even;

14. $2\left(\frac{\sin x}{1} - \frac{\sin 2x}{2} + \frac{\sin 3x}{3} - \cdots\cdots\right)$;

15. $f(x) = \pi + \frac{4}{\pi}\left\{\frac{\cos x}{1^2} + \frac{\cos 3x}{3^2} + \cdots\cdots\right\} + 4\left\{\frac{\sin x}{1} + \frac{\sin 3x}{3} + \cdots\cdots\right\}$;

16. $\frac{\pi}{2} + 2\left\{\sin x + \frac{\sin 3x}{3} + \frac{\sin 5x}{5} + \cdots\cdots\right\}$;

17. $\frac{1}{2} + \frac{2}{\pi}\left\{\sin\frac{\pi x}{2} + \frac{1}{3}\sin\frac{3\pi x}{2} + \frac{1}{5}\sin\frac{5\pi x}{2} + \cdots\cdots\right\}$

$f(0) = \frac{1}{2}$, $f(-2) = f(2) = \frac{1}{2}$

18. $x^2 = \frac{\pi^2}{3} + 4 \sum_{n=1}^{\infty} (-1)^n \frac{\cos nx}{n^2}$

(i) $\frac{\pi^2}{6}$ (ii) $\frac{\pi^2}{12}$ (iii) $\frac{\pi^2}{8}$

19. $\frac{\pi^2}{6}$; **20.** $f(x) = \frac{4}{\pi}\left(\sin x + \frac{\sin 3x}{3} + \frac{\sin 5x}{5} + \cdots\cdots\right)$; $\frac{\pi}{4}$;

21. $\frac{4k}{\pi}\left(\sin x + \frac{1}{3}\sin 3x + \frac{1}{5}\sin 5x + \cdots\cdots\right)$;

22. $\frac{k}{2} + \frac{2k}{\pi}\left(\cos\frac{\pi}{2}x - \frac{1}{3}\cos\frac{3\pi}{2}x + \frac{1}{5}\cos\frac{5\pi}{2}x - \cdots\cdots\right)$; square waveform;

23. $f(t) = \frac{E}{\pi} + \frac{E}{2}\sin wt - \frac{2E}{\pi}\left(\frac{1}{1.3}\cos 2wt + \frac{1}{3.5}\cos 4wt + \cdots\cdots\right)$;

24. Saw-toothed waveform; $\pi + 2\left(\sin x - \frac{1}{2}\sin 2x + \frac{1}{3}\sin 3x \cdots\cdots + \cdots\cdots\right)$;

25. $f(x) = -\frac{\pi}{4} - \frac{2}{\pi}\left(\cos x + \frac{\cos 3x}{3^2} + \frac{\cos 5x}{5^2} + \cdots\cdots\right)$

$$+3\sin x - \frac{\sin 2x}{2} + \frac{3\sin 3x}{3} - \frac{\sin 4x}{4} + \cdots\cdots;$$

26. $e^{-x} = \sin x 4\left\{\frac{1}{4} - 8\left(\frac{1}{16+\pi^2}\cos\frac{\pi x}{4} - \frac{1}{16+2^2\pi^2}\cos\frac{2\pi x}{4}\right.\right.$

$$\left.\left.+\frac{1}{16+3^2\pi^2}\cos\frac{3\pi x}{4} - \cdots\cdots\right)\right\} - 2\pi\left\{\frac{1}{16+\pi^2}\sin\frac{\pi x}{4} - \frac{2}{16+2^2\pi^2}\sin\frac{2\pi x}{4}\right.$$

$$\left. + \frac{3}{16+3^2\pi^2}\sin\frac{3\pi x}{4} - \cdots\cdots\right\};$$

27. $\frac{8}{\pi^2}\left(\frac{\cos x}{1^2} + \frac{\cos 3x}{3^2} + \frac{\cos 5x}{5^2} + \cdots\cdots\right)$;

28. $\frac{\pi}{4}-\frac{2}{\pi}\sum_{n=1}^{\infty}\frac{\cos(2n-1)x}{(2n-1)^2}+\sum_{n=1}^{\infty}\frac{(-1)^n \sin nx}{n}$;

29. $|\sin x|=\frac{2}{\pi}-\frac{4}{\pi}\sum_{n=1}^{\infty}\frac{\cos 2nx}{4n^2-1}$;

30. $\sum_{n=1}^{\infty}\frac{4}{\pi n^2}\sin\frac{n\pi}{2}\sin nx$

31. $\sin x+\frac{1}{3}\sin 3x+\frac{1}{5}\sin 5x+\cdots\cdots$

32. $\frac{1}{2}+\frac{2}{\pi}\left(\sin x+\frac{1}{3}\sin 3x+\frac{1}{5}\sin 5x+\cdots\cdots\right),\ \frac{1}{2}$

33. Yes

34. $f(x)=\frac{1}{\pi}-\frac{2}{\pi}\sum_{n=1}^{\infty}\frac{1}{4n^2-1}\cos 2nx+\frac{1}{2}\sin x$

36. $f(x)=\frac{\pi}{4}-\frac{2}{\pi}\left(\frac{\cos x}{1^2}+\frac{\cos 3x}{3^2}+\frac{\cos 5x}{5^2}+\cdots\cdots\right)$

$+\left(\frac{\sin x}{1}-\frac{\sin 2x}{2}+\frac{\sin 3x}{3}-\cdots\cdots\right);\ \frac{\pi}{2}$

38. $\frac{\pi^2}{6}-2\left(\cos x-\frac{\cos 2x}{2^2}+\frac{\cos 3x}{3^2}-\cdots\cdots\right)$

$-\frac{1}{\pi}\left\{\left(\frac{2}{1^3}-\frac{\pi^2}{1}\right)\sin x-\left(\frac{2}{2^3}-\frac{\pi^2}{2}\right)\sin 2x+\cdots\cdots\right\}$

39. $2\left(\pi-\frac{4}{\pi}\right)\sin x-\pi\sin 2x+\frac{2}{3}\left(\pi-\frac{4}{9\pi}\right)\sin 3x-\frac{\pi}{2}\sin 4x+\cdots\cdots$

40. $f(t)=\frac{2}{3}+\frac{4}{\pi^2}\left(\cos\pi t-\frac{\cos 2\pi t}{2^2}+\frac{\cos 3\pi t}{3^2}+\cdots\cdots\right)$

41. $\sin ax=\frac{2\sin a\pi}{\pi}\left(\frac{\sin x}{1^2-a^2}-\frac{2\sin 2x}{2^2-a^2}+\frac{3\sin 3x}{3^2-a^2}-\cdots\cdots\right)$

42. $\frac{1}{2}\pi-\frac{4}{\pi}\left(\cos x+\frac{\cos 3x}{3^2}+\frac{\cos 5x}{5^2}+\cdots\cdots\right)$

43. $\frac{2}{\pi}+\frac{4}{\pi}\left\{\frac{1}{3}\cos 2x-\frac{1}{15}\cos 4x+\cdots\cdots\right\}$

44. $\frac{2}{3}\pi^2$;

45. $\frac{\pi^2}{6} = a_0$ **46.** $-1+\pi\sin x-\frac{1}{2}\cos x+2\sum_{n=2}^{\infty}\frac{\cos nx}{n^2-1}$;

47. $x^2 = \frac{4\pi^2}{3}+\sum_{n=1}^{\infty}\left(\frac{4}{n^2}\cos nx-\frac{4\pi}{n}\sin nx\right)$; **48.** $f(x)=\sum_{n=1}^{\infty}\frac{\sin nx}{n}$

49. $\pi - x = \frac{\pi}{2}+\frac{4}{\pi}\left(\frac{\cos x}{1^2}+\frac{\cos 3x}{3^2}+\cdots\cdots\right)$; $\frac{\pi^2}{8}$;

50. (i) $f(x)=\frac{\pi^2}{16}-2\left\{\frac{\cos 2x}{2^2}+\frac{\cos 6x}{6^2}+\cdots\cdots\right\}$, $0 \le x \le \pi$

(ii) $f(x)=\sin x-\frac{\sin 3x}{3^2}+\frac{\sin 5x}{5^2}-\cdots\cdots$,,;

51. $1-\frac{8}{\pi^2}\left(\cos\frac{\pi x}{2}+\frac{1}{3^2}\cos\frac{3\pi x}{2}+\frac{1}{5^2}\cos\frac{5\pi x}{2}+\cdots\cdots\right)$;

52. Triangular pulse; $f(x)=\frac{8k}{\pi^2}\left(\frac{1}{1^2}\sin\frac{\pi x}{l}-\frac{1}{3^2}\sin\frac{3\pi}{l}x+\frac{1}{5^2}\sin\frac{5\pi}{l}x-\cdots\cdots\right)$;

53. $f(x)=\frac{2}{\sqrt{3}}\left(\cos x-\frac{1}{5}\cos 5x+\frac{1}{7}\cos 7x-\frac{1}{11}\cos 11x+\cdots\cdots\right)$

57. $\sin x-\sin 2x+\frac{1}{3}\sin 3x+\frac{1}{5}\sin 5x-\frac{1}{3}\sin 6x+\cdots\cdots$

$f(\pi/2)=\frac{\pi}{4}$, $f(\pi)=0$

58. $\sin 2x+\frac{1}{2}\sin 4x+\frac{1}{4}\sin 8x+\cdots\cdots$

$f(0)=f(\pi)=0$; $f(\pi/3)=\pi/6$, $f(2\pi/3)=-\frac{1}{6}\pi$

62. $\frac{1}{2}\pi-\frac{4}{\pi}\left(\cos x+\frac{\cos 3x}{3^2}+\frac{\cos 5x}{5^2}+\cdots\cdots\right)$; $\frac{\pi^2}{8}$

63. $-\frac{2}{\pi}\sum_{n=1}^{\infty}\frac{\sin n\pi x}{n}$ **64.** $\frac{1}{3}+\frac{4}{\pi^2}\sum_{n=1}^{\infty}\frac{1}{n^2}\cos n\pi x$; π^2

65. $\frac{8}{\pi^3}\left(\frac{\sin \pi t}{1^3}+\frac{\sin 3\pi t}{3^3}+\frac{\sin 5\pi t}{5^3}+\cdots\cdots\right)$

66. $\sum_{n=1}^{\infty}\frac{2n\pi}{1+n^2\pi^2}(1-e\cos n\pi)\sin n\pi x$

67. $f(x)=\frac{al}{3}-\frac{8al}{\pi^2}\left\{\frac{1}{2^2}\cos\frac{2\pi x}{l}+\frac{1}{6^2}\cos\frac{6\pi x}{l}+\frac{1}{10^2}\cos\frac{10\pi x}{l}+\cdots\cdots\right\};\ \frac{\pi^2}{8}$

68. $\frac{8}{\pi}\cos\frac{\pi}{4}\left\{\frac{\sin 2x}{1.3}-\frac{\sin 6x}{5.7}+\frac{\sin 10x}{9.11}+\cdots\cdots\right\}$

69. $\frac{16}{\pi^2}\left\{\cos\frac{\pi x}{4}+\frac{1}{3^2}\cos\frac{3\pi x}{4}+\frac{1}{5^2}\cos\frac{5\pi x}{4}+\cdots\cdots\right\}$

70. $\frac{\pi}{2}-\frac{4}{\pi}\left\{\frac{\cos x}{1^2}+\frac{\cos 3x}{3^2}+\frac{\cos 5x}{5^2}+\cdots\cdots\right\};$

71. $x=2\left\{\sin x-\frac{1}{2}\sin 2x+\frac{1}{3}\sin 3x-\cdots\cdots\right\}$ which shows this is expansion

73. (i) no, (ii) no.

1.2 FOURIER TRANSFORM

1.2.1. Introduction

In the previous chapter we have seen if a function $f(x)$ satisfies Dirichlet's condition on the interval $(-T,\ T)$ then it can be expanded in the Fourier series

$$f(x)=\frac{a_0}{2}+\sum_{n=1}^{\infty}\left(a_n\cos\frac{n\pi x}{T}+b_n\sin\frac{n\pi x}{T}\right)$$

where a_n, b_n are Fourier coefficients. Using the Euler's identity $e^{i\theta}=\cos\theta+i\sin\theta$ the Fourier series of $f(x)$ can be written as

$$f(x)=\sum_{n=-\infty}^{\infty}c_n e^{\frac{i n\pi x}{T}} \qquad \dots (1)$$

where $c_n=\frac{1}{2T}\int_{-T}^{T}f(x)e^{-i\frac{n\pi x}{T}}dx$ are also known as Fourier coefficients.

If now this hold for all values of T then as $T\to\infty$, the expansion (1) takes the form $f(x)=\int_{-\infty}^{\infty}\left\{\frac{1}{2\pi}F(s)\right\}e^{-isx}\,ds$

where $F(s)=\int_{-\infty}^{\infty}f(t)e^{ist}dt$ which is known as Fourier Transform.

This is a particular type of Integral transform. Its theory affords mathematical devices through which solutions of different boundary value problems of engineering can be obtained, as for example, transverse vibrations of string, transverse oscillation of an elastic beam, conduction of heat, transmission of lines etc.

1.2.2. Fourier Transforms

It transforms an integrable function to an another function defined as follow.

Let $f(x)$ be an integrable function on any interval $(-T, T)$. Then the improper integral

$$\mathcal{F}\left(f(x)\right)=F(s)=\int_{-\infty}^{\infty}f(x)e^{isx}dx$$

is called the *Fourier Transform* of the function *f*. This is a function of *s*.

Fourier Sine & Cosine Transforms

The Fourier sine tranform of $f(x)$, $0 < x < \infty$ is defined as

$$\mathcal{F}_s(f(x)) = F_s(s) = \int_0^\infty f(x) \sin sx \, dx$$

The Fourier cosine transform of $f(x)$, $0 < x < \infty$ is defined as

$$\mathcal{F}_c(f(x)) = F_c(s) = \int_0^\infty f(x) \cos sx \, dx$$

provided the improper integrals are convergent.

Illustration.

(i) Let us consider the function

$$f(x) = 1, \quad -1 \le x \le 1$$
$$= 0, \quad \text{otherwise}$$

Its Fourier Transform,

$$F(s) = \int_{-\infty}^{\infty} f(x) e^{isx} dx = \int_{-\infty}^{-1} o e^{isx} dx + \int_{-1}^{1} 1 \cdot e^{isx} dx + \int_{1}^{\infty} 0 \, e^{isx} dx = \int_{-1}^{1} e^{isx} dx$$

$$= \left[\frac{e^{isx}}{is} \right]_{-1}^{1} = \frac{e^{is} - e^{-is}}{is} = \frac{(\cos s + i \sin s) - (\cos s - i \sin s)}{is}$$

$$= \frac{2 \sin s}{s} \quad \text{for } s \neq 0$$

For, $s = 0$, $F(0) = \int_{-1}^{1} 1 \, e^{0 \cdot x} dx = \int_{-1}^{1} dx = 2$

So, the Fourier Transform of $f(x)$ is

$$F(s) = \frac{2 \sin s}{s}, \quad s \neq 0$$
$$= 2, \quad s = 0$$

(ii) Consider the function $f(x) = e^{-x}$, $x \ge 0$.

Its Sine Transform,

$$F_s(s) = \int_0^\infty f(x) \sin sx \, dx$$

$$= \int_0^\infty e^{-x} \sin sx \, dx = \lim_{X \to \infty} \int_0^X e^{-x} \sin sx \, dx \quad \ldots (1)$$

Now, $I = \int_0^X e^{-x} \sin sx \, dx$

$$= \left[-e^{-x} \frac{\cos sx}{s} \right]_{x=0}^{X} - \int_0^X e^{-x} \cdot \frac{\cos sx}{s} dx$$

$$= -\left(\frac{\cos s X}{se^X} - \frac{1}{s} \right) - \frac{1}{s} \left\{ \left[e^{-x} \frac{\sin s x}{s} \right]_{x=0}^{X} + \int_0^X e^{-x} \frac{\sin s x}{s} dx \right\}$$

$$= \frac{1}{s} - \frac{\cos s X}{se^X} - \frac{1}{s} \left\{ \frac{e^{-X} \sin s X}{s} + \frac{1}{s} I \right\}$$

$$= \frac{1}{s} - \frac{\cos sX}{se^X} - \frac{\sin sX}{s^2 e^X} - \frac{I}{s^2}$$

or, $I \frac{1+s^2}{s^2} = \frac{1}{s} - \frac{\cos sX}{s\, e^X} - \frac{\sin sX}{s^2 e^X}$

or, $I \frac{1+s^2}{s} = 1 - \frac{\cos sX}{e^X} - \frac{\sin sX}{s\, e^X}$

or, $I = \frac{s}{1+s^2} \left[1 - \frac{\cos sX}{e^X} - \frac{\sin sX}{s\, e^X} \right]$

$\therefore \int_0^X e^{-x} \sin s\, x \, dx = \frac{s}{1+s^2} \left[1 - \frac{\cos sX}{e^X} - \frac{\sin sX}{se^X} \right]$

$\therefore \lim_{X \to \infty} \int_0^X e^{-x} \sin sx \, dx = \frac{s}{1+s^2} \left[1 - \lim_{X \to \infty} \frac{\cos sX}{e^X} - \lim_{X \to \infty} \frac{\sin sX}{se^X} \right]$

$$= \frac{s}{1+s^2} (1 - 0 - 0) \qquad [\because \cos sX \text{ and } \sin sX \text{ are bounded }]$$

$$= \frac{s}{1+s^2}.$$

So, the Fourier sine Transform of the given function is $F_s(s) = \frac{s}{1+s^2}$.

Theorem 1.

If $f(x), -\infty < x < \infty$ is an even function then its Fourier transform,

$F(s) = 2F_c(s)$ where $F_c(s)$ is the Fourier cosine transform of $f(x)$.

Proof. $F(s) = \int_{-\infty}^{\infty} f(x) e^{isx} dx$,

$$= \int_{-\infty}^{\infty} f(x)(\cos sx + i \sin sx) dx$$

$$= \int_{-\infty}^{\infty} f(x) \cos sx \, dx + i \int_{-\infty}^{\infty} f(x) \sin sx \, dx$$

$$= 2\int_{0}^{\infty} f(x) \cos s\, x \, dx + i \times 0$$

[$\because f(x)\cos sx$ is even function and $f(x)\sin sx$ is odd]

$\therefore F(s) = 2F_c(s)$

Theorem 2.

If $f(x)$, $-\infty < x < \infty$ is an odd funtion then its Fourier Transform,

$F(s) = 2i\, F_s(s)$

where $F_s(s)$ is the Fourier sine transform of $f(x)$.

Proof. As the previous.

Illustration.

In the previous illustration (i) the function $f(x) = 1, -1 < x < 1$
$= 0$, otherwise

is an even function. So its Fourier Transform, $F(s) = 2 \times F_c(s)$

or, $F_c(s) = \frac{1}{2} F(s)$

So its Fourier Cosine Transform,

$$F_c(s) = \frac{1}{2} \cdot \frac{2 \sin s}{s}, \quad s \neq 0$$

$$= \frac{1}{2} \cdot 2, \quad s = 0$$

i.e., $F_c(s) = \frac{\sin s}{s} \quad s \neq 0$

$= 1 \quad s = 0$.

Transforms of some Standard Functions

Following can be treated as formulae :

The Fourier Transforms are

(1) $F(s)=\sqrt{2\pi}\, e^{-\frac{s^2}{2}}$ of $f(x)=e^{-\frac{x^2}{2}}$

(2) $F(s)=\dfrac{2a}{a^2+s^2}$ of $f(x)=e^{-a|x|}$

(3) $F(s)=\dfrac{2\sin sa}{s}$, $s\neq 0$ of $f(x)=1$, $|x|<a$

$=2a$, $s=0$ $=0$ $|x|>a$

(4) $F_s(s)=\dfrac{s}{a^2+s^2}$ of $f(x)=e^{-ax}$ $(a>0)$

(5) $F_c(s)=\dfrac{a}{a^2+s^2}$ of $f(x)=e^{-ax}$ $(a>0)$

(6) $F_s(s)=\dfrac{a\cos sa-b\cos sb}{s}+\dfrac{\sin sb-\sin sa}{s^2}$

of $f(x)=0$, $0<x<a$

$=x$, $a\leq x\leq b$

$=0$, $x>b$

Proofs.

(1) The Fourier Transform corresponding to $f(x)$ is

$$F(s)=\int_{-\infty}^{\infty} f(x)e^{isx}dx$$

$$=\int_{-\infty}^{\infty} e^{-\frac{x^2}{2}}e^{isx}dx=\int_{-\infty}^{\infty} e^{\frac{-x^2+2isx}{2}}dx$$

$$=\int_{-\infty}^{\infty} e^{-\frac{x^2-2isx}{2}}dx=\int_{-\infty}^{\infty} e^{-\frac{x^2-2\cdot x\cdot is+(is)^2+s^2}{2}}dx$$

$$= \int_{-\infty}^{\infty} e^{-\frac{(x-is)^2+s^2}{2}} dx = e^{-\frac{s^2}{2}} \int_{-\infty}^{\infty} e^{-\frac{(x-is)^2}{2}} dx$$

[Putting $t = x - is$ i.e. $dt = dx$]

$$= e^{-\frac{s^2}{2}} \int_{-\infty}^{\infty} e^{-\frac{t^2}{2}} dt \qquad = 2e^{-\frac{s^2}{2}} \int_{0}^{\infty} e^{-\frac{t^2}{2}} dt$$

$$= 2\sqrt{2}\, e^{-\frac{s^2}{2}} \int_{0}^{\infty} e^{-z^2} dz$$

[Putting $\frac{t}{\sqrt{2}} = z$, $\frac{1}{\sqrt{2}} dt = dz$] $\quad = 2\sqrt{2}\, e^{-\frac{s^2}{2}} \cdot \frac{\sqrt{\pi}}{2} = \sqrt{2\pi}\, e^{-\frac{s^2}{2}}$

(2) The Fourier Transform,

$F(s) = 2F_c(s)$ [since $e^{-a|x|}$ is even function]

$$= 2\int_{0}^{\infty} e^{-a|x|} \cos sx \, dx$$

$$= 2\int_{0}^{\infty} e^{-ax} \cos sx \, dx \qquad [\because x > 0 \text{ in } (0, \infty)]$$

$$= 2 \lim_{X \to \infty} \int_{0}^{X} e^{-ax} \cos sx \, dx$$

Let $I = \int_{0}^{X} e^{-ax} \cos sx \, dx$

or, $I = \left[e^{-ax} \frac{\sin sx}{s} \right]_{x=0}^{X} + \int_{0}^{X} a\, e^{-ax} \frac{\sin sx}{s} dx$

$$= \frac{\sin sX}{s\, e^{aX}} + \frac{a}{s} \int_{0}^{X} e^{-ax} \sin sx \, dx$$

$$= \frac{\sin sX}{s\, e^{aX}} + \frac{a}{s} \left\{ \left[-e^{-ax} \frac{\cos sx}{s} \right]_0^X - \int_{0}^{X} ae^{-ax} \cdot \frac{\cos sx}{s} dx \right\}$$

$$= \frac{\sin sX}{s\, e^{aX}} + \frac{a}{s} \left\{ -\left(\frac{\cos sX}{e^{aX} \cdot s} - \frac{1}{s} \right) - \frac{a}{s} \int_{0}^{X} e^{-ax} \cos sx \, dx \right\}$$

$$= \frac{\sin sX}{se^{ax}} + \frac{a}{s}\left\{\frac{1}{s} - \frac{\cos sX}{se^{aX}} - \frac{a}{s}I\right\}$$

or, $$\left(1+\frac{a^2}{s^2}\right)I = \frac{\sin sX}{se^{aX}} + \frac{a}{s^2} - \frac{a\cos sX}{s^2 e^{aX}}$$

$$\therefore\ I = \frac{s^2}{s^2+a^2}\left\{\frac{\sin sX}{se^{aX}} + \frac{a}{s^2} - \frac{a\cos sX}{s^2 e^{aX}}\right\}$$

$$\therefore\ \lim_{X\to\infty}\int_0^X e^{-ax}\cos sx\,dx = \frac{s^2}{s^2+a^2}\lim_{X\to\infty}\left\{\frac{\sin sX}{s\,e^{aX}} + \frac{a}{s^2} - \frac{a\cos sX}{s^2 e^{aX}}\right\}$$

$$= \frac{s^2}{s^2+a^2}\cdot\frac{a}{s^2} = \frac{a}{s^2+a^2}$$

$$\therefore\ F(s) = \frac{2a}{s^2+a^2}$$

(3) The Fourier Transform of the given function $f(x)$ is

$$F(s) = \int_{-\infty}^{\infty} f(x)e^{isx}dx$$

$$= \int_{-\infty}^{-a} 0\cdot e^{isx}dx + \int_{-a}^{a} 1\cdot e^{isx}dx + \int_{a}^{\infty} 0\cdot e^{isx}dx$$

$$= \int_{-a}^{a} e^{isx}dx = \left[\frac{e^{isx}}{is}\right]_{x=-a}^{a} \quad = \frac{1}{is}\left(e^{isa} - e^{-isa}\right)$$

$$= \frac{1}{is}(\cos sa + i\sin sa - \cos sa + i\sin sa)$$

$$= \frac{2\sin sa}{s} \quad \text{for } s \neq 0$$

For $s = 0$, $F(s) = \int_{-a}^{a} e^0 dx = 2a$

So the required Fourier Transform,

$$F(s) = \frac{2\sin sa}{s}, s \neq 0$$

$$= 2a, \qquad s = 0.$$

(4) The Fourier sine Transform of the given function $f(x)$ is

$$F_s(s)=\int_0^{\infty} f(x)\sin sx\,dx=\int_0^{\infty} e^{-ax}\sin sx\,dx=\lim_{X\to\infty}\int_0^{X} e^{-ax}\sin sx\,dx$$

Now, $I=\int_0^{X} e^{-ax}\sin sx\,dx$

After doing integration by parts repeatedly twice we get

$$I=\frac{s}{a^2+s^2}\left[1-\frac{\cos sX}{e^{aX}}-\frac{\sin sX}{e^{ax}\cdot s}\right]$$

$$\therefore\ F_s(s)=\lim_{X\to\infty}\frac{s}{a^2+s^2}\left[1-\frac{\cos sX}{e^{aX}}-\frac{\sin sX}{e^{aX}\cdot s}\right]$$

$$=\frac{s}{a^2+s^2},\ \text{since } a>0$$

(5) Left as exercise.

(6) $F_s(s)=\int_{-\infty}^{a} 0\sin sx\,dx+\int_a^b x\sin sx\,dx+\int_b^{\infty} 0\sin sx\,dx$

or, $F_s(s)=\int_a^b x\sin sx\,dx=-\left[\frac{x\cos sx}{s}\right]_{x=a}^{b}+\int_a^b\frac{\cos sx}{s}dx$

$$=-\left(\frac{b\cos sb}{s}-\frac{a\cos sa}{s}\right)+\frac{1}{s}\left[\frac{\sin sx}{s}\right]_{x=a}^{b}$$

$$=\frac{a\cos sa}{s}-\frac{b\cos sb}{s}+\frac{1}{s}\left(\frac{\sin sb}{s}-\frac{\sin sa}{s}\right)$$

$$=\frac{a\cos sa-b\cos sb}{s}+\frac{\sin sb-\sin sa}{s^2}.$$

1.2.3. Properties of Fourier Transform.

(1) Linear Property. If $f(x)$ and $g(x)$ are two functions having Fourier Transforms then $\mathcal{F}(c_1 f+c_2 g)=c_1\,\mathcal{F}(f)+c_2\,\mathcal{F}(g)$
where c_1, c_2 are constants.

Proof. $\mathcal{F}\,(c_1 f+c_2 g)=\int_{-\infty}^{\infty}\{c_1 f(x)+c_2 g(x)\}e^{isx}\,dx$

$$= c_1 \int_{-\infty}^{\infty} f(x) e^{isx} dx + c_2 \int_{-\infty}^{\infty} g(x) e^{isx} dx$$

$$= c_1 \mathcal{F}(f) + c_2 \mathcal{F}(g).$$

(2) Change of Scale Property. If the Fourier Transform of $f(x)$, $\mathcal{F}\{f(x)\} = F(s)$ then the Fourier transform of $f(ax)$ $\mathcal{F}\{f(ax)\} = \frac{1}{a} F\left(\frac{s}{a}\right)$

Proof. $\because F\{f(x)\} = F(s)$ therefore $F(s) = \int_{-\infty}^{\infty} f(x) e^{isx}$ ... (1)

Now, $\mathcal{F}\{f(ax)\} = \int_{-\infty}^{\infty} f(ax) e^{isx} dx$ [Putting $t = ax$ i.e. $dt = a\,dx$]

$$= \frac{1}{a} \int_{-\infty}^{\infty} f(t) e^{\frac{ist}{a}} dt$$

$$= \frac{1}{a} \int_{-\infty}^{\infty} f(t) e^{i \cdot \frac{s}{a} \cdot x} dx = \frac{1}{a} F\left(\frac{s}{a}\right) \quad \text{by (1)}$$

(3) Shifting Property. If the Fourier Transform of $f(x)$, $\mathcal{F}\{f(x)\} = F(s)$ then the Fourier transform of $f(x-a)$,

$$\mathcal{F}\{f(x-a)\} = e^{ias} F(s)$$

Proof. $\mathcal{F}\{f(x-a)\} = \int_{-\infty}^{\infty} f(x-a) e^{isx} dx$

$$= \int_{-\infty}^{\infty} f(t) e^{is(a+t)} dt \qquad [\text{put } x - a = t \therefore dx = dt]$$

$$= e^{isa} \int_{-\infty}^{\infty} f(t) e^{ist} dt$$

$$= e^{isa} \int_{-\infty}^{\infty} f(x) e^{isx} dx = e^{ias} F(s).$$

(4) Modulation Property. If the Fourier Transform of $f(x)$, $\mathcal{F}\{f(x)\} = F(s)$ then the Fourier Transform of

$\mathcal{F}\{f(x) \cos a\,x\} = \frac{1}{2}\{F(s-a) + F(s+a)\}$,

Proof. $\mathcal{F}\{f(x) \cos a\,x\} = \int_{-\infty}^{\infty} f(x) \cos ax\, e^{isx} dx$

$$= \int_{-\infty}^{\infty} f(x)\, e^{isx}\, \frac{e^{iax} + e^{-iax}}{2}\, dx$$

$$= \frac{1}{2}\left\{ \int_{-\infty}^{\infty} f(x) e^{i(s+a)x} dx + \int_{-\infty}^{\infty} f(x) e^{i(s-a)x} dx \right\}$$

$$= \frac{1}{2}\{F(s+a) + F(s-a)\} = \frac{1}{2}\{F(s-a) + F(s+a)\}.$$

Illustration. Let we are to evaluate the Fourier Transform of the function $f(x) = 4e^{-|x|} - 5e^{-3|x+2|}$

$$\mathcal{F}\left(4e^{-|x|} - 5e^{-3|x+2|}\right) = 4\,\mathcal{F}\left(e^{-|x|}\right) - 5\,\mathcal{F}\left(e^{-3|x+2|}\right) \quad \text{... (1)}$$

Now, by formula, $\mathcal{F}\left(e^{-|x|}\right) = \dfrac{2}{s^2+1}$

By shifting property, $\mathcal{F}\left(e^{-3|x+2|}\right) = \mathcal{F}\left(e^{-3|x-(-2)|}\right)$

$$= e^{-2is}\,\mathcal{F}\left(e^{-3|x|}\right) = e^{-2is}\,\mathcal{F}\left(e^{-|3x|}\right)$$

$$= e^{-2is} \cdot \frac{1}{3} F\left(\frac{s}{3}\right), \text{ by change of scale property}$$

where $F(s) = \mathcal{F}\left(e^{-|x|}\right) = \dfrac{2}{s^2+1}$

$$\therefore\ F\left(\frac{s}{3}\right) = \frac{2}{\left(\frac{s}{3}\right)^2 + 1} = \frac{18}{s^2+9} = \frac{2}{s^2+1}$$

$$\therefore\ \mathcal{F}\left(e^{-3|x+2|}\right) = \frac{e^{-2is}}{3} \cdot \frac{18}{s^2+9} = \frac{6e^{-2is}}{s^2+9}$$

So, form (1) we get, $\mathcal{F}\left(4e^{-|x|} - 5e^{-3|x+2|}\right)$

$$= 4 \times \frac{2}{s^2+1} - 5 \times \frac{6e^{-2is}}{s^2+9} = \frac{8}{s^2+1} - \frac{30^{-2is}}{s^2+9}.$$

(5) On Derivative of a Function.

If $f(x), f'(x), f''(x), \cdots \cdots f^{(n-1)}(x)$ all tend to 0 as $x \to \pm\infty$ and $\int_{-\infty}^{\infty} \left| f^{(j)}(x) \right| dx$ converges for all j then $\mathcal{F}\left\{ f^{(n)}(x) \right\} = (-is)^n \, \mathcal{F}\left\{ f(x) \right\}$

Proof. Beyond the scope of this text.

(6) On Multiplication by *x*.

If $\mathcal{F}\{f(x)\} = F(s)$

then $\mathcal{F}\{xf(x)\} = -i\, F'(s)$

Proof. $\mathcal{F}\{xf(x)\} = \int_{-\infty}^{\infty} xf(x)\, e^{isx}\, dx$

$$\text{Now} \quad F'(s) = \frac{d}{ds} \int_{-\infty}^{\infty} f(x) e^{isx}\, dx$$

$$= \int_{-\infty}^{\infty} ixf(x) e^{isx}\, dx$$

$$= i \int_{-\infty}^{\infty} xf(x) e^{isx}\, dx$$

$$= i\, \mathcal{F}\{xf(x)\}$$

$$\therefore \quad \mathcal{F}\{xf(x)\} = -i \frac{d}{ds} \mathcal{F}\{f(x)\}$$

Illustration.

Suppose we are to find the Fourier Transform $\mathcal{F}\left(x\, e^{-4x^2}\right)$

$$\text{Now,} \quad \mathcal{F}\left(e^{-4x^2}\right) = \mathcal{F}\left(e^{-\frac{(\sqrt{8}x)^2}{2}}\right)$$

$$\text{Now,} \quad \mathcal{F}\left(e^{-\frac{x^2}{2}}\right) = \sqrt{2\pi}\, e^{-\frac{s^2}{2}}$$

$$\therefore \quad \mathcal{F}\left(e^{-\frac{(\sqrt{8}x)^2}{2}}\right) = \frac{1}{\sqrt{8}} \cdot \sqrt{2\pi} \cdot e^{-\frac{\left(\frac{s}{\sqrt{8}}\right)^2}{2}} \quad \text{(by change of scale property)}$$

$$= \frac{1}{2}\sqrt{\pi}\; e^{-\frac{s^2}{16}}$$

$$\therefore \quad \mathcal{F}\left(e^{-4x^2}\right) = \frac{\sqrt{\pi}}{2} e^{-\frac{s^2}{16}} \qquad \ldots (1)$$

or, $\mathcal{F}\left(\frac{d}{dx}\left(e^{-4x^2}\right)\right) = (-is)\,\mathcal{F}\left(e^{-4x^2}\right)$

or, $\mathcal{F}\left(-8xe^{-4x^2}\right) = -is\,\mathcal{F}\left(e^{-4x^2}\right)$

or, $-8\,\mathcal{F}\left(x\,e^{-4x^2}\right) = -is \cdot \frac{\sqrt{\pi}}{2}\, e^{-\frac{s^2}{16}}$ by (1)

or, $\mathcal{F}\left(x\,e^{-4x^2}\right) = \frac{is\sqrt{\pi}}{16}\, e^{-\frac{s^2}{16}}$

(7) Convolution and Convolution Theory.

If $f(x)$ and $g(x)$ are two functions defined on the interval $(-\infty, \infty)$ then

$$f * g = \int_{-\infty}^{\infty} f(u)\, g(x-u)\, du$$

is called the convolution of the two functions f and g

Convolution Theorem.

The Fourier Transform of the convolution of $f(x)$ and $g(x)$ is the product of their Fourier Transform, i.e.

$$\mathcal{F}\,(f * g) = \mathcal{F}\,\{f(x)\}\,\mathcal{F}\,\{g(x)\}$$

Proof. Beyond the scope of the book.

Note. $\mathcal{F}^{-1}(F(s)\,G(s)) = f * g$. In fact the inverse $\mathcal{F}^{-1}$ is going to be defined in next article.

1.2.4. Inverse Fourier Transform or Fourier Integral

In this article we discuss how a function can be found if its Fourier Transform is known. This is due to the following theorem.

Theorem 1. (Fourier Integral Theorem)

If a function $f(x)$

(i) satisfies Dirichlet's condition in every finite interval $[-T, T]$

(ii) $\int_{-\infty}^{\infty} |f(x)|\, dx$ exists finitely

then $\frac{1}{2\pi}\int_{-\infty}^{\infty} F(s)\, e^{-isx} ds = f(x)$ where $f(x)$ is continuous

$$= \frac{1}{2}\left\{\lim_{t\to x+} f(t) + \lim_{t\to x-} f(t)\right\} \text{ when } f(x) \text{ is discontinuous at } x$$

where $F(s)$ is the Fourier Transform of $f(x)$.

Proof. Beyond the scope of the book.

Definition. The integral $\mathcal{F}^{-1}(F(s)) = \frac{1}{2\pi}\int_{-\infty}^{\infty} F(s)\, e^{-isx}\, ds$ is called **Inverse Fourier Transform** or **Fourier Integral** of the function $f(x)$, where $F(s)$ is the Fourier Transform of $f(x)$

So, $\mathcal{F}^{-1}(F(s)) = f(x)$ at the point of continuity

Theorem 2. (Fourier Sine and Cosine Integral Theorem)

If a function $f(x)$ (i) satisfies Dirichlet's condition in every finite interval $[0, T]$

(ii) $\int_0^{\infty} |f(x)|\, dx$ exists finitely

then

(a) $\frac{2}{\pi}\int_0^{\infty} F_s(s) \sin sx\, ds = f(x)$ where $f(x)$ is continuous

$$= \frac{1}{2}\left\{\lim_{t\to x+} f(t) + \lim_{t\to x-} f(t)\right\} \text{ where } f(x) \text{ is discotinuous.}$$

where $F_s(s)$ is the Fourier Sine Transform of $f(x)$.

(b) $\frac{2}{\pi}\int_0^{\infty} F_c(s) \cos sx\, ds = f(x)$ where $f(x)$ is continuous

$= \frac{1}{2}\left\{\lim_{t\to x+} f(t) + \lim_{t\to x-} f(t)\right\}$ where $f(x)$ is discontinuous where $F_c(s)$ is the Fourier Cosine Transform of $f(x)$

Proof. Beyond the scope of the book.

Definition :

(a) The integral $\mathcal{F}_s^{-1}(F_s(s)) = \frac{2}{\pi}\int_0^{\infty} F_s(s) \sin sx\, ds$ is called **Inverse Fourier Sine Transform** or **Fourier-sine-Integral** of $f(x)$, where $F_s(s)$ is Fourier sine Transform of $f(x)$

(b) The integral $\mathcal{F}_c^{-1}(F_c(s)) = \frac{2}{\pi}\int_0^{\infty} F_c(s)\cos sx\,ds$ is called **Inverse Fourier Cosine Transform** or **Fourier-Cosine-Integral** of $f(x)$, where $F_c(s)$ is Fourier Cosine Transform of $f(x)$.

Note. With the help of above Integral Theorem we can evaluate many complicated improper integral.

Illustration. Consider the function

$$f(x) = 1, \quad -1 < x < 1$$
$$= 0 \quad \text{otherwise.}$$

Its Fourier Transform,

$$F(s) = \int_{-\infty}^{-1} 0\, e^{isx}dx + \int_{-1}^{1} 1\cdot e^{isx}dx + \int_{1}^{\infty} 0\cdot e^{isx}dx$$

$$= \left[\frac{e^{isx}}{is}\right]_{-1}^{1} = \frac{1}{is}\left(e^{is} - e^{-is}\right)$$

$$= \frac{1}{is}(\cos s + i\sin s - \cos s + i\sin s) = \frac{2\sin s}{s}$$

Then its Inverse Fourier Transform is $\int_{-\infty}^{\infty} F(s)e^{-isx}ds$

$$= \int_{-\infty}^{\infty} \frac{2\sin s}{s} e^{-isx}ds = 2\int_{-\infty}^{\infty} \frac{\sin s(\cos sx - i\sin sx)}{s}ds$$

$$= 2\int_{-\infty}^{\infty} \frac{\sin s\cos sx}{s}ds - 2i\int_{-\infty}^{\infty} \frac{\sin s\sin sx}{s}ds = 4\int_0^{\infty} \frac{\sin s\cos sx}{s}ds$$

$$\left[\because \frac{\sin s\cos sx}{s} \text{ is even and } \frac{\sin s\sin sx}{s} \text{ is odd}\right]$$

So, by Fourier Integral Theorem

$$\frac{1}{2\pi}\cdot 4\int_0^{\infty} \frac{\sin s\cos sx}{s}ds = f(x), \quad x \neq -1, 1$$

$$= \frac{1}{2}\left\{\lim_{t\to 1-} f(t) + \lim_{t\to 1+} f(t)\right\} \text{ at } x = 1$$

$$= \frac{1}{2}\left\{\lim_{t\to -1-} f(t) + \lim_{t\to -1+} f(t)\right\} \text{ at } x = -1$$

$$\therefore \frac{2}{\pi}\int_0^\infty \frac{\sin s\cos sx}{s}\,ds = f(x) \qquad x \neq -1, 1$$

$$= \frac{1}{2}(0+1) \quad \text{at } x = 1$$

$$= \frac{1}{2}(1+0) \quad \text{at } x = -1$$

$$\therefore \frac{2}{\pi}\int_0^\infty \frac{\sin s\cos sx}{s}\,ds = f(x), \quad \text{at } x \neq -1, 1$$

$$= \frac{1}{2}, \qquad \text{at } x = -1, 1$$

$$\therefore \frac{2}{\pi}\int_0^\infty \frac{\sin s\cos sx}{s}\,ds = 1, \qquad -1 < x < 1$$

$$= 0, \qquad x > 1 \text{ or } x < -1$$

$$= \frac{1}{2}, \qquad x = -1$$

$$= \frac{1}{2}, \qquad x = 1$$

$$\therefore \int_0^\infty \frac{\sin s\cos sx}{s}\,ds = \frac{\pi}{2}, \qquad |x| < 1$$

$$= 0, \qquad |x| > 1$$

$$= \frac{\pi}{4}, \qquad |x| = 1$$

Thus the improper integral $\int_0^\infty \frac{\sin s\cos sx}{s}\,ds$ is evaluated.

Linear Property of Inverse Fourier Transform

If $F(s)$ and $G(s)$ are Fourier transforms of the functions $f(x)$ and $g(x)$ repectively then

$$\mathcal{F}^{-1}\{c_1F(s)+c_2G(s)\} = c_1\,\mathcal{F}^{-1}\{F(s)\} + c_2\,\mathcal{F}^{-1}\{G(s)\}$$

where c_1, c_2 are constants.

Proof. Left as exercise.

This property is also valid on Inverse Fourier sine or cosine transform.

Illustrative Examples.

Ex. 1. *Find the Fourier Transform of the function*

$$f(x) = 1 - x^2 , \ |x| \le 1$$
$$= 0, \quad |x| > 1$$

Hence evaluate $\int_0^\infty \frac{x\cos x - \sin x}{x^3} \cos\frac{x}{2} dx$

The Fourier Transform,

$$F(s) = \int_{-\infty}^{\infty} f(x) e^{isx} dx = \int_{-1}^{1} (1 - x^2) e^{isx} dx$$

$$= \int_{-1}^{1} (1 - x^2)(\cos sx + i \sin sx) dx$$

$$= \int_{-1}^{1} (1 - x^2) \cos sx \, dx + i \int_{-1}^{1} (1 - x^2) \sin sx \, dx$$

$$= 2\int_{0}^{1} (1 - x^2) \cos sx \, dx$$

[$\because$ $(1 - x^2)\cos sx$ is even and $(1 - x^2)\sin sx$ is odd function]

$$= 2\left\{ \left[(1 - x^2)\frac{\sin sx}{s} \right]_0^1 + \int_0^1 2x \frac{\sin sx}{s} dx \right\}$$

$$= \frac{4}{s} \int_0^1 x \sin sx \, dx = \frac{4}{s} \left\{ -\left[x \frac{\cos sx}{s} \right]_0^1 + \int_0^1 \frac{\cos sx}{s} dx \right\}$$

$$= \frac{4}{s} \left\{ -\frac{\cos s}{s} + \frac{1}{s} \left[\frac{\sin sx}{s} \right]_0^1 \right\}$$

$$= \frac{4}{s} \left\{ -\frac{\cos s}{s} + \frac{\sin s}{s^2} \right\} = -\frac{4\{s\cos s - \sin s\}}{s^3}.$$

Since $f(x)$ is an even function therefore $F(s) = 2F_c(s)$

or, $F_c(s) = \frac{1}{2} F(s) = -\frac{2\{s\cos s - \sin s\}}{s^3}$.

By Fourier Cosine Integral Theroem,

$$\frac{2}{\pi} \int_0^\infty F_c(s) \cos sx \, ds = f(x)$$ [$\because$ $f(x)$ is continuous everywhere]

or, $\frac{2}{\pi}\int_0^\infty -\frac{2\{s\cos s-\sin s\}}{s^3}\cos sx\,ds = f(x)$

or, $-\frac{4}{\pi}\int_0^\infty \frac{(s\cos s-\sin s)}{s^3}\cos sx\,ds = 1-x^2\,,\quad |x|\le 1$

$$= 0\,,\quad |x|>1$$

Putting $x=\frac{1}{2}$ we get

$$-\frac{4}{\pi}\int_0^\infty \frac{(s\cos s-\sin s)\cos\frac{s}{2}}{s^3}ds = 1-\frac{1}{4}=\frac{3}{4}$$

or, $\int_0^\infty \frac{(s\cos s-\sin s)}{s^3}\cos\frac{s}{2}\,ds = -\frac{3\pi}{16}$

i.e., $\int_0^\infty \frac{x\cos x-\sin x}{x^3}\cos\frac{x}{2}\,dx = -\frac{3\pi}{16}$.

Ex. 2. *Find the Fourier Sine Transform of the function* $f(x)=1,\ 0<x<l$.

Its Fourier Sine Transform,

$$F_s(s)=\int_0^\infty f(x)\sin sx\,dx=\int_0^l 1\cdot\sin sx\,dx$$

$$=\left[-\frac{\cos sx}{s}\right]_{x=0}^{l} = -\frac{1}{s}\left(\cos sl-\cos 0^\circ\right)=\frac{1}{s}(1-\cos sl).$$

Ex. 3. *Find the Fourier Sine Transform of the function*

$$f(x)=1,\quad 0\le x\le\pi$$
$$=0,\quad x>\pi$$

Hence evaluate the integral $\int_0^\infty \frac{1-\cos\pi s}{s}\sin xs\,ds$.

The required Fourier Sine Transform,

$$F_s(s)=\int_0^\infty f(x)\sin sx\,dx=\int_0^\pi \sin sx\,dx=-\left[\frac{\cos sx}{s}\right]_0^\pi=-\frac{1}{s}(\cos s\pi-1)=\frac{1-\cos\pi s}{s}.$$

By Fourier Sine Integral Theorem, we have

$\frac{2}{\pi}\int_0^\infty F_s(s)\sin(sx)ds = f(x)$ where $f(x)$ is continuous

$$=\frac{1}{2}\left\{\lim_{t\to x+} f(t)+\lim_{t\to x-} f(t)\right\}$$ where $f(x)$ is discontinuous

$$\therefore \frac{2}{\pi}\int_0^\infty \frac{1-\cos\pi s}{s}\sin(sx)ds = 1, \qquad 0 \le x < \pi$$

$$= \frac{1}{2}\left\{\lim_{t\to\pi+} f(t) + \lim_{t\to\pi-} f(t)\right\}, x = \pi$$

$$= 0. \qquad x > \pi$$

$$\text{or, } \frac{2}{\pi}\int_0^\infty \frac{1-\cos\pi s}{s}\sin(sx)ds = 1, \qquad 0 \le x \le \pi$$

$$= \frac{1}{2}, \qquad x = \pi$$

$$= 0, \qquad x > \pi$$

$$\therefore \int_0^\infty \frac{1-\cos\pi s}{s}\sin(sx)ds = \frac{\pi}{2}, \qquad 0 \le x < \pi$$

$$= \frac{\pi}{4}, \qquad x = \pi$$

$$= 0, \qquad x > \pi.$$

Ex. 4. *FInd the Fourier Cosine Transform of the function*

$f(x) = \cos x$, $0 < x < a$

$= 0$, $x > a$

The required transform,

$$F_c(s) = \int_0^\infty f(x)\cos sx\, dx = \int_0^a \cos x \cos sx\, dx = \frac{1}{2}\int_0^a \{\cos(1+s)x + \cos(1-s)x\}dx$$

$$= \frac{1}{2}\left\{\frac{\sin(1+s)a}{1+s} + \frac{\sin(1-s)a}{1-s}\right\}.$$

Ex. 5. *Find the Fourier Sine Transform of* $f(x) = \frac{1}{x}$.

The required transform,

$$F_s(s) = \int_0^\infty f(x)\sin sx\, dx = \int_0^\infty \frac{\sin sx}{x}dx$$

$$= \frac{1}{s}\int_0^\infty \frac{s\sin t}{t}dt \qquad \text{putting } sx = t \;\therefore dx = \frac{1}{s}dt$$

$$= \int_0^\infty \frac{\sin t}{t}dt = \frac{\pi}{2}.$$

Ex. 6. *Find Fourier sine transform of* $\frac{e^{-ax}}{x}$

Let $f(x)=\frac{e^{-ax}}{x}$. Then the Fourier sine transform,

$$F_s(s)=\int_0^\infty \frac{e^{-ax}}{x}\sin sx\,dx \quad \ldots (1)$$

or, $\frac{d}{ds}F_s(s)=\int_0^\infty \frac{e^{-ax}}{x}\cdot x\cos sx\,dx$

or, $\frac{dF_s}{ds}=\int_0^\infty e^{-ax}\cos sx\,dx = \left[\frac{e^{-ax}}{a^2+s^2}(-a\cos sx+s\sin sx)\right]_0^\infty = \frac{a}{a^2+s^2}$

$\therefore\ dF_s=\frac{a}{a^2+s^2}ds$

or, $F_s(s)=a\int\frac{ds}{a^2+s^2}=\tan^{-1}\frac{s}{a}+C$

Now when $s=0$, $F_s(s)=0$ from (1). So form (1), putting $s=0$, we get $C=0$

$\therefore\ F_s(s)=\tan^{-1}\frac{s}{a}$.

Ex. 7. *If* $f(x)$ *is an even function show that* $F(\lambda)=2\int_0^\infty f(u)\cos\lambda u\,du$ *and* $f(x)=\frac{1}{\pi}\int_0^\infty F(\lambda)\cos\lambda x\,d\lambda$ *where* $f(x)$ *is continuous.*

The Fourier Transform of f is

$$F(s)=\int_{-\infty}^{\infty} f(x)e^{isx}dx=\int_{-\infty}^{\infty} f(x)\cos sx\,dx+i\int_{-\infty}^{\infty} f(x)\sin sx\,dx$$

$$=2\int_{-\infty}^{\infty} f(x)\cos sx\,dx$$

[$\because$ $f(x)\cos sx$ is even and $f(x)\sin sx$ is odd function]

i.e., $F(\lambda)=2\int_0^\infty f(u)\cos\lambda u\,du$

Now, $F(-\lambda)=2\int_0^\infty f(u)\cos(-\lambda)u\,du=2\int_0^\infty f(u)\cos\lambda u\,du=F(\lambda)$. So, F is an even function

Again by Fourier Integral Theorem

$$\frac{1}{2\pi}\int_{-\infty}^{\infty} F(s)e^{-isx}ds = f(x) \text{ where } f(x) \text{ is continuous}$$

or, $$\frac{1}{2\pi}\left\{\int_{-\infty}^{\infty} F(s)(\cos sx - i\sin sx)ds\right\} = f(x)$$

or, $$\frac{1}{2\pi}\left\{\int_{-\infty}^{\infty} F(s)\cos sx\, ds - i\int_{-\infty}^{\infty} F(s)\sin sx\, ds\right\} = f(x)$$

or, $$\frac{1}{2\pi}\left\{2\int_{0}^{\infty} F(s)\cos sx\, ds - i\times 0\right\} = f(x)$$

[$\because$ $F(s)\cos sx$ is even and $F(s)\sin sx$ is odd function]

or, $$\frac{1}{\pi}\int_{0}^{\infty} F(s)\cos sx\, ds = f(x)$$

i.e. $$\frac{1}{\pi}\int_{0}^{\infty} F(\lambda)\cos\lambda x\, d\lambda = f(x) \text{ where } f(x) \text{ is continuous.}$$

Ex. 8. *With the help of Fourier Transform find the value of* $\int_0^\infty \frac{\sin x}{x}dx$.

We know the Fourier Transform $F(s) = \frac{2\sin s}{s}$, $s \neq 0$

corresponding to the function $f(x) = 1$, $\quad |x| < 1$

$= 0$, $\quad |x| > 1$

Since this function is even so $F(s) = 2F_c(s)$ or, $F_c(s) = \frac{\sin s}{s}$, $s \neq 0$

where $F_c(s)$ is Fourier Cosine Transform. Clearly $f(x)$ satisfies Dirichlet's condition. So by Fourier Cosine Integral Theorem

$$\frac{2}{\pi}\int_0^\infty F_c(s)\cos sx\, ds = f(x) \text{ where } f(x) \text{ is continuous.}$$

Since $f(x)$ is continuous at $x = 0$ so we get

$$\frac{2}{\pi}\int_0^\infty \frac{\sin s}{s}\cos 0\, ds = f(0)$$

or, $$\int_0^\infty \frac{\sin s}{s}ds = \frac{\pi}{2} \quad \text{or,} \quad \int_0^\infty \frac{\sin x}{x}dx = \frac{\pi}{2}.$$

Ex. 9. *Prove that* $\lim_{s\to\infty}\int_0^T \frac{\sin sv}{v}\,dv = \frac{\pi}{2}$

Put $sv = y \therefore dv = \frac{dy}{s}$

$$\therefore \lim_{s\to\infty}\int_0^T \frac{\sin sv}{v}\,dv = \lim_{s\to\infty}\int_0^{sT} \frac{s\sin y}{y}\frac{dy}{s} = \int_0^{\infty}\frac{\sin y}{y}\,dy = \frac{\pi}{2}.$$

Ex. 10. *Evaluate* $\int_0^{\infty}\frac{\cos\lambda x}{\lambda^2+1}\,d\lambda$ *using Fourier Transforms.*

We see the part $\frac{1}{\lambda^2+1}$ of the whole integrant $\frac{\cos\lambda x}{\lambda^2+1}$ is the Fourier Cosine Transforms $F_c(\lambda)$ corresponding to the function $f(x)=e^{-x}$

$\therefore F_c(\lambda)=\frac{1}{\lambda^2+1}$ corresponding to $f(x)=e^{-x}$.

By Fourier Cosine Integral Theorem

$\frac{2}{\pi}\int_0^{\infty}F_c(\lambda)\cos\lambda x\,d\lambda = f(x)$ where $f(x)$ is continuous. Since $f(x)=e^{-x}$ is every where continuous so

$$\frac{2}{\pi}\int_0^{\infty}\frac{1}{\lambda^2+1}\cos\lambda x\,d\lambda = e^{-x} \qquad \text{or,} \quad \int_0^{\infty}\frac{\cos\lambda x}{\lambda^2+1}\,d\lambda = \frac{\pi}{2}e^{-x}.$$

Ex. 11. *Given* $\int_0^{\infty} f(x)\cos\alpha x\,dx = 1-\alpha, \quad 0\le\alpha\le 1$

$$= 0, \quad \alpha > 1$$

Find $f(x)$. *Hence evaluate* $\int_0^{\infty}\frac{\sin^2 x}{x^2}\,dx$.

Since $\int_0^{\infty} f(x)\cos\alpha x\,dx = 1-\alpha, \quad 0\le\alpha\le 1$

$$= 0, \quad \alpha > 1,$$

so the Fourier Cosine Transform,

$$\mathcal{F}_c(f(x)) = F_c(\alpha) = 1-\alpha, \qquad 0\le\alpha\le 1$$
$$= 0, \qquad \alpha > 1$$

where $f(x)$ is defined on $0<x<\infty$. We extend $f(x)$ in $-\infty<x<0$ such that it becomes even. So,

$F(\alpha)=2F_c(\alpha)$ for $f(x)$

or, $F(\alpha)=2(1-\alpha), \quad 0\le\alpha\le 1$

$\qquad =0 \qquad \alpha>1$

By Fourier Cosine Integral Theorem,

$\frac{2}{\pi}\int_0^\infty F_c(\alpha)\cos x\alpha\, d\alpha = f(x)$ where $f(x)$ is continuous.

or, $\frac{2}{\pi}\left\{\int_0^1 (1-\alpha)\cos x\alpha\, d\alpha + \int_1^\infty 0\cdot\cos s\alpha\, d\alpha\right\} = f(x)$

or, $\frac{2}{\pi}\int_0^1 (1-\alpha)\cos x\alpha\, d\alpha = f(x)$ where $f(x)$ is continuous.

or, $\frac{2(1-\cos x)}{\pi x^2} = f(x)$.

Now, $\int_0^\infty f(x)\cos(sx)dx = F_c(s)$

or, $\int_0^\infty \frac{2(1-\cos x)}{\pi x^2}\cos(sx)dx = 1-s, \quad 0\le s\le 1$

$\qquad = 0, \quad s>1$.

Putting $s=0$ we get $\int_0^\infty \frac{2(1-\cos x)}{\pi x^2} dx = 1$

or, $\int_0^\infty \frac{1-\cos x}{x^2} = \frac{\pi}{2}$ $\qquad$ or, $\int_0^\infty \frac{2\sin^2\frac{x}{2}}{x^2} dx = \frac{\pi}{2}$

Putting $t=\frac{x}{2}$ we get $\int_0^\infty \frac{\sin^2 t}{t^2} dt = \frac{\pi}{2}$ i.e., $\int_0^\infty \frac{\sin^2 x}{x^2} dx = \frac{\pi}{2}$.

Ex. 12. *Find the function whose cosine transform is* $\frac{\sin as}{s}$.

Let $f(x)$ be the required function. Then $F_c(s)=\frac{\sin as}{s}$ corresponding to $f(x)$.

By Fourier cosine integral theorem,

$\frac{2}{\pi}\int_0^\infty F_c(s)\cos sx\, ds = f(x)$ where $f(x)$ is continuous or, $\frac{2}{\pi}\int_0^\infty \frac{\sin as}{s}\cos sx\, ds = f(x)$

or, $f(x) = \frac{1}{\pi}\int_0^\infty \frac{2\sin as\cos sx\, ds}{s} = \frac{1}{\pi}\int_0^\infty \frac{\sin(a+x)s + \sin(a-x)s}{s}ds$

$$= \frac{1}{\pi}\left\{\int_0^\infty \frac{\sin(a+x)s}{s}ds + \int_0^\infty \frac{\sin(a-x)s}{s}ds\right\}$$

or, $f(x) = \frac{1}{\pi}\left(\frac{\pi}{2}+\frac{\pi}{2}\right) = 1$ if $x<a$

$= \frac{1}{\pi}\left(\frac{\pi}{2}-\frac{\pi}{2}\right) = 0$ if $x>a$ $\quad \left[\because \int_0^\infty \frac{\sin sx}{x}dx = \frac{\pi}{2}\right]$

Ex. 13. *Find the function* $f(x)$ *from the following integral equation* $f(x) = g(x) + \int_{-\infty}^{\infty} f(u)h(x-u)du$.

Taking Fourier Transform on both side we get

$$\mathcal{F}\{f(x)\} = \mathcal{F}\{g(x)\} + \mathcal{F}\left\{\int_{-\infty}^{\infty} f(u)h(x-u)du\right\} = \mathcal{F}\{g(x)\} + \mathcal{F}(f*h)$$

$= G(s) + \mathcal{F}\{f(x)\}\,\mathcal{F}\{h(x)\}$, by Convolution Theorem

or, $F(s) = G(s) + F(s)H(s)$

or, $F(s) = \frac{G(s)}{1-H(s)}$ $\quad \therefore\ \mathcal{F}(f(x)) = \frac{G(s)}{1-H(s)}$

So by Fourier Integral Theorem

$\therefore\ \frac{1}{2\pi}\int_{-\infty}^{\infty}\frac{G(s)}{1-H(s)}e^{-isx}ds = f(x)$ where $f(x)$ is continuous

$\therefore\ f(x) = \frac{1}{2\pi}\int_{-\infty}^{\infty}\frac{G(s)}{1-H(s)}e^{-isx}ds$ at the point of continuity.

Ex. 14. *Using Fourier sine inversion formula find* $f(x)$ *whose Fourier sine transform is* $F_s(s) = \frac{s}{1+s^2}$.

Using Fourier sine inversion formula we have,

(1) $f(x) = \frac{2}{\pi}\int_0^\infty \frac{s}{1+s^2}\sin sx\, ds$ (where $f(x)$ has continuity)

$$= \frac{2}{\pi}\int_0^\infty \frac{s^2+1-1}{s(s^2+1)}\sin sx\, ds = \frac{2}{\pi}\int_0^\infty \frac{\sin sx}{s}ds - \frac{2}{\pi}\int_0^\infty \frac{\sin sx}{s(s^2+1)}ds$$

$$= \frac{2}{\pi}\int_0^\infty \frac{\sin sx}{sx}d(sx) - \frac{2}{\pi}\int_0^\infty \frac{\sin sx}{s(s^2+1)}ds$$

$$= \frac{2}{\pi}\cdot\frac{\pi}{2} - \frac{2}{\pi}\int_0^\infty \frac{\sin sx}{s(s^2+1)}ds \text{ from an earlier problem (Ex. 8)}$$

(2) or, $f(x) = 1 - \frac{2}{\pi}\int_0^\infty \frac{\sin sx}{s(s^2+1)}ds$

(3) $\therefore$ $\frac{df}{dx} = -\frac{2}{\pi}\int_0^\infty \frac{\cos sx}{s^2+1}ds$ [differentiating both side w.r.t.x]

or, $\frac{d^2 f}{dx^2} = -\frac{2}{\pi}\int_0^\infty \frac{(-\sin sx)s}{s^2+1}ds = \frac{2}{\pi}\int_0^\infty \frac{s\sin sx}{s^2+1}ds = f(x)$ from (1)

or, $\frac{d^2 f}{dx^2} - f = 0$. Solution of this second order differential equaiton

is $f(x) = Ae^x + Be^{-x}$ where A and B are constants.

or, $\frac{df}{dx} = Ae^x - Be^{-x}$

Now, from (2) when $x = 0, f(x) = 1$

and $\frac{df}{dx} = -\frac{2}{\pi}\int_0^\infty \frac{ds}{s^2+1}$ from (3)

$$= -\frac{2}{\pi}\left[\tan^{-1} s\right]_0^\infty = -1$$

So, $A + B = 1$ and $A - B = -1$ $\therefore$ $A = 0,\ B = 1$

$\therefore$ $f(x) = e^{-x}$.

Ex. 15. *Evaluate* $\mathcal{F}^{-1}\left(\frac{1}{s^2+4s+13}\right)$.

By definition,

$$\mathcal{F}^{-1}\left(\frac{1}{s^2+4s+13}\right) = \frac{1}{2\pi}\int_{-\infty}^\infty \frac{1}{s^2+4s+13}e^{isx}ds = \frac{1}{2\pi}\int_{-\infty}^\infty \frac{1}{(s+2)^2+9}e^{isx}ds$$

$$= \frac{1}{2\pi}\int_{-\infty}^{\infty} \frac{1}{w^2+9} e^{i(w-2)x} dw \quad [s+2 = w \therefore ds = dw$$

$$= \frac{e^{-2ix}}{6} \cdot \frac{1}{2\pi}\int_{-\infty}^{\infty} \frac{2\times 3}{w^2+3^2} e^{iwx} dw$$

$$= \frac{e^{-2ix}}{6} \mathcal{F}^{-1}\left(\frac{2\times 3}{w^2+3^2}\right) = \frac{e^{-2ix}}{6} \mathcal{F}^{-1}\left(\frac{2\times 3}{w^2+3^2}\right)$$

$$= \frac{e^{-2ix}}{6} e^{-3|x|} = \frac{e^{-2ix} \cdot e^{-3|x|}}{6}$$

Ex. 16. *Find the Fourier integral of* $f(x) = e^{-kx}$ *when* $x > 0$ *and* $f(-x) = f(x)$

Since $f(x)$ is even therefore its Fourier transform,

$$F(s) = 2F_c(s) = 2\int_0^{\infty} f(x) \cos sx \, dx = 2\int_0^{\infty} e^{-kx} \cos sx \, dx.$$

Now, by integration by parts

$$\int_0^{\infty} e^{-kx} \cos sx \, dx = \frac{e^{-kx}}{s^2+k^2}(-k\cos sx + s\sin sx)$$

$$\therefore \int_0^{\infty} e^{-kx} \cos sx \, dx = \left[\frac{e^{-kx}}{s^2+k^2}(-k\cos sx + s\sin sx)\right]_0^{\infty} = \frac{k}{s^2+k^2}$$

$$\therefore F(s) = \frac{2k}{s^2+k^2}.$$

Ex. 17. *Find* $f(x)$ *if* $F_s(s) = \frac{e^{-as}}{s}$. *Hence deduce* $\mathcal{F}_s^{-1}\left(\frac{1}{s}\right)$.

Using Fourier sine inversion formula, we have,

$$f(x) = \frac{2}{\pi}\int_0^{\infty} \frac{e^{-as}}{s} \sin sx \, ds \qquad \cdots \quad (1)$$

Diffrentiating both side w.r.t. x we get

$$\frac{df}{dx} = \frac{2}{\pi}\int_0^{\infty} \frac{e^{-as}}{s}(\cos sx) s \, ds = \frac{2}{\pi}\int_0^{\infty} e^{-as} \cos sx \, ds$$

$$= \frac{2}{\pi} \frac{a}{a^2+x^2} \qquad \text{[detail calculation is not shown]}$$

$$\text{or, } df = \frac{2a}{\pi} \frac{dx}{a^2+x^2}$$

or, $f=\frac{2a}{\pi}\int\frac{dx}{x^2+a^2}=\frac{2a}{\pi}\frac{1}{a}\tan^{-1}\frac{x}{a}+C$

or, $f(x)=\frac{2}{\pi}\tan^{-1}\frac{x}{a}+C$

From (1) we get when $x=0,\ f(x)=0$

$\therefore$ From above we get $C=0\quad\therefore\ f(x)=\frac{2}{\pi}\tan^{-1}\frac{x}{a}$

We see $\mathcal{F}_s^{-1}\left(\frac{e^{-as}}{s}\right)=\frac{2}{\pi}\tan^{-1}\frac{x}{a}$

Putting $a=0$ we get $\mathcal{F}_s^{-1}\left(\frac{1}{s}\right)=\frac{2}{\pi}\tan^{-1}\infty=1$.

Ex. 18. *Find the inverse Fourier transform of the function* $e^{-|s|y}$.

The Inverse Fourier transform,

$$f(x)=\frac{1}{2\pi}\int_{-\infty}^{\infty}e^{-|s|y}\cdot e^{-isx}ds=\frac{1}{2\pi}\int_{-\infty}^{0}e^{sy}\cdot e^{-isx}ds+\frac{1}{2\pi}\int_{0}^{\infty}e^{-sy}\cdot e^{-isx}ds$$

$$=\frac{1}{2\pi}\int_{-\infty}^{0}e^{(y-ix)s}ds+\frac{1}{2\pi}\int_{0}^{\infty}e^{-(y+is)s}ds=\frac{1}{2\pi}\left[\frac{e^{(y-ix)s}}{y-ix}\right]_{-\infty}^{0}+\frac{1}{2\pi}\left[\frac{e^{-(y+ix)s}}{-(y+ix)}\right]_{0}^{\infty}$$

$$=\frac{1}{2\pi}\frac{1}{y-ix}(1-0)-\frac{1}{2\pi}\frac{1}{y+ix}(0-1)=\frac{1}{2\pi}\frac{1}{y-ix}+\frac{1}{2\pi}\frac{1}{y+ix}=\frac{y}{\pi(y^2+x^2)}.$$

EXERCISES

1. (i). Find the Fourier Transform of

$f(x)=x^2,\quad |x|<a$

$=0,\quad |x|\geq a$

(ii) Find the Fourier Transform of the function

$f(x)=a^2-x^2,\quad |x|<a$

$=0,\quad |x|>a$

2. If the Fourier transform, $F(s)=e^{-2|s|}$ evaluate $f(x)$

3. Find the Fourier sine and cosine transforms of $f(x)=1,\ 0\leq x<a$

$=0,\ x\geq a$

4. Find the Fourier cosine transform of

$$f(x)=x, \quad 0<x<1$$
$$=2-x, \quad 1<x<2$$
$$=0, \quad x>2$$

5. Find the Fourier sine transform of $e^{-|x|}$

Hence show that $\int_0^{\infty}\frac{x\sin mx}{1+x^2}dx=\frac{\pi e^{-m}}{2}$, $m>0$

6. Find Fourier Cosine Transform of $f(x)=\frac{1}{1+x^2}$ and hence derive Fourier sine transform of the function $\frac{x}{1+x^2}$.

7. (i) Find the Fourier sine and Cosine Transform of $f(x)=x^2$, $0<x<k$.

(ii) Find the Fourier Cosine Transform of $f(x)=\frac{1}{\sqrt{x}}$.

(iii) Find the Fourier cosine transform of $f(x)=e^{-x^2}$

8. Find the Fourier Transform of the function $f(x)=1, \quad |x|\le a$

$$=0, \quad |x|>a$$

Hence evaluate

(i) $\int_{-\infty}^{\infty}\frac{\sin sa\cos sx}{s}ds$ [*W.B.U.Tech,2004*]

(ii) $\int_0^{\infty}\frac{\sin x}{x}dx$

Hint : See the illustration given in a previous Art.

9. Find the Fourier Transform of

$$f(x)=1 \quad |x|<1$$
$$=0 \quad |x|>1$$

Hence evaluate $\int_0^{\infty}\frac{\sin x}{x}dx$

10. Find the Fourier Transform of

$$f(x)=\frac{\sqrt{2\pi}}{2\in} \quad |x|\le\in$$
$$=0 \quad |x|>\in$$

11. Find the sine and cosine transform of $\dfrac{e^{ax}+e^{-ax}}{e^{\pi x}-e^{-\pi x}}$.

12. If $f(x)=1, \quad 0 \le x < 1$

$\quad = 0; \quad x \ge 1$

find the (i) Fourier sine transform of $f(x)$

(ii) Fourier Cosine transform of $f(x)$

Use this to show that

(i) $\displaystyle\int_0^\infty \left(\frac{1-\cos x}{x}\right)^2 dx = \frac{\pi}{2}$ (ii) $\displaystyle\int_0^\infty \frac{\sin^4 x}{x^2} dx = \frac{\pi}{2}$

13. Evaluate $\mathcal{F}\left\{4x^2 e^{-3|x|}\right\}$

[Hint : $\mathcal{F}\left\{4x^2 e^{-3|x|}\right\} = 4i^2 \dfrac{d^2}{ds^2} \mathcal{F}\left\{e^{-3|x|}\right\}$]

14. Evaluate $\mathcal{F}\left\{\dfrac{\cos 3x}{x^2+2}\right\}$

15. Evaluate $\mathcal{F}\left\{\dfrac{3}{2x^2+1} - 5e^{-|x|}\right\}$

16. Evaluate the integral $\displaystyle\int_0^\infty \frac{s \sin sx}{1+s^2}$

[Hint : Consider $F_s(s) = \dfrac{s}{1+s^2}$ corresponding to the function $f(x)=e^{-x}$]

17. Evaluate : $\displaystyle\int_0^\infty \frac{\cos xs + s \sin xs}{1+s^2} ds$.

18. Prove that $\displaystyle\int_0^\infty \frac{s^3 \sin xs}{s^4+4} ds = \frac{\pi}{2} e^{-x} \cos x; \; x > 0$

19. Prove that $\displaystyle\int_0^\infty \frac{\sin \pi s \sin xs}{1-s^2} ds = \frac{\pi}{2} \sin x, \; 0 \le x \le \pi$

$\quad = 0, \quad x > \pi$

20. (i) Prove that $\int_0^\infty \frac{1-\cos \pi s}{s} \sin xs \, ds = \frac{\pi}{2}, \; 0 < x < \pi$

$= 0, \; x > \pi$

(ii) Show that $\int_0^\infty \frac{(x\cos x - \sin x)^2}{x^6} = \frac{\pi}{15}$.

21. Prove that $\lim_{\lambda \to \infty} \int_{-l}^{0} \frac{\sin \lambda v}{v} dv = \frac{\pi}{2}$

22. Find Fourier sine and cosine transform of e^{-x} and using the inversion formulae Recover the original function, in both the cases.

[*W.B.U.Tech,2003*]

23. Find $f(x)$ whose Fourier sine transform is $s e^{-as}$

24. Find $f(x)$ whose Fourier cosine transform is $s e^{-as}$.

25. Find $f(x)$ if its Fourier cosine transform is

$F_c(s) = a - \frac{s}{2}, \quad s < 2a$

$= 0, \quad s \geq 2a$

26. Given $\int_0^\infty f(x) \sin \lambda x \, dx = 1, \; 0 \leq \lambda < 1$

$= 2, 1 \leq \lambda < 2$

$= 0, \; \lambda \geq 2$

find $f(x)$ [or, solve the integral equation

$\int_0^\infty f(x) \sin \lambda x \, dx = 1, \quad 0 \leq \lambda < 1$

$= 2 \quad 1 \leq \lambda < 2$

$= 0] \quad \lambda \geq 2$

27. Solve the integral equation $\int_0^\infty f(x) \cos sx \, dx = e^{-s}$

28. Verify the Convolution theorem for the function

$f(x) = 1, \; |x| < 1$ and $g(x) = 1, \; |x| < 1$

$= 0, \; |x| > 1$ $\quad = 0, \; |x| > 1$

29. If $0 < x < \pi$ prove that $\mathcal{F}_s^{-1}\left(\frac{1-\cos n\pi}{n^3}\right) = \frac{x}{2}(\pi - x)$

30. Evaluate $\mathcal{F}^{-1}\left\{\dfrac{4\sin s}{s}-\dfrac{1}{\sqrt{|s|}}\right\}$

31. Evaluate $\mathcal{F}^{-1}\left\{e^{-2|s-3|}\right\}$.

32. Evaluate $\mathcal{F}^{-1}\left\{\dfrac{9}{2s+i}\right\}$

33. Evaluate $\mathcal{F}^{-1}\left\{e^{-|s|}\cos s\right\}$

34. Show that $e^{-\frac{x^2}{2}}$ is its own Fourier Transform.

35. Find the Fourier integral of the function

$$\begin{aligned} f(t) &= 1, \quad |t| \le 1 \\ &= 0, \quad |t| > 1 \end{aligned}$$

36. Apply Fourier Integral Theorem to show that

$$f(x)=\frac{1}{\pi}\int_0^{\infty}\frac{\cos\alpha x+\cos\alpha(\pi-x)}{1-\alpha^2}\,d\alpha$$

where $f(x)=\sin x,\ 0\le x\le\pi$

$= 0,$ otherwise

Hence show that $\displaystyle\int_0^{\infty}\frac{\cos\frac{\alpha\pi}{2}}{1-\alpha^2}\,d\alpha=\frac{\pi}{2}$

[Hint : put $x=\pi/2$]

37. Prove that $\displaystyle\frac{2}{\pi}\int_0^{\infty}e^{-bx}\sin\alpha x\,dx=\frac{2\alpha}{\pi\left(\alpha^2+b^2\right)}$

Hence prove that $\displaystyle e^{-bx}=\frac{2}{\pi}\int\frac{\alpha\sin\alpha x}{\alpha^2+b^2}\,d\alpha\ (b>0, x>0)$

ANSWERS

1. (i) $2\left\{\left(a^2s^2-2\right)\sin as+2as\cos as\right\}/s^3$ **2.** $\dfrac{1}{x^2+4}$ **3.** $\dfrac{1-\cos as}{s}$; $\dfrac{\sin as}{s}$

4. $\dfrac{2\cos s(1-\cos s)}{s^2}$ **5.** $\dfrac{s}{1+s^2}$ **6.** $\dfrac{\pi}{2}e^{-s}$; $-\dfrac{\pi}{2}e^{-s}$

7.(i) $\dfrac{2k}{s^2}\sin sk+\dfrac{2}{s^3}\cos sk-\dfrac{2}{s^3}-\dfrac{k^2\cos sk}{s}$ (iii) $\dfrac{\sqrt{\pi}}{2}e^{-\frac{s^2}{4}}$

8. (i) $=\pi$ when $|x|<a$ The Fourier trans is $F(s)=2\dfrac{\sin sa}{s}$, $s\neq 0$

$=\dfrac{\pi}{2}$ when $|x|=a$ $=2a$, $s=0$

$=0$ when $|x|>a$

(ii) Putting $x=0,\ a=1$ in (1) we get the result $\dfrac{\pi}{2}$

9. $\dfrac{\pi}{2}$ **10.** $\sqrt{2\pi}\,\dfrac{\sin s\varepsilon}{s\varepsilon}$ **11.** $\dfrac{\cos\frac{a}{2}\left(e^{s/2}+e^{-s/2}\right)}{2\cos a+e^{s}+e^{-s}}$

12. (i) $\dfrac{1-\cos s}{s}$, (ii) $\dfrac{\sin s}{s}$, **13.** $48\left\{\dfrac{1}{\left(s^2+9\right)^2}-\dfrac{4s^2}{\left(s^2+9\right)^3}\right\}$

14. $\dfrac{\pi}{2\sqrt{2}}\left(e^{-\sqrt{2}|s-3|}+e^{-\sqrt{2}|s+3|}\right)$ **15.** $\dfrac{3\pi}{\sqrt{2}}e^{-\frac{|s|}{\sqrt{2}}}-\dfrac{10}{s^2+1}$ **16.** $\dfrac{\pi}{2}e^{-x}$

17. 0 when $x<0$, $\pi/2$ when $x=0, \pi e^{-x}$ when $x>0$.

23. $\dfrac{2}{\pi}\dfrac{\sin 2\theta}{\left(a^2+x^2\right)}$ where $\tan\theta=\dfrac{x}{a}$,

24. $\dfrac{2}{\pi}\dfrac{\cos 2\theta}{\left(a^2+x^2\right)}$ where $\tan\theta=\dfrac{x}{a}$

25. $\dfrac{\sin^2 ax}{\pi x^2}$ **26.** $f(x) = (2 + 2\cos x - 4\cos 2x)/\pi x$ **27.** $2/\pi(1 + x^2)$

30. $2\{H(x+1) - H(x-1)\} - \dfrac{1}{\sqrt{2\pi|x|}}$ **32.** $-\dfrac{9i}{2} H(-x) e^{x/2}$

33. $\dfrac{1}{2\pi}\left\{\dfrac{1}{(x+1)^2+1} + \dfrac{1}{(x-1)^2+1}\right\}$ **36.** $\dfrac{2}{\pi}\displaystyle\int_0^\infty \dfrac{\sin y \cos xy}{y}\, dy$

Module -II

Calculus of Complex Variable

2.1 FUNCTION, LIMIT, CONTINUITY AND ANALYTIC FUNCTIONS

2.1.1. Introduction

Student are very well familiar about functions. limit and continuity for a real variable. In this chapter we discuss about that analogy for functions of complex variable. For this firstly we define and explain some related basic concepts as follow :

Complex Point Set or (simply a set)

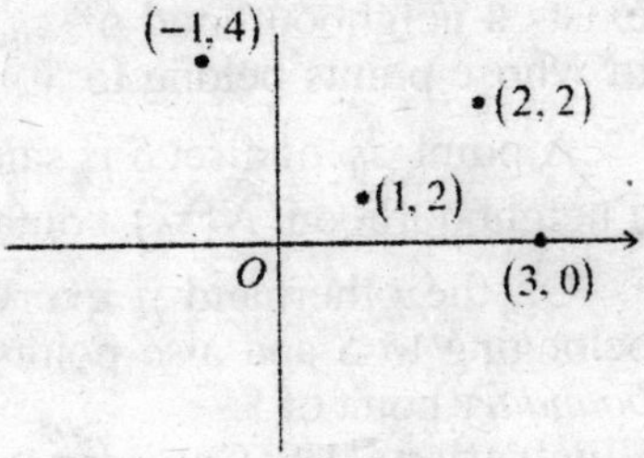

A collection of points in the complex plane i.e. in the Argand plane is called a point set. So the complex point set is nothing but the set of some complex numbers.

For example : the set $\{1+2i, -1+4i, 3, 2+2i\}$ is a complex point set. The points of this set are shown by dots in the adjacent diagram.

Neighbourhood of a point.

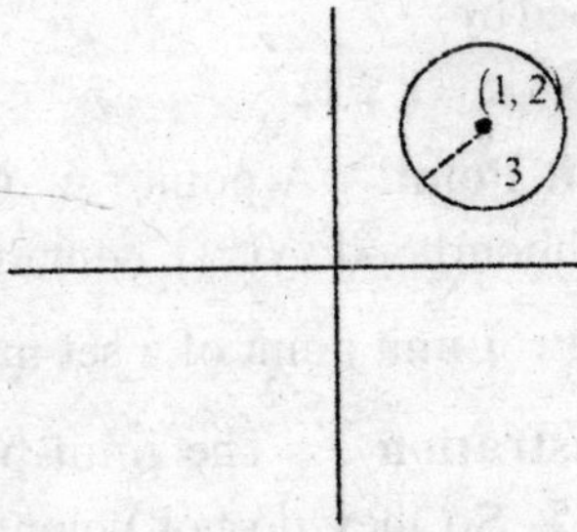

Let z_0 be a point of the complex plane. The set of all points z satisfying the inequality $|z-z_0|<\varepsilon$ is called the *neighbourhood* of z_0 and is denoted by $N(z_0,\varepsilon)$ or simply $N(z)$.

For Example : $N(1+2i,\ 0\cdot 3)$ is a neighbourhood as shown by the circle in the adjacent argand diagram. Note that this is a circular disc with centre (1, 2) and radius 0.3.

If from the neighbourhood of a point z_0 we exclude the point z_0, then such a neighbourhood is called the ***deleted neighbourhood*** of z_0 and is reperesented by $0<|z-z_0|<\varepsilon$ or by $N^0(z_0,\varepsilon)$ or simply $N^0(z_0)$.

Bounded and Unbounded sets.

A set of points S is said to be *bounded* if there exists a finite positive number M such that $|z|\le M \quad \forall z \in S$. If there exists no such number M, then the set S is called *unbounded.*

For example : We consider the set S of complex points (shown by dots) in the adjacent diagram. These are contained within the circle with centre origin, radius 10 (say). So this set $S = \{z : |z| \leq 10\}$ is bounded.

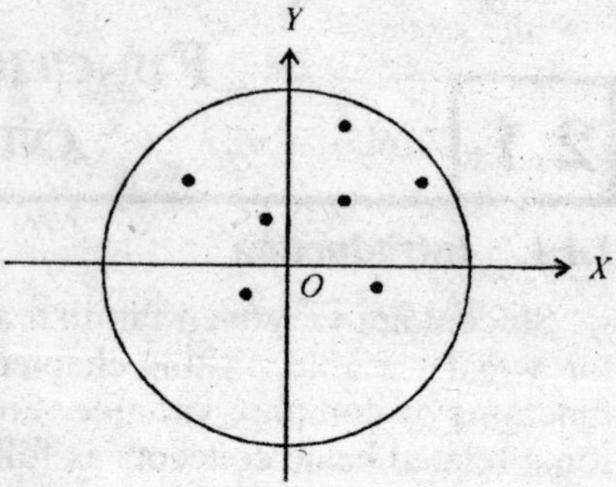

Interior, Exterior and Boundary Points.

A point z_0 is said to be an *interior* point of a set S if there exists a neighbourhood of z_0 all of whose points belong to S.

A point z_0 of a set S is said to an *exterior or isolated* point if it has a neighbourhood $N(z_0)$ containing no other points of S.

On the otherhand if every neighbourhood of z_0 contains points belonging to S and also points not belonging to S, then z_0 is called a *boundary* point of S.

Illustration. (1) Consider the set S defined by $S : |z| < 3$

(i) Every point of S is an interior point.

(ii) The point for which $|z| = 3$ are boundary points

(2) The point $z = 2 - 7i$ is an exterior or isolated point of the set S defined by

$S : |z - 1| \leq 1$,

Limit Point. A point z_0 is called a limit point of a set S if every neighbourhood $N(z_0)$ contain at least one point of S other than z_0.

Note : Limit point of a set may or may not belong to the set.

Illustration . The limit points of the set $S : |z| < 5$ are given by $|z| \leq 5$. So they do not belong to the set. But the limit points of the set $S : |z| \leq 5$ are given by $|z| \leq 5$. So they all belong to the set S.

This is further illustrated by the adjacent diagram. The set $S = \{z : |z| < 5\}$ is represented by the points within the circle with centre at origin and radius 5 (not

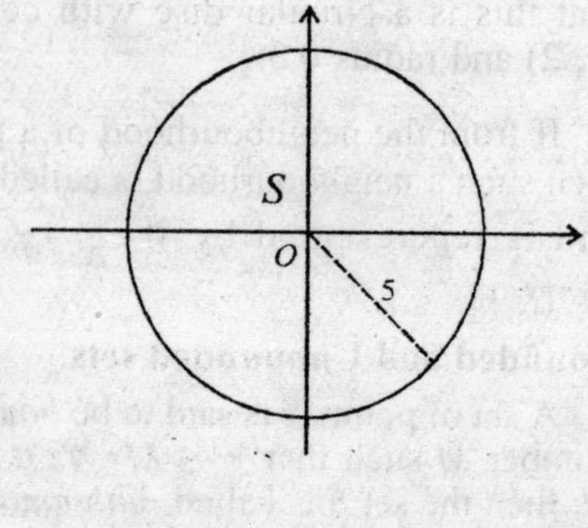

by the points on the circumference). The limiting points of S are the points within the circle as well as those on the circumference.

2.1.2. Function of a Complex Variable.

When a symbol z takes any one of the values of a set of complex numbers then z is called a *complex variable.*

Let D and R be any two non-empty point sets in the complex plane. A complex variable $w \in R$ is said to be a *function* of the complex variable $z \in D$, if, to every value of z there correspond one or more values of w. Thus if w is a function of z, it is written as $w = f(z)$.

If $z = x + iy$ and $w = u + iv$, then u and v are both functions of two real variable x and y and we may write

$$w = f(z) = u(x, y) + iv(x, y)$$

Here D is called the *domain* and R is called the *range* of f.

Illustration. (i) $f(z) = z^4 + 5z - 3$ is a function of z and is called the *rational integral function* of z.

(ii) $f(z) = \cos z$ is a *trigonometric function* of z.

(iii) $w = f(z) = \dfrac{z-2}{z+3}$ is known as *rational function* of z whose domain is entire complex plane except $z = -3$ and the range is the entire complex plane except $w = 1$.

(iv) For the function $w = f(z) = z^2 + 5$, we have

$$u + iv = (x + iy)^2 + 5 = (x^2 - y^2 + 5) + i \cdot 2xy$$

$$\therefore u(x, y) = x^2 - y^2 + 5 \text{ and } v(x, y) = 2xy.$$

2.1.3 Limit of a Function

Let D be the domain of a function $f(z)$ and $z_0 \in D$. Then we say that $f(z)$ tends to the limit l, (a complex number) as z tends to z_0 if to each $\varepsilon(>0)$ small at pleasure there exists a number $\delta > 0$ such that

$$|f(z) - l| < \varepsilon \ \forall z \in N^0(z_0, \delta) \cap D$$

i.e. for all pts of D satisfying $0 < |z - z_0| < \delta$.

Symbolically we write $\underset{z \to z_0}{Lt} f(z) = l$

Illustration. (i) Let $f(z)=\begin{cases} z^2, & z \neq z_0 \\ 0, & z = z_0 \end{cases}$

Then for $\varepsilon > 0$ we can find $\delta \le 1$ such that

$$\left|f(z)-z_0^2\right| = \left|z^2 - z_0^2\right| = |z-z_0||z+z_0|$$

$$= |z-z_0||z-z_0+2z_0|$$

$$\le |z-z_0|\{|z-z_0|+|2z_0|\}$$

$$< \delta(\delta+2|z_0|) \text{ if } |z-z_0| < \delta$$

$$< \delta(1+2|z_0|) \text{ as } \delta \le 1$$

$$< \varepsilon \text{ if } \delta = \frac{\varepsilon}{1+2|z_0|}.$$

$\therefore \left|f(z)-z_0^2\right| < \varepsilon$ whenever $|z-z_0| < \delta$.

Hence $\underset{z \to z_0}{Lt}\, f(z) = z_0^2$.

Existence of Limit. The conception of left hand limit and right hand limit for a real valued function is generalized for complex valued functions. Here the complex variable z can tend to the complex number a along any path. In the adjacent figure we show two of these, say $C_1 : y=\phi(x)$, $C_2 : y=\psi(x)$

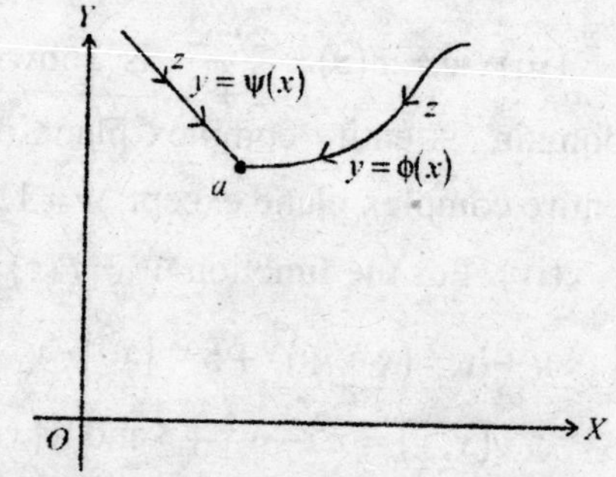

If there exist two paths say C_1, C_2 *such that*

$\lim\limits_{\substack{z \to a \\ \text{along } C_1}} f(z) \neq \lim\limits_{\substack{z \to a \\ \text{along } C_2}} f(z)$ *then the limit* $\lim\limits_{z \to a} f(z)$ *does not exist.*

Example can be found in the subsequent Illustrative Examples.

Theorems on Limit

Theorem 1. The limit of a function, when exists, is unique.

Proof: Left as an exercise.

Theorem 2. A necessary and sufficient condition that the function $f(z)=u+iv$ tends to $l=\alpha+i\beta$ as $z=x+iy$ tends to $z_0=x_0+iy_0$ is that $u\to\alpha$ and $v\to\beta$ as $(x,y)\to(x_0,y_0)$.

Proof: Beyond the scope of the book.

Theorem 3. If $\underset{z\to z_0}{Lt} f(z)=l_1$ and $\underset{z\to z_0}{Lt} g(z)=l_2$

then (i) $\underset{z\to z_0}{Lt}\left[pf(z)\pm qg(z)\right]=pl_1\pm ql_2$ (*p*, *q* are constants)

(ii) $\underset{z\to z_0}{Lt}\left[f(z)g(z)\right]=l_1l_2$

(iii) $\underset{z\to z_0}{Lt}\dfrac{f(z)}{g(z)}=\dfrac{l_1}{l_2}$, provided $l_2\neq 0$.

Proof: Left as an exercise.

Like real valued function here also the following formulae can be presented.

Formulae : (1) $\lim\limits_{z\to a} z^n=a^n$

(2) $\lim\limits_{z\to a}\sin z=\sin a$ **(3)** $\lim\limits_{z\to a}\cos z=\cos a$

(4) $\lim\limits_{z\to a} b^z=b^a$ **(5)** $\lim\limits_{z\to a}\log z=\log a$

Proof: (1) is shown for $n=2$ in the illustration (i) of Art 4.3.

Illustrative Examples.

Ex. 1. *Evaluate* $\underset{z\to -2i}{Lt}\dfrac{2z^2+z-3}{(z^2-2z+4)(z-1)}$

$$\underset{z\to -2i}{Lt}\frac{2z^2+z-3}{(z^2-2z+4)(z-1)}=\underset{z\to -2i}{Lt}\frac{(2z+3)(z-1)}{(z^2-2z+4)(z-1)}$$

$$=\underset{z\to -2i}{Lt}\frac{2z+3}{z^2-2z+4}=\frac{-4i+3}{-4+4i+4}=-1-\frac{3}{4}i\text{. by the above fuormulae.}$$

Ex. 2. *Prove that* $\lim\limits_{z\to a}\dfrac{\bar z}{z}$ *does not exist.* [*W.B.U.Tech 2002*]

Let $z=x+iy$ $\quad\therefore \bar z=x-iy\quad\therefore \dfrac{\bar z}{z}=\dfrac{x-iy}{x+iy}$.

Let $z \to 0$ along the x-axis. Then $y = 0 \therefore \underset{z\to 0}{Lt} \frac{\bar{z}}{z} = \underset{x\to 0}{Lt} \frac{x}{x} = 1$.

Next let $z \to 0$ along the y-axis. Then $x = 0$.

$$\therefore \underset{z\to 0}{Lt} \frac{\bar{z}}{z} = \underset{y\to 0}{Lt} \frac{-iy}{iy} = -1.$$

Thus $\underset{z\to 0}{Lt} \frac{\bar{z}}{z}$ has different values along different path.

Hence by the previous discussion $\underset{z\to 0}{Lt} \frac{\bar{z}}{z}$ does not exist.

Ex. 3. *Show that* $\underset{z\to 0}{Lt} \frac{xy}{x^2 + 2y^2}$ *does not exist.*

Let $z \to 0$ i.e. $(x, y) \to 0$ along the st. line $y = mx$.

Then $\underset{z\to 0}{Lt} \frac{xy}{x^2 + 2y^2} = \underset{(x,y)\to 0}{Lt} \frac{xy}{x^2 + 2y^2}$

$= \underset{x\to 0}{Lt} \frac{x \cdot mx}{x^2 + 2(mx)^2} = \frac{m}{1 + 2m^2}$ which has different values for different values of m i.e. for different path. Hence, by a previous discussion , $\underset{z\to 0}{Lt} \frac{xy}{x^2 + 2y^2}$ does not exist.

2.1.4. Continuity

Continuity at a Point.

Let D be the domain of a function $f(z)$ and $z_0 \in D$. Then $f(z)$ is said to be continuous at z_0 if to each $\varepsilon (> 0)$ small at pleasure there exists a number $\delta > 0$ such that $|f(z) - f(z_0)| < \varepsilon$, $\forall z \in N(z_0, \delta) \cap D$ i.e. for all pts of D satisfying $|z - z_0| < \delta$.

From the above definition it is clear that $f(z)$ will be continuous at z_0 iff $\underset{z\to z_0}{Lt} f(z) = f(z_0)$.

Continuity in a Region.

A function $f(z)$ is said to be continuous in a certain region D if it is continuous at every point of D.

Illustration. **(i)** The function $f(z)=z^2$ is continuous at $z=z_0$, as by previous formulae, we have $\underset{z\to z_0}{Lt} f(z)=z_0^2=f(z_0)$ because $f(z_0)=0$.

But the function $f(z)= z^2, \quad z\neq z_0$

$\qquad\qquad\qquad = 0, \quad z=z_0$

is not continuous at $z=z_0$, as $\underset{z\to z_0}{Lt} f(z)=z_0^2\neq f(z_0)$.

(ii) Also the function $f(z)=z^2$ is continuous in the region $|z|\leq 1$, as the function $f(z)$ is continuous at any point $z=z_0$ in the region $|z|\leq 1$, by **(i)**.

Theorems on Continuity

Theorem 1. A necessary and sufficient condition of a function $f(z)=u+iv$ to be continuous at $z_0=x_0+iy_0$ is that $u(x,y)$ and $v(x,y)$ be continuous at (x_0,y_0).

Proof. Beyond the scope of the book.

Theorem 2. If the function $f(z)$ and $g(z)$ are defined in D and continuous at $z=z_0$, then

(i) $pf(z)\pm qg(z)$, (p, q are constants)

(ii) $f(z)g(z)$

(iii) $\dfrac{f(z)}{g(z)}$, if $g(z_0)\neq 0$ are also continuous at $z=z_0$.

Proof: Left as an exercise.

Illustrative Examples.

Ex. 1. *For what value of z, the following functions are continuous?*

(i) $f(z)=\dfrac{2z-3}{z^2+2z+2}$ *(ii)* $f(z)=\dfrac{1}{z}-\sec z$.

(i) $f(z) = \dfrac{2z-3}{z^2+2z+2}$ is not continuous when $z^2+2z+2=0$

i.e. $z = \dfrac{-2 \pm \sqrt{4-8}}{2} = -1 \pm i$.

$z = \dfrac{-b \pm \sqrt{b^2-4ac}}{2a}$

So $f(z)$ is continuous everywhere except $z = -1 \pm i$.

(ii) $f(z) = \dfrac{1}{z} - \sec z = \dfrac{\cos z - z}{z \cos z}$ is not continuous when $z \cos z = 0$

i.e. $z = 0$ or $\cos z = 0$ or, $z = (2n+1)\dfrac{\pi}{2}$, $n = 0, \pm 1, \pm 2, \ldots$

So $f(z)$ is continuous everywhere except at $z = 0$, $(2n+1)\dfrac{\pi}{2}$, $n = 0, \pm 1, \pm 2, \ldots$.

Ex. 2. *Discuss the continuity of the function* $f(z) = |z|^2$.

Let $z = x + iy \quad \therefore |z|^2 = x^2 + y^2$.

Also let $f(z) = u(x,y) + iv(x,y)$

$\therefore u(x,y) = x^2 + y^2$, $v(x,y) = 0$.

Now $v(x,y) = 0$ is continuous everywhere.

Also $u(x,y) = x^2 + y^2$ is continuous everywhere.

Hence by Theorem (1), $f(z) = |z|^2$ is continuous everywhere.

Ex. 3. *Test the continuity at origin of the following function*

$$f(z) = \frac{xy^3}{x^2+y^6}, \quad \textit{for } z \neq 0$$

$$= 0, \qquad \textit{for } z = 0.$$

Let $f(z) = u(x,y) + iv(x,y)$. Then

$$u(x,y) = \frac{xy^3}{x^2+y^6}, \quad (x,y) \neq (0,0)$$

$$= 0, \quad (x,y) = (0,0)$$

and $v(x,y) = 0$, $\forall (x,y)$.

Let $(x,y) \to (0,0)$ along the curve $x = my^3$.

$$\therefore \quad \underset{(x,y)\to(0,0)}{Lt} u(x,y) = \underset{y\to 0}{Lt} \frac{my^3 \cdot y^3}{m^2 y^6 + y^6} = \frac{m}{1+m^2}.$$

$\therefore \quad \underset{(x,y)\to(o,o)}{Lt} u(x,y)$ has different values for different values of m i.e. along different path. Hence $\underset{(x,y)\to(0,0)}{Lt} u(x,y)$ does not exist. Therefore $u(x,y)$ is not continuous at $(0,0)$. Consequently $f(z)$ is not continuous at $z=0$, by Theorem1.

2.1.5. Analytic Functions.

Differentiability.

A function $w = f(z)$ defined in a certain domain D is said to be differentiable at $z = z_0$ if the limit $\lim\limits_{z\to z_0} \frac{f(z)-f(z_0)}{z-z_0}$ exists, i.e., if $\frac{f(z)-f(z_0)}{z-z_0}$ tends to a unique finite limit as z tends to z_0 along any path of the domain D. Then this unique limit is called the derivative or differential co-efficient of $f(z)$ at $z = z_0$ and is denoted by $f'(z_0)$ or $\left(\frac{dw}{dz}\right)_{z=z_0}$. Thus

$$f'(z_0) = \underset{z\to z_0}{Lt} \frac{f(z)-f(z_0)}{z-z_0}.$$

Note : We may also write the derivative of $f(z)$ at $z = z_0$ as follows :

$$f'(z_0) = \underset{\Delta z\to 0}{Lt} \frac{f(z_0+\Delta z)-f(z_0)}{\Delta z}$$

$$\text{or, } f'(z_0) = \underset{h\to 0}{Lt} \frac{f(z_0+h)-f(z_0)}{h}.$$

Theorem 1. If a function is differentiable at a point, then it is continuous at that point.

Proof : Left as an exercise.

...se of the above theorem is not true i.e. a function ...as is not necessarily differentiable.

...ve this by the following examples.

Ex... *Consider the function* $f(z)=\bar{z}$.

Then $|f(z)-f(0)|=|\bar{z}-0|=|\bar{z}|=|z|<\varepsilon$ when $|z-0|<\delta$ where $\delta=\varepsilon$.

So $f(z)=\bar{z}$ is continuous at $z=0$.

But $f'(0)=\underset{z\to 0}{Lt}\dfrac{f(z)-f(0)}{z-0}=\underset{z\to 0}{Lt}\dfrac{\bar{z}}{z}$ does not exist (for details see Ex 2 of art 4.3)

Example 2. *Consider the function* $f(z)=|z|^2$.

Here $f(z)=|z|^2=x^2+y^2$ is continuous everywhere (See Ex 2 of art 4.4). But

$$f'(z_0)=\underset{\Delta z\to 0}{Lt}\frac{f(z_0+\Delta z)-f(z_0)}{\Delta z}=\underset{\Delta z\to 0}{Lt}\frac{|z_0+\Delta z|^2-|z_0|^2}{\Delta z}$$

$$=\underset{\Delta z\to 0}{Lt}\frac{(z_0+\Delta z)(\bar{z}_0+\overline{\Delta z})-z_0\bar{z}_0}{\Delta z}=\underset{\Delta z\to 0}{Lt}\left(\bar{z}_0+\overline{\Delta z}+z_0\frac{\overline{\Delta z}}{\Delta z}\right)$$

$$=\bar{z}_0+z_0\underset{\Delta z\to 0}{Lt}\frac{\overline{\Delta z}}{\Delta z}.$$

Now at $z_0=0$, $f'(0)=0$

When $z_0\neq 0$, let $\Delta z=r(\cos\theta+i\sin\theta)=re^{i\theta}$

$\therefore\ \overline{\Delta z}=r(\cos\theta-\sin\theta)=re^{-i\theta}\quad \therefore\ \dfrac{\overline{\Delta z}}{\Delta z}=e^{-2i\theta}=\cos 2\theta-i\sin 2\theta$

Hence $\underset{\Delta z\to 0}{Lt}\dfrac{\overline{\Delta z}}{\Delta z}$ have different values for different values of θ.

$\therefore\ \underset{\Delta z\to 0}{Lt}\dfrac{\overline{\Delta z}}{\Delta z}$ does not exist.

$\therefore\ f'(z_0)$ does not exist except $z_0=0$.

Analytic at a Point.

A function $f(z)$ in a domain D is said to be analytic at a point $z = z_0$ if there exists a neighbourhood $|z - z_0| < \delta$ at all points of which the function is differentiable i.e. $f'(z)$ exists. [i.e., there exists a circle with centre at z_0 within which the function is differentiable at every point]

Analytic Function. If a function $f(z)$ be such that $f'(z)$ exists at every point of the domain D then $f(z)$ is said to be analytic in D.

Note : The terms regular and homomorphic are also sometimes used as synonyms for analytic.

Singular Point. If a function $f(z)$ is not analytic at a point z_0, then z_0 is called a singular point or a singularity of $f(z)$.

Zero of a Function. A point z_0 is called a zero of the function $f(z)$ if $f(z)$ is analytic at $z = 0$ and $f(z_0) = 0$.

Illustration. **(i)** $f(z) = e^z$ is analytic in every finite region of the complex plane.

(ii) $f(z) = \dfrac{2z}{z + 3i}$ is analytic in every finite region except at $z = -3i$.

So $z = -3i$ is a singularity of $f(z)$ whereas $z = 0$ is a zero of $f(z)$.

2.1.6. Cauchy-Riemann Conditions.

Theorem 1. (Necessary conditions for $f(z)$ to be analytic.)

The neceassary conditions for $w = f(z) = u(x, y) + iv(x, y)$ to be analytic (i.e. differentiable) at any point $z = x + iy$ of its domain D is that the four partial derivatives $\dfrac{\partial u}{\partial x}, \dfrac{\partial u}{\partial y}, \dfrac{\partial v}{\partial x}$ and $\dfrac{\partial v}{\partial y}$ should exist and

$$\frac{\partial u}{\partial x} = \frac{\partial v}{\partial y} \text{ and } \frac{\partial u}{\partial y} = -\frac{\partial v}{\partial x} \quad \ldots (1)$$

Proof : Beyond the scope of this book.

Note : (i) The relations (1) are called the Cauchy-Riemann Equations (in brief C-R equations).

Corollary 1. If $w = f(z)$ is differentiable at any point z then

$$\frac{dw}{dz} = \frac{\partial w}{\partial x} = \frac{1}{i}\frac{\partial w}{\partial y}$$

Corollary 2. If $f'(z)$ exists then $f'(z)$ is given by

$$f'(z)=\frac{\partial u}{\partial x}+i\frac{\partial v}{\partial x} \text{ or } f'(z)=\frac{1}{i}\frac{\partial u}{\partial y}+\frac{\partial v}{\partial y}=\frac{\partial v}{\partial y}-i\frac{\partial u}{\partial y}.$$

Theorem 2. (Sufficient Conditions for $f(z)$ to be analytic.)

The function $w=f(z)=u(x,y)+iv(x,y)$ is analytic in a domain D if

(i) four partial derivatives u_x, u_y, v_x and v_y exist and $u_x=v_y, u_y=-v_x$

(ii) the partial derivatives u_x, u_y, v_x and v_y are all continuous in D.

Proof: Beyond the scope of the book.

Illustrative Examples.

Ex. 1. *Show that the function* $f(z)=\sin x\cosh y+i\cos x\sinh y$ *is continuous as well as analytic everywhere and find its derivative.*

Let $f(z)=u(x,y)+iv(x,y)$.

$\therefore u(x,y)=\sin x\cosh y$, $v(x,y)=\cos x\sinh y$

Clearly $u(x,y)$ and $v(x,y)$ are continuous for all values of x and y. Hence $f(z)$ is continuous everywhere.

Again $\dfrac{\partial u}{\partial x}=\cos x\cosh y$, $\dfrac{\partial u}{\partial y}=\sin x\sinh y$, $\dfrac{\partial v}{\partial x}=-\sin x\sinh y$,

$\dfrac{\partial v}{\partial y}=\cos x\cosh y$

$\therefore \dfrac{\partial u}{\partial x}=\dfrac{\partial v}{\partial y}$ and $\dfrac{\partial u}{\partial y}=-\dfrac{\partial v}{\partial x}$.

Thus Cauchy-Riemann conditions are satisfied.

Also it can be easily verified that all the partial derivatives $\dfrac{\partial u}{\partial x},\dfrac{\partial u}{\partial y},\dfrac{\partial v}{\partial x},\dfrac{\partial v}{\partial y}$ are continuous for all values of x and y.

Hence the function $f(z)$ is analytic everywhere, by Theorem 2.

$$\therefore f'(z)=\frac{\partial u}{\partial x}+i\frac{\partial v}{\partial x}=\cos x\cosh y-i\sin x\sinh y$$

Ex. 2. *Verify whether* $f(z) = \operatorname{Im}(z)$ *is differentiable everywhere in the finite complex plane.*

Let $z = x + iy$ $\therefore f(z) = \operatorname{Im}(z) = iy$

Also let $f(z) = u(x,y) + iv(x,y)$ $\therefore u(x,y) = 0$, $v(x,y) = y$

$$\therefore \frac{\partial u}{\partial x} = 0,\ \frac{\partial u}{\partial y} = 0,\ \frac{\partial v}{\partial x} = 0,\ \frac{\partial v}{\partial y} = 1. \qquad \therefore \frac{\partial u}{\partial x} \neq \frac{\partial v}{\partial y}.$$

So Cauchy-Riemann equations are not satisfied unless $x = y = 0$. Hence $f(z)$ is not differentiable at any point other than origin.

Ex. 3. *Show that the function* $f(z) = xy + iy$ *is everywhere continuous but is not analytic.*

Let $f(z) = u(x,y) + iv(x,y)$ $\therefore u(x,y) = xy,\ v(x,y) = y$

Thus we see that both u and v being polynomials in x and y are continuous everywhere. Hence $f(z)$ is continuous everywhere.

$$\text{Here } \frac{\partial u}{\partial x} = y, \frac{\partial u}{\partial y} = x, \frac{\partial v}{\partial x} = 0, \frac{\partial v}{\partial y} = 1.$$

$$\therefore \frac{\partial u}{\partial x} \neq \frac{\partial v}{\partial y}, \frac{\partial u}{\partial y} \neq -\frac{\partial v}{\partial x}.$$

Hence Cauchy-Riemann equations are not satisfied anywhere. Thus $f(z)$ is not analytic at any point in the complex plane.

Ex. 4. *Prove that the function* $f(z) = \dfrac{x^3(1+i) - y^3(1-i)}{x^2 + y^2}$, $z \neq 0$

$= 0$, $z = 0$

is continuous everywhere but $f'(0)$ *does not exist though Cauchy-Riemann Equations are satisfied at the origin.* [W.B.U.Tech 2002]

Let $f(z) = u(x,y) + iv(x,y)$

$$\therefore u(x,y) = \frac{x^3 - y^3}{x^2 + y^2},\quad v(x,y) = \frac{x^3 + y^3}{x^2 + y^2} \quad \text{when } z \neq 0$$

$$= 0, \qquad = 0 \qquad \text{when } z = 0.$$

As u and v are both rational and finite for all values of $z \neq 0$, so u and v are continuous at all these points for which $z \neq 0$. Hence $f(z)$ is continuous when $z \neq 0$. Again u and v are both continuous at the origin. Consequently $f(z)$ is continuous at the origin. Thus $f(z)$ is continuous everywhere.

Now at the origin.

$$\frac{\partial u}{\partial x} = \underset{x\to 0}{Lt} \frac{u(x,0)-u(0,0)}{x} = \underset{x\to 0}{Lt}\left(\frac{x}{x}\right) = 1$$

$$\frac{\partial u}{\partial y} = \underset{y\to 0}{Lt} \frac{u(0,y)-u(0,0)}{y} = \underset{y\to 0}{Lt}\left(\frac{-y}{y}\right) = -1$$

$$\frac{\partial v}{\partial x} = \underset{x\to 0}{Lt} \frac{v(x,0)-v(0,0)}{x} = \underset{x\to 0}{Lt}\left(\frac{x}{x}\right) = 1$$

$$\frac{\partial v}{\partial y} = \underset{y\to 0}{Lt} \frac{v(0,y)-v(0,0)}{y} = \underset{y\to 0}{Lt}\left(\frac{y}{y}\right) = 1.$$

Thus at the origin, $\frac{\partial u}{\partial x} = \frac{\partial v}{\partial y}, \frac{\partial u}{\partial y} = -\frac{\partial v}{\partial x}$.

Hence Cauchy-Riemann Equations are satisfied at the origin.

Again $f'(0) = \underset{z\to 0}{Lt} \frac{f(z)-f(0)}{z} = \underset{(x,y)\to(0,0)}{Lt} \frac{(x^3-y^3)+i(x^3+y^3)}{(x^2+y^2)(x+iy)}$

Now let $z \to 0$ along $y = x$. Then

$$f'(0) = \underset{x\to 0}{Lt} \frac{x^3-x^3+i(x^3+x^3)}{(x^2+x^2)(x+ix)} = \frac{2i}{2(1+i)} = \frac{1-i}{2}$$

Again let $z \to 0$ along $y = 0$. Then $f'(0) = \underset{x\to 0}{Lt} \frac{x^3+ix^3}{x^2 \cdot x} = 1+i$

Thus the values of $f'(0)$ are not same as $z \to 0$ along different paths.

Hence $f'(0)$ does not exist.

Ex. 5. *For the function defined by* $f(z) = \sqrt{|xy|}$, *show that the Cauchy-Riemann Equations are satisfied at* $(0,0)$ *but the function is not differentiable at that point.* [W.B.U.Tech 2003]

Let $f(z)=u(x,y)+iv(x,y)$.

Then $u(x,y)=\sqrt{|xy|}$, $v(x,y)=0$. Now at the origin,

$$\frac{\partial u}{\partial x}=\underset{x\to 0}{Lt}\frac{u(x,0)-u(0,0)}{x}=\underset{x\to 0}{Lt}\frac{0-0}{x}=0$$

$$\frac{\partial u}{\partial y}=\underset{y\to 0}{Lt}\frac{u(0,y)-u(0,0)}{y}=\underset{y\to 0}{Lt}\frac{0-0}{y}=0.$$

Similarly $\frac{\partial v}{\partial x}=0$, $\frac{\partial v}{\partial y}=0$.

Hence Cauchy-Riemann Equations are satisfied at the origin.

Again $f'(0)=\underset{z\to 0}{Lt}\frac{f(z)-f(0)}{z}=\underset{(x,y)\to 0}{Lt}\frac{\sqrt{(|xy|)}}{x+iy}$.

Let $z\to 0$ along the straight line $y=mx$. Then

$$f'(0)=\underset{x\to 0}{Lt}\frac{\sqrt{|mx^2|}}{x+imx}=\frac{\sqrt{|m|}}{1+im}$$

which have different values for different m i.e. along different straight line $y=mx$. Hence $f'(0)$ does not exist. Thus the function is not differentiable at the origin.

2.1.7. Harmonic Function.

Laplace's Equation. A partial differential equation of the form $\frac{\partial^2 f}{\partial x^2}+\frac{\partial^2 f}{\partial y^2}=0$ is called the Laplace's Equation.

Harmonic Function. A function $f(x,y)$ which possesses continuous partial derivatives of first and second orders and satisfies Laplace's Equation is called the *Harmonic Function.*

Conjugate Harmonic Functions. If two harmonic functions $u(x,y)$ and $v(x,y)$ satisfy the Cauchy-Riemann Equations $\frac{\partial u}{\partial x}=\frac{\partial v}{\partial y}$, $\frac{\partial u}{\partial y}=-\frac{\partial v}{\partial x}$, then they are called the *Conjugate Harmonic Functions.*

Theorem 1. If $f(z)=u+iv$ is an analytic function, then u and v are conjugate harmonic function.

Proof: Beyond the scope of this book.

Illustrative Examples.

Ex. 1. *Show that the function* $u=\cos x\cosh y$ *is harmonic and find its harmonic conjugate.*

Here $u=\cos x\cosh y$

$$\therefore \frac{\partial u}{\partial x}=-\sin x\cosh y,\quad \frac{\partial^2 u}{\partial x^2}=-\cos x\cosh y$$

$$\frac{\partial u}{\partial y}=\cos x\sinh y,\quad \frac{\partial^2 u}{\partial y^2}=\cos x\cosh y.$$

$$\therefore \frac{\partial^2 u}{\partial x^2}+\frac{\partial^2 u}{\partial y^2}=-\cos x\cosh y+\cos x\cosh y=0$$

Hence u is a harmonic function.

Let v be the conjugate harmonic of u. Then

$$\frac{\partial u}{\partial x}=\frac{\partial v}{\partial y},\quad \frac{\partial u}{\partial y}=-\frac{\partial v}{\partial x} \qquad \ldots (1)$$

Now $dv=\frac{\partial v}{\partial x}\cdot dx+\frac{\partial v}{\partial y}\cdot dy=-\frac{\partial u}{\partial y}\cdot dx+\frac{\partial u}{\partial x}\cdot dy$, by (1)

$$=-\cos x\sinh y\,dx-\sin x\cosh y\,dy=-d(\sin x\sinh y).$$

Integrating this we get $v=-\sin x\sinh y+c$ where c is a real constant.

Ex. 2. *Prove that* $u=2x-2xy$ *is a harmonic function. Determine its harmonic conjugate and find the corresponding analytic function* $f(z)$ *in terms of* z.

Here $u=2x-2xy$

$$\therefore \frac{\partial u}{\partial x}=2-2y,\quad \frac{\partial^2 u}{\partial x^2}=0,\quad \frac{\partial u}{\partial y}=-2x,\quad \frac{\partial^2 u}{\partial y^2}=0.$$

$$\therefore \frac{\partial^2 u}{\partial x^2}+\frac{\partial^2 u}{\partial y^2}=0.$$

So u is a harmonic function.

Let v be the corresponding conjugate of u. Then

$$\frac{\partial u}{\partial x}=\frac{\partial v}{\partial y} \text{ and } \frac{\partial u}{\partial y}=-\frac{\partial v}{\partial x} \qquad \cdots \quad (1)$$

Now $dv=\frac{\partial v}{\partial x}\cdot dx+\frac{\partial v}{\partial y}\cdot dy=-\frac{\partial u}{\partial y}\cdot dx+\frac{\partial u}{\partial x}\cdot dy$

$=2xdx+2dy-2ydy$

Integrating this we get $v=x^2+2y-y^2+c$, where c is a real constant.

Hence the corresponding analytic function $f(z)$ is given by

$$f(z)=u+iv=(2x-2xy)+i(x^2+2y-y^2+c)$$

$$=2(x+iy)+i(x+iy)^2+ic=2z+iz^2+ic.$$

Ex. 3. *If* $u=e^{-x}(x\sin y-y\cos y)$, *show that there exists a function* $v(x,y)$ *such that* $w=u+iv$ *is analytic.* [*W.B.U.Tech 2002*]

Here $u=e^{-x}(x\sin y-y\cos y)$

$$\therefore \frac{\partial u}{\partial x}=e^{-x}\sin y-xe^{-x}\sin y+ye^{-x}\cos y \qquad \cdots \quad (1)$$

$$\frac{\partial^2 u}{\partial x^2}=-2e^{-x}\sin y+xe^{-x}\sin y-ye^{-x}\cos y$$

$$\frac{\partial u}{\partial y}=xe^{-x}\cos y+ye^{-x}\sin y-e^{-x}\cos y \qquad \cdots \quad (2)$$

$$\frac{\partial^2 u}{\partial y^2}=-xe^{-x}\sin y+2e^{-x}\sin y+ye^{-x}\cos y$$

$$\therefore \frac{\partial^2 u}{\partial x^2}+\frac{\partial^2 u}{\partial y^2}=0.$$

So u is a harmonic function.

As $w=u+iv$ is analytic, so v is harmonic conjuate of u, by Theorem 1 of art 4.7.

$$\therefore \frac{\partial u}{\partial x}=\frac{\partial v}{\partial y} \qquad \cdots \quad (3)$$

$$\frac{\partial u}{\partial y}=-\frac{\partial v}{\partial x} \qquad \cdots \quad (4)$$

Now $dv = \frac{\partial v}{\partial x}\cdot dx + \frac{\partial v}{\partial y}\cdot dy = -\frac{\partial u}{\partial y}\cdot dx + \frac{\partial u}{\partial x}\cdot dy$, by (3), (4)

$$= -\left(xe^{-x}\cos y + ye^{-x}\sin y - e^{-x}\cos y\right)dx$$

$$+\left(e^{-x}\sin y - xe^{-x}\sin y \quad + ye^{-x}\cos y\right)dy$$

$$= d\left(xe^{-x}\right)\cos y - ye^{-x}\sin y\,dx + e^{-x}d\left(y\sin y\right) - xe^{-x}\sin y\,dy$$

$$= d\left(xe^{-x}\cos y\right) + d\left(y\sin ye^{-x}\right).$$

Integrating this we get $v = xe^{-x}\cos y + y\sin ye^{-x} + c$, c is a real constant.

$$= e^{-x}\left(x\cos y + y\sin y\right) + c$$

$$\therefore w = u + iv = e^{-x}\left(x\sin y - y\cos y\right) + ie^{-x}\left(x\cos y + y\sin y\right) + ic$$

$$= ie^{-x}\left(x + iy\right)\cos y + e^{-x}\sin y\left(x + iy\right) + ic$$

$$= i\left(x + iy\right)e^{-x}\left(\cos y - i\sin y\right) + ic$$

$$= i\left(x + iy\right)e^{-x}e^{-iy} + ic = ize^{-(x+iy)} + ic = ize^{-z} + ic.$$

Alternative method for finding v.

From (1) and (3)

$$\frac{\partial v}{\partial y} = e^{-x}\sin y - xe^{-x}\sin y + ye^{-x}\cos y$$

which on integration, keeping x constant, gives

$$v = -e^{-x}\cos y + xe^{-x}\cos y + e^{-x}\left(y\sin y + \cos y\right) + \phi(x)$$

$$= ye^{-x}\sin y + xe^{-x}\cos y + \phi(x)$$

where $\phi(x)$ is an arbitrary real function of x.

$$\therefore \frac{\partial v}{\partial x} = -ye^{-x}\sin y + e^{-x}\cos y - xe^{-x}\cos y + \phi'(x)$$

$$\therefore -\frac{\partial u}{\partial y} = -ye^{-x}\sin y + e^{-x}\cos y - xe^{-x}\cos y + \phi'(x)\text{, by (4)}$$

$$\therefore -xe^{-x}\cos y - ye^{-x}\sin y + e^{-x}\cos y =$$

$$-ye^{-x}\sin y + e^{-x}\cos y - xe^{-x}\cos y + \phi'(x)$$

by (2).

$\therefore \phi'(x) = 0$.

$\therefore \phi(x) = c$, a real constant.

$\therefore v = ye^{-x} \sin y + xe^{-x} \cos y + c$.

Ex. 4. *Prove that an analytic function with constant modulus is constant.*

Let $f(z) = u + iv$ be an analytic function with constant modulus. Then

$$\frac{\partial u}{\partial x} = \frac{\partial v}{\partial y}, \quad \frac{\partial u}{\partial y} = -\frac{\partial v}{\partial x} \qquad \dots \qquad (1)$$

and $|f(z)|^2 = u^2 + v^2 = c^2$, a constant, by hypothesis (2)

From (2), we have,

$$u\frac{\partial u}{\partial x} + v\frac{\partial v}{\partial x} = 0 \qquad \dots \qquad (3)$$

and $u\dfrac{\partial u}{\partial y} + v\dfrac{\partial v}{\partial y} = 0$

i.e. $-u\dfrac{\partial v}{\partial x} + v\dfrac{\partial u}{\partial x} = 0 \qquad \dots \qquad (4)$

by (1).

Squaring and adding (3) and (4), we get

$$\left(u^2 + v^2\right)\left[\left(\frac{\partial u}{\partial x}\right)^2 + \left(\frac{\partial v}{\partial x}\right)^2\right] = 0$$

or, $c^2\left[\left(\dfrac{\partial u}{\partial x}\right)^2 + \left(\dfrac{\partial v}{\partial x}\right)^2\right] = 0$

or, $\left(\dfrac{\partial u}{\partial x}\right)^2 + \left(\dfrac{\partial v}{\partial x}\right)^2 = 0$. $\quad (\because c \neq 0)$

$\therefore \dfrac{\partial u}{\partial x} = \dfrac{\partial v}{\partial x} = 0$. $\quad \therefore \dfrac{\partial u}{\partial y} = \dfrac{\partial v}{\partial y} = 0$, by (1)

$\therefore du = \dfrac{\partial u}{\partial x}dx + \dfrac{\partial u}{\partial y}dy = 0 \quad \therefore u = \text{constant} = A$, say

$$dv = \frac{\partial v}{\partial x}dx + \frac{\partial v}{\partial y}dy = 0 \text{, } v = \text{constant} = B \text{ (say).}$$

$\therefore f(z) = A + iB = \text{constant}.$

Ex. 5. *If* $f(z)$ *is analytic, prove that* $\left(\frac{\partial^2}{\partial x^2}+\frac{\partial^2}{\partial y^2}\right)|f(z)|^2 = 4|f'(z)|^2$.

[*W.B.U.Tech 2004*]

Let $f(z) = u + iv \qquad \therefore f'(z) = u_x + iv_x$.

$\therefore |f(z)|^2 = u^2 + v^2, \ |f'(z)|^2 = u_x^2 + v_x^2$.

Now $\frac{\partial}{\partial x}(u^2) = 2uu_x$ and $\frac{\partial^2}{\partial x^2}(u^2) = 2(uu_{xx} + u_x^2)$.

Similarly $\frac{\partial^2}{\partial y^2}(u^2) = 2(uu_{yy} + u_y^2)$.

$$\therefore \left(\frac{\partial^2}{\partial x^2}+\frac{\partial^2}{\partial y^2}\right)u^2 = 2u(u_{xx}+u_{yy}) + 2(u_x^2+u_y^2) = 0 + 2(u_x^2+u_y^2)$$

[$\because u$ is harmonic]

$$= 2(u_x^2 + u_y^2).$$

Similarly $\left(\frac{\partial^2}{\partial x^2}+\frac{\partial^2}{\partial y^2}\right)v^2 = 2(v_x^2 + v_y^2)$.

$$\therefore \left(\frac{\partial^2}{\partial x^2}+\frac{\partial^2}{\partial y^2}\right)|f(z)|^2 = \left(\frac{\partial^2}{\partial x^2}+\frac{\partial^2}{\partial y^2}\right)(u^2+v^2) = 2(u_x^2+u_y^2+v_x^2+v_y^2)$$

$= 4(u_x^2 + v_x^2)$, using C-R equations $u_x = v_y, u_y = -v_x$

$= 4|f'(z)|^2$.

2.1.8. Construction of Analytic Functions.

Milne Thomson method.

Let $f(z) = u + iv$ be an analytic function where u, v are conjugate harmonic. If one of these say u is given then to determine $f(z)$ directly without finding v is due to Milne Thomson and is given below :

Let $z = x + iy$ $\quad \therefore \bar{z} = x - iy$.

$\therefore x = \frac{1}{2}(z + \bar{z})$, $y = \frac{1}{2i}(z - \bar{z})$.

Now $f(z) = u(x, y) + iv(x, y) = u\left(\frac{z + \bar{z}}{2}, \frac{z - \bar{z}}{2i}\right) + iv\left(\frac{z + \bar{z}}{2}, \frac{z - \bar{z}}{2i}\right)$,

When $z = \bar{z}$, $y = 0$, $x = z$ so that

$f(z) = u(z, 0) + iv(z, 0)$.

$\therefore f'(z) = \frac{\partial u}{\partial x} + i\frac{\partial v}{\partial x} = \frac{\partial u}{\partial x} - i\frac{\partial u}{\partial y}$, by C-R equations.

Now we choose $\frac{\partial u}{\partial x} = \phi_1(x, y)$, $\frac{\partial u}{\partial y} = \phi_2(x, y)$.

$\therefore f'(z) = \phi_1(x, y) - i\phi_2(x, y)$

$= \phi_1(z, 0) - i\phi_2(z, 0)$, as $x = z, y = 0$

Integrating this we get $f(z) = \int[\phi_1(z, 0) - i\phi_2(z, 0)]dz + c$ where c is an arbitrary constant.

Similarly when v is given, $f(z)$ is given by

$f(z) = \int[\psi_1(z, 0) + i\psi_2(z, 0)]dz + c$ where

$\frac{\partial v}{\partial y} = \psi_1(x, y)$, $\frac{\partial v}{\partial x} = \psi_2(x, y)$.

Illustrative Examples. (In the following Examples we find $f(z)$ directly by using Milne's method).

Ex. 1. *Find* $f(z)$ *where* $u = e^{-x}(x\sin y - y\cos y)$

Here $u = e^{-x}(x\sin y - y\cos y)$.

$\therefore \frac{\partial u}{\partial x} = e^{-x}\sin y - xe^{-x}\sin y + ye^{-x}\cos y = \phi_1(x, y)$, say.

$\frac{\partial u}{\partial y} = xe^{-x}\cos y + ye^{-x}\sin y - e^{-x}\cos y = \phi_2(x, y)$, say.

By Milne's method, we have

$f'(z) = \phi_1(z, 0) - i\phi_2(z, 0) = 0 - i(ze^{-z} - e^{-z})$.

Integrating, we get

$$f(z) = -i\int ze^{-z}dz + i\int e^{-z}dz = -i\left(-ze^{-z} - e^{-z}\right) - ie^{-z} + c$$

$= ize^{-z} + c$, where c is an arbitrary constant.

Ex. 2. *Prove that* $u = x^3 - 3xy^2 + 3x^2 - 3y^2 + 1$ *is a harmonic function and determine the corresponding analytic function* $u + iv$.

[*W.B.U.Tech 2003*]

Here $u = x^3 - 3xy^2 + 3x^2 - 3y^2 + 1$.

$$\therefore \frac{\partial u}{\partial x} = 3x^2 - 3y^2 + 6x, \quad \frac{\partial^2 u}{\partial x^2} = 6x + 6$$

$$\frac{\partial u}{\partial y} = -6xy - 6y, \quad \frac{\partial^2 u}{\partial y^2} = -6x - 6.$$

$\therefore \dfrac{\partial^2 u}{\partial x^2} + \dfrac{\partial^2 u}{\partial y^2} = 0$ and hence u is a harmonic function.

Let $\dfrac{\partial u}{\partial x} = \phi_1(x,y)$, $\dfrac{\partial u}{\partial y} = \phi_2(x,y)$.

$$\therefore \phi_1(x,y) = 3x^2 - 3y^2 + 6x$$

$$\phi_2(x,y) = -6xy - 6y$$

$\therefore$ By Milne's method, we have

$$f'(z) = \phi_1(z,0) - i\phi_2(z,0)$$

$$= \left(3z^2 + 6z\right) - i \cdot 0.$$

Integrating we get $f(z) = z^3 + 3z^2 + c$, where c is an arbitrary constant.

Ex. 3. *If* $u - v = (x - y)(x^2 + 4xy + y^2)$ *and* $f(z) = u + iv$ *is an analytic function of* $z = x + iy$, *find* $f(z)$ *in terms of* z. [*W.B.U. Tech 2004*]

As $f(z) = u + iv$ is an analytic function, so

$$\frac{\partial u}{\partial x} = \frac{\partial v}{\partial y}, \frac{\partial u}{\partial y} = -\frac{\partial v}{\partial x} \quad \cdots \quad (1)$$

Also $u-v=(x-y)(x^2+4xy+y^2)$.

$$\therefore \frac{\partial u}{\partial x}-\frac{\partial v}{\partial x}=x^2+4xy+y^2+(x-y)(2x+4y)$$

$$\therefore \frac{\partial u}{\partial x}+\frac{\partial u}{\partial y}=3x^2+6xy-3y^2 \qquad \text{... (2),}$$

by (1)

$$\frac{\partial u}{\partial y}-\frac{\partial v}{\partial y}=-(x^2+4xy+y^2)+(x-y)(4x+2y).$$

$$\therefore \frac{\partial u}{\partial y}-\frac{\partial u}{\partial x}=3x^2-6xy-3y^2 \qquad \text{...(3),}$$

by (1).

Solving (2) and (3) we get

$$\frac{\partial u}{\partial x}=6xy=\phi_1(x,y), \text{ (say)}$$

$$\frac{\partial u}{\partial y}=3x^2-3y^2=\phi_2(x,y), \text{ (say).}$$

$\therefore$ by Milne's Method we have

$$f'(z)=\phi_1(z,0)-i\phi_2(z,0) \;=0-i3z^2.$$

Integrating we get $f(z)=-iz^3+c$, where c is an arbitrary constant.

Ex. 4. *Find the analytic function* $f(z)=u+iv$ *of which the imaginary part* $v=e^x\sin y$.

Here $v=e^x\sin y$.

$$\therefore \frac{\partial v}{\partial x}=e^x\sin y=\psi_2(x,y), \text{ say}$$

$$\frac{\partial v}{\partial y}=e^x\cos y=\psi_1(x,y), \text{ say.}$$

By Milne's method, we have

$$f'(z)=\psi_1(z,0)+i\psi_2(z,0) \;=e^z+i0.$$

Integrating we get $f(z)=e^z+c$, where c is an arbitrary constant.

Ex. 5. *Determine the analytic function* $f(z)=u+iv$ *whose imaginary part is* $v=x^4-6x^2y^2+y^4$. [*W.B.U. Tech 2002*]

Here $v=x^4-6x^2y^2+y^4$.

$$\therefore \frac{\partial v}{\partial x}=4x^3-12xy^2=\psi_2(x,y), \text{ say.}$$

$$\frac{\partial v}{\partial y}=-12x^2y+4y^3=\psi_1(x,y), \text{ say.}$$

By Milne's method we have

$$f'(z)=\psi_1(z,0)+i\ \psi_2(z,0)=0+i4z^3=4iz^3.$$

Integrating we get $f(z)=i\,z^4+c$, where c is an arbitrary constant.

EXERCISES

1. Using the definition of limit, prove the following :

(i) $\underset{z\to i}{Lt}\left(z^2+2z\right)=2i-1$ (ii) $\underset{z\to(2+3i)}{Lt}\left\{x+i\left(x+y^2\right)\right\}=2+11i$

(iii) $\underset{z\to -3i}{Lt}\dfrac{z^2+9}{z+3i}=-6i$.

2. Evaluate the following using theorems on limits :

(i) $\underset{z\to\frac{i}{2}}{Lt}\dfrac{(2z-3)(4z+i)}{(iz-1)^2}$ (ii) $\underset{z\to 1+i}{Lt}\left(\dfrac{z-1-i}{z^2-2z+2}\right)^2$

(iii) $\underset{z\to e^{\frac{\pi i}{3}}}{Lt}\left(z-e^{\frac{\pi i}{3}}\right)\left(\dfrac{z}{z^3+1}\right)$.

3. Prove that if $f(z)=4z^3+3z$, then $\underset{z\to z_0}{Lt}\dfrac{f(z)-f(z_0)}{z-z_0}=12z_0^2+3$.

4. Find all points of discontinuity for the following functions :

(i) $\dfrac{z}{z+\bar{z}}$, (ii) $\dfrac{1}{1-|z|^2}$ (iii) $\dfrac{3z^2+4}{z^4-16}$.

5. Show that $f(z)=\dfrac{z^2+1}{z^2-3z+2}$ is continuous for all z outside $|z|=2$.

6. Prove that $f(z)=\dfrac{z^2+1}{z^3+9}$ is continuous in the region $|z|\le 2$.

7. Is the function $f(z)=\dfrac{z^2+(2-i)z-2i}{z-i}$ continuous at $z=i$? If not, can it be made continuous by redefining at $z=i$?

8. Discuss the continuity of the function $f(z)=\arg z$ where $-\pi<\arg z\le\pi$ in the entire complex plane.

9. Show that $\underset{z\to 0}{Lt}\, f(z)$ does not exist where

(i) $f(z)=\dfrac{x^2+5y^2}{(x-y)^2}$ (ii) $f(z)=\dfrac{xy(x-2y)}{x^3+y^3}$.

10. Test the continuity at the origin of the following functions :

(i) $f(z)=\dfrac{xy}{x^2+y^2}$, for $z\ne 0$

$=0$, for $z=0$.

(ii) $f(z)=\dfrac{x^2y^2}{x^2y^2+(x-y)^2}$, for $z\ne 0$

$=0$, for $z=0$.

(iii) $f(z)=\dfrac{x^2y}{x^4+y^2}$, for $z\ne 0$

$=0$, for $z=0$.

(iv) $f(z)=\dfrac{2x^2y}{x^2+y^2}$, for $z\ne 0$

$=0$, for $z=0$.

(v) $f(x,y)=\dfrac{x^4y(y-x)}{(x^8+y^2)(y+x)}$, $(x,y)\ne(0,0)$

$=0$, $(x,y)=(0,0)$.

11. Show that $f(z)=\log z$ is analytic every where in the complex plane except at the origin. Find its derivative.

12. Prove that the function $f(z)=z^n$ where n is a positive integer, is an analytic function. Find its derivative.

13. Show that $\sinh z$ is an analytic function. Also find its derivative.

14. Show that if $w=f(z)$ is an analytic function of z,

$$\left(\frac{\partial^2}{\partial x^2}+\frac{\partial^2}{\partial y^2}\right)\log|f'(z)|=0.$$

15. Test whether the following functions are analytic or not

(i) $z^2-\dfrac{1}{2z}$. (ii) $e^{-x}(\sin y+i\cos y)$.

(iii) $\dfrac{x+iy}{x^2+y^2}$. (iv) $e^{\sin z}$.

16. Show that the Cauchy-Riemann equations are not satisfied for $f(z)=(\bar{z})^2$.

17. Show that (i) an analytic function with constant real part is a constant and (ii) an analytic function with constant modulus is also a constant.

18. (a) Show that $f(z)=\bar{z}$ is nowhere analytic. [*W.B.U. Tech 2004*]

(b) Show that $f(z)=\dfrac{x-iy}{x-iy+a}$ where $a\neq 0$ is not an analytic function of z.

(c) Show that $f(z)=\sin\bar{z}$ is not an analytic function of z.

19. Show that the function $f(z)=x^2+iy^3$ is everywhere continuous but is not analytic.

20. (a) Show that the function $f(z)$ defined by

$$f(z)=\frac{(\bar{z})^2}{z},\ z\neq 0$$
$$=0,\qquad z=0.$$

is not differentiable at the origin though the C-R equations are satisfied at that point.

(b) Show by considering the function $f(z)$ defined as

$$f(z)=\frac{xy(y-ix)}{x^2+y^2}, \text{ for } z\neq 0$$
$$=0, \qquad z=0$$

that the C-R equations are not the sufficient conditious for a function to be analytic.

21. Show that the following functions are not differentiable at origin though continuous everywhere :

(i) $f(z)=|z|$ (ii) $f(z)=z\,\mathrm{Re}(z)$.

22. Show that the following functions are not differentiable at the origin :

(i) $$f(z)=\frac{x^2y^2(y-ix)}{x^2y^2+(x-y)^2}, \text{ for } z\neq 0$$
$$=0, \qquad \text{for } z=0.$$

(ii) $$f(z)=\frac{x^3y(y-ix)}{x^6+y^2}, \ z\neq 0$$
$$=0, \qquad z=0.$$

23. Given that $u(x,y)=x^2-y^2$ and $v(x,y)=-\frac{y}{x^2+y^2}$ prove that both u and v are harmonic functions but $u+iv$ is not an analytic function of z.

24. Prove that the following functions are harmonic. Find its harmonic conjugate and the corresponding analytic function $f(z)$ in terms of z.

(i) $u=x^3-3xy^2$

(ii) $u=\frac{1}{2}\log\left(x^2+y^2\right)$ [*W.B.U. Tech 2004 M*]

(iii) $u=2x-x^3+3xy^2$ (iv) $u=x^2-y^2$ (v) $3x^2y+x^2-y^3-y^2$

25. Find the analytic function $f(z)=u+iv$ of which the real part is

(a) $u=e^x(x\cos y-y\sin y)$ (b) $u=e^x\cos y$

(c) $u=\frac{\sin 2x}{\cosh 2y+\cos 2x}$ (d) $u=x^3-3xy^2+3x^2-3y^2+1$

(e) $u=-\sin x\sinh y$.

26. Show that the following functions are harmonic. Find the corresponding analytic function $f(z)=u+iv$ if

(i) $v=x^2-y^2-2xy+2x-3y$ (ii) $e^{-x}(x\cos y+y\sin y)$

(iii) $e^x \sin y-x$

(iv) $e^x(x\sin y+y\cos y)$. [*W.B.U.Tech 2004*]

(v) $2xy-\dfrac{y}{x^2+y^2}$

27. If $f(z)=u+iv$ is an analytic function of $z=x+iy$ and $u-v=e^x(\cos y-\sin y)$, find $f(z)$ in terms of z.

28. (i) If $u+v=\dfrac{2\sin 2x}{e^{2y}+e^{-2y}-2\cos 2x}$ and $f(z)=u+iv$ is an analytic function of $z=x+iy$. find $f(z)$ in terms of z.

(ii) If $f(z)=u+iv$ is an analytic function such that $u-v=\dfrac{\cos x+\sin x-e^{-y}}{2\cos x-e^{y}-e^{-y}}$ and $f(\pi/2)=0$, find the function $f(z)$.

29. If $u=x^2-y^2$, $v=\dfrac{-y}{x^2+y^2}$, prove that both u and v satisfy Laplace's equation but $u+iv$ is not an analytic function of z.

30. If $f(z)=\dfrac{xy^2(x+iy)}{x^2+y^4}$, $z\neq 0$, $f(0)=0$, prove that $\dfrac{f(z)-f(0)}{z}\to 0$ as $z\to 0$ along any radius vector but not as $z\to 0$ in any manner.

Answers

2. (i) $-\dfrac{4}{3}-4i$ (ii) $-\dfrac{1}{4}$ (iii) $\dfrac{1}{6}-i\dfrac{\sqrt{3}}{6}$

4. (i) All pts on imaginary axis i.e. where $x=0$

(ii) All pts on the circle $|z|=1$ (iii) $\pm 2, \pm 2i$

7. Discontinuous at $z=i$, define $f(z)=2+i$ to make continuous at $z=i$.

8. Continuous everywhere except at the origin and the negative half of the real axis.

10. (i) Discontinuous (ii) Discontinuous (iii) Discontinuous

(iv) Continuous (v) Discontinuous.

11. $\frac{1}{z}$ **12.** nz^{n-1} **13.** $\cosh z$

15. (i) analytic for all $z \neq 0$ (ii) analytic for all z

(iii) not analytic (iv) analytic for all z.

24. (i) $v = 3x^2 y - y^3 + c, z^3 + ic$ (ii) $v = \tan^{-1}\left(\frac{y}{x}\right) + c, \log z + ic$

(iii) $v = 2y + y^3 - 3x^2 y + c, 2z - z^3$

(iv) $v = 2xy + c, z^2$ (v) $3y^2 + 2xy - x^2 + c, -iz^3 + z^2 + ic$

25. (a) $ze^z + c$ (b) $e^z + c$ (c) $\tan z + c$

(d) $z^3 + 3z^2 + c$ (e) $-i\cos z + c$.

26. (i) $z^2(i-1) + 2iz - 3z + c$ (ii) $ize^{-z} + c$

(iii) $e^z - iz + ic$ (iv) $ze^z + ic$ (v) $z^2 + \frac{1}{z} + c$

27. $e^z + c$ **28.** (i) $\frac{1}{2}(1+i)\cot z + c$. **28.** (ii) $1 - \cot\frac{z}{2}$

2.2 COMPLEX INTEGRATION

2.2.1. Curve Represented by a Complex Variable

Let $z = x + iy$ be a complex variable. Since a fixed z represents a point (x, y) in Argand Plane, so as z varies the point (x, y) moves on the plane and makes a locus or a curve.

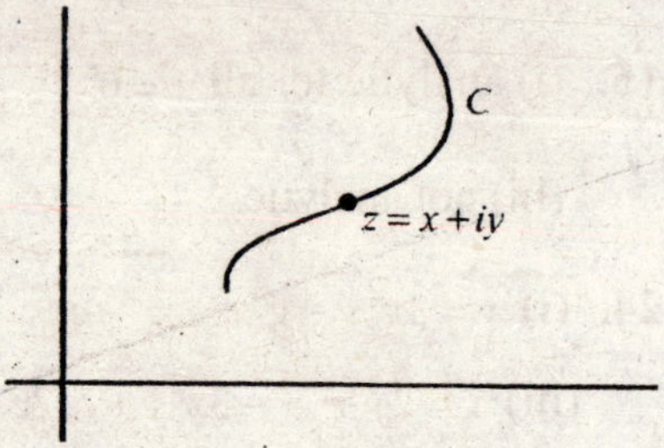

We say this curve, say C is represented by the complex variable z.

Parametric Representation

Let $z = x + iy$ be a complex variable. If $x = \phi(t)$ and $y = \psi(t)$ where t is a real variable then there is a relation between x and y. This relation gives a curve, C (say).

Then we say $z = \phi(t) + i\psi(t)$

gives the curve C ; t is parameter.

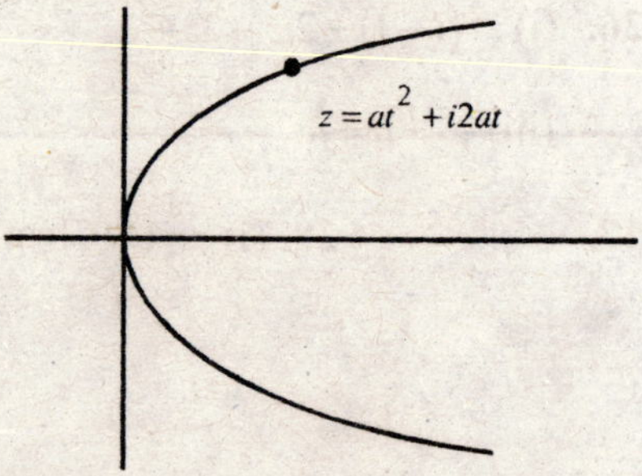

For example let $z = x + iy$ and $x = at^2, y = 2at$. (x, y) traces the parabola $y^2 = 4ax$. (x, y) traces the parabola $y^2 = 4ax$. Then we see $z = at^2 + i2at$ represents the parabola $y^2 = 4ax$ on the Argand plane.

Simple Curve.

A curve C is called simple if it does not intersect itself. So a curve $C : z = \phi(t) + i\psi(t)$ is simple if $t_1 \neq t_2$ implies $z(t_1) \neq z(t_2)$.

The curve shown in Fig1 is a simple curve. But the curve shown in Fig 3 is not so.

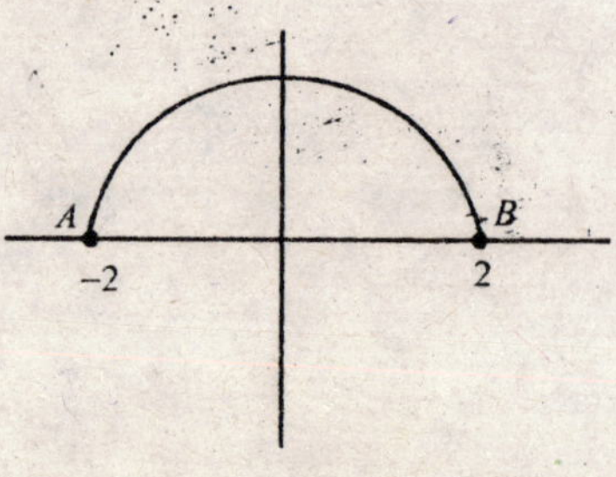

Fig 1

Closed Curve.

A simple curve is called closed if the two end points of the curve coincide.

The curve shown in Fig 2 is a closed curve.

Smooth Curve.

A curve C is called smooth if it possess unique tangent at every point. The curves shown in Fig 1 Fig 2 are smooth.

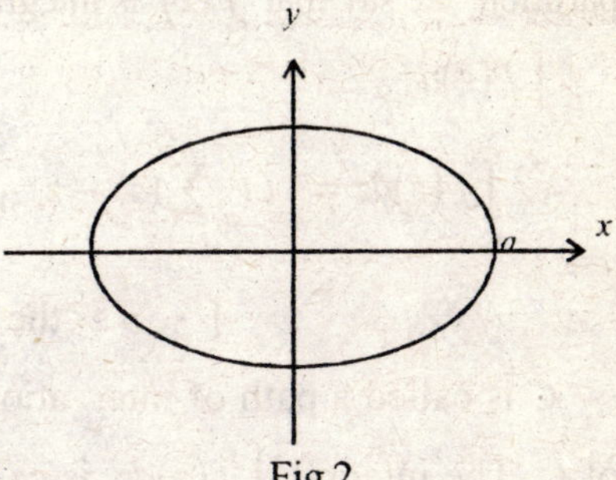

Fig 2

Contour or Piecewise smooth curve

A curve is called contour or piecewise smooth if it is comprised of a finite number of smooth curves. For example the triange ABC in Fig 4 represents a piecewise smooth curve. Here the three pieces AB, BC and CA are smooth.

Fig 3

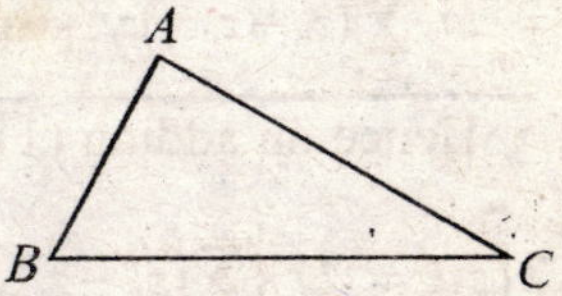

Fig 4

2.2.2. Riemann's Definition of Integration.

Let $f(z)$ be a function where z varies over a piecewise smooth and simple curve C.

Let the initial of C are A and B respectively. If C is closed then $A = B$.

Divide C into n arcs by means of the points $P : z_0 = A,\ z_1, z_2, \cdots z_{n-1}, z_n = B$

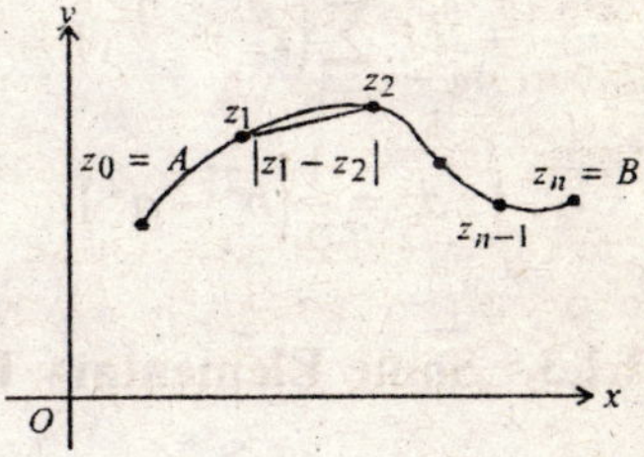

along C. On each arc, between z_{r-1} and z_r, choose some point ξ_r. Now $|z_{r-1} - z_r|$ is the length of the chord joining the two points z_{r-1} and z_r. Let $\|P\| =$ maximum of all such chords $= \max\limits_{1 \le r \le n} |z_{r-1} - z_r|$. Form the sum

$$S_p = \sum_{r=1}^{n} (z_r - z_{r-1}) f(\xi_r),$$

where ξ_r is a point lying on the arc joining z_{r-1} and z_r.

Nowif $\underset{\|P\| \to 0}{Lt}\ S_p = \underset{\|P\| \to 0}{Lt} \sum_{r=1}^{n} (z_r - z_{r-1}) f(\xi_r)$ exists finitely and for any partition, we say that $f(z)$ is integrable along the curve C and is written as $\int_C f(z)dz$.

$$\therefore \int_C f(z)dz = \underset{\|P\|\to 0}{Lt} \sum_{r=1}^{n}(z_r - z_{r-1})f(\xi_r) = \underset{n\to\infty}{Lt} \sum_{r=1}^{n}(z_r - z_{r-1})f(\xi_r).$$

[$\because$ n is the number of points in the partition P]

C is called a path of integration or **contour.** *P is called a* **partition** of C. The integral $\int_C f(z)dz$ is called **Contour Integral.**

Illustration. **(i)** By definition $\int_C z\,dz = \underset{n\to\infty}{Lt} \sum_{r=1}^{n}(z_r - z_{r-1})\xi_r$, $z_{r-1} \le \xi_r \le z_r$

[$\because f(z) = z$] [C is shown in the figure]

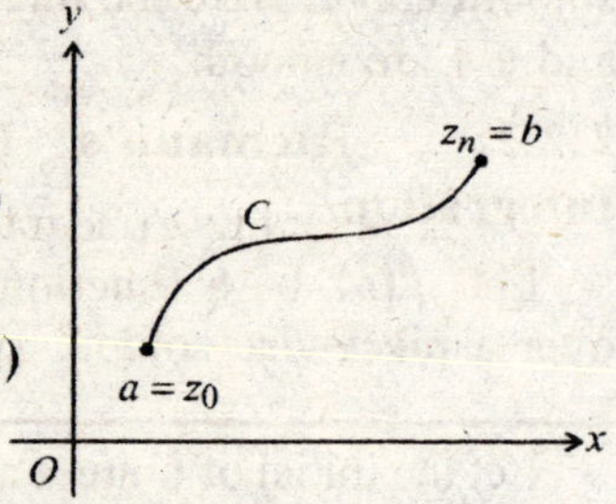

$$= \underset{n\to\infty}{Lt} \sum_{r=1}^{n}(z_r - z_{r-1})z_r \text{ taking } \xi_r = z_r \quad (1)$$

$$= \underset{n\to\infty}{Lt} \sum_{r=1}^{n}(z_r - z_{r-1})z_{r-1}, \text{ taking } \xi_r = z_{r-1} \quad (2)$$

Hence on adding (1) and (2), we have

$$2\int_C z\,dz = \underset{n\to\infty}{Lt} \left\{\sum_{r=1}^{n}(z_r - z_{r-1})z_r + \sum_{r=1}^{n}(z_r - z_{r-1})z_{r-1}\right\}$$

$$= \underset{n\to\infty}{Lt} \sum_{r=1}^{n}\left(z_r^2 - z_{r-1}^2\right) = \underset{n\to\infty}{Lt}\left(z_n^2 - z_0^2\right) = b^2 - a^2$$

$$\therefore \int_C z\,dz = \frac{1}{2}\left(b^2 - a^2\right).$$

2.2.3. Some Elementary Properties of Complex Integrals.

(a) $\int_C [f(z) + \phi(z)]dz = \int_C f(z)dz + \int_C \phi(z)dz$

(b) $\int_C f(z)dz = -\int_{-C} f(z)dz$

where $-C$ indicates the curve C traversed in the opposite direction.

(c) $\int_C k\,f(z)dz = k\int_C f(z)dz$ where k is a constant.

(d) $\int_C f(z)dz = \int_{C_1} f(z)dz + \int_{C_2} f(z)dz$

where C_1, C_2 are two parts of C

Proof. Beyond the scope of the book.

2.2.4. Line Integral along a Piecewise Smooth Curve.

Let $P(x,y)$ and $Q(x,y)$ be two real functions of x and y which are defined and continuous at all points of a piecewise smooth simple arc

$C: z(t) = \phi(t) + i\psi(t)$, $\alpha \le t \le \beta$.

Then the real line integral of $Pdx + Qdy$ along C can be defined in a similar manner as Art 5.2 and is denoted by

$\int_C [P(x,y)dx + Q(x,y)dy]$. Since $dx = \phi'(t)dt$, $dy = \psi'(t)dt$.

The value of this line integral is given by

$$\int_C (P\,dx + Q\,dy) = \int_\alpha^\beta P(\phi(t),\psi(t))\phi'(t)dt + \int_\alpha^\beta Q(\phi(t),\psi(t))\psi'(t)dt.$$

Theorem. (Complex integral as the sum of two line integrals).

Let $f(z) = u(x,y) + iv(x,y)$ be continuous on a piecewise smooth simple arc C. Then $f(z)$ is integrable along C and is given by

$$\int_C f(z)dz = \int_C (u\,dx - v\,dy) + i\int_C (v\,dx + u\,dy).$$

Proof. Beyond the scope of the book.

Corollary. When the equation of the arc C is

$z(t) = \phi(t) + i\psi(t)$, $\alpha \le t \le \beta$, then

$$\int_C f(z)dz = \int_\alpha^\beta \left[u(\phi(t),\psi(t))\phi'(t) - v(\phi(t),\psi(t))\psi'(t)\right]dt$$

$$+i\int_\alpha^\beta \left[v(\phi(t),\psi(t))\phi'(t) + u(\phi(t),\psi(t))\psi'(t)\right]dt.$$

Illustrative Example.

Ex. 1. *Evaluate* $\int_{i}^{2-i} \left(3xy + iy^2\right) dz$

(*a*) *along the straight line joining* $z = i$ *and* $z = 2 - i$

(*b*) *along the curve* $x = 2t - 2$, $y = 1 + t - t^2$.

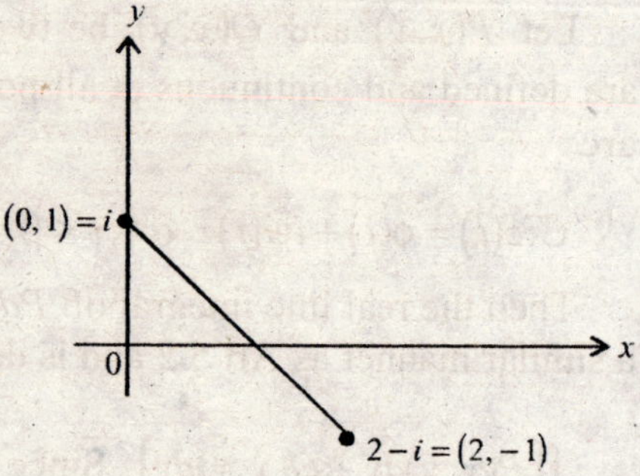

(*a*) The equation of the straight line joining $z = i$ and $z = 2 - i$ is

$$\frac{y-1}{x-0} = \frac{1+1}{0-2} \text{ or, } x = -y + 1$$

whose parametric equation is $x = -t + 1$, $y = t$.

$$\therefore \int_{i}^{2-i} \left(3xy + iy^2\right) dz$$

$$= \int_{i}^{2-i} \left(3xy + iy^2\right)(dx + i\,dy)$$

$$= \int_{1}^{-1} \left\{3(-t+1)\cdot t + it^2\right\}(-dt + i\,dt)$$

[At $z = i$, $x = 0$, $y = 1$ i.e., $t = 1$; Similarly at $z = 2 - i$, $t = -1$]

$$= (-1+i)\int_{1}^{-1} \left(-3t^2 + 3t + it^2\right) dt = (-1+i)\left[-t^3 + \frac{3t^2}{2} + i\frac{t^3}{3}\right]_1^{-1}$$

$$= (-1+i)\left(2 - \frac{2i}{3}\right) = \frac{-4}{3} + i\frac{8}{3}.$$

(*b*) The points i and $2 - i$ on the curve $x = 2t - 2$, $y = 1 + t - t^2$ correspond to $t = 1$ and $t = 2$ respectively.

$$\therefore \int_{i}^{2-i} \left(3xy + iy^2\right) dz = \int_{i}^{2-i} \left(3xy + iy^2\right)(dx + i\,dy)$$

$$=\int_1^2\left\{3\cdot(2t-2)\left(1+t-t^2\right)+i\left(1+t-t^2\right)\right\}\left\{2dt+i(1-2t)dt\right\}$$

$$=\int_1^2\left(-13+29t^2-12t^3-5t^4+2t^5\right)dt+i\int_1^2\left(-4+16t+10t^2-34t^3+14t^4\right)dt$$

$$=-\frac{1}{3}+\frac{79}{30}i.$$

Ex. 2. *Evaluate* $\int_C \bar{z}dz$ *from* $z=0$ *to* $z=4+2i$ *along the curve C given by (a) the straight line joining* $z=0$ *and* $z=4+2i$ *(b) the straight line from* $z=0$ *to* $z=2i$ *and the straight line from* $z=2i$ *to* $z=4+2i$.

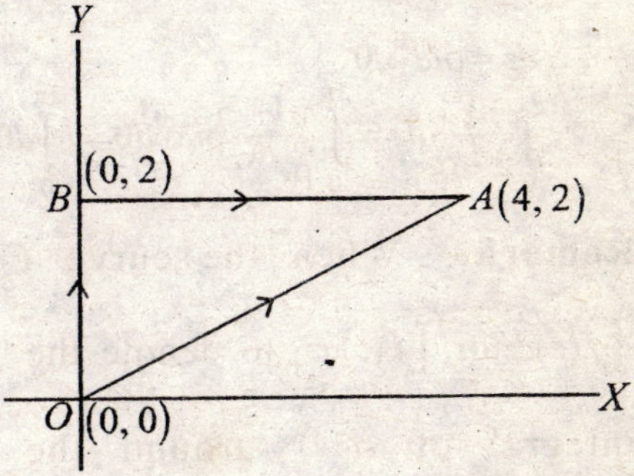

(*a*) The equation of the straight line joining $z=0$ and $z=4+2i$ is

$$\frac{y-0}{x-0}=\frac{2-0}{4-0} \text{ i.e. } x=2y$$

whose parametric equation is $x=2t$, $y=t$

$$\therefore \int_C \bar{z}dz=\int_{OA}(x-iy)(dx+idy)$$

$$=\int_0^2(2t-it)(2dt+i\,dt)$$

$$=(2+i)\int_0^2(2t-it)dt=(2+i)(4-2i).$$

[$\because$ At $z=0$, $t=0$, at $z=4+2i$, $t=2$]

(*b*) $\int_C \bar{z}dz=\int_{0B}(x-iy)(dx+i\,dy)+\int_{BA}(x-iy)(dx+idy)$

On $0B$, $x=0$ (constant), $0\le y\le 2$. $\therefore dx=0$

and on BA, $y=2$ (constant), $0\le x\le 4$. $\therefore dy=0$.

$$\therefore \int_C \bar{z}\,dz=\int_0^2(0-yi)(0+idy)+\int_0^4(x-2i)(dx+i\cdot 0)$$

$$= \int_0^2 y\,dy + \int_0^4 (x-2i)\,dx = 2 + \left[\frac{x^2}{2} - 2xi\right]_0^4 = 10 - 8i\,.$$

Ex. 3. *Evaluate* $\int_C \frac{1}{z-\alpha}dz$ *where C is a circle whose equation is* $|z-\alpha| = \rho$.

On the circle $|z-\alpha| = \rho$. If θ be *amp* $(z-\alpha)$, we have

$$z - \alpha = \rho(\cos\theta + i\sin\theta) = \rho e^{i\theta}$$

i.e. $z = \alpha + \rho e^{i\theta}$ where θ varies from 0 to 2π.

$\therefore dz = \rho i e^{i\theta} d\theta$

$$\because \int_C \frac{1}{z-\alpha}dz = \int_0^{2\pi} \frac{1}{\rho e^{i\theta}} \cdot \rho i e^{i\theta} d\theta = i\int_0^{2\pi} d\theta = 2\pi i\,.$$

Remark. When the curve C is closed, we use symbol $\oint_C f(z)dz$ or, $\int f(z)dz$ to denote the integral of $f(z)$ around the boundary C in positive sense i.e. in the counter clockwise direction. The integral around C is of called a *Contour integral.*

2.2.5. Cauchy's Theorem.

Theorem. 1. (Green's Theorem in the plane).

Let $M(x,y)$ and $N(x,y)$ be continuous and have continuous partial derivatives in a region R bounded by a closed curve C. Then

$$\oint_C (M dx + N dy) = \iint_R \left(\frac{\partial N}{\partial x} - \frac{\partial M}{\partial y}\right) dx\,dy\,.$$

Proof. Beyond the scope of this book.

Theorem. 2. (Cauchy's Theorem) :

Let $f(z)$ be an analytic function and $f'(z)$ is continuous at each point within the domain D bounded by a closed contour C. Then $\oint_C f(z)dz = 0$.

Proof. Let $f(z) = u(x, y) + i\,v(x, y)$.

As $f(z)$ in an analytic function, so by Cauchy-Riemann condition, we have

$$\frac{\partial u}{\partial x} = \frac{\partial v}{\partial y},\ \frac{\partial u}{\partial y} = -\frac{\partial v}{\partial x} \qquad \ldots \qquad (1)$$

Also $f'(z)$ is continuous and as $f'(z) = \frac{\partial u}{\partial x} + i\frac{\partial v}{\partial x} = \frac{\partial v}{\partial y} - i\frac{\partial u}{\partial y}$, by (1). So u, v and their partial derivatives $\frac{\partial u}{\partial x}, \frac{\partial u}{\partial y}, \frac{\partial v}{\partial x}, \frac{\partial v}{\partial y}$ are all continous within and on C. So Green's theorem can be applied.

$$\therefore \oint_C f(z)dz = \oint_C (u + iv)(dx + i\,dy)$$

$$= \oint_C \{(udx - vdy) + i(vdx + udy)\} = \oint_C (udx - vdy) + i\oint_C (vdx + udy)$$

$$= \iint_D \left(-\frac{\partial v}{\partial x} - \frac{\partial u}{\partial y}\right)dx\,dy + i\iint_D \left(\frac{\partial u}{\partial x} - \frac{\partial v}{\partial y}\right)dx\,dy \text{ by Green's theorem}$$

$$= \iint_D \left(\frac{\partial u}{\partial y} - \frac{\partial u}{\partial y}\right)dx\,dy + i\iint_D \left(\frac{\partial u}{\partial x} - \frac{\partial u}{\partial x}\right)dx\,ds \text{ by (1)}$$

$$= 0 + i \cdot 0 = 0.$$

Remark. Famous mathematician Goursat first established that the condition of continuity of $f'(z)$ is unnecessary and proved the theorem without using the continuity condition of $f'(z)$ and established a revised form of the above theorem known as *Cauchy Goursat Theorem* which is given below.

Theorem. 3. (Cauchy-Goursat theorem).

Let $f(z)$ be an analytic function within and on a simple closed contour C. Then $\oint_C f(z)dz = 0$.

Proof. Beyond the scope of this book.

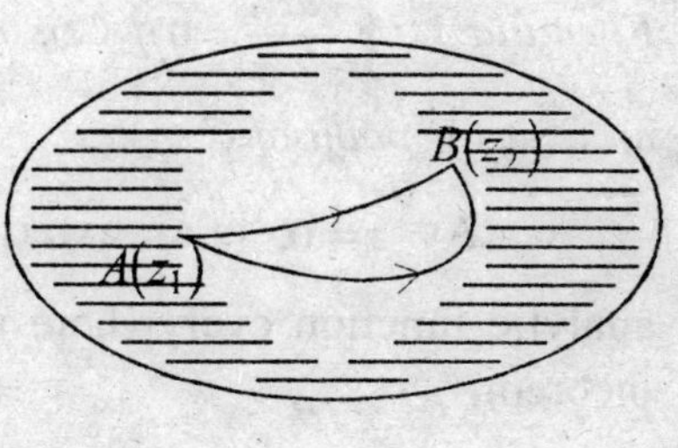

Note. (i) The above theorem is often called *Cauchy's theorem.*

(ii) The above theorem is valid

for both simply and multi-connected domain.

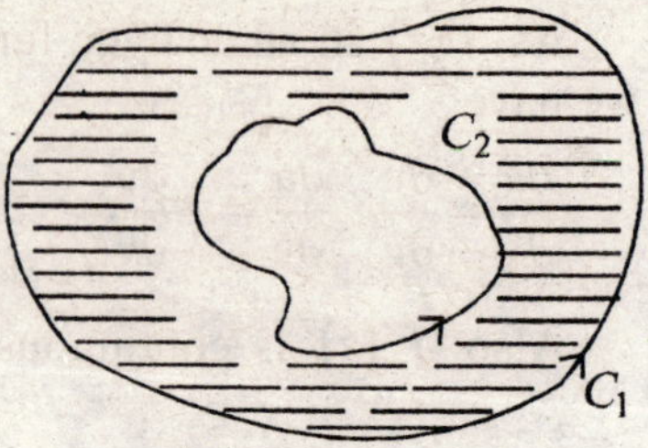

Some Consequences of Cauchy's theorem.

Theorem. 4. Let $f(z)$ be analytic in a simply connected domain D. Then the integral of $f(z)$ along any path joining any two points of D is same i.e.

$$\int_{C_1} f(z)dz = \int_{C_2} f(z)dz$$

where C_1 and C_2 are two curves in D joining by two points $A(z_1)$ and $B(z_2)$ as shown in the fig.

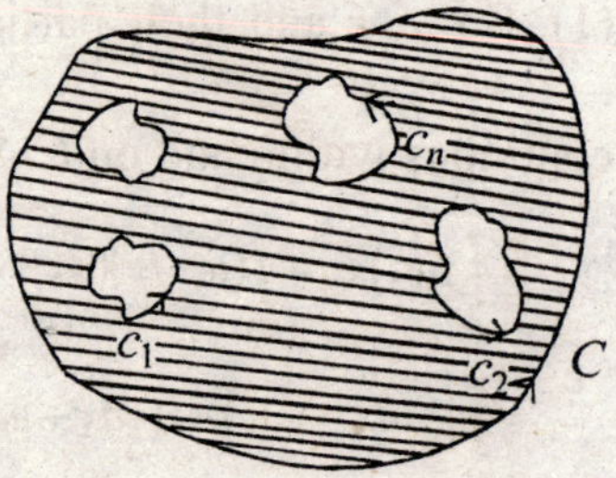

Proof. Beyond the scope of the book.

Theorem. 5. Let $f(z)$ be analytic in the annular region bounded by two closed curves C_1 and C_2 where C_2 lies wholly within C_1. Then

$$\oint_{C_1} f(z)dz = \oint_{C_2} f(z)dz$$

Proof. Beyond the scope of the book.

Theorem. 6. Let $f(z)$ be analytic within and on the boundary of the region bounded by the non-overlapping simple closed curves $C, C_1, C_2 \cdots C_n$ where $C_1, C_2 \cdots C_n$ lie wholly within C. Then

$$\oint_{C} f(z)dz = \oint_{C_1} f(z)dz + \int_{C_2} f(z)dz + \cdots + \int_{C_n} f(z)dz$$

Proof .Beyond the scope of the book.

Formula 1. $\oint_C \frac{dz}{z-\alpha} = 0$ *if C is any simple closed curve and* $z = \alpha$ *is an exterior point of C.*

As $z = \alpha$ is an exterior point of C so $f(z) = \frac{1}{z-\alpha}$ is an analytic function everywhere within and on C. Hence by Cauchy's theorem,

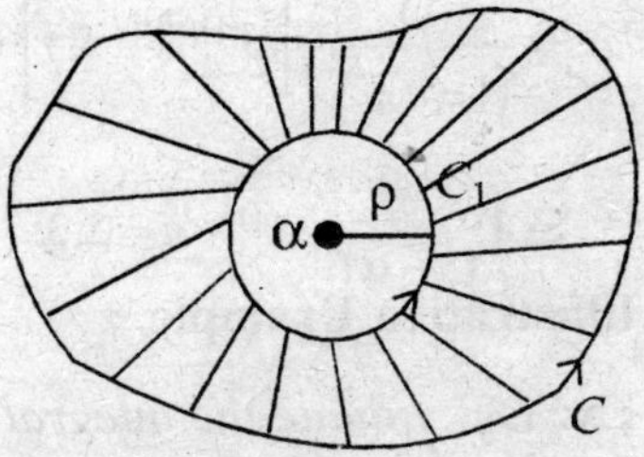

$\oint_C f(z)dz = 0$ i.e. $\oint_C \frac{1}{z-\alpha} dz = 0$.

Formula 2. $\oint_C \frac{dz}{z-\alpha} = 2\pi i$ *if C is any simple closed curve and* $z = \alpha$ *is an interior point of C.*

Let C_1 be a circle with centre at $z = \alpha$ and radius ρ so that C_1 lies wholly within C. Then by Theorem 5 of art 5.5, we have

$\oint_C f(z)dz = \oint_{C_1} f(z)dz$, as

$f(z) = \frac{1}{z-\alpha}$ is analytic in the annular region bounded by C and C_1.

But $\oint_{C_1} f(z)dz = \oint_{C_1} \frac{1}{z-\alpha} dz = 2\pi i$, by Ex-3 of Art 5.4

$\therefore \oint_C \frac{dz}{z-\alpha} = 2\pi i$.

Formula 3. $\oint_C \frac{dz}{(z-\alpha)^n} = 0$, $n = 2,3,4,\cdots$

where α *is an interior point of any simple closed curve C.*

As in Formula 1. we can write

$\oint_C \frac{dz}{(z-\alpha)^n} = \oint_{C_1} \frac{dz}{(z-\alpha)^n}$ where C_1 is a circle lying within C and the equation of the circle C_1 is $(z-\alpha) = \rho$.

On the circle $|z-\alpha| = \rho$, we have $z - \alpha = \rho(\cos\theta + i\sin\theta) = \rho e^{i\theta}$

i.e. $z = \alpha + \rho e^{i\theta}$ where θ varies from 0 to 2π.

$\therefore dz = \rho i e^{i\theta} d\theta$.

$$\therefore \oint_{C_1} \frac{dz}{(z-\alpha)^n} = \int_0^{2\pi} \frac{i\rho e^{i\theta} d\theta}{\rho^n e^{in\theta}} = \frac{i}{\rho^{n-1}} \int_0^{2\pi} e^{(1-n)i\theta} d\theta = \frac{i}{\rho^{n-1}} \left[\frac{e^{(1-n)i\theta}}{(1-n)i} \right]_0^{2\pi}$$

$$= \frac{1}{(1-n)\rho^{n-1}}\left[e^{2(1-n)\pi i} - 1\right] = \frac{1}{(1-n)\rho^{n-1}}(1-1) = 0$$

$$\therefore \oint_C \frac{dz}{(z-\alpha)^n} = 0, \quad n = 2, 3, \cdots.$$

Illustrative Example :

Ex. 1. *Evaluate the integral* $\int_C \frac{z}{z^2-25} dz$ *on the circle* $C: |z-1| = 2$.

$(1, 0)$ *is the centre and 2 is radius of the circle C.*

$$\text{Let } f(z) = \frac{z}{z^2-25} = \frac{1}{2}\left(\frac{1}{z-5} + \frac{1}{z+5}\right)$$

which is analytic within and on the circle C, as $z = \pm 5$ lies outside the circle $|z-1| = 2$. Hence by Cauchy's theorem,

$$\int_C \frac{z}{z^2-25} dz = 0.$$

Ex. 2. *Evaluate the integral* $\int_C \frac{4z-9}{z(z-3)} dz$ *where C is a closed contour containing 0 and 3 in its interior.*

$$\text{As } \frac{4z-9}{z(z-3)} = \frac{3}{z} + \frac{1}{z-3},$$

$$\int_C \frac{4z-9}{z(z-3)} = 3\int \frac{dz}{z-0} + \int \frac{dz}{z-3}$$

$= 3 \cdot 2\pi i + 2\pi i$, (by Ex. 2) as 0,1 are interior points of C.

$= 8\pi i$.

Ex. 3. *Evaluate* $\int \frac{2z+1}{z(2z+i)^2} dz$ *on the circle* $C: |z| = 1$.

$\frac{2z+1}{z(2z+i)^2}$ can be decomposed into the sum of partial fraction

$$\frac{-1}{z} + \frac{2}{2z+i} + \frac{2(1+i)}{(2z+i)^2}$$

Now $2z + i = 0 \Rightarrow z = -\frac{i}{2} = 0 - \frac{1}{2}i$ which is a point lying within C because C is a circle with centre origin and radius 1.

$$\therefore \int_C \frac{dz}{(2z+i)^2} = \frac{1}{4}\int_C \frac{dz}{\left(z+\frac{i}{2}\right)^2} = 0, \text{ by Formula 3.}$$

$$\text{Again} \int_C \frac{dz}{(2z+i)} = \frac{1}{2}\int_C \frac{dz}{z+\frac{i}{2}} = \frac{1}{2}\cdot 2\pi i = \pi i \text{ by Formula 2.}$$

$$\text{And} \int_C \frac{dz}{z} = 2\pi i \quad \therefore z = 0 \text{ lies within } C$$

$$\therefore \int_C \frac{2z+1}{z(2z+i)^2}dz = -\int_C \frac{dz}{z} + 2\int_C \frac{dz}{2z+i} + 2(1+i)\int_C \frac{dz}{(2z+i)^2}$$

$$= -2\pi i + 2\cdot \pi i + 2(1+i)\times 0 = 0$$

2.2.6. Cauchy's Integral Formula.

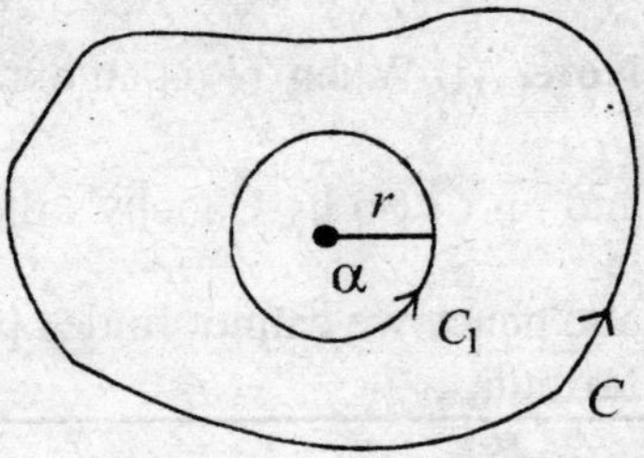

Theorem 1. If $f(z)$ is analytic within and on a simple closed curve C and α is any point within C, then

$$f(\alpha) = \frac{1}{2\pi i}\oint_C \frac{f(z)}{z-\alpha}dz.$$

Proof : As $\dfrac{f(z)}{z-\alpha}$ is analytic everywhere within C except at $z=\alpha$, we draw a circle C_1 with centre at $z=\alpha$ and radius r so that C_1 lies wholly within C. Then

$\dfrac{f(z)}{z-\alpha}$ is analytic within the annular region bounded by C and C_1.

Hence.

$$\oint_C \frac{f(z)}{z-\alpha}dz = \oint_{C_1} \frac{f(z)}{z-\alpha}dz, \text{ by Theorem 5 of Art 5.5.}$$

Now the equation of the circle C_1 is $|z-\alpha| = r$

$\therefore z-\alpha = re^{i\theta}$, when θ varies from 0 to 2π.

$\therefore dz = rie^{i\theta}d\theta$

$$\therefore \oint_{C_1} \frac{f(z)}{z-\alpha} dz = \int_0^{2\pi} \frac{f\left(\alpha + re^{i\theta}\right)}{re^{i\theta}} rie^{i\theta} d\theta = i \int_0^{2\pi} f\left(\alpha + re^{i\theta}\right) d\theta \quad \dots (2)$$

Now taking limit, as $r \to 0$, on both side of (2) and using the continuity of $f(z)$, we have

$$\oint_{C_1} \frac{f(z)}{z-\alpha} dz = i \underset{r\to 0}{Lt} \int_0^{2\pi} f\left(\alpha + re^{i\theta}\right) d\theta = i \int_0^{2\pi} \underset{r\to 0}{Lt}\, f\left(\alpha + re^{i\theta}\right) d\theta.$$

$$= i \int_0^{2\pi} f(\alpha) d\theta = if(\alpha) \int_0^{2\pi} d\theta = 2\pi i f(\alpha)$$

$\therefore$ From (1), $\oint_C \frac{f(z)}{z-\alpha} dz = 2\pi i f(\alpha)$. $\quad \therefore f(\alpha) = \frac{1}{2\pi i} \oint_C \frac{f(z)}{z-\alpha} dz$.

Note. (i) When α is an exterior point of *C*, $\frac{f(z)}{z-\alpha}$ is analytic within and on *C*. So by Cauchy's theorem $\oint_C \frac{f(z)}{z-\alpha} dz = 0$, and hence we cannot find $f(\alpha)$ in this case by using Cauchy's integral formula.

(ii) We can extend Cauchy's integral formula for multi-connected domain.

Theorem 2. ***(Cauchy's integral formula for the derivative of an analytic function).***

If $f(z)$ is analytic within and on *a* closed curve *C*, then the derivative of $f(z)$ at any interior point α of *C* is given by

$$f'(\alpha) = \frac{1}{2\pi i} \oint_C \frac{f(z)}{(z-\alpha)^2} dz.$$

Proof: As $z = \alpha$ is an interior point of C. So by Theorem 1,

$$f(\alpha) = \frac{1}{2\pi i} \oint_C \frac{f(z)}{(z-\alpha)} dz.$$

Now let $z = \alpha + h$ be a point in the neighbourhood of α. Then

$$f(\alpha+h)=\frac{1}{2\pi i}\oint_C \frac{f(z)}{z-\alpha-h}dz$$

$$\therefore\ f(\alpha+h)-f(\alpha)=\frac{1}{2\pi i}\oint_C\left[\frac{1}{z-\alpha-h}-\frac{1}{z-\alpha}\right]f(z)dz$$

$$=\frac{1}{2\pi i}\oint_C \frac{h}{(z-\alpha-h)(z-\alpha)}f(z)dz$$

$$\therefore\ \frac{f(\alpha+h)-f(\alpha)}{h}=\frac{1}{2\pi i}\oint_C \frac{f(z)}{(z-\alpha-h)(z-\alpha)}dz.$$

Now taking limit, as $h\to 0$, on both side we get

$$\underset{h\to 0}{Lt}\frac{f(\alpha+h)-f(\alpha)}{h}=\frac{1}{2\pi i}\underset{h\to 0}{Lt}\oint_C \frac{f(z)}{(z-\alpha-h)(z-\alpha)}dz$$

i.e. $$f'(\alpha)=\frac{1}{2\pi i}\oint_C \underset{h\to 0}{Lt}\frac{f(z)}{(z-\alpha-h)(z-\alpha)}dz$$

$$\therefore\ f'(\alpha)=\frac{1}{2\pi i}\oint_C \frac{f(z)}{(z-\alpha)^2}dz.$$

Theorem 3. (Cauchy's integral formula for the successive derivatives of an analytic function).

If $f(z)$ is analytic within and on a closed curve C, then the nth order derivatives of $f(z)$ at any interior point α of C is given by

$$f^{(n)}(\alpha)=\frac{n!}{2\pi i}\oint_C \frac{f(z)}{(z-\alpha)^{n+1}}dz.$$

Proof: Left as an exercise.

Illustrative Examples.

Ex. 1. *Evaluate* $\oint_C \frac{e^z}{(z+1)(z+2)}dz$ *where C is the circle* $|z-1|=4$.

Here $f(z)=e^z$ is analytic within and on the circle $|z-1|=4$ and $z=-1$, -2 are the interior points of C. Now $\frac{1}{(z+1)(z+2)}=\frac{1}{z+1}-\frac{1}{z+2}$

$$\therefore\ \oint_C \frac{e^z}{(z+1)(z+2)}dz=\int_C \frac{e^z}{z+1}dz-\int_C \frac{e^z}{z+2}dz$$

$=2\pi i\times e^{-1}-2\pi i\times e^{-2}$, by Cauchy's integral formula

$= 2\pi i\left(e^{-1} - e^{-2}\right)$.

Ex. 2. *Evaluate* $\oint_C \frac{z^2+6z-1}{z-4}dz$ *if C is (a) the circle* $|z|=5$ *(b)* $|z| = \frac{1}{2}$.

(*a*) Here $f(z) = z^2 + 6z - 1$ is analytic within and on the circle $|z| = 5$ and 4 is an interior point of C. Hence by Cauchy's integral formula

$$\oint_C \frac{z^2+6z-1}{z-4}dz = 2\pi i \times \left(4^2 + 6\times 4 - 1\right) = 78\pi i.$$

(*b*) As $z = 4$ is the exterior point of $|z| = \frac{1}{2}$, so $\frac{z^2+6z-1}{z-4}$ is analytic within and on the circle $|z| = \frac{1}{2}$. Hence by Cauchy's theorem

$$\oint_C \frac{z^2+6z-1}{z-4}dz = 0.$$

Ex. 3. *Evaluate* $\oint_C \frac{\sin 3z + 2\cos z}{z + \pi/2}dz$ *if C is the circle* $|z| = 5$.

Here $f(z) = \sin 3z + 2\cos z$ is analytic within and and on the circle $|z| = 5$ and $z = -\pi/2$ lies inside the circle *C*. So by Cauchy's integral formula

$$\oint_C \frac{\sin 3z + 2\cos z}{z+\pi/2}dz = 2\pi i\left\{\sin 3\left(-\pi/2\right) + 2\cos\left(-\pi/2\right)\right\} = 2\pi i(1 + 2.0) = 2\pi i.$$

Ex. 4. *Evaluate* $\oint_C \frac{\cos^3 z}{\left(z - \pi/4\right)^3}dz$ *where C is the circle* $|z| = 1$.

Let $f(z) = \cos^3 z$ which is analytic within and on *C*. Also $z = \pi/4$ lies inside the circle *C*. Hence by Cauchy's integral formula

$$f''(\alpha) = \frac{2!}{2\pi i}\oint_C \frac{f(z)}{(z-\alpha)^3}dz$$

We get $f''\left(\pi/4\right) = \frac{1}{\pi i}\oint_C \frac{\cos^3 z}{\left(z-\pi/4\right)^3}dz$.

$$\therefore \oint_C \frac{\cos^3 z}{\left(z-\pi/4\right)^3}dz = \pi i\left(6\cos\pi/4 \sin^2\pi/4 - 3\cos^3\pi/4\right) \quad [\because f''(z) = 6\cos z\sin^2 z - 3\cos^3 z]$$

$$= \frac{3\sqrt{2}\pi i}{4}$$

Ex. 5. *Evaluate* $\frac{1}{2\pi i}\oint_C \frac{e^{zt}}{(z^2+1)^2}dz$ *if* $t>0$ *and C is the circle* $|z|=3$.

Here $f(z)=e^{zt}$ is analytic within and on the circle $|z|=3$ and $z=\pm i$ are the interior points of C.

Now let $\frac{1}{(z^2+1)^2} = \frac{1}{(z+i)^2(z-i)^2} = \frac{A}{z+i}+\frac{B}{(z+i)^2}+\frac{C}{z-i}+\frac{D}{(z-i)^2}$

$\therefore \ 1 = A(z-i)^2(z+i)+B(z-i)^2+C(z-i)(z+i)^2+D(z+i)^2$.

Putting $z=i,-i,0,2i$, we get

$A=\frac{i}{4}, B=-\frac{1}{4}, C=-i/4, D=-\frac{1}{4}$.

$$\therefore \frac{1}{(z^2+1)^2} = \frac{i}{4}\cdot\frac{1}{z+i}-\frac{1}{4}\cdot\frac{1}{(z+i)^2}-\frac{i}{4}\cdot\frac{1}{z-i}-\frac{1}{4}\cdot\frac{1}{(z-i)^2}$$

$$\therefore \frac{1}{2\pi i}\oint_C \frac{e^{zt}}{(z^2+1)^2}dz$$

$$= \frac{1}{2\pi i}\left\{\frac{i}{4}\oint_C \frac{e^{zt}}{z+i}dz - \frac{1}{4}\oint_C \frac{e^{zt}}{(z+i)^2}dz - \frac{i}{4}\oint_C \frac{e^{zt}}{z-i}dz - \frac{1}{4}\oint_C \frac{e^{zt}}{(z-i)^2}dz\right\}.$$

$$= \frac{1}{2\pi i}\left\{\frac{i}{4}\times 2\pi i\times e^{-it} - \frac{1}{4}\times 2\pi i\times te^{-it} - \frac{i}{4}\times 2\pi i\times e^{it} - \frac{1}{4}\times 2\pi i\times te^{it}\right\},$$

by using Cauchy's integral formula

$$= \frac{i}{4}\left(e^{-it}-e^{it}\right)-\frac{1}{4}t\left(e^{-it}+e^{it}\right) = \frac{i}{4}\times(-)2i\sin t - \frac{1}{4}t\cdot 2\cos t = \frac{1}{2}(\sin t - t\cos t)$$

Ex. 6. *If* $F(z_0)=\oint_C \frac{2z^2-z-2}{z-z_0}dz$ *where C is the circle* $|z|=3$, *show that* $F(2)=8\pi i$. *What is the value of* $F(z_0)$ *when* $|z_0|>3$?

Now $F(2)=\oint_C \frac{2z^2-z-2}{z-2}dz$ where C: $|z|=3$.

Let $f(z)=2z^2-z-2$. Then $f(z)$ is analytic within and on the circle $|z|=3$. Also $z=2$ is an interior point of C. So by Cauchy's integral formula, we have

$$f(2)=\frac{1}{2\pi i}\oint_C \frac{f(z)}{z-2}dz \quad \therefore \oint_C \frac{f(z)}{z-2}dz=2\pi i f(2).$$

$$\therefore \oint_C \frac{2z^2-z-2}{z-2}dz=8\pi i \quad [\because f(2)=4]$$

$$\therefore F(2)=8\pi i.$$

When $|z_0|>3$, then z_0 lies outside the circle $|z|=3$. So $\frac{2z^2-z-2}{z-z_0}$ is analytic within and on the circle $|z|=3$. Hence by Cauchy's integral formula, we have

$$\oint_C \frac{2z^2-z-2}{z-z_0}dz=0. \qquad \therefore F(z_0)=0.$$

2.2.7. Taylor's Series

Taylor's Theorem. If a function $f(z)$ is analytic within a circle C with centre at α, then at each point z inside C,

$$f(z)=f(\alpha)+\frac{(z-\alpha)}{1!}f'(\alpha)+\frac{(z-\alpha)^2}{2!}f''(\alpha)+\cdots\cdots+\frac{(z-\alpha)^n}{n!}f^{(n)}(\alpha)+\cdots\cdots \quad (1)$$

i.e. $f(z)=\sum_{n=0}^{\infty}\frac{(z-\alpha)^n}{n!}f^n(\alpha)$.

Proof: Beyond the scope of the book.

Note. (i) The above series (1) is called the *Taylor's Series* of $f(z)$ about the point $z=\alpha$ or at the point $z=\alpha$ or about the point $z=\alpha$.

(ii) Putting $\alpha = 0$ in the above series (1) we get

$$f(z) = f(0) + \frac{z}{1!} f'(0) + \frac{z^2}{2!} f''(0) + \cdots + \frac{z^n}{n!} f^n(0) + \cdots$$

$$= \sum_{n=0}^{\infty} \frac{z^n}{n!} f^n(\alpha)$$

which is the Taylor's Series of $f(z)$ about the origin.

(iii) Putting $z - \alpha = h$, i.e. $z = \alpha + h$ in the series (1), we get

$$f(\alpha + h) = f(\alpha) + \frac{h}{1!} f'(\alpha) + \frac{h^2}{2!} f''(\alpha) + \cdots + \frac{h^n}{n!} f^{(n)}(\alpha) + \cdots\cdots$$

which is the another form of Taylor's series.

Illustrative Example.

Ex. 1. *Expand* $f(z) = \dfrac{1}{1+2z}$ *in Taylor's Series about the origin.*

Here $f(z) = \dfrac{1}{1+2z}$ is analytic in the neighbourhood of the point $z = 0$. So the Taylor's series expansion of $f(z)$ about $z = 0$ is

$$f(z) = f(0) + z f'(0) + \frac{z^2}{2!} f''(0) + \cdots + \frac{z^n}{n!} f^n(0) + \cdots\cdots.$$

Now $f^n(z) = \dfrac{(-1)^n 2^n n!}{(1+2z)^{n+1}}$ $\quad \therefore f^n(0) = (-1)^n 2^n n!$

$\therefore f'(0) = -2$, $\therefore f''(0) = 2^2 \cdot 2!$, $f'''(0) = -2^3 \cdot 3!$ and so on.

Also $f(0) = 1$.

Hence $f(z) = 1 - z \cdot 2 + z^2 \cdot 2^2 - z^3 \cdot 2^3 + \cdots + (-1)^n z^n 2^n + \cdots$

$$= 1 - 2z + 2^2 z^2 - \cdots + (-1)^n \cdot 2^n z^n + \cdots.$$

Ex. 2. *Expand* $f(z) = \sin z$ *in a Taylor's series about* $z = \pi/4$

[*W.B.U.Tech 2003*]

As $f(z) = \sin z$ is analytic for all finite values of z, so $f(z) = \sin z$ is analytic in the neighbourhood of the point $z = \pi/4$. So the Taylor's series expansion of $f(z)$ about $z = \pi/4$ is

$$f(z) = f(\pi/4) + (z - \pi/4) f'(\pi/4) + \frac{(z - \pi/4)^2}{2!} f''(\pi/4) + \cdots$$

$$\cdots + \frac{(z - \pi/4)^n}{n!} f^{(n)}(\pi/4) + \cdots\cdots$$

Now $f^n(z) = \sin\left(\frac{n\pi}{2} + z\right)$

$$\therefore f^n\left(\frac{\pi}{4}\right) = \sin\left(\frac{n\pi}{2} + \frac{\pi}{4}\right)$$

$$\because f'\left(\frac{\pi}{4}\right) = \sin\left(\frac{\pi}{2} + \frac{\pi}{4}\right) = \cos\frac{\pi}{4} = \frac{1}{\sqrt{2}}$$

$$f''\left(\frac{\pi}{4}\right) = \sin\left(2\cdot\frac{\pi}{2} + \frac{\pi}{4}\right) = -\frac{1}{\sqrt{2}}$$

$$f'''\left(\frac{\pi}{4}\right) = \sin\left(3\cdot\frac{\pi}{2} + \frac{\pi}{4}\right) = -\frac{1}{\sqrt{2}}$$

and so on.

Hence

$$f(z) = \frac{1}{\sqrt{2}} + \left(z - \frac{\pi}{4}\right)\cdot\frac{1}{\sqrt{2}} + \frac{\left(z-\frac{\pi}{4}\right)^2}{2!}\left(\frac{1}{-\sqrt{2}}\right) + \frac{\left(z-\frac{\pi}{4}\right)^3}{3!}\left(-\frac{1}{\sqrt{2}}\right) + \cdots\cdots$$

$$= \frac{\sqrt{2}}{2}\left\{1 + \left(z-\frac{\pi}{4}\right) - \frac{\left(z-\frac{\pi}{4}\right)^2}{2!} - \frac{\left(z-\frac{\pi}{4}\right)^3}{3!} + \cdots\right\}.$$

Ex. 3. *Expand* $\log(1+z)$ *in power of* z *and indicate the region of convergence.*

Let $f(z) = \log(1+z)$. Now the nearest singularity of $f(z) = \log(1+z)$ is $z = -1$. So $f(z)$ is analytic within the region $|z| < 1$. Then the Taylor's series expansion of $f(z)$ about $z = 0$ is

$$f(z) = f(0) + z f'(0) + \frac{z^2}{2!} f''(0) + \cdots + \frac{z^n}{n!} f^n(0) + \cdots\cdots$$

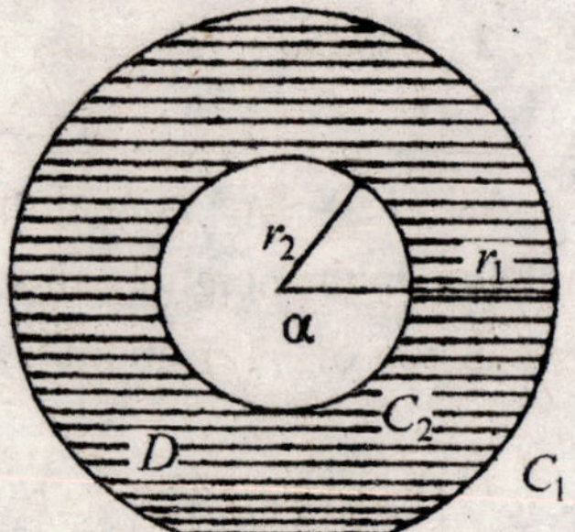

Now $f^n(z) = \dfrac{(-1)^{n-1}(n-1)!}{(1+z)^n}$

$$\therefore f^n(0) = (-1)^{n-1}(n-1)!$$

$$\therefore f'(0) = 1, \quad f''(0) = -1, \quad f'''(0) = 2!$$

and so on.

Also $f(0) = \log 1 = 0$.

Hence $f(z) = 0 + z\cdot 1 + \dfrac{z^2}{2!}(-1) + \dfrac{z^3}{3!}2! + \cdots\cdots$

$$\therefore \log(1+z) = z - \frac{z^2}{2} + \frac{z^3}{3} - \frac{z^4}{4} + \cdots$$

2.2.8. Laurent Series.

Laurent's Theorem. If a function $f(z)$ is analytic in the annulus (ring shaped) region D bounded by two concentric circles C_1 and C_2 with centre at the point $z = \alpha$ and radius r_1 and r_2 $(r_1 > r_2)$, then for all z in D,

$$f(z) = \sum_{n=0}^{\infty} a_n (z-\alpha)^n + \sum_{n=1}^{\infty} b_n (z-\alpha)^{-n} \quad \ldots \quad (1)$$

$$\text{where } a_n = \frac{1}{2\pi i} \oint_{C_1} \frac{f(z)}{(z-\alpha)^{n+1}} dz, \; n = 0, 1, 2, \cdots$$

$$b_n = \frac{1}{2\pi i} \oint_{C_2} \frac{f(z)}{(z-\alpha)^{-n+1}} dz, \; n = 1, 2, 3, \cdots.$$

Proof: Beyond the scope of this book.

Note. (i) The above series (1) is called the **Laurent's Series** of $f(z)$ at the point $z = \alpha$ or about the point $z = \alpha$.

(ii) Replacing n by $-n$ in a_n we have

$$a_{-n} = \frac{1}{2\pi i} \oint_C \frac{f(z)}{(z-\alpha)^{-n+1}} dz = b_n.$$

Hence Laurent's Series can be written as

$$f(z) = \sum_{n=0}^{\infty} a_n (z-\alpha)^n + \sum_{n=1}^{\infty} a_{-n} (z-\alpha)^{-n} = \sum_{-\infty}^{\infty} a_n (z-\alpha)^n \quad \ldots \quad (2)$$

where $a_n = \frac{1}{2\pi i} \oint_C \frac{f(z)}{(z-\alpha)^{n+1}} dz$ where C is any circle of radius r with centre at α such that $r_2 < r < r_1$,

(iii) If $f(z)$ is analytic within C_1, then

$$a_n = \frac{1}{2\pi i} \oint_{C_1} \frac{f(z)}{(z-\alpha)^{n+1}} dz = \frac{f^n(\alpha)}{n!}$$, by Cauchy's integral formula for successive derivative

and $b_n = \dfrac{1}{2\pi i}\oint_{C_1} \dfrac{f(z)}{(z-\alpha)^{-n+1}}dz = \dfrac{1}{2\pi i}\oint_{C_1}(z-\alpha)^{n-1} f(z)dz$

$= 0$, by Cauchy's Theorem.

Hence, in this case Laurent's Series (1) reduces to Taylor's Series,

$$f(z) = \sum_{n=0}^{\infty} \frac{(z-\alpha)^n}{n!} f^n(\alpha)$$

(iv) When $\alpha = 0$, the Laurent Series (2) becomes

$$f(z) = \sum_{n=-\infty}^{\infty} a_n z^n$$

where $a_n = \dfrac{1}{2\pi i}\oint_C \dfrac{f(z)}{z^{n+1}} dz$.

(v) The working procedure for finding the Laurent series of a function $f(z)$ would be more simpler. If $f(z)$ can be expanded into an infinite series (using Binomial law or other means) subject to the restriction on z the obtained infinite series is nothing but the required Laurent's series. See the following examples.

Illustrative Examples.

Ex. 1. *Expand the function* $f(z) = \dfrac{z^2-1}{(z+2)(z+3)}$ *when* (*i*) $|z| < 2$ (*ii*) $2 < |z| < 3$ (*iii*) $|z| > 3$.

Let $f(z) = 1 + \dfrac{A}{z+2} + \dfrac{B}{z+3}$.

Then $z^2 - 1 = (z+2)(z+3) + A(z+3) + B(z+2)$.

Putting $z = -2, -3$, we get $A = 3$, $B = -8$.

$\therefore f(z) = 1 + \dfrac{3}{z+2} - \dfrac{8}{z+3}$.

(i) When $|z| < 2$,

$$f(z) = 1 + \frac{3}{z+2} - \frac{8}{z+3} = 1 + \frac{3}{2}\left(1+\frac{z}{2}\right)^{-1} - \frac{8}{3}\left(1+\frac{z}{3}\right)^{-1}$$

$$=1+\frac{3}{2}\left(1-\frac{z}{2}+\frac{z^2}{2^2}-\cdots\right)-\frac{8}{3}\left(1-\frac{z}{3}+\frac{z^2}{3^2}-\cdots\right),$$

which holds for $\left|\frac{z}{2}\right|<1$ and $\left|\frac{z}{3}\right|<1$ i.e. for $|z|<2$

$$\therefore f(z)=1+\frac{3}{2}\sum_{n=0}^{\infty}(-1)^n\left(\frac{z}{2}\right)^n-\frac{8}{3}\sum_{n=0}^{\infty}(-1)^n\left(\frac{z}{3}\right)^n.$$

(ii) When $2<|z|<3$,

$$f(z)=1+\frac{3}{z}\left(1+\frac{2}{z}\right)^{-1}-\frac{8}{3}\left(1+\frac{z}{3}\right)^{-1}$$

$$=1+\frac{3}{z}\sum_{n=0}^{\infty}(-1)^n\left(\frac{2}{z}\right)^n-\frac{8}{3}\sum_{n=0}^{\infty}(-1)^n\left(\frac{z}{3}\right)^n$$,which holds for $\left|\frac{2}{z}\right|<1$ i.e.

$|z|>2$ and for $\left|\frac{z}{3}\right|<1$ i.e. $|z|<3$. i.e. for $2<|z|<3$.

(iii) When $|z|>3$,

$$f(z)=1+\frac{3}{z}\left(1+\frac{2}{z}\right)^{-1}-\frac{8}{z}\left(1+\frac{3}{z}\right)^{-1}$$

$$=1+\frac{3}{z}\sum_{n=0}^{\infty}(-1)^n\left(\frac{2}{z}\right)^n-\frac{8}{z}\sum_{n=0}^{\infty}(-1)^n\left(\frac{3}{z}\right)^n,$$

which holds for $\left|\frac{2}{z}\right|<1$ and $\left|\frac{3}{z}\right|<1$ i.e. for $|z|>2$, $|z|>3$.

Ex. 2. *Expand the function* $f(z)=\dfrac{1}{(z-1)(z-2)}$ *between the annular region* $|z|=1$ *and* $|z|=2$. [*W.B.U.Tech 2004*]

The given function can be written as

$$f(z)=\frac{1}{z-2}-\frac{1}{z-1}=-\frac{1}{2}\left(1-\frac{z}{2}\right)^{-1}-\frac{1}{z}\left(1-\frac{1}{z}\right)^{-1}$$

$$=-\frac{1}{2}\sum_{n=0}^{\infty}\left(\frac{z}{2}\right)^n-\frac{1}{z}\sum_{n=0}^{\infty}\left(\frac{1}{z}\right)^n$$, which holds for $\left|\frac{z}{2}\right|<1$ i.e. for $|z|<2$

and for $\left|\frac{1}{z}\right| < 1$ i.e. for $|z| > 1$. i.e. for $1 < |z| < 2$.

$\therefore \quad f(z) = -\sum_{n=0}^{\infty} \frac{z^n}{2^{n+1}} - \sum_{n=0}^{\infty} \frac{1}{z^{n+1}}$ which is the required Laurent Series.

Ex. 3. *Represent the function* $f(z) = \frac{z}{(z-1)(z-3)}$ *by a series of positive and negative powers of* $(z-1)$ *which converges to* $f(z)$ *when* $0 < |z-1| < 2$.

Here $f(z) = \frac{z}{(z-1)(z-3)}$ can be written as

$$f(z) = -\frac{1}{2} \cdot \frac{1}{z-1} + \frac{3}{2} \frac{1}{z-3} = -\frac{1}{2} \cdot \frac{1}{z-1} - \frac{3}{2} \frac{1}{2-(z-1)}$$

$$= -\frac{1}{2} \cdot \frac{1}{z-1} - \frac{3}{4}\left(1 - \frac{z-1}{2}\right)^{-1} = -\frac{1}{2} \cdot \frac{1}{z-1} - \frac{3}{4} \sum_{n=0}^{\infty} \left(\frac{z-1}{2}\right)^n,$$

which holds for $|z-1| < 2$ i.e. $0 < |z-1| < 2$.

Ex. 4. Find the Laurent's Series of the function $\frac{e^z}{(z-2)^3}$ about the point $z = 2$.

Let $z - 2 = u$ then $z = u + 2$.

$$\therefore \quad \frac{e^z}{(z-2)^3} = \frac{e^{u+2}}{(u+2-2)^3} = \frac{e^2 \cdot e^u}{u^3} = \frac{e^2}{u^3}\left(1 + u + \frac{u^2}{2!} + \frac{u^3}{3!} + \cdots\cdots\right)$$

$$= e^2\left(\frac{1}{u^3} + \frac{1}{u^2} + \frac{1}{2!} \cdot \frac{1}{u} + \frac{1}{3!} + \frac{1}{4!} u + \cdots\cdots\right)$$

$$= \frac{e^2}{(z-2)^3} + \frac{e^2}{(z-2)^2} + \frac{1}{2!} \cdot \frac{1}{z-2} + \frac{1}{3!} + \frac{1}{4!}(z-2) + \cdots\cdots$$

EXERCISES

1. Find the value of the integral $\int_0^{1+i} \left(x - y + ix^2\right) dz$

(*a*) along the straight line from $z = 0$ to $z = 1 + i$.

(*b*) along the real axis from $z = 0$ to $z = 1$ and then along a line parallel to the imaginary axis from $z = 1$ to $z = 1 + i$.

2. Evaluate $\oint_C |z|^2 dz$ around the square with vertices at $(0,0)$, $(1,0)$, $(1,1)$, $(0,1)$.

3. (*a*) Evaluate $\int_{1-i}^{2+i}(2x+iy+1)dz$, along the two paths :

(i) $x=t+1$, $y=2t^2-1$ (ii) the straight line joining $1-i$ and $2+i$.

(*b*) Evaluate $\int_{(0,3)}^{(2,4)}\left(2y+x^2\right)dx+(3x-y)dy$ along the paths

(i) the parabola $x=2t$, $y=t^2+3$ (ii) the straight line joining $(0,3)$ and $(2,4)$. [*W.B.U.Tech 2002*]

4. Evaluate $\int_C\left(z-z^2\right)dz$ where *C* is the (i) upper half of the circle $|z|=1$ (ii) lower half of the circle $|z|=1$.

5. (a) Evaluate $\int_C\left(z^2+3z\right)dz$ along (*a*) the straight line from $(2,0)$ to $(2,2)$ (*b*) the straight lines from $(2,0)$ to $(2,2)$ and then from $(2,2)$ to $(0,2)$.

(*b*) Evalute $\int_C\left(z^2+3z+2\right)dz$ where *C* is the arc of the cycloid $x=a(\theta+\sin\theta)$ $y=a(1-\cos\theta)$ between the points $(0,0)$ and $(\pi a,2a)$.

[*W.B.U.Tech 2004*]

6. Verify Cauchy's theorem for the function $3z^2+iz-4$ if *C* is the square with vertices at $1\pm i,-1\pm i$.

7. Evaluate the following integrals using Cauchy's theorem and Cauchy's integral Formula

(i) $\oint_C\frac{dz}{z^2+8}$ on circle $C:|z|=2$.

(ii) $\oint_C\frac{z}{(2z+1)(z-3)}dz$ on the circle $C:|z-1|=5$.

(iii) $\oint\frac{6z}{z^2+2z-115}dz$ on the circle $C:|z|=1$.

(iv) $\int_C\frac{e^z}{\left(z^2+\pi^2\right)^2}dz$ on the circle $C:|z|=4$.

(v) $\int_C \frac{4z^2-3z+2}{(z+1)(z-2)^2}dx$ where $C:|z|=3$

[Hint. write $\frac{4z^2-3z+2}{(z+1)(z-2)^2}=\frac{1}{z+1}+\frac{3}{z-2}+\frac{4}{(z-2)^2}$]

8. Evaluate $\oint_{|z|=1} \frac{dz}{z+2}$ and deduce that $\int_0^{\pi} \frac{1+2\cos\theta}{5+4\cos\theta}d\theta=0$.

9. Evauate $\oint_C \frac{2z-1}{z^2-z}dz$ where C is closed contour containing 0 and 1 as its interior points.

10. Using Cauchy's Integral formula, evaluate the following integrals :

(i) $\oint \frac{\sin 3z}{z+\pi/2}dz$ where C is the circle $|z|=5$.

(ii) $\int \frac{e^{2z}}{(z-1)(z-2)}dz$ where C is the circle $|z|=3$.

(iii) $\int_C \frac{zdz}{z^2-1}$ where C is the positively oriented circle $|z|=2$.

[*W.B.U.Tech 2004,2003*]

(iv) $\oint_C \frac{3z^2+7z+1}{z+1}dz$ where C is the circle $|z|=1/2$.

(v) $\int_C \frac{\cos\pi z}{z^2-1}$ where C is a rectangle with vertices $(2,1)$, $(2,-1)$, $(-2,1)$, $(-2,-1)$.

(vi) $\oint_C \frac{\cos^3(3z)+\cosh^4(4z)}{z}dz$ where C is any closed contour having $z=0$ as its interior.

(vii) $\oint_C \frac{\sin\pi z^2+\cos\pi z^2}{(z-1)(z-2)}dz$ where C is the circle $|z|=5$.

[*W.B.U.Tech 2003*]

(viii) $\int_C \frac{z+7}{z^2+2z+5}dz$ if C is $|z-i|=3/2$.

(ix) $\int_C \frac{dz}{z^3(z+4)}$ if C is $|z+2|=3$.

(x) $\int_C \frac{e^z}{(z^2+\pi^2)^2}dz$ if C is $|z|=4$.

(xi) $\int_C \frac{2z-1}{z(z+6)}dx$ where C is $|z|=4$

11. If $g(z_0)=\oint_C \frac{2z^2-z+1}{z-z_0}dz$ where C is the circle $|z|=2$, show that $g(1)=4\pi i$. What is the value of $g(z_0)$ when $|z_0|>2$?

12. Evaluate the following integrals :

(i) $\oint_C \frac{\sin^6 z}{\left(z-\pi/6\right)^3}dz$ if C is the circle $|z|=1$.

(ii) $\oint_C \frac{e^{2z}}{(z+1)^4}dz$ if C is the circle $|z|=3$. [*W.B.U.Tech 2002*]

(iii) $\oint_C \frac{z^3-2z+1}{(z-i)^2}$ if C is the circle $|z|=2$.

(iv) $\oint_C \frac{\cos z}{(z+\pi i)^4}$ where C is the closed square bounded by the lines $x=\pm 4$, $y=\pm 4$.

13. Expand each of the following functions in a Taylor's Series :

(i) $\cos z$ about $z=\pi/2$.

(ii) $\frac{1}{1+z}$ about $z=1$.

(iii) $\frac{2z^3+1}{z^2+z}$ about the point $z=i$.

(iv) $\frac{z-1}{z+1}$ about $z=0$.

(v) $\frac{z+1}{(z-3)(z-4)}$ about $z=2$; find also the region of convergency.

[*W.B.U.Tech 2002*]

14. Expand $f(z)=\dfrac{z}{(z-1)(2-z)}$ in a Laurent's series valid for

(*a*) $|z|<1$ (*b*) $1<|z|<2$ (*c*) $|z|>2$ (*d*) $|z-1|>1$ (*e*) $0<|z-2|<1$.

15. Find the Laurent's expansion of $\dfrac{z^2-6z-1}{(z-1)(z-3)(z+2)}$ in the region $3<|z+2|<5$.

16. Find the Taylor's or Laurent's series which represents the function

$\dfrac{1}{(1+z^2)(z+2)}$ when (i) $|z|<1$ (ii) $1<|z|<2$ (iii) $|z|>2$.

ANSWERS

1. (a) $\dfrac{-1+i}{3}$ (b) $-\dfrac{1}{2}+\dfrac{5i}{6}$. **2.** $-1+i$.

3. (a) (i) $4+\dfrac{25}{3}i$ (ii) $4+8i$ · (b) (i) $\dfrac{33}{2}$ (ii) $\dfrac{47}{3}$.

5. (a) $\dfrac{-44}{3}-\dfrac{8}{3}i$ (b) $\dfrac{-49}{3}-\dfrac{8}{3}i$

7. (i) 0 (ii) πi (iii) 0 (iv) $\dfrac{1}{2}$ (v) $8\pi i$.

8. 0. **9.** $4\pi i$.

10. (i) $2\pi i$ (ii) $2\pi i(e^4-e^2)$ (iii) 0 (iv) 0

(v) 0 (vi) $4\pi i$ (vii) $4\pi i$ (viii) $\pi(3+i)$

(ix) 0 (x) i/π. (xi) $-\dfrac{\pi i}{3}$

11. 0.

12. (i) $\dfrac{21\pi i}{16}$ (ii) $\dfrac{8\pi i}{3e^2}$ (iii) $-10\pi i$ (iv) $\dfrac{\pi}{3}\sinh\pi = -\dfrac{\pi}{3}i\sin\pi i$

13. (i) $\sum_{n=0}^{\infty}(-1)^{n+1}\frac{\left(z-\pi/2\right)^{2n+1}}{n!}$ (ii) $\frac{1}{2}-\frac{z-1}{2^2}+\frac{(z-1)^2}{2^3}-\cdots$

(iii) $\left(\frac{i}{2}-\frac{3}{2}\right)+\left(3+\frac{i}{2}\right)(z-i)+\sum_{n=2}^{\infty}(-1)^n\left\{\frac{1}{i^{n+1}}+\frac{1}{(1+i)^{n+1}}\right\}(z-i)^n$

(iv) $-1+2\left(z-z^2+z^3-z^4+\cdots\cdots\right)$

(v) $\left(4-\frac{5}{2}\right)+\left(4-\frac{5}{4}\right)(z-2)+\left(4-\frac{5}{2^3}\right)(z-2)^2+\left(4-\frac{5}{2^4}\right)(z-2)^3+\cdots\cdots\infty$

where $|z-2|<1$ is region convergence

14. (a) $-\frac{1}{2}z-\frac{3}{4}z^2-\frac{7}{8}z^3-\cdots\ \cdots$ (b) $\cdots\cdots\frac{1}{z^2}+\frac{1}{z}+1+\frac{1}{2}z+\frac{1}{4}z^2+\frac{1}{8}z^3+\cdots\cdots$

(c) $-\frac{1}{2}-\frac{3}{z^2}-\frac{7}{z^3}-\frac{15}{z^4}\cdots$ (d) $-\frac{1}{z-1}-\frac{2}{(z-1)^2}-\frac{2}{(z-1)^3}\cdots$

(e) $1-\frac{2}{z-2}-(z-2)+(z-2)^2-(z-2)^3+\cdots\cdots$.

15. $\frac{2}{z+2}+\frac{3}{(z+2)^2}+\frac{3^2}{(z+2)^3}+\cdots+\frac{1}{5}\left(1+\frac{z+2}{5}+\frac{(z+2)^2}{5^2}+\cdots\right)$

16. (i) $\frac{1}{10}\sum_{n=0}^{\infty}(-1)^n\left(\frac{z}{2}\right)^n+\frac{2-z}{5}\sum_{n=0}^{\infty}(-1)^n z^{2n}$

(ii) $\frac{1}{10}\sum_{n=0}^{\infty}(-1)^n\left(\frac{z}{2}\right)^n+\frac{2-z}{5z^2}\sum_{n=0}^{\infty}(-1)^n\left(\frac{1}{z^2}\right)^n$

(iii) $\frac{1}{5z}\sum_{n=0}^{\infty}(-1)^n\left(\frac{2}{z}\right)^n-\frac{1}{5}\left(\frac{1}{z}-\frac{2}{z^2}\right)\sum_{n=0}^{\infty}(-1)^n\frac{1}{z^{2n}}$.

2.3 ZERO & SINGULARITIES OF AN ANALYTIC FUNCTION : RESIDUE THEOREM

2.3.1. Zero of an Analytic Function .

Definition. A point $z=\alpha$ is said to be *zero* of an analytic function $f(z)$ if $f(\alpha)=0$

Order of Zero. If $f(z)$ is analytic in a domain D and $\alpha \in D$, then α is called *zero of* $f(z)$ *of order m* if $f(\alpha)=f'(\alpha)=f^2(\alpha)=\cdots=f^{m-1}(\alpha)=0$ but $f^m(\alpha)\neq 0$; $f^n(\alpha)$ stands for nth derivative of f.

Thus from Taylor's Theorem $f(z)=\sum_{n=m}^{\infty} a_n(z-\alpha)^n$ if α is mth order 0 of $f(z)$

Simple zero. A zero of order one of an analytic function is called a *simple zero.*

Illustration. $z=0$ is a zero of order 2 of the analytic function $f(z)=\sin^2 z$, as $f(0)=0, f'(z)=0$, but $f''(0)=2\neq 0$.

Theorem. The zero of an analytic function are isolated points

Proof. Beyond the scope of this book.

2.3.2. Singularities of an Analytic Function

Singularity. If a function $f(z)$ is not analytic at the point $z=\alpha$,then α is called the *singularity or the singular point* of $f(z)$.

Isolated and Non -Isolated Singularity.

A singularity $z=\alpha$ of a function $f(z)$ is said to be an *isolated singularity* if there is no other singularity within a small neighbourhood of $z=\alpha$

If a singularity $z=\alpha$ of a function $f(z)$ is not isolated, then it is called a *non-isolated singularity.*

Illustration. (i) The function $f(z)=\frac{1}{z-1}$ is analytic everywhere except at $z=1$. So $z=1$ is the only singularity of $f(z)$.As the function $f(z)$ has no other singularity, so $z=1$ is an isolated singularity .

(ii) The function $f(z)=\cot\frac{\pi}{z}=\frac{\cos\frac{\pi}{z}}{\sin\frac{\pi}{z}}$ is not analytic at $z=0$ and at the points where $\sin\frac{\pi}{z}=0$ i.e. $\frac{\pi}{z}=n\pi, n=0,\pm1,\pm2\cdots$

i.e. $z=\frac{1}{n}, n=\pm1,\pm2,\ldots$

So the singularities of $f(z)$ are $\pm1,\pm\frac{1}{2},\pm\frac{1}{3},\ldots$. Here $z=0$ is a non-isolated singularity as in every neighbourhood of $z=0$ there is an infinite number of singularities.

2.3.3. Essential Singularity and Pole

Principal Part of an Analytic Function.

Let α be an isolated singularity of an analytic function $f(z)$ in a domain D. Now we draw a circle C_1 with center at $z=\alpha$ and radius as small as we please and another large concentric circle C_2 if any radius lying wholly within the domain D. Then $f(z)$ is analytic within the annular region between these two circles C_1 and C_2. Hence by Laurent's Theorem, we have

$$f(z)=\sum_{n=0}^{\infty}a_n(z-\alpha)^n+\sum_{n=1}^{\infty}b_n(z-\alpha)^{-n}$$

The second term in R.H.S i.e. $\sum_{n=1}^{\infty}b_n(z-\alpha)^{-n}$ is called the principal part of $f(z)$ at the point $z=\alpha$.

Essential Singularities. If the principal part of $f(z)$ (at the isolated singularity α) contains an infinite number of terms, then the singularity $z=\alpha$ is called an *essential singularity.*

Pole : If the principal part of $f(z)$ contains a finite number of terms, say m, then the singularity $z=\alpha$ is called a *pole* of $f(z)$ *of order m.* In this case the Laurent Series of $f(z)$ is of the form

$$f(z)=\sum_{n=0}^{\infty}a_n(z-\alpha)^n+\frac{b_1}{z-\alpha}+\frac{b_2}{(z-\alpha)^2}+\cdots+\frac{b_m}{(z-\alpha)^m}, b_m\neq0$$

That is, in *L-series* the heighest power of $\dfrac{1}{z-\alpha}$ is finite and this finite highest power is the order of the pole.

When $m=1$, the pole is said to be **Simple Pole.**

Illustration. (i) As $f(z)=e^{\frac{1}{z}}=1+\dfrac{1}{z}+\dfrac{1}{2!}\cdot\dfrac{1}{z^2}+\dfrac{1}{3!}\cdot\dfrac{1}{z^3}+\cdots$ contains an infinite number terms in negative power of z, so $z=0$ is an essential singularity.

(ii) As the function

$$f(z)=\frac{\sin(z-a)}{(z-a)^4}=\frac{1}{(z-a)^4}\left[(z-a)-\frac{1}{3!}(z-a)^3+\frac{1}{5!}(z-a)^5-\cdots\right]$$

$$=\frac{1}{(z-a)^3}-\frac{1}{3!}\frac{1}{z-a}+\frac{1}{5!}(z-a)-\frac{1}{7!}(z-a)^3+\cdots$$ contains only two terms in negative power of $z-a$, so $z=a$ is a pole of order 2.

Theorem 1. If an analytic function $f(z)$ has a pole of order m at $z=a$, then $\dfrac{1}{f(z)}$ has a zero of order m at $z=a$ and conversely.

Proof. Left as an exercise.

Theorem 2. The limit point of the set of all poles of a function $f(z)$ is a non-isolated essential singularity.

Proof. Beyond the scope of this book.

Theorem 3. The limit point of the zeros of a function $f(z)$ is an isolated essential singularity.

Proof. Beyond the scope of this book.

Illustrative Examples

Ex. 1. *Prove that the function* $e^{-\frac{1}{z^2}}$ *has no singularities but has a zero of order two.*

Let $f(z)=e^{-\frac{1}{z^2}}=\dfrac{1}{e^{\frac{1}{z^2}}}$.

Poles of $f(z)$ are given by equating to zero the denominator of $f(z)$ (by Th. 1) i.e. $e^{\frac{1}{z^2}} = 0$ which does not give any value of z. Hence the function has no singularities.

As $e^{-\infty} = 0$, so the zeros of $f(z)$ are given by $z^2 = 0$ i.e. $z = 0, 0$. Thus $z = 0$ is a zero of order two.

Ex. 2. *Find out the zeros and discuss the nature of the singularities of*

$$f(z) = \frac{z-2}{z^2} \sin\frac{1}{z-1}.$$

Poles of $f(z)$ are given by putting the denominator equal to zero (by Th1 above) i.e. by $z^2 = 0$. i.e. $z = 0, 0$. So $z = 0$ is a pole of $f(z)$ of order two

Again the zeros of $f(z)$ are given by equating to zero the numerator of $f(z)$ (by Th.1) i.e. $(z-2)\sin\frac{1}{z-1} = 0$. or, $z - 2 = 0$ and $\sin\frac{1}{z-1} = 0$.

Now $z - 2 = 0$ gives $z = 2$ and $\sin\frac{1}{z-1} = 0$ gives

$$\frac{1}{z-1} = n\pi, (n = \pm 1, \pm 2, \cdots) \quad \text{i.e. } z = 1 + \frac{1}{n\pi}$$

Thus all the zeros of $f(z)$ are given by $z = 2, 1 + \frac{1}{n\pi} (n = \pm 1, \pm 2, \cdots)$.

Also the limit point of the zeros, given by $z = 1 + \frac{1}{n\pi}$ $(n = \pm 1, \pm 2, \cdots \cdots)$ is $z = 1$. Hence $z = 1$ is an *isolated essential singularity.*

Ex. 3. *Find the singularities of the followings function and determine their nature (i)* $\dfrac{1}{\sin z - \cos z}$

(ii) $\sec\dfrac{1}{z}$ [*W.B.U.Tech 2003*]

(iii) $\dfrac{1 - e^z}{1 + e^z}$

(i) Here poles of the given function are given by $\sin z - \cos z = 0$ i.e. $\tan z = 1$.

i.e. $z = n\pi + \pi/4 \ (n = 0,1,2\cdots)$

All these poles are simple pole.

(ii) Pole of the given function are given by $\cos\frac{1}{z} = 0$ (by Th 1) i.e. $\frac{1}{z} = n\pi + \pi/2$ i.e. $z = \frac{1}{n\pi + \pi/2}$, $(n = 0, \pm 1, \pm 2, \cdots)$. Obviously the limit point of these poles is $z = 0$. Hence $z = 0$ is a non-isolated essential singularity of the function. (By Th 2)

(iii) Poles of the function are given by $1 + e^z = 0$, i.e. $e^z = -1$

i.e. $e^z = e^{i\pi} = e^{i\pi + 2n\pi i}$ $[e^{2n\pi i} = \cos 2n\pi + i \sin 2n\pi = 1 + 0.i = 1]$

$\therefore \ z = (2n+1)\pi i, (n = 0, \pm 1, \pm 2, \cdots)$

Obviously $z = \infty$ is the limit point of these poles. Hence $z = \infty$ is a non isolated essential singularity . (By Th 2)

Ex. 4. *Find Laurent Series about the indicated singularity for each of the following functions.Determine the nature of the singularities in each case.*

(i) $\frac{e^z}{(z-1)^3}; z = 1$ *(ii)* $\frac{e^{z^2}}{z^3} : z = 0$ *(iii)* $e^{z/z-2} : z = 2.$

(i) Let $f(z) = \frac{e^z}{(z-1)^3} = \frac{e^{1+u}}{u^3}$, by putting $z - 1 = u$

$$= \frac{e}{u^3}\left(1 + u + \frac{u^2}{2!} + \frac{u^3}{3!} \cdots\right).$$

$$= e\left(\frac{1}{u^3} + \frac{1}{u^2} + \frac{1}{2u} + \frac{1}{3!} + \frac{u}{4!} + \frac{u^2}{5!} + \cdots\right)$$

$$= \frac{e}{(z-1)^3} + \frac{e}{(z-1)^2} + \frac{e}{2(z-1)} + \frac{e}{3!} + \frac{z-1}{4!} + \frac{(z-1)^2}{5!} + \cdots$$

which contains only three terms in negative power of $z-1$. Hence $z=1$ is a pole of order 3 .

(ii) Let $f(z)=\dfrac{e^{z^2}}{z^3}$

$$=\frac{1}{z^3}\left(1+z^2+\frac{(z^2)^2}{2!}+\frac{(z^2)^3}{3!}+\cdots\right)$$

$$=\frac{1}{z^3}+\frac{1}{z}+\frac{z}{2!}+\frac{z^3}{3!}+\frac{z^5}{5!}+\cdots$$

As the highest power of $\frac{1}{z}$ in the Laurent's Series is 3, so $z=0$ is a pole of order 3.

(iii) Let $f(z)=e^{\frac{z}{z-2}}=e^{\frac{u+2}{u}}$, by putting $z-2=u$

$$= e.e^{\frac{2}{u}}$$

$$=e\left(1+\frac{2}{u}+\frac{1}{2!}\left(\frac{2}{u}\right)^2+\frac{1}{3!}\left(\frac{2}{u}\right)^3+\cdots\right)$$

$$=e\{1+\frac{2}{z-2}+\frac{2^2}{2!(z-2)^2}+\frac{2^3}{3!(z-2)^3}+\cdots\}$$

Since there are infinite number of terms in negative power of $z-2$ in the Laurent's Series , so $z=2$ is the essential singularity.

2.3.4. Residue Theorem

Residue. Let α be an isolated singularity of an analytic function $f(z)$. Then by Laurent's Theorem ,

we have

$$f(z)=\sum_{n=0}^{\infty}a_n(z-\alpha)^n+\sum_{n=1}^{\infty}b_n(z-\alpha)^{-n} \text{ where}$$

$$a_n=\frac{1}{2\pi i}\oint_C\frac{f(z)}{(z-\alpha)^{n+1}}dz,$$

$$b_n=\frac{1}{2\pi i}\oint_C\frac{f(z)}{(z-\alpha)^{-n+1}}dz .$$

The coefficient b_1 of $\frac{1}{z-\alpha}$ in the above infinite series is called the residue of $f(z)$ at the singularity $z=\alpha$ and is denoted by $\text{Res}(\alpha)$. Thus

$$\text{Res}(\alpha)=b_1=\frac{1}{2\pi i}\oint_C f(z)dz \quad \therefore \oint_C f(z)dz=2\pi i\,\text{Res}(\alpha)\cdots(1).$$

Illustration . Consider the function $f(z)=\frac{1}{(z+1)^2(z-2)}$.

Here $z=-1, 2$ are the singularities of $f(z)$.

Now $f(z)=\frac{1}{t^2(t-1-2)}$, by putting $z+1=t$

$$=\frac{1}{-3t^2}\left(1-\frac{t}{3}\right)^{-1}$$

$$=\frac{1}{-3t^2}\left(1+\frac{t}{3}+\frac{t^2}{9}+\cdots\right).$$

$$=-\frac{1}{3t^2}-\frac{1}{9}\cdot\frac{1}{t}-\frac{1}{27}-\cdots$$

$$=-\frac{1}{3(z+1)^2}-\frac{1}{9}\cdot\frac{1}{z+1}-\frac{1}{27}-\cdots$$

So the coefficient of $\frac{1}{z+1}$ is $-\frac{1}{9}$

Hence the residue at $z=-1$ is $-\frac{1}{9}$.

Theorem 1. Let α be a pole of $f(z)$ of order m. Then the residue of $f(z)$ at $z=\alpha$ is given by $\frac{1}{(m-1)!}\lim_{z\to\alpha}\frac{d^{m-1}}{dz^{m-1}}[(z-\alpha)^m f(z)]$.

Proof. Beyond the scope of this book .

Corollary . When α is a simple pole of $f(z)$, then $\text{Res}(\alpha)= \lim_{z\to\alpha}(z-\alpha)f(z)$

Theorem 2 . (Cauchy's Residue Theorem)

Let $f(z)$ be analytic within and on a closed contour C ,except at a finite number of singularities $\alpha_1, \alpha_2, \cdots, \alpha_n$ and let $R_1, R_2 \cdots R_n$ be respectively the residues of $f(z)$ at these poles .

Then $\oint_C f(z)dz = 2\pi i(R_1 + R_2 \cdots + R_n)$.

$= 2\pi i$ (sum of residues at the poles within C) .

Proof . Beyond the scope of this book .

Illustrative Examples

Ex. 1. Find the residue of $\dfrac{z^3}{(z-1)^4(z-2)(z-3)}$ at $z=1$.

Let $f(z) = \dfrac{z^3}{(z-1)^4(z-2)(z-3)}$

$$= \frac{(1+t)^3}{t^4(t-1)(t-2)}, \text{ by putting } z-1=t$$

$$= \frac{(1+t)^3}{2t^4}\left(1+\frac{t^2-3t}{2}\right)^{-1}$$

$$= \frac{(1+t)^3}{2t^4}\left\{1-\frac{t^2-3t}{2}+\left(\frac{t^2-3t}{2}\right)^2+\cdots\cdots\right\}$$

$$= \frac{1}{2t^4}\left(1+3t+3t^2+t^3\right)\left(1+\frac{3}{2}t+\frac{7}{4}t^2+\frac{15}{8}t^3+\cdots\right).$$

So the coefficient of $\dfrac{1}{z-1}$ i.e of $\dfrac{1}{t}$ is $\dfrac{1}{2}\left(1+\dfrac{9}{2}+\dfrac{21}{4}+\dfrac{15}{8}\right) = \dfrac{101}{16}$

Hence the residue at $z=1$ is $\dfrac{101}{16}$.

Alternative method : Since $z=1$ is a pole of order 4, so by Theorem 1.

$$\text{Res}(1) = \frac{1}{(4-1)!}\lim_{z\to 1}\frac{d^3}{dz^3}\left[(z-1)^4 f(z)\right]$$

$$= \frac{1}{3!}\lim_{z\to 1}\frac{d^3}{dz^3}\left[(z-1)^4\cdot\frac{z^3}{(z-1)^4(z-2)(z-5)}\right]$$

$$=\frac{1}{6}\lim_{z\to 1}\frac{d^3}{dz^3}\left[\frac{z^3}{(z-2)(z-3)}\right]$$

$$=\frac{1}{6}\lim_{z\to 1}\frac{d^3}{dz^3}\left(z+5-\frac{8}{z-2}+\frac{27}{z-3}\right)$$

$$=\frac{1}{6}\lim_{z\to 1}\left\{0-8\frac{(-6)}{(z-2)^4}+27\frac{(-6)}{(z-3)^4}\right\}=\frac{101}{16}.$$

Ex. 2. *Show that the sum of residue of* $\frac{e^{-z}}{z^2-4}$ *at its pole is* $-\frac{1}{2}\sinh 2$

Let $f(z)=\frac{e^{-z}}{z^2-4}$.

So the poles of $f(z)$ are given by $z^2-4=0$ i.e. $z=\pm 2$ which are the simple poles of $f(z)$.

$\therefore$ $\text{Res}(2)=\lim_{z\to 2}(z-2)f(z)$, by corollary of Theorem 1

$$=\lim_{z\to 2}(z-2)\frac{e^{-z}}{z^2-4}=\lim_{z\to 2}\frac{e^{-z}}{z+2}=\frac{e^{-2}}{4}.$$

Similarly $\text{Res}(-2)=-\frac{e^2}{4}$. Thus the sum of residue of $f(z)$ is $\frac{e^{-2}}{4}-\frac{e^2}{4}=-\frac{1}{4}\left(e^2-e^{-2}\right)=-\frac{1}{4}\sinh 2$.

Ex. 3. *Find the residues of the function* $\frac{\cot\pi z}{(z-a)^2}$. [*W.B.U.Tech,2004*]

Let $f(z)=\frac{\cot\pi z}{(z-a)^2}=\frac{\cos\pi z}{(z-a)^2\sin\pi z}$

So the poles of $f(z)$ are given by $(z-a)^2\sin\pi z=0$ i.e. by $(z-a)^2=0$, or by $\sin\pi z=0$. Now $(z-a)^2=0$ gives $z=a,a$.

So $z=a$ is a pole of order 2

Again $\sin\pi z=0$ gives $\pi z=n\pi$ $(n=0,\pm 1,\pm 2,\cdots)$

$\therefore$ $z=n,(n=0,\pm 1,\pm 2,\cdots)$

These are simple poles if n is finite ,

$$\therefore \text{Res}(a) = \frac{1}{(2-1)!} \lim_{z \to a} \frac{d}{dz} [(z-a)^2 f(z)]$$

$$= \lim_{z \to a} \frac{d}{dz} (\cot \pi z) = \lim_{z \to a} \left(-\pi \operatorname{cosce}^2 \pi z \right) = -\pi \operatorname{cosec}^2 \pi a .$$

So residue of $f(z)$ at $z = a$ is $-\pi \operatorname{cosce}^2 \pi a$.

Now $\text{Res}(n) = \lim_{z \to n} [(z-n) f(z)]$

$$= \lim_{z \to n} [(z-n) \frac{\cos \pi z}{\sin \pi z . (z-a)^2}] \left(\frac{0}{0} \, form \right)$$

$$= \lim_{z \to n} \frac{\cos \pi z - \pi (z-n) \sin \pi z}{\pi \cos \pi z (z-a)^2 + 2(z-a) \sin \pi z} \quad \text{(Using L'Hospital rule)}$$

$$= \frac{\cos \pi n - 0}{\pi \cos \pi n (n-a)^2 + 2(n-a) \sin \pi n} = \frac{(-1)^n}{\pi (-1)^n (n-a)^2} = \frac{1}{\pi (n-a)^2}$$

So the residue of $f(z)$ at $z = n$ is $\dfrac{1}{\pi (n-a)^2}$

Ex. 4. *Evaluate* $\oint_C \frac{z+1}{z^2 - 2z} dz$ *where* C *is the circle* $|z| = 5$.

Let $f(z) = \dfrac{z+1}{z^2 - 2z}$. Then the poles of $f(z)$ are given by $z^2 - 2z = 0$ i.e. by $z = 0,2$. Here $z = 0,2$ are both simple pole of $f(z)$ and both lie within the circle $|z| = 5$. Hence by Residue Theorem we have

$$\oint_C f(z) dz = 2\pi i \ (\text{Res } (0) + \text{Res} (2)) \qquad \ldots (1)$$

Now $\text{Res}(0) = \lim_{z \to 0} z \cdot f(z) = \lim_{z \to 0} \dfrac{z+1}{z-2} = -\dfrac{1}{2}$

$\text{Res}(2) = \lim_{z \to 2} (z-2) f(z) = \lim_{z \to 2} \dfrac{z+1}{z} = \dfrac{3}{2}$

$\therefore$ From (1), $\oint_C f(z) dz = 2\pi i \left(-\dfrac{1}{2} + \dfrac{3}{2} \right)$

$$\therefore \oint_C \frac{z+1}{z^2 - 2z} dz = 2\pi i .$$

Ex. 5. *Evaluate* $\oint_C \frac{z\,dz}{(z-1)(z-2)^2}$ *where* C *is* $|z-2| = \frac{1}{2}$.

Let $f(z) = \frac{z}{(z-1)(z-2)^2}$.

Then the poles of $f(z)$ are given by $(z-1)(z-2)^2 = 0$ i.e. $z = 1, 2$. Here $z = 1$ is a simple pole and $z = 2$ is a double pole. But only $z = 2$ lies within the circle C. Hence by Residue Theorem,

$$\oint_C f(z)dz = 2\pi i \times \text{Res}(2) \qquad \dots \quad (1)$$

$$\text{Res}(2) = \lim_{z\to 2}\frac{d}{dz}(z-2)^2 \cdot f(z) = \lim_{z\to 2}\frac{d}{dz}\left(\frac{z}{z-1}\right) = \lim_{z\to 2} -\frac{1}{(z-1)^2} = -1.$$

$\therefore$ From (1), $\oint_C f(z)dz = 2\pi i(-1)$

$$\therefore \oint_C \frac{z\,dz}{(z-1)(z-2)^2} = -2\pi i.$$

Ex. 6. *Using Cauchy's Residue theorem, prove that*

$$\oint_C \frac{z\cos z}{\left(z-\frac{\pi}{2}\right)^3}dz = -2\pi i \quad \textit{where C is the circle } |z-1| = 1.$$

Let $f(z) = \frac{z\cos z}{\left(z-\frac{\pi}{2}\right)^3}$.

The poles of $f(z)$ are given by $\left(z-\frac{\pi}{2}\right)^3 = 0$ i.e. $z = \frac{\pi}{2}$ which is a pole of order 3 and lies within the circle $|z-1| = 1$. Hence by Cauchy's Residue theorem, we have $\oint_c f(z)dz = 2\pi i \,\text{Res}\left(\frac{\pi}{2}\right)$.

Now $\text{Res}\left(\frac{\pi}{2}\right) = \frac{1}{(3-1)!}\lim_{z\to\frac{\pi}{2}}\frac{d^2}{dz^2}\left(z-\frac{\pi}{2}\right)^3 f(z) = \frac{1}{2}\lim_{z\to\frac{\pi}{2}}\frac{d^2}{dz^2}(z\cos z)$

$$= \frac{1}{2}\lim_{z\to\frac{\pi}{2}}(-2\sin z - z\cos z) = \frac{1}{2}(-2-0) = -1.$$

$$\therefore \oint_C f(z)dz = 2\pi i(-1)$$

$$\therefore \oint_C \frac{z\cos z}{\left(z-\pi/2\right)^3} dz = -2\pi i .$$

Ex. 7. *Evalute* $\int_0^{2\pi} \frac{d\theta}{1+a^2-2a\cos\theta}$, $0<a<1$.

Let $I = \int_0^{2\pi} \frac{d\theta}{1+a^2-2a\cos\theta} = \int_C \frac{1}{1+a^2-a\left(z+\frac{1}{z}\right)} \frac{dz}{iz}$, putting $z=e^{i\theta}$ so that

$d\theta = \frac{dz}{iz}$ and $\cos\theta = \frac{1}{2}\left(z+\frac{1}{z}\right)$ where C is the circle $|z|=1$.

$$= -\frac{1}{ia}\int_C \frac{dz}{(z-a)\left(z-\frac{1}{a}\right)}.$$

Let $f(z) = \frac{1}{(z-a)\left(z-\frac{1}{a}\right)}$.

Then poles of $f(z)$ are $a, \frac{1}{a}$ which are simple poles. As $0<a<1$, so only the pole $z=a$ lies within the circle C. Hence by residue theorem

$$\int_C f(z)dz = 2\pi i \times \text{Res}(a).$$

Now $\text{Res}(a) = \lim_{z\to a}(z-a)f(z) = \lim_{z\to a}\frac{1}{z-\frac{1}{a}} = \frac{1}{a-\frac{1}{a}} = \frac{a}{a^2-1}$.

$$\therefore \int_C f(z)dz = \frac{2\pi i a}{a^2-1}$$

i.e. $\int_C \frac{dz}{(z-a)\left(z-\frac{1}{a}\right)} = \frac{2\pi i a}{a^2-1}$.

$$\therefore I = -\frac{1}{ia}\times\frac{2\pi i a}{a^2-1} = \frac{2\pi}{1-a^2}$$

EXERCISES

1. Find the zeros of the following functions and indicate their order :

(i) $z^2 \sin 2z$ (ii) $\frac{(z+1)^2}{3z^2+5z-2}$ (iii) $\frac{z^3-1}{2z^2+5}$ (iv) $\cot z$ (v) $\frac{z^5-1}{z^2+5}$

2. Examine the nature of singularities of the following functions :

(a) $\sin\frac{1}{z-a}$ (b) $\frac{1}{1-e^z}$

(c) $\frac{z^2+4}{e^z}$ (d) $\sin\frac{1}{z}$ (e) $ze^{\frac{1}{z}}$.

3. Find the singularities and their nature of the following function :

(a) $\frac{1}{(z+1)^2(3-z)}$ (b) $\frac{z}{(z^2+4)^2}$ (c) $\frac{\tan z}{z}$

(d) $z \operatorname{cosec} z$ (e) $\frac{e^{2z}}{(z-1)^3}$ (*f*) $\frac{1}{z(1-e^z)}$

4. Expand each of the following functions in a Laurent Series about $z=0$. Determine the nature of the singularities in each case :

(a) $\frac{1-\cos z}{z}$, (b) $\frac{z-2}{z(z-1)}$, (c) $(z-5)\sin\frac{1}{z}$.

5. Determine and classify all the singularities of the function $\frac{z}{e^{\frac{1}{z}}-1}$.

6. Show that the singularities of $f(z)=\cot \pi z$ are all simple poles and find the residues of the function at these poles.

7. Find the residues of the following function at each pole

(a) $\frac{1-e^{2z}}{z^4}$ (b) $\frac{z^2-2z}{(z+1)^2(z^2+4)}$ (c) $\frac{e^z}{\sin^2 z}$

(d) $\operatorname{sech} z$ (e) $\frac{z}{\cos z}$ [*W.B.U.Tech 2004*]

(f) $\frac{ze^z}{(z-a)^3}$ (g) $\frac{z^3+2z+5}{z^2(z^2+1)}$.

8. Evaluate the following integrals :

(a) $\oint_C \frac{e^z}{\cosh z}dz$, $C:|z|=5$

(b) $\oint_C \frac{z+1}{z^2-2z}dz$, $C:|z|=1$

(c) $\oint_C \frac{4z^2-4z+1}{(z-2)(4+z^2)}dz$, $C:|z|=1$

(d) $\oint_C \frac{3z^2+2}{(z-1)(z^2+9)}dz$, $C:|z-2|=2$

(e) $\oint_C \frac{e^z}{z^2+1}dz$, $C:|z|=2$

(f) $\int_C \frac{3z^2+z+1}{(z^2-1)(z-3)}dz$ around the circle $C:|z|=2$. [W.B.U.Tech 2002]

(g) $\int_C \frac{dz}{z^8(z+4)}$ where C is the circle $|z+2|=3$.

(h) $\int_C \frac{e^{2z}}{(z+1)^3}dz$ where C is the ellipse $4x^2+4y^2=36$.

9. Evaluate $\oint_C \frac{2+3\sin\pi z}{z(z-1)^2}dz$ where C is a square having vertices at $(3,3)$, $(3,-3)$, $(-3,3)$ $(-3,-3)$.

10. Using Cacuhy's Residue theorem, prove that

(i) $\int_0^{2\pi} \frac{d\phi}{a^2+\sin^2\phi} = \frac{\pi}{a\sqrt{1+a^2}}$, $(a>0)$

(ii) $\int_0^{2\pi} \frac{\sin^2\theta}{a+b\cos\theta}d\theta = \frac{2\pi}{b^2}\left[a-\sqrt{a^2-b^2}\right]$; $a>b>0$

(iii) $\int_0^{2\pi} \frac{d\theta}{2+\cos\theta} = \frac{2\pi}{\sqrt{3}}$ [W.B.U.Tech 2004]

(iv) $\int_0^{2\pi} \frac{d\theta}{5+4\cos\theta} = \frac{2\pi}{3}$ [W.B.U.Tech 2004]

(v) $\int_0^{2\pi} \frac{\cos 2\theta}{5+4\cos\theta} d\theta = \frac{\pi}{6}$

ANSWERS

1. (i) $z=0$ is a zero of order 3, $\pm\frac{\pi}{2}, \pm\frac{2\pi}{2}, \pm\frac{3\pi}{2}$...... are zero of order 1

(ii) $z=-1$ is a zero of order 2

(iii) simple zeros at 1, $e^{\frac{2\pi i}{3}}$, $e^{\frac{4\pi i}{3}}$

(iv) simple zero at $(2n+1)\pi/2$, $n=0,\pm 1,\pm 2,\cdots$

2. (a) a is an isolated essential singularity.

(b) $2\pi i$ is a simple pole.

(c) $z=\infty$ is an isolated essential singularity.

(d) $z=0$ is an isolated essential singularity.

(e) $z=0$ is an essential singular point.

3. (a) $z=-1$ is a pole of order 2 and $z=3$ is a simple pole.

(b) $z=\pm 2i$, pole of order 2.

(c) $z=0,\pm\pi/2, \frac{3\pi}{2},\cdots$ are simple poles.

(d) $z=n\pi(0,\pm 1,\pm 2,\cdots)$ are simple poles.

(e) $z=1$ is a pole of order 3.

(f) $z=0$ is a double pole.

4. (a) $\frac{z}{2!}-\frac{z^3}{4!}+\frac{z^5}{6!}\cdots$, ordinary point.

(b) $\frac{2}{z}+1+z+z^2+\cdots$, simple pole.

(c) $1-\frac{5}{z}-\frac{1}{3!z^2}+\frac{5}{3!z^3}+\frac{1}{5!z^4}\cdots$, essential singularity.

5. $\frac{i}{2n\pi}$, $(n=\pm1,\pm2,\cdots)$; simple poles $z=0$, essential singularity $z=\infty$, pole of order 2.

6. $\frac{1}{\pi}$

7. (a) $\text{Res}(0)=-4/3$

(b) $\text{Res}(-1)=-\frac{14}{25}$, $\text{Res}(2i)=\frac{7+i}{25}$, $\text{Res}(-2i)=\frac{7-i}{25}$.

(c) $\text{Res}(n\pi)=e^{n\pi}$

(d) $\text{Res}\left(\frac{2k+1}{2}\pi i\right)=(-1)^{k+1}i$.

(e) $\text{Res}\left\{(2n+1)\pi/2\right\}=(-1)^n(2n+1)\pi/2$, $n=0,\pm1,\pm2,\cdots$

(f) $\text{Res}(a)=e^a\left(1+\frac{a}{2}\right)$

(g) $\text{Res}(\pm i)=\frac{1}{2}(-1\pm5i)$ $\text{Res}(0)=2$.

8. (a) $8\pi i$ (b) $-\pi i$ (c) 0 (d) πi (e) $\pi\left(e^i-e^{-i}\right)$

(f) $-\pi i/4$ (g) 0 (h) $4\pi i/e^2$.

9. $-6\pi^2 i$.

2.4 CONFORMAL MAPPINGS

2.4.1. Transformation or Mapping

Consider the complex valued function $w = f(z)$. Then corresponding to each point z_0, we have a point w_0 in the complex plane such that $f(z_0) = w_0$. Let $z = x + iy$ and $w = f(z) = u + iv$. Then $u = u(x, y)$, $v = v(x, y)$ represent a mapping which establish a correspondence between the points (x, y) in the *xy*-plane i.e. *z*-plane and the points (u, v) in the *uv*-plane i.e. *w*-plane. Here using the above mapping, a curve or a region of *z*-plane transformed into another curve or a region of *w*-plane.

2.4.2. Conformal Mapping

Isogonal Mapping. A mapping is said to be *isogonal* if two curves in the *z*-plane intersecting at the point z_0 at an angle θ are transformed into two corresponding curves in the *w*-plane intersecting at the point w_0, which corresponds to z_0, at the same angle θ under the mapping (see Fig. 1).

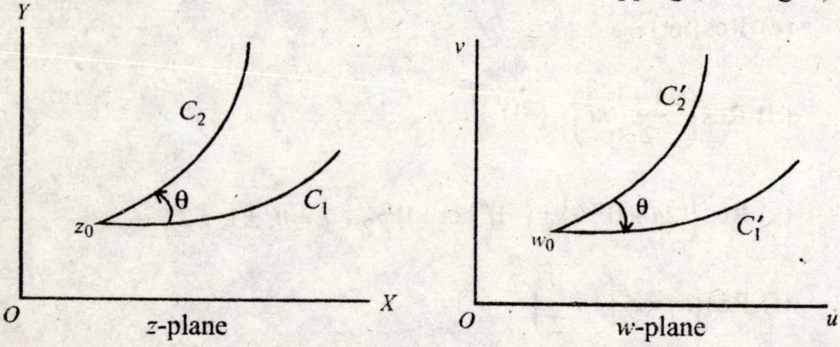

Fig. 1

Conformal Mapping. A isogonal mapping $w = f(z) = u + iv$ is said to be *conformal* if the sense of the rotation as well as the magnitude of the angle is preserved, (see adjoining Fig.2).

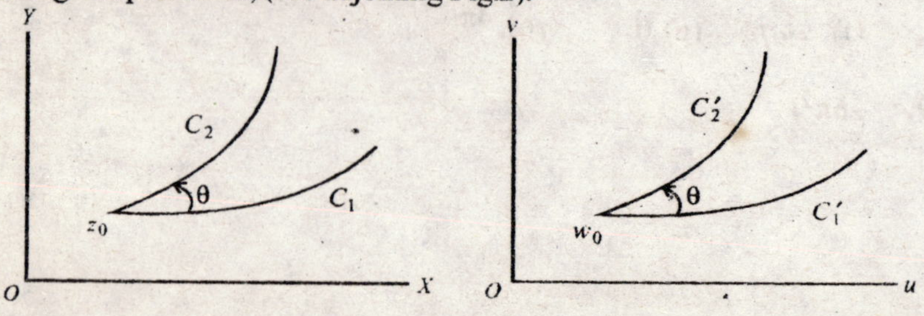

Fig. 2

Theorem 1.

If a function $f(z)$ is analytic and $f'(z) \neq 0$ in a region D of the z-plane, then the mapping $w = f(z)$ is conformal at all points of D.

Proof: Beyond the scope of the book.

Note. The points where $f'(z) = 0$ are called the ***Critical Points*** of the mapping.

Illustration. **(i)** Consider the function $f(z) = e^z$. e^z is analytic everywhere in the finite z-plane and $f'(z) \neq 0$ for all z. So the mapping $w = e^z$ is conformal everywhere in the finite z-plane.

(ii) Consider the function $f(z) = 3z^2 - 2z + 1$ which is analytic everywhere in the finite z-plane but $f'(z) = 6z - 2 = 0$ when $z = 1/3$. So the mapping $w = 3z^2 - 2z + 1$ is conformal everywhere in the finite z-plane except at $z = 1/3$. Hence $z = 1/3$ is the critical point of this mapping.

Theorem 2.

Let $w = f(z)$ be a conformal mapping at z_0. Then under this mapping the tangent at z_0 to any curve C_1 in the z-plane passing through z_0 is rotated through an angle arg $f'(z_0)$.

Proof: Let $C_1 : z(t) = x(t) + i\,y(t),\ a \le t \le b$ be a smooth curve. So $x'(t)$ and $y'(t)$ are continuous and $z'(t) = x'(t) + i\,y'(t) \neq 0$ for all $t \in (a,b)$.

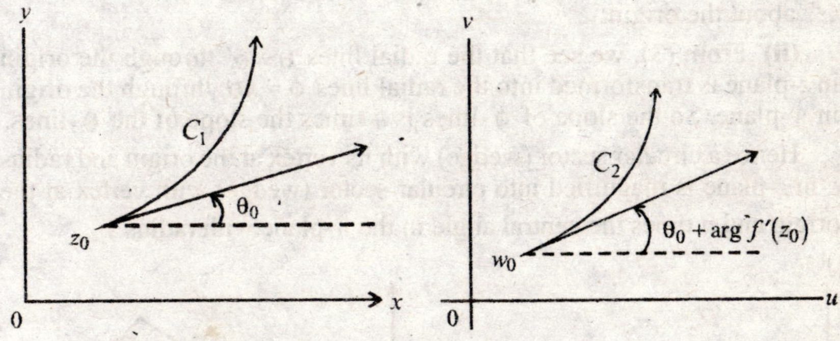

Let $f(z)$ be defined on C_1. Then the parametric representation of the image of C_1 i.e. C_2 (as shown in above fig) under the mapping $w = f(z)$ is given by $w(t) = f(z(t)) = u(t) + i\,v(t)$.

As $w = f(z)$ is a conformal mapping at z_0, so by Th-1 $f(z)$ is analytic at $z = z_0$ and $f'(z_0) \neq 0$. Hence by Chain Rule,

$w'(t_0)=f'(z_0)z'(t_0)$, $t_0 \in (a,b)$ and $z_0 = z(t_0)$.

Since $f'(z_0) \neq 0$, it follows that

$\arg w'(t_0) = \arg f'(z_0) + \arg z'(t_0)$. i.e. $\phi_0 = \theta_0 + \arg f'(z_0)$.

Where θ_0, ϕ_0 are the inclination of the directed tangents to C_1 and C_2 at z_0 and $w_0 = f(z_0)$ respectively.

Note. This angle of rotation $\arg f'(z)$ is approximately same for all points in the neighbourhood of z_0.

2.4.3. Some Special Transformation.

1. The Transformation $w = z^n$ (n being a positive integer).

Here $\frac{dw}{dz} = nz^{n-1}$. So $\frac{dw}{dz} = 0$ at $z = 0$. Thus the transformation $w = z^n$ (1)

is conformal at every point in the finite z-plane except at the point $z = 0$. (by a previous theorem)

Let $w = Re^{i\phi}$, $z = re^{i\theta}$ then (1) becomes $Re^{i\phi} = r^n e^{in\theta}$.

$\therefore R = r^n$ ... (2)

and $\phi = n\theta$ ... (3)

Using the equations (2) and (3) we may conclude the following facts about the transformation :

(i) When $r = \text{constant} = c$, say, from (2) $R = c^n$. So the circle of radius c about the origin in z-plane is mapped on the circle of radius c^n about the origin.

(ii) From (3), we see that the radial lines $\theta = \alpha$ through the origin in z-plane is transformed into the radial lines $\phi = n\alpha$ through the origin in w-plane. So the slope of ϕ-lines is n times the slope of the θ-lines.

Hence a circular sector (wedge) with its vertex at the origin and radius c in z-plane is magnified into circular sector (wedge) with vertex at the origin and n times the central angle in the w-plane with radius c^n.

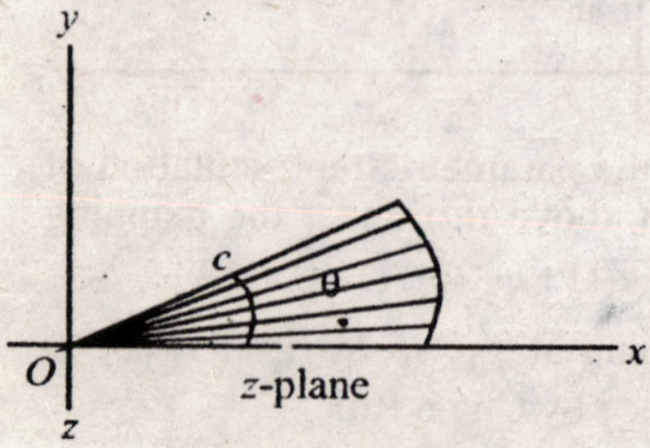

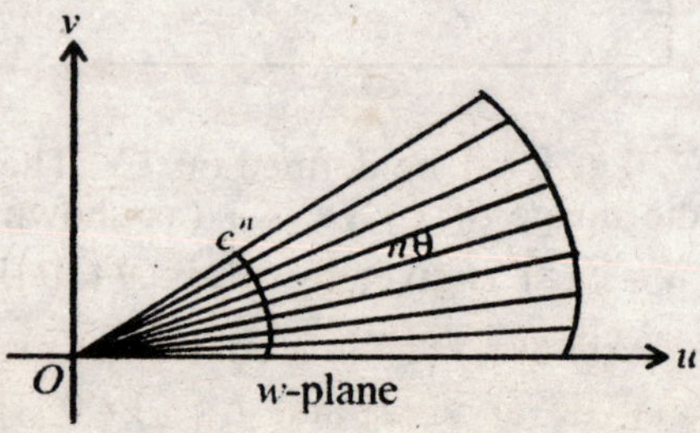

(iii) The interior of the circular sector with central angle $\frac{\pi}{n}$ and radius c is mapped conformally on to the upper half of w-plane for which Im $(w)>0$ and height c^n as shown in the following fig.

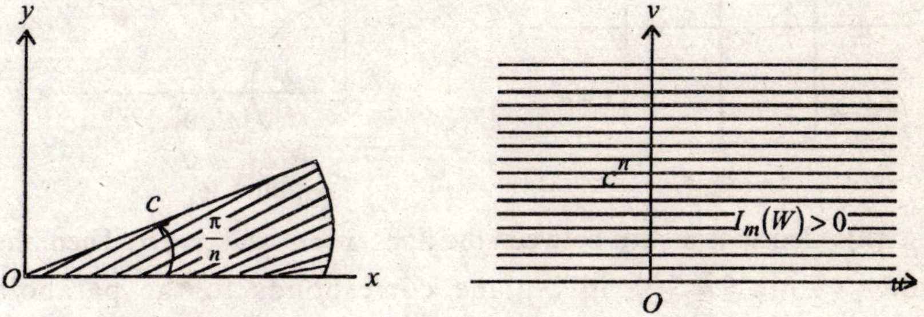

z-plane w-plane

2. The Transformation $w=z^2$.

This transformation is a particular case of the above Transformation 1 and can be discussed similarly by converting it to the polar co-ordinate. But we discuss this transformation by converting it to the cartesian co-ordinates as given below.

Here $\frac{dw}{dz}=2z$. So $\frac{dw}{dz}=0$ at $z=0$.

Thus the transformation $w=z^2$ ... (1)

is conformal everywhere except at $z=0$.

Let $w=u+iv$, $z=x+iy$. Then (1) becomes $u+iv=(x+iy)^2$.

$\therefore u=x^2-y^2$ ... (2)

and $v=2xy$ ... (3)

Using the equation (2) and (3), we may conclude the following facts about the transformation :

(i) When $x=\text{constant}=a$, from (2) and (3), we have $u=a^2-y^2$, $v=2ay$.

Eliminating y between these we get $v^2=-4a^2\left(u-a^2\right)$.

Thus the line $x=a$ in the z-plane corresponds to the parabola $v^2=-4a^2\left(u-a^2\right)$ in w-plane as shown in the Fig. Replaceing a by $-a$ in the above we see that the line $x=-a$ in z-plane corresponds to the same parabola in the w-plane.

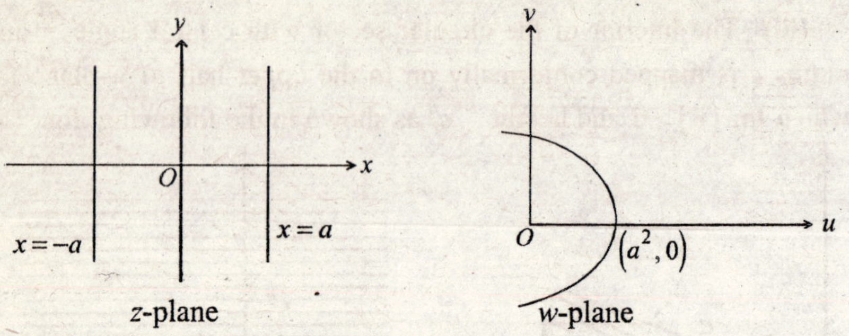

(ii) Consider a strip between the lines $x = a$ and $x = b$. Then, the line $x = \lambda(a \le \lambda \le b)$ in z-plane corresponds to the parabola $v^2 = -4\lambda^2(u - \lambda^2)$. in w-plane, as discussed in **(i)**. Hence the area described by $x = \lambda$ lying between the lines $x = a$ and $x = b$ in z-plane is transformed to the area enclosed between the parabola $v^2 = -4a^2(u - a^2)$ and $v^2 = -4b^2(u - b^2)$ in w-plane (as shown in the second figure).

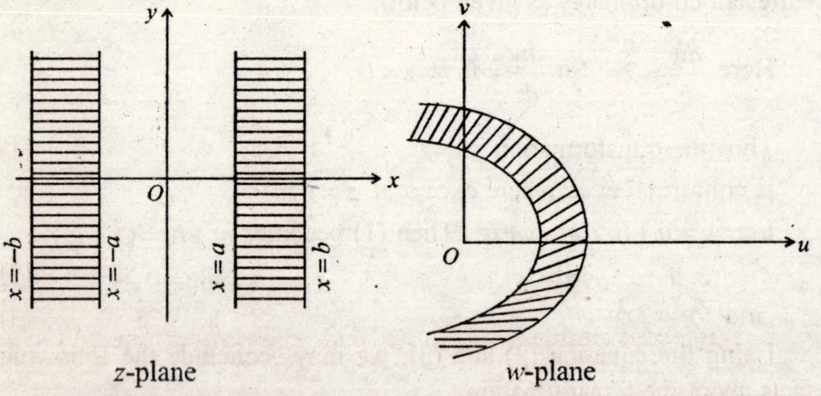

Thus the strip between $x = a$, $x = b$ in z-plane is transformed to the domain between the parabola $v^2 = -4a^2(u - a^2)$, $v^2 = -4b^2(u - b^2)$ in w-plane.

Similarly another strip between $x = -a$, $x = -b$ in z plane corresponds to the same domain enclosed between the same two parabolas.

(iii) It can be easily shown that the both domain in the z-plane to the right of the line $x = a$ and also to the left of $x = -a$ correspond to the exterior (outside) of the parabola $v^2 = -4a^2(u - a^2)$ in w-plane as shown in the 2nd fig. given below

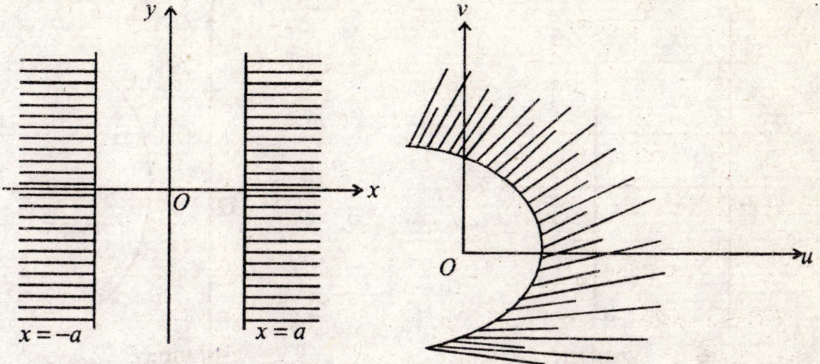

z-plane. w-plane.

(iv) Also it can be easily shown that the domain in z-plane enclosed between the lines $x = 0$ and $x = a$ corresponds to the interior of the parabola $v^2 = -4a^2(u - a^2)$ enclosed by v-axis in w-plane as shown in the 2nd fig. given below.

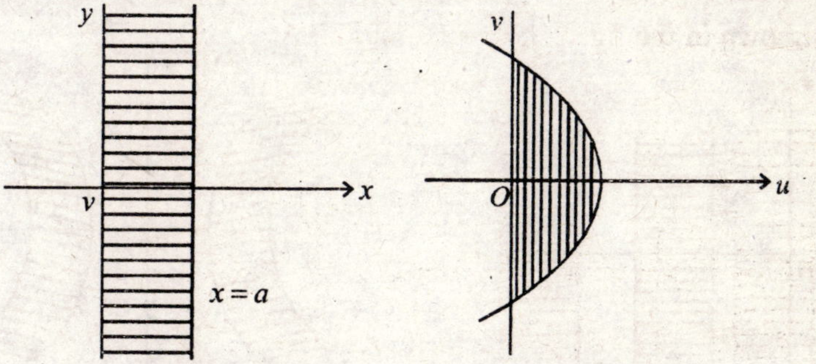

z-plane w-plane

3. The Transformation $z = \sqrt{w}$ (inverse mapping of $w = z^2$)

As (2), we have $u = x^2 - y^2$ ... (1)

$$v = 2xy \quad \text{... (2)}$$

Here we discuss the images of the lines $u = a$, $v = b$ in the w-plane as follows :

(i) when $u = a$ where $a > 0$, we have from (1)

$x^2 - y^2 = a$ which is a rectangular hyperbola. Then the line $u = a$ in w-plane is transformed to the rectangular hyperbola $x^2 - y^2 = a$ in the z-plane as shown in following fig.

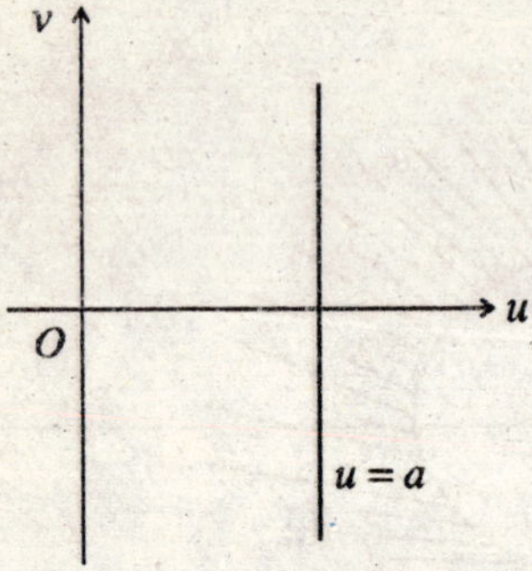

w-plane.

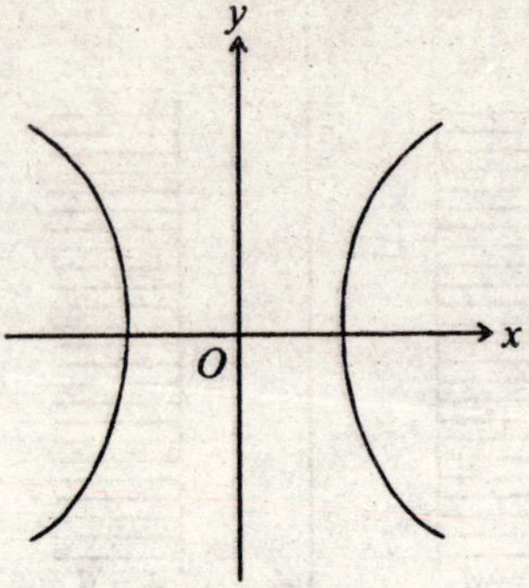

z-plane.

(ii) Consider a strip between the lines $u = a$, $u = b$ in the w-plane. As $u = a$ and $u = b$ in w-plane correspond to the rectangular hyperbolas $x^2 - y^2 = a$, $x^2 - y^2 = b$ respectively in z-plane, by (i), so the strip between $u = a$, $u = b$ in w-plane is transformed to the domain enclosed between the hyperbolas $x^2 - y^2 = a$, $x^2 - y^2 = b$ in z-plane as shown in the fig.

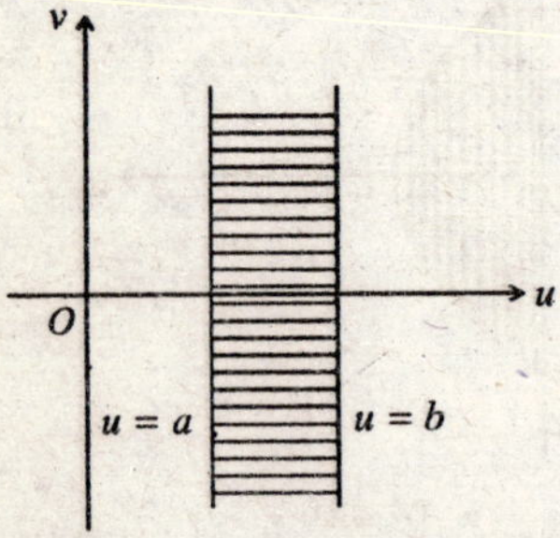

w-plane

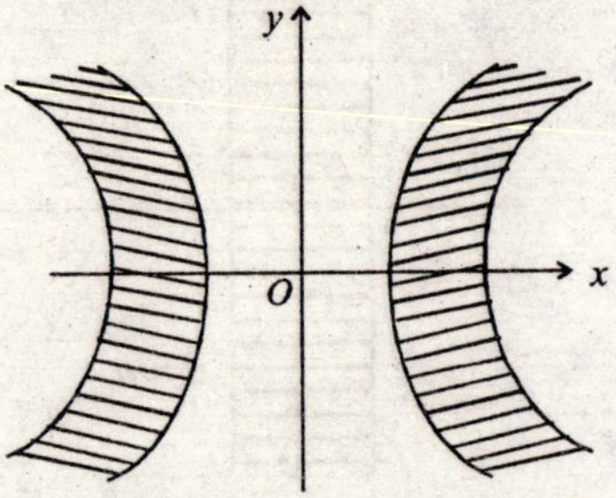

z-plane

(iii) It can be easily shown that the domain in the w-plane to the right of the line $u = a$ corresponds to the interior of the rectangular hyperbola $x^2 - y^2 = a$ in z-plane as shown in following fig.

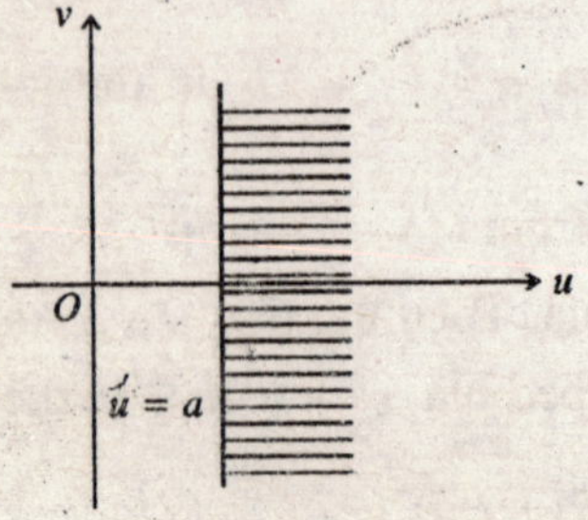

w-plane

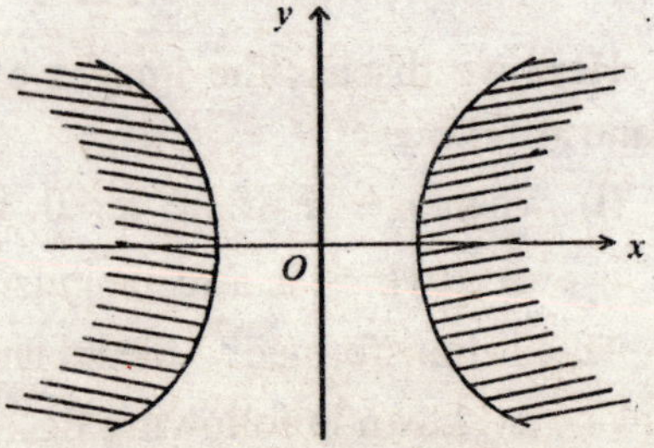

z-plane

(iv) When $v = a$, where $a > 0$, then from (2), $2xy = a$ which is rectangular hyperbola. (Shown in the follownig figure)

So the line $v = a$ in w-plane corresponds to the rectangular hyperbola $2xy = a$ in z-plane as shown in fig.

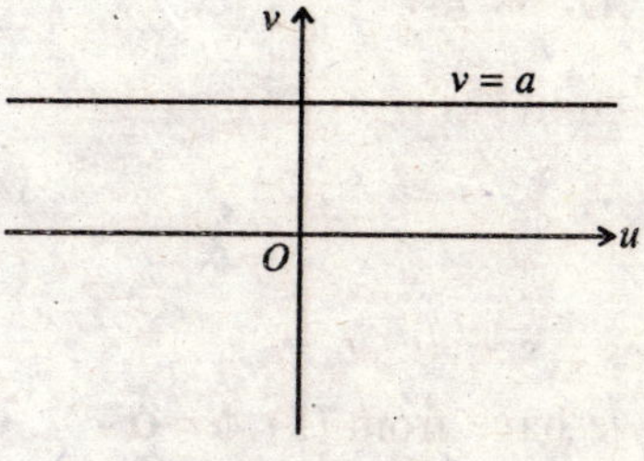

w-plane

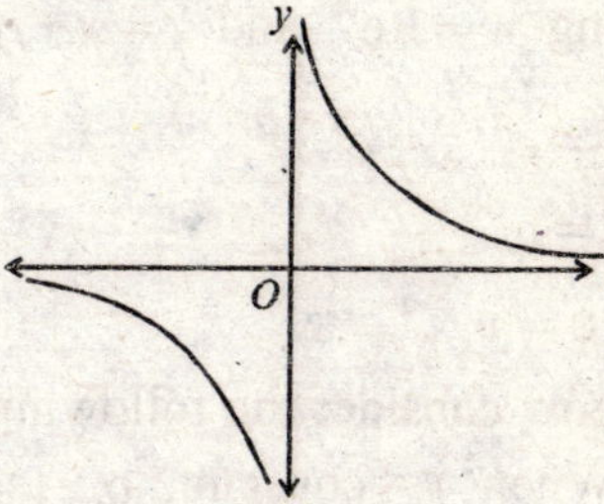

z-plane

(v) The strip between the lines $v = a$, $v = b$ in w-plane corresponds to the area between the hyperbolas $2xy = a$, $2xy = b$ in z-plane as shown in the following figure (only one branch of the hyperbola is shown) :

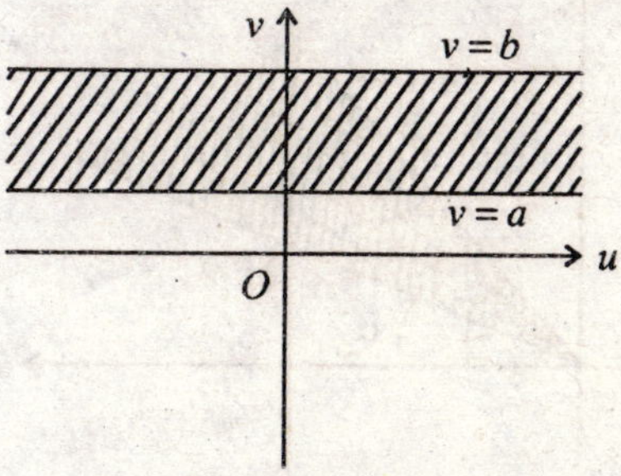

w-plane

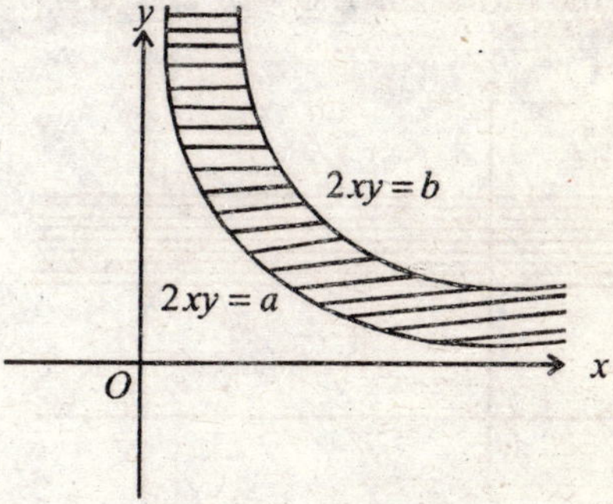

z-plane

(vi) The domain in the w-plane above the line $v = a$ corresponds to the interior of the hyperbola $2xy = a$ in z-plane as shown in fig (only one branch of the hyperbola is shown).

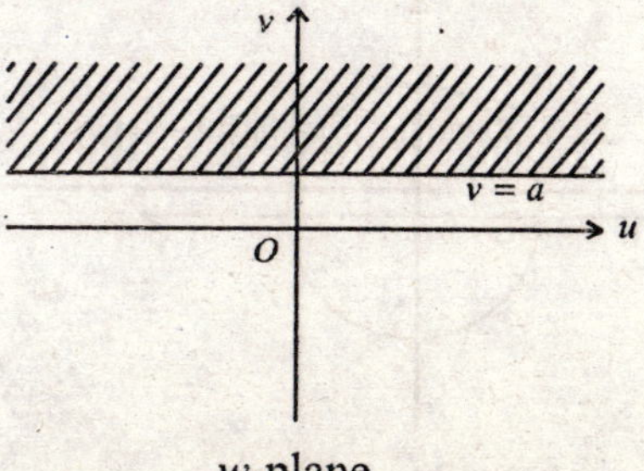

w-plane

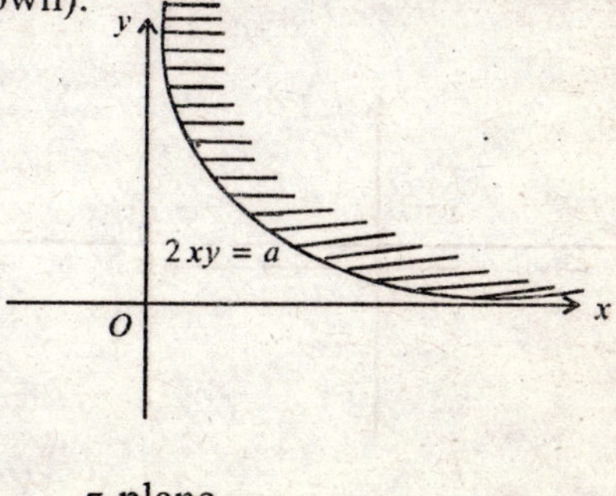

z-plane.

4. The Transformation $w = e^z$

Here $\frac{dw}{dz} = e^z \neq 0$ for all z. So the transformation $w = e^z$ ... (1)
is conformal at every point in the z-plane.

Putting $w = Re^{i\phi}$ and $z = x + iy$ in (1), we get

$Re^{i\phi} = e^{x+iy} = e^x \cdot e^{iy}$

$\therefore \; R = e^x$ ... (2)

and $\phi = y$ ... (3)

We now consider the following cases :

(i) When $y =$ constant, α, (say), we have from (3) $\phi = \alpha$.

So the line $y = \alpha$ in z-plane is transformed to the radial line $\phi = \alpha$ in w-plane. Hence the region bounded by the lines $y = \alpha$ and $y = \beta$ in z-plane is transformed to wedge shaped region in w-plane bounded by the radial lines $\phi = \alpha$ and $\phi = \beta$ as shown in the fig.

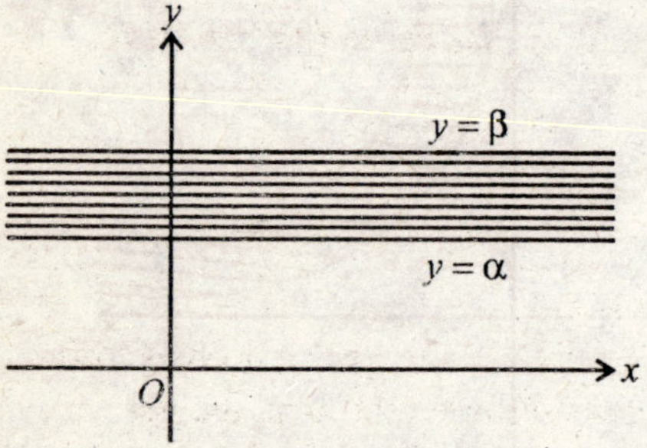

z-plane

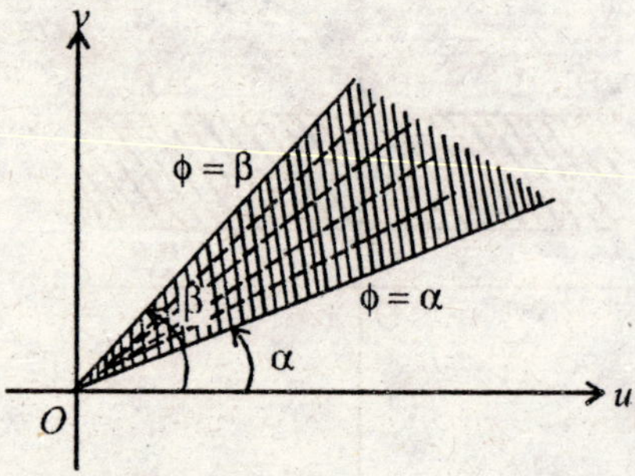

w-plane

(ii) When $x = 0$, from (2), we have $R = 1$ i.e. $|w| = 1$.

Thus the imaginary axis i.e. $x = 0$ in z-plane is transformed to the unit circle $|w| = 1$ in the w-plane.

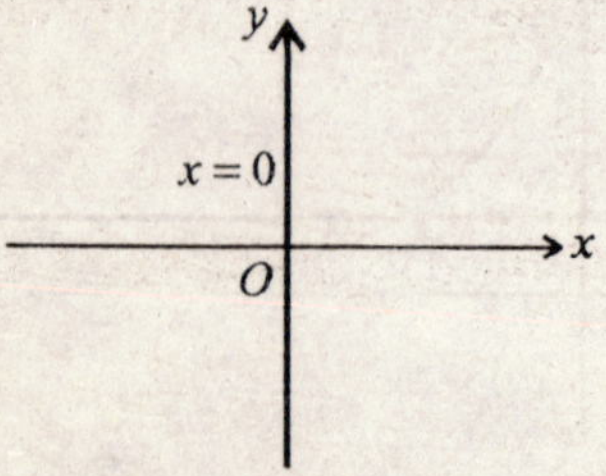

z-plane

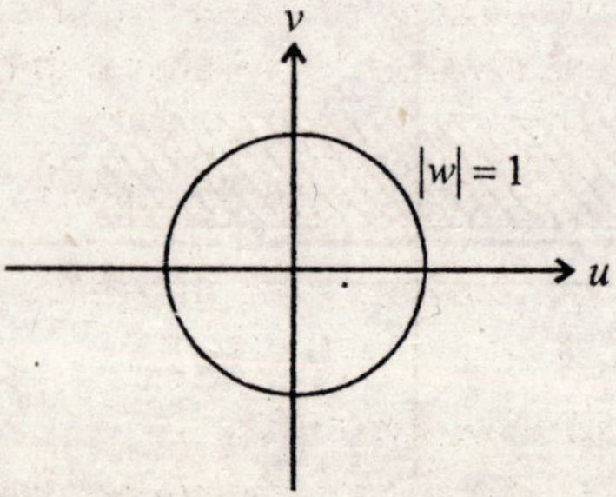

w-plane

(iii) By (i) and (ii) we conclude that the region bounded by $y=0$, $y=\pi$ and $x=0$ in *z*-plane. Corresponds to $\phi=0$, $\phi=\pi$ and $|w|=1$ in *w*-plane as shown in following fig.

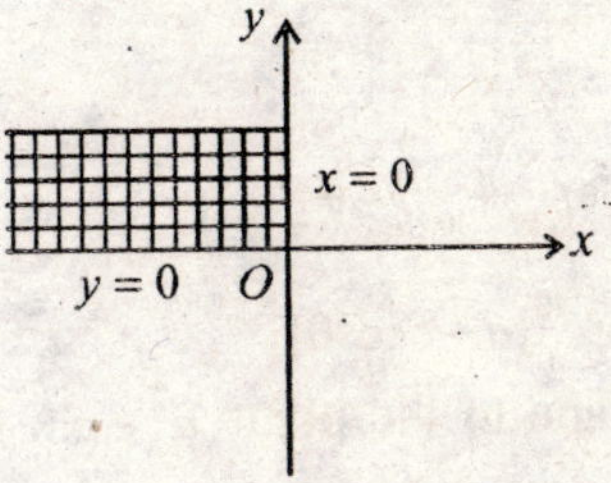

z-plane

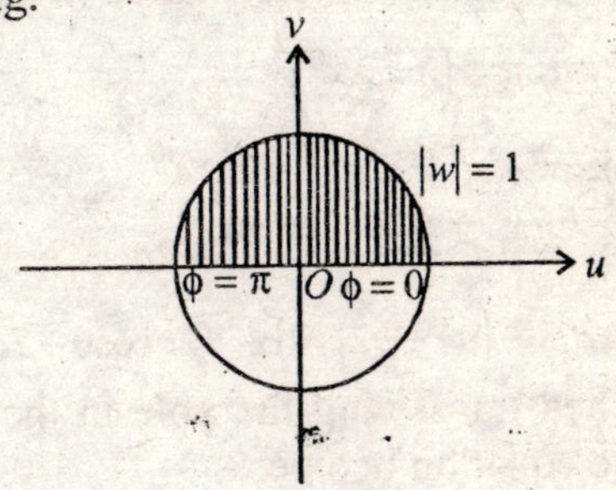

w-plane

5. The Transformation $w = \cos z$

Here $\dfrac{dw}{dz} = -\sin z = 0$ at $z=0$. So the transformation $w=\cos z$ ... (1)

is conformal everywhere except at $z=0$. From (1), we have.

$$u+iv=\cos(x+iy)=\cos x\cosh y - i\sin x\, hy$$

$$\therefore u=\cos x\cosh y \qquad \ldots (2)$$

$$v=-\sin x\sinh y \qquad \ldots (3)$$

We now consider the following cases :

(i) When $x=a$, from (2) and (3) we get $u=\cos a\cosh y$, $v=-\sin a\sinh y$. Eliminating *y* between these relations, we get

$$\frac{u^2}{\cos^2 a}-\frac{v^2}{\sin^2 a}=1 \qquad \ldots (4)$$

Thus the line $x=a$ in *z*-plane is transformed to the hyperbola (4) in w-plane.

(ii) When $y=$ constant, *a* say, then from (2) and (3) we get $u=\cos x\cosh a$, $v=-\sin x\sinh a$. Eliminating *x* between these relations we get $\dfrac{u^2}{\cosh^2 a}+\dfrac{v^2}{\sinh^2 a}=1$ ... (5)

Thus the line $y=a$ in *z*-plane is transformed to an ellipse (5) in *w*-plane.

Illustrative Examples.

Ex. 1. *Show that the circles $|z-a|=c$ (a, c being real) in z-plane correspond to the limacons in the w-plane by the transformation $w=z^2$.*

On the circle $|z-a|=c$, we have

$z-a=ce^{i\theta}$ i.e. $z=a+ce^{i\theta}$

So from the transformation $w=z^2$ we have

$$w=\left(a+ce^{i\theta}\right)^2$$

or, $w-a^2=c^2e^{2i\theta}+2ace^{i\theta}$

or, $w-a^2+c^2=c^2e^{2i\theta}+2ace^{i\theta}+c^2=ce^{i\theta}\left(ce^{i\theta}+ce^{-i\theta}+2a\right)$

or, $w-\left(a^2-c^2\right)=ce^{i\theta}(2c\cos\theta+2a)=2ce^{i\theta}(a+c\cos\theta)$.

Then transfering the pole in the w-plane to the point a^2-c^2, the above equation bcomes

$$W=2ce^{i\theta}(a+c\cos\theta)$$

Let $W=\text{Re}^{i\phi}$. Then $\text{Re}^{i\phi}=2ce^{i\theta}(a+c\cos\theta)\ =2c(a+c\cos\theta)e^{i\theta}$

$\therefore R=2c(a+c\cos\theta)$ and $\phi=\theta$.

$\therefore R=2c(a+c\cos\phi)$ which is the polar equation of limacon.

Hence the circle $|z-a|=c$ in the z-plane corresponds to the limacon $R=2c(a+c\cos\phi)$ in the w-plane.

Note. When $c=a$, then the circle $|z-a|=a$ maps into the Cardiod $R=2a^2(1+\cos\phi)$.

Ex. 2. *Find the image of the circle* $|w-1|=1$ *under the transformation* $z=\sqrt{w}$.

The transformation is $z=\sqrt{w}$ i.e. $w=z^2$

$\therefore z^2-1=w-1$

or, $(z-1)(z+1)=w-1$ or, $|z-1||z+1|=|w-1|$

$\therefore |z-1||z+1|=1$ when $|w-1|=1$.

Thus the image of the circle $|w-1|=1$ under the given transformation is $|(z-1)||(z+1)|=1$.

We shall now show that $|(z-1)||(z+1)|=1$ represents Lemniscate.

Now $|(z-1)||(z+1)|=1$

or, $\left|z^2-1\right|=1$

or, $\left|\left(x^2-y^2-1\right)+2xyi\right|=1$ $\quad [\because z=x+iy]$

$\therefore \left(x^2-y^2-1\right)^2+4x^2y^2=1$.

or, $\left(x^2-y^2\right)^2-2\left(x^2-y^2\right)+1+4x^2y^2=1$

$\therefore \left(x^2+y^2\right)^2=2\left(x^2-y^2\right)$.

Which in polar co-ordinate gives the standard form of Lemniscate $r^2=2\cos 2\theta$ [by putting $x=r\cos\theta \quad y=r\sin\theta$].

Ex. 3. *Show that the circle* $|w|=1$ *corresponds to the circle* $x^2+y^2\pm 2y-1=0$ *under the transformation* $w=\frac{1}{2}\left(z+\frac{1}{z}\right)$.

The transformation is $w=\frac{1}{2}\left(z+\frac{1}{z}\right)$.

$\therefore \ |w|=\frac{1}{2}\left|z+\frac{1}{z}\right|$.

So $|w|=1$ gives $\frac{1}{2}\left|z+\frac{1}{z}\right|=1$.

$\frac{1}{4}\left(z+\frac{1}{z}\right)^2=1$ or, $\left(z+\frac{1}{z}\right)\overline{\left(z+\frac{1}{z}\right)}=4$ or, $\left(z+\frac{1}{z}\right)\left(\bar{z}+\frac{1}{\bar{z}}\right)=4$

or, $z\bar{z}+\frac{1}{z\bar{z}}+\frac{z}{\bar{z}}+\frac{\bar{z}}{z}=4$ or, $(z\bar{z})^2+1+z^2+(\bar{z})^2=4z\bar{z}$

or, $(z\bar{z}-1)^2+(z-\bar{z})^2=0$ or, $\left(|z|^2-1\right)^2+(z-\bar{z})^2=0$

or, $\left(x^2+y^2-1\right)^2-4y^2=0$ [$\because z-\bar{z}=2iy$ and $|z|^2=x^2+y^2$]

or, $\left(x^2+y^2-1\right)\pm 2y=0$. $\therefore\ x^2+y^2\pm 2y-1=0$.

Thus the circle $|w|=1$ corresponds to the circles $x^2+y^2\pm 2y-1=0$ under the transformation $w=\frac{1}{2}\left(z+\frac{1}{z}\right)$.

Ex. 4. *Show that the transformation* $w(z+i)^2=1$ *maps the inside of the circle* $|z|=1$ *in the z-plane on the exterior of a parabola.*

The transformation is $w(z+i)^2=1$. or, $z=\frac{1}{\sqrt{w}}-i$.

Let $w=\mathrm{R}e^{i\phi}$. Then $z=\frac{1}{\sqrt{R}}e^{\frac{-i\phi}{2}}-1 \ =\frac{1}{\sqrt{R}}\cos\frac{\phi}{2}-i\left(\frac{1}{\sqrt{R}}\sin\frac{\phi}{2}+1\right)$

$\therefore |z|^2=\left(\frac{1}{\sqrt{R}}\cos\frac{\phi}{2}\right)^2+\left(\frac{1}{\sqrt{R}}\sin\frac{\phi}{2}+1\right)^2=\frac{1}{R}+\frac{2}{\sqrt{R}}\sin\frac{\phi}{2}+1$.

When $|z| \le 1$, $\frac{1}{R} + \frac{2}{\sqrt{R}} \sin\frac{\phi}{2} + 1 \le 1$.

i.e. $\frac{1}{R} + \frac{2}{\sqrt{R}} \sin\frac{\phi}{2} \le 0$ i.e. $1 + 2\sqrt{R}\sin\frac{\phi}{2} \le 0$

i.e. $-2\sqrt{R}\sin\frac{\phi}{2} \ge 1$ i.e. $4R\sin^2\frac{\phi}{2} \ge 1$ i.e. $\frac{1}{2R} \le 1 - \cos\phi$.

Thus the interior of the circle $|z| = 1$ in z-plane corresponds the exterior of the parabola $\frac{1}{2R} = 1 - \cos\phi$ in w-plane because in polar co-ordinate system the equation of a parabola is $\frac{1}{r} = 1 - \cos\theta$.

2.4.4. The Bilinear Transformation or Mobius Transformation.

A transformation of the form

$$w = \frac{az + b}{cz + d} \quad \ldots \quad (1)$$

where a, b, c, d are constants and $ad - bc \ne 0$ is called bilinear transformation. The transformation (1) can also be written as

$$cwz - az + wd - b = 0 \cdot$$

which is linear both in w and z and hence the name bilinear.

Now $\frac{dw}{dz} = \frac{ad - bc}{(cz + d)^2} \ne 0$ everywhere except at $z = \infty$ and $f(z) = \frac{az + b}{cz + d}$ is analytic except when $cz + d = 0$ i.e. $z = -d/c$. So the transformation

(i) is Conformal everywhere in the finite z-plane except at $z = -d/c$.

Illustration. The transformation $w = \frac{2z - 5}{z + 4}$ is bilinear, as $2 \cdot 4 - 1 \cdot (-5) = 13 \ne 0$. Now $\frac{dw}{dz} = \frac{13}{(z + 4)^2} \ne 0$ everywhere except at $z = \infty$ and $f(z) = \frac{2z - 5}{z + 4}$ is analytic everywhere except at $z = -4$. So the transformation is conformal everywhere in finite z-plane except at $z = -4$.

Theorem 1. Every bilinear transformation maps circles and lines into circles and lines (a line is also called a circle with infinite radius).

Proof: Beyond the scope of this book.

Fixed Point of a Bilinear Transformation.

When a point coincides with its image under a bilinear transformation, the point is called a *fixed point or an invariant point.*

Theorem 2. A bilinear transformation having exactly one fixed point is of the form $\frac{1}{w-p}=\frac{1}{z-p}+k$ ... (1)

where $k \neq 0$ and p is the fixed point.

Proof: Beyond the scope of the book.

Note. This transformaton is called *parabola transformation.*

Theorem 3. A bilinear transformation having exactly two fixed points is of the form $\frac{w-p}{w-q}=k\frac{z-p}{z-q}$ ... (2)

where $k \neq 0$, 1 and p, q are the fixed points.

Proof: Beyond the scope of the book.

Note. (i) The above transformation is called ***elliptic* if** $|k|=1$ and ***hyperbolic*** if k is real.

(ii) The transformaton (1) and (2) are known as the ***normal form*** or ***canonical form*** of a bilinear transformation.

Illustration.

(i) Consider the transformation $w=\frac{3z-4}{z-1}$.

For a fixed point, we have $w=z$

$\therefore z=\frac{3z-4}{z-1}$ or, $z^2-z=3z-4$

or, $(z-2)^2=0$ $\therefore z=2$.

So $z=2$ is the only fixed point.

Now $w-2=\frac{3z-4}{z-1}-2=\frac{z-2}{z-1}$.

$$\therefore \frac{1}{w-2}=\frac{z-1}{z-2}=\frac{z-2+1}{z-2}=\frac{1}{z-2}+1$$

which is the normal form of the transformation and is called parabolic.

(ii) Consider the transformation $w=\frac{z-1}{z+1}$.

Then for a fixed point, we have $w=z$.

$\therefore z=\frac{z-1}{z+1}$ or, $z^2+z=z-1$ or, $z^2=-1$ $\therefore z=\pm i$.

So the fixed points are i and $-i$.

Now $w-i=\dfrac{z-1}{z+1}-i=\dfrac{z-1-i(z+1)}{z+1}=\dfrac{(z-i)(1-i)}{z+1}$

and $w+i=\dfrac{z-1}{z+1}+i=\dfrac{(z-1)+i(z+1)}{z+1}=\dfrac{(z+i)(1+i)}{z+1}$

$\therefore \quad \dfrac{w-i}{w+i}=\dfrac{z-i}{z+i}\dfrac{1-i}{1+i}=k\dfrac{z-i}{z+i}$ where $k=\dfrac{1-i}{1+i}$ and is the normal form of the transformation.

Now $|k|=\left|\dfrac{1-i}{1+i}\right|=\dfrac{\sqrt{2}}{\sqrt{2}}=1$. Hence the transformation is elliptic.

Theorem. 4. The bilinear transformation which maps the poits z_1, z_2, z_3, of the z-plane into the points w_1, w_2, w_3 of the w-plane respectively is

$$\frac{(w-w_1)(w_2-w_3)}{(w-w_3)(w_2-w_1)}=\frac{(z-z_1)(z_2-z_3)}{(z-z_3)(z_2-z_1)}.$$

Proof: Left as exercise.

Illustration. The bilinear transformaton which maps the points $z=2, i, -2$ into the points $w=1, i, -1$ is given by

$$\frac{(w-1)(i+1)}{(w+1)(i-1)}=\frac{(z-2)(i+2)}{(z+2)(i-2)} \quad \text{or,} \quad \frac{w-1}{w+1}\frac{(i+1)^2}{i^2-1}=\frac{z-2}{z+2}\cdot\frac{(i+2)^2}{i^2-4}$$

$$\text{or,} \quad \frac{w-1}{w+1}(-i)=\frac{z-2}{z+2}\times\frac{3+4i}{-5} \quad \text{or,} \quad \frac{w-1}{w+1}=\frac{z-2}{z+2}\times\frac{3i-4}{-5}.$$

Using componendo and dividendo we get $w=\dfrac{3z+2i}{iz+6}$.

2.4.5. Some Special Bilinear Transformations.

Theorem 1. The bilinear transformation $w=\dfrac{az+b}{cz+d}$ where a, b, c, d are real and $ad-bc>0$, maps the upper half of the z-plane into the upper half of w-plane and conversely.

Proof: Beyond the scope of this book.

Note. When $ad-bc<0$, the upper half of the z-plane is mapped into the lower half of the w-plane.

Theorem 2. The bilinear transformation $w = e^{i\alpha}\dfrac{z - z_1}{z - \bar{z}_1}$ where α is real and imaginary part of z_1 maps the upper half of the z-plane into the interior of the unit circle in w-plane i.e. $|w| \le 1$ (as shown in following fig.) and conversely.

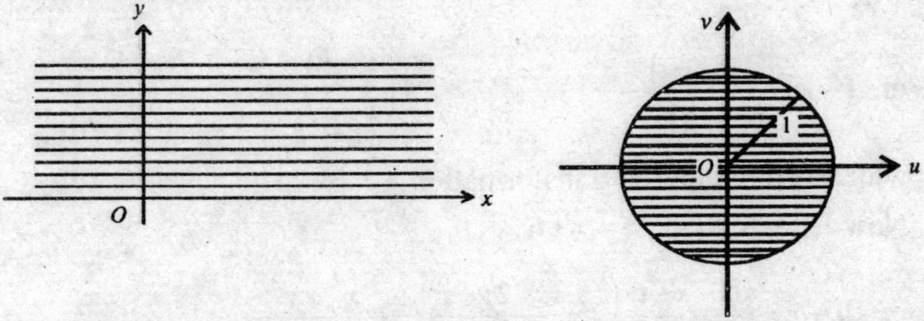

z-plane *w*-plane

Proof: Beyond the scope of this book.

Theorem 3. The bilinear transformation $w = e^{i\lambda}\dfrac{z - \alpha}{\alpha z - 1}$ where λ is real and $|\alpha| < 1$ maps the unit circle $|z| \le 1$ in z-plane into the unit circle $|w| \le 1$ in w-plane and conversely (as shown in following fig.)

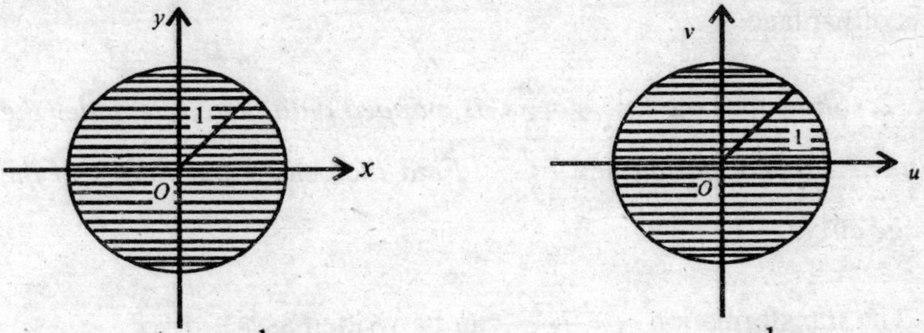

z-plane *w*-plane

Proof: Beyond the scope of this book.

Illustrative Examples.

Ex. 1. *Find the bilinear transformation which maps* $z = i, 1, -1$ *onto* $w = 1, 0, \infty$ *respectively. Also show that the unit circle* $|z| = 1$ *in z-plane maps into the real-axis of w-plane.*

Let the bilinear transformation be

$$\frac{(w - w_1)(w_2 - w_3)}{(w - w_3)(w_2 - w_1)} = \frac{(z - z_1)(z_2 - z_3)}{(z - z_3)(z_2 - z_1)} \quad \ldots \quad (1)$$

where $z_1 = i$, $z_2 = 1$, $z_3 = -1$, $w_1 = 1$, $w_2 = 0$, $w_3 = \infty$.

As $w_3=\infty$, so (1) can be written as

$$\frac{(w-w_1)\left(\frac{w_2}{w_3}-1\right)}{\left(\frac{w}{w_3}-1\right)(w_2-w_1)}=\frac{(z-z_1)(z_2-z_3)}{(z-z_3)(z_2-z_1)} \text{ i.e. } \frac{(w-1)(0-1)}{(0-1)(0-1)}=\frac{(z-i)(1+1)}{(z+1)(1-i)}$$

or, $1-w=\dfrac{2(z-i)}{(z+1)(1-i)}$ $\qquad \therefore w=i\left(\dfrac{1-z}{1+z}\right)$

which is the required transformation.

Now let $w=u+iv$, $z=x+iy$

$$\therefore \quad u+iv=i\frac{(1-x-iy)}{(1+x+iy)}=\frac{2y}{(1+x)^2+y^2}-i\frac{x^2+y^2-1}{(1+x)^2+y^2}$$

$\therefore v=-\dfrac{x^2+y^2-1}{(1+x)^2+y^2}$. Therefore when $x^2+y^2=1$ i.e. $|z|=1$, then $v=0$.

Hence the transformation maps the unit circle $|z|=1$ into the real axis of w-plane.

Ex. 2. *Show that the line* $y=\frac{1}{3}x$ *is mapped onto the circle under the bilinear transformation* $w=\dfrac{iz+2}{4z+i}$. *Find the centre and radius of the image circle.*

The transformation $w=\dfrac{iz+2}{4z+i}$ can be written as $z=\dfrac{2-iw}{4w-i}$.

Putting $z=x+iy$, $w=u+iv$, we get

$$x+iy=\frac{2-iu+v}{4u+4iv-i}=\frac{(v+2-iu)\{4u+i(4v-1)\}}{(4u)^2+(4v-1)^2}$$

$$=\frac{9u-i\left(4u^2+4v^2+7v-2\right)}{16u^2+(4v-1)^2}$$

$$\therefore \quad x=\frac{9u}{16u^2+(4v-1)^2}, \qquad y=\frac{-\left(4u^2+4v^2+7v-2\right)}{16u^2+(4v-1)^2}.$$

So the line $y=\frac{1}{3}x$ corresponds to

$$\frac{-\left(4u^2+4v^2+7v-2\right)}{16u^2+(4v-1)^2}=\frac{1}{3}\cdot\frac{9u}{16u^2+(4v-1)^2}$$

i.e. $u^2+v^2+\frac{3}{4}u+\frac{7}{4}v-\frac{1}{2}=0$.

which is the equation of a circle with centre $\left(-\frac{3}{8},-\frac{7}{8}\right)$ and radius $\frac{3\sqrt{10}}{8}$ in uv plane.

Ex. 3. *Show that the transformation* $w=\frac{z+i}{z-i}$ *maps the interior of the circle* $|w|=1$ *i.e.* $|w|\le 1$ *into the lower half plane* $I(z)\le 0$.

The transformation is $w=\frac{z+i}{z-i}$

$$\therefore\ \overline{w}=\frac{\overline{z}+\overline{i}}{\overline{z}-\overline{i}}=\frac{\overline{z}-i}{\overline{z}+i}$$

$$\therefore\ w\overline{w}-1=\frac{z+i}{z-i}\cdot\frac{\overline{z}-i}{\overline{z}+i}-1\quad=-\frac{2i(z-\overline{z})}{|z-i|^2}$$

$$=-\frac{2i(2iy)}{|z-i|^2}=\frac{4I(z)}{|z-i|^2}=\frac{4y}{|x+i(y-1)|^2}=\frac{4y}{x^2+(y-1)^2}$$

$$\therefore |w|^2-1=\frac{4y}{x^2+(y-1)^2}$$

When $|w|=1,\ y=0$
and when $|w|<1$, when $y<0$.

$\therefore |w|\le 1$ corresponds to $y\le 0$.

Hence the interior of the circle $|w|=1$ maps into the lower half plane $I(z)\le 0$ under the given transformation.

Ex. 4. *Find the bilinear transformation which transforms the circle* $|z|=1$ *onto* $|w|=1$ *and makes the points* $z=1,-1$ *correspond to* $w=1$ *and* $w=-1$ *respectively.*

As Th-3 of Art 7.6 let the transformation which transforms the circle $|z|=1$ onto $|w|=1$ be

$$w=e^{i\lambda}\frac{z-\alpha}{\bar{\alpha}z-1} \text{ where } \lambda \text{ is real and } |\alpha|<1.$$

The point $z=1$ maps onto the point $w=1$

$$\therefore \quad 1=e^{i\lambda}\frac{1-\alpha}{\bar{\alpha}-1} \qquad \dots \qquad (1)$$

Also the point $z=-1$ maps onto the point $w=-1$

$$\therefore \quad -1=e^{i\lambda}\frac{-1-\alpha}{-\bar{\alpha}-1}=e^{i\lambda}\frac{1+\alpha}{1+\bar{\alpha}} \qquad \dots \qquad (2)$$

$\therefore$ From (1) and (2), we have,

$$\frac{1+\alpha}{1+\bar{\alpha}}=-\frac{1-\alpha}{\bar{\alpha}-1} \quad \text{i.e.} \quad \frac{1+\alpha}{1+\bar{\alpha}}=\frac{1-\alpha}{1-\bar{\alpha}} \quad \therefore \quad \frac{1+\alpha}{1-\alpha}=\frac{1+\bar{\alpha}}{1-\bar{\alpha}}$$

Hence $\alpha=\bar{\alpha}$.

So the required transformation is $w=e^{i\lambda}\dfrac{z-\alpha}{\alpha z-1}$.

EXERCISES

1. Find the critical points of the following mappings

   (i) $w=e^{2z}-2iz+3$, (ii) $w=\sin z$.

2. Find the Jacobian of the following transformations :

   (i) $w=z^2$, (ii) $w=2z^2-iz+3-i$.

3. (a) Find the image of the staright line $x+y=1$ under the transformation

   (i) $w=z^2$ (ii) $w=\dfrac{1}{z}$.

   (b) Find the image of the circle $|z|=\lambda$ by the transformation

   (i) $w=z+3+4i$, (ii) $w=\sqrt{2}(1+i)z$

(c) Find the image of the circle $|z-1|=1$ in the complex plane under the mapping $w=\frac{1}{z}$.

4. Show that by means of the transformation $w=\frac{1}{z}$ the circle C given by $|z-3|=5$ is mapped into the circle $\left|w+\frac{3}{16}\right|=\frac{5}{16}$.

5. Show that the transformation $w=\left(\frac{1+z}{1-z}\right)^{2/3}$ maps the unit circle on to a wedge shaped region.

6. Show that the mapping $w=\sqrt{z}$ transforms the family of circles $|z-1|=c$ into the family of lemniscates $|w-1||w+1|=c$.

7. Show that the transformation $w=\frac{2}{\sqrt{z}}-1$ transforms the region outside the parabola $y^2=4(1-x)$ into the interior of the unit circle in w-plane.

8. Show that the transformation $w=\tan^2\left(\frac{\pi}{4}\sqrt{z}\right)$ transforms the interior of unit circle $|w|=1$ into the interior of a parabola.

9. If $w=\frac{1}{2}\left(z+\frac{1}{z}\right)$, then prove that in general circles $|z|=$ constant and lines $\arg(z)=$ constant correspond to conics with foci at $w=\pm 1$ in the w-plane.

10. Prove that the transformation $w=\sin z$ maps the families of lines $x=$ constant and $y=$ constant into two families of confocal conics.

11. Show that the transformation $w=z+\frac{a^2}{z}$ transforms circle with origin at the centre in the z-plane into co-axial concentric, confocal ellipses in w-plane.

12. Find the fixed points (invariant point) of the following transformation :

(i) $w=z^3$ (ii) $w=\frac{3z-4}{z-1}$ (iii) $w=\frac{2z+6}{z+7}$

13. (a) Find the bilinear transformation which maps the points $1,-i,2$ in z-plane to the points $0,2,-i$ in w-plane.

(b) Find the bilinear transformation that maps the points $0,1,2$ of the z-plane onto $-1,i,1$ of the w-plane.

14. Find the bilinear transformation which maps the points $1,-1,\infty$ in z-plane to the points $1+i,1-i,1$ in w-plane.

15. Find the bilinear transformation which maps $z=1,i,-1$ respectively to $w=i,0,-i$. Also find the image of $|z|\le 1$ under the transformation. [*W.B.U.Tech 2002*]

16. Show that the transformation $w=\dfrac{5-4z}{4z-2}$ transforms the circle $|z|=1$ into a circle of radius unity in w-plane.

17. Find the bilinear transformation which transforms the points $z=1,-i,-1$ into the points $w=0,i,\infty$ and show further that the area of the circle $|z|=1$ is transformed into the half plane above the real axis in w-plane.

18. Show that the transformation $w=\dfrac{2z+3}{z-4}$ maps the circle $x^2+y^2-4x=0$ onto the straight line $4u+3=0$.

19. Find the image of the line $y=x+1$ under the transformation $w=\dfrac{z+i}{z-1}$.

20. Find the bilinear transformation which maps the points $z=1,0,-1$ into the points $w=i,0,-i$ and find the fixed points of the transformation. [*W.B.U.Tech 2003*]

21. Find the fixed points of the bilinear transformation $w=\dfrac{13iz+75}{3z-5i}$. Hence show that the transformation is elliptic.

22. Determine all bilinear transformation which have fixed points $-i$ and i.

ANSWERS

1. (i) $\left(n+\dfrac{1}{2}\right)\pi,\ n\in z,$ (ii) $i\left(n\pi+\dfrac{\pi}{4}\right),\ n\in z.$

2. (i) $4\left(x^2+y^2\right)$ (ii) $(4z-i)^2$.

3. (a) (i) $u^2+2v=1$, (ii) $u^2+2uv+2v^2=u+v$

(b) (i) $(u-3)^2+(v-4)^2=\lambda^2$ (ii) $u^2+v^2=4\lambda^2$ (c) $u=\frac{1}{2}$

12. (i) $z=0,1,-1$ (ii) $z=2$ (iii) $z=-6,1$

13. (a) $w=2\frac{z-1}{(1+i)z-2}$, (b) $w=\frac{z+i-1}{z-i+1}$

15. $w=\frac{i-z}{i+z}$ Right half of w-plane

19. $u^2+v^2-2u+v+1=0$, **20.** $w=zi,0$

21. $-4+3i, 4+3i, k=\frac{-1}{5}(4+3i)$ **22.** $w=\frac{9z+b}{-bz+a}, a^2+b^2\neq 0$.

Module -III
Probability

3.1 FUNDAMENTAL OF THE PROBABILITY THEORY

3.1.1. Introduction : In Science and Technology we have to be concerned with every phenomena whose future behaviour is not predictable in a deterministic fashion. We have to depend on 'Chance' in every field. In theory of probability we are very much concerned with 'Chance'. In fact 'Probability' is nothing but a numerical measurement of this 'Chance'. Future can be guessed if this measurement is deduced. In this chapter we are going to deal with this measurement 'Probability'. Conception on Set-theory and Combinatiories Theory are the only prerequisistes for this chapter.

3.1.2. Introductory Definitions :

Random Experiment : An experiment or observation which may be repeated a large number of times under very nearly identical conditions and the possible outcome of any particular observation is unpredictable but all possible outcomes can be described prior to its performance, is known as Random Experiment.

For example, the experiment of tossing a coin is a random experiment, as the possible outcomes are 'tails' or 'heads' but the outcome of a particular tossing cannot be predicted.

Sample Points/ Event Point. The outcomes of a random experiment are called sample points or event points.

For example, the sample points in the experiment 'tossing a coin' are Head and Tail, in symbol H and T.

Sample Space/Event Space. The set of all sample points i.e. the set of all possible outcomes of a Random Experiment is called the sample space. It is denoted by S.

For example if we throw two coins once, then $S = \{HH, HT, TH, TT\}$; if we roll a die once, we have $S = \{1, 2, 3, 4, 5, 6\}$ as the event space.

Event. Any subset of the sample space S of a random experiment is called an Event.

For example, in the experiment of 'throwing two coins', $A = \{TH, HT\}$ is an event because $A \subset S$.

Certain Event. Since every set is a subset of itself, so the sample space is a subset of itself. So this is an event. This event is called a certain event.

Impossible Event. An event that contains no sample points is called impossible event. It is denoted by ϕ. For example in the experiment 'throwing a die' the event 'Face 7' $=\phi$.

Complementary Event. For any event A, there is an event containing all the sample points in the sample space which are not in A. This event is called the complementary event of A and is denoted by A' or $\overline{A}$ or A^c. Obviously $A'=$ 'Not A'.

For example, if $A=\{TH, HT\}$ where $S=\{HH, TT, TH, HT\}$, then $\overline{A}=\{HH, TT\}$.

Note. $\overline{S}=\phi$; $\overline{\phi}=S$; $\overline{(\overline{A})}=A$

Simultaneous Occurence of two Events.

Let A_1 and A_2 be two events. Then the set $A_1 \cap A_2$ represents the simultaneous occurence of the two events A_1 and A_2. This event is also denoted by $A_1 A_2$.

For example, in the experiment 'rolling a die' let $A_1=$ 'Even face' $A_2=$ 'Multiple of three'. Then $A_1 \cap A_2=\{6\}$ is the event whose occurence shows the simultaneous occurence of A_1 and A_2.

At least one of Two Events.

Let A_1 and A_2 be two events. Then the set $A_1 \cup A_2$ represents 'at least one of A_1 and A_2'. This event is also denoted by A_1+A_2.

For example, in the experiment rolling a die let $A_1=$ Even face $=\{2,4,6\}$, $A_2=$ Multiple of three $=\{3,6\}$. Then $A_1 \cup A_2 =\{2,4,6,3\}$ is the event whose occurence shows the occurence of at least one of 'even face' and 'Multiple of 3'.

Disjoint or Mutually Exclusive (m.e) Events. If two events A_1, A_2 have no common sample points i.e. if $A_1 \cap A_2=\phi$, they are called Mutually Exclusive Events.

For example, in a previous example if $A_1=\{HH, TT\}$ $A_2=\{HT, TH\}$, then $A_1 \cap A_2=\phi$. So, A_1 and A_2 are mutually exclusive events. Two m.e events cannot occur simultaneously.

Pairwise Disjoint Events. Let $A_1, A_2, \cdots, A_n$ be n number of events. Events $A_i (i=1,2,\cdots,n)$ are said to be pairwise disjoint if no two of them have any common event points i.e. if $A_i \cap A_j=\phi, i \neq j$ and $i, j=1,2,\cdots,n$.

Exhaustive Events. Two or more events are said to be exhaustive if at least one of them neccessarily occurs or in other words the events $A_1, A_2, \cdots \cdots$ are exhaustive if $A_1 \cup A_2 \cup A_3 \cup \cdots \cdots = S$.

For example, in the experiment of throwing two coins once, the events $A_1 = \{HH\}$, $A_2 = \{TT\}$ and $A_3 = \{HT, TH\}$ are exhaustive.

Elementary or Simple Event. An event containing exactly one sample point is called Elementary Event.

For example, $A_1 = \{2\}$, $A_2 = \{5\}$, $A_3 = \{3\}$ etc are simple event of an experiment of rolling a die.

Composite Event. The event which can be decomposed into simple events i.e. which can be expressed as union of two or more simple events is called Composite Event.

For example, in the experiment of 'rolling a die' $A_1 = \{2, 3, 4\}$, $A_2 = \{1, 5\}$ etc are composite events.

Equally Likely Sample Points. The sample points of a sample space are said to be equally likely if one of them may not be expected rather than the other.

3.1.3. Classical Definition of Probability.

Let us suppose that a random experiment E is such that its sample space S contains a finite number $n(S)$ of sample points, all of which are equally likely. Then the probability of an event A which contains $n(A)$ sample points, is defined by

$$P(A) = \frac{n(A)}{n(S)}. \qquad \ldots (1)$$

Illustration. A perfect die is rolled once and observed whether an odd number appears. If A denotes this event, then $A = \{1, 3, 5\}$ and the sample space $S = \{1, 2, 3, 4, 5, 6\}$

$$\therefore\ n(A) = 3 \text{ and } n(S) = 6$$

$$\therefore\ P(A) = \frac{n(A)}{n(S)} = \frac{3}{6} = \frac{1}{2}$$

This tells that chance of occuring 'odd face' is one per two throws.

Criticism of Classical Definition.

(i) To define the probability, firstly we presumes that the sample points are equally likely which means equally probable i.e., probability of each sample point is same. So this definition is circular.

(ii) This defintion does not provide any criterion of deciding whether the possible outcomes of an experiment are equally likely.

(iii) In many experiments, the no. of possible outcomes is infinite and so this defintion is not suitable in those cases.

(iv) This definition can be used only in very simple and unimportant cases like games of chance.

(v) In some complicated problems, the calculation of possible outcomes and favourable cases are difficult, for example the sex of a newly born child and the throw of an untrue coin.

3.1.4. Theorems on Probability.

Some important properties of Probability are presented as the following theorems. Proofs are given considering the Frequency definition.

Theorem 1. $0 \leq P(A) \leq 1$ for any event A.

Proof : Let a random experiment E be repeated n times under identical conditions and A be an event which occurs $n(A)$ times. Then we have

$$0 \leq n(A) \leq n \text{ or, } 0 \leq \frac{n(A)}{n} \leq 1$$

$$\therefore \; 0 \leq P(A) \leq 1.$$

Theorem 2. $P(S) = 1$ and $P(\phi) = 0$ where S is certain event, ϕ is impossible event.

Proof: Now $P(S) = \frac{n(S)}{n} = \frac{n}{n} = 1$

$$P(\phi) = \frac{n(\phi)}{n} = \frac{0}{n} = 0.$$

Theorem 3. If A_1 and A_2 be two mutually exclusive events, then $P(A_1 \cup A_2) = P(A_1) + P(A_2)$.

Proof : Let a random experiment E be repeated n times under identical condition and A_1, A_2 be two events which occurs $n(A_1)$ and $n(A_2)$ times. Since A_1 and A_2 are mutually exclusive, so $n(A_1 \cup A_2) = n(A_1) + n(A_2)$

$$\therefore \; \frac{n(A_1 \cup A_2)}{n} = \frac{n(A_1)}{n} + \frac{n(A_2)}{n}$$

$$\therefore \; P(A_1 \cup A_2) = P(A_1) + P(A_2)$$

Theorem 4. If the events $A_1, A_2, \cdots, A_n$ are mutually exclusive then $P(A_1 \cup A_2 \cup \cdots \cup A_n) = P(A_1) + P(A_2) + \cdots + P(A_n)$

i.e., $P\left(\bigcup_{i=1}^{n} A_i\right) = \sum_{i=1}^{n} P(A_i)$

Proof: Left as an exercise.

(Proceeding as Th. 3 and using induction theorem can be proved)

Theorem 5. (Total Probability Theorem). For any two events A_1 and A_2 (may not be mutually exclusive),

$P(A_1 \cup A_2) = P(A_1) + P(A_2) - P(A_1 \cap A_2)$ [*W.B.U.Tech 2007*]

Proof: $P(A_1 \cup A_2) = P\left(A_1 \cup \left(\overline{A}_1 \cap A_2\right)\right)$, when $\overline{A}$ is complement of A.

$= P(A_1) + P(\overline{A}_1 \cap A_2)$, since A_1 and $\overline{A}_1 \cap A_2$ are m.e

Again, $A_2 = (A_1 \cap A_2) \cup (\overline{A}_1 \cap A_2)$

$\therefore \; P(A_2) = P\left((A_1 \cap A_2) \cup (\overline{A}_1 \cap A_2)\right)$

$= P(A_1 \cap A_2) + P(\overline{A}_1 \cap A_2)$ $\quad\because A_1 \cap A_2$ and $\overline{A}_1 \cap A_2$ are m.e

or, $P(\overline{A}_1 \cap A_2) = P(A_2) - P(A_1 \cap A_2)$

Therefore, from above

$P(A_1 \cup A_2) = P(A_1) + P(A_2) - P(A_1 \cap A_2)$

Theorem 6. If A_1, A_2, A_3 are any three events (not necessarily m.e.) then

$$P(A_1 \cup A_2 \cup A_3) = P(A_1) + P(A_2) + P(A_3) - P(A_1 \cap A_2)$$
$$-P(A_2 \cap A_3) - P(A_3 \cap A_1) + P(A_1 \cap A_2 \cap A_3)$$

Proof: Left as an exercise.

Theorem 7. If $A_1, A_2, A_3, \cdots \cdots, A_n$ are any n events then

$$P(A_1 \cup A_2 \cup A_3 \cup \cdots \cup A_n) = \sum P(A_1) - \sum P(A_1)P(A_2)$$
$$+\sum P(A_1)P(A_2)P(A_3) - \cdots \cdots + (-1)^n P(A_1)P(A_2)\cdots P(A_n)$$

Proof: Omitted.

Theorem 8. For any event A, $P(\overline{A}) = 1 - P(A)$ where $\overline{A}$ is the complementary event of A.

Proof: We have A and $\overline{A}$ are mutually exclusive events and $A \cup \overline{A} = S$, the certain event.

$\therefore\ P(A \cup \overline{A}) = P(S)$

or, $P(A) + P(\overline{A}) = 1$, since A and $\overline{A}$ are mutually exclusive.

$\therefore\ P(\overline{A}) = 1 - P(A)$.

Theoem 9. If $A_1, A_2, \cdots, A_n$ are mutually exclusive and exhaustive events, then $\sum_{i=1}^{n} P(A_i) = 1$.

Proof : Since the events are exhaustive, so $\bigcup_{i=1}^{n} A_i = S$, the certain event.

Hence by Theorem (4) we have

$$P\left(\bigcup_{i=1}^{n} A_i\right) = \sum_{i=1}^{n} P(A_i)$$

i.e., $\sum_{i=1}^{n} P(A_i) = P(S)$

$\therefore\ \sum_{i=1}^{n} P(A_i) = 1$, by Theorem 2.

Theorem 10. For any two events A_1 and A_2 where $A_1 \subset A_2$

(i) $P(A_1) \le P(A_2)$ (ii) $P(A_2 - A_1) = P(A_2) - P(A_1)$

Proof:

(i) $A_1 \subset A_2 \Rightarrow n(A_1) \le n(A_2) \Rightarrow \dfrac{n(A_1)}{n} \le \dfrac{n(A_2)}{n} \Rightarrow P(A_1) \le P(A_2)$

(ii) $A_1 \subset A_2 \Rightarrow A_2 = A_1 \cup (A_2 - A_1)$, where $A_1, A_2 - A_1$ become m.e.

$\therefore\ P(A_2) = P(A_1) + P(A_2 - A_1)\ \therefore\ P(A_2 - A_1) = P(A_2) - P(A_1)$

3.1.5. Axiomatic definition of probability. [*W.B.U.Tech 2005*]

Let E be a random experiment and S be its sample space ; Σ be the class of all events (i.e., subsets of S). Let P be a function from Σ to the set of all real numbers satisfying the following axioms :

Axiom I. $P(A) \ge 0$, for every event A in Σ

Axiom II. $P(S) = 1$.

Axiom III. If $A_1, A_2, \cdots$ be a finite or infinite sequence of pairwise mutually exclusive events, then

$P(A_1 \cup A_2 \cup \cdots) = P(A_1) + P(A_2) + \cdots$

Then for any event A the real number $P(A)$ is called its probability.

Illustration. Let a biased coin be tossed. The event space $S=\{H,T\}$.

It has four events $\{H\},\{T\}$ ϕ and S itself.

We define P such that $P(H)=1/3, P(T)=2/3,$ $P(\phi)=0,$ $P(\{H,T\})=1$

Then we see $P(A)\geq 0$ for all event A i.e., Axiom I is satisfied. $P(S)=P(\{H,T\})=1$ i.e., Axiom II is satisfied. Consider the two m.e. events $\{H\}$ and $\{T\}$.

Then $\{H\}\cup\{T\}=\{H,T\}$

$$\therefore\ P(\{H\}\cup\{T\})=P(\{H,T\})=1=\frac{1}{3}+\frac{2}{3}=P(\{H\})+P(\{T\}),$$

i.e., Axiom III is satisfied.

Thus this function P represents probability. In particular we can say probability of Head is $\frac{1}{3}$.

All the Theorems which were deduced for the Frequency definition of probability can be deduced for Axiomatic definition also. Some of thems are shown in the next page.

Some Deductions from the Axioms :

(i) ***For any*** $A\subset S,\ P(A)\leq 1$

Let $\overline{A}=S-A$. Then $A,\overline{A}$ are mutually exclusive and $A\cup\overline{A}=S$.

$\therefore\ P(A\cup\overline{A})=P(S)$

or, $P(A)+P(\overline{A})=1$, by Axiom II and III

or, $P(A)=1-P(\overline{A})$

... (1)

Again by axiom I, $P(\overline{A})\geq 0$

Hence, $P(A)=1-P(\overline{A})\leq 1$

(ii) $P(\phi)=0$

Since $\overline{\phi}=S$. So from (1)

$P(\phi)=1-P(\overline{\phi})=1-P(S)=1-1$

by Axiom (II)

$=0$

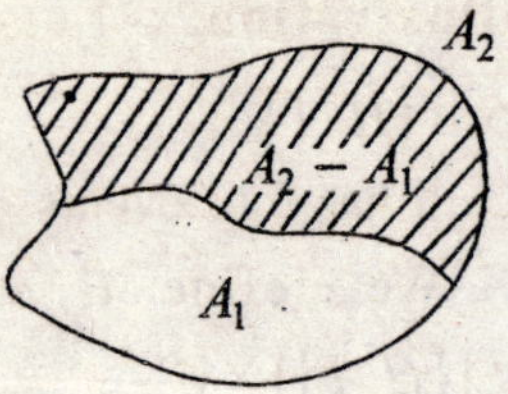

(iii) For any two events $A_1, A_2 \subset S$ and $A_1 \subset A_2$. We have $P(A_1) \le P(A_2)$

and $P(A_2 - A_1) = P(A_2) - P(A_1)$

Since $A_1 \subset A_2, A_2 = A_1 \cup (A_2 - A_1)$.

Also since A_1 and $A_2 - A_1$ are mutually exclusive, so by Axiom III, we have,

$$P(A_2) = P(A_1) + P(A_2 - A_1)$$

$$\therefore P(A_2 - A_1) = P(A_2) - P(A_1)$$

By Axiom I, $P(A_2 - A_1) \ge 0 \quad \therefore P(A_2) - P(A_1) \ge 0$

i.e., $P(A_2) \ge P(A_1)$.

(iv) (Boole's inequality)

For n events $A_1, A_2, \cdots, A_n$

$$P(A_1 \cup A_2 \cup \cdots \cup A_n) \le P(A_1) + P(A_2) + \cdots + P(A_n)$$

i.e., $P(\bigcup_{i=1}^{n} A_i) \le \sum_{i=1}^{n} P(A_i)$

Proof: For $n = 1$, the theorem is obviously true. We have

$$P(A_1 \cup A_2) = P(A_1) + P(A_2) - P(A_1 \cap A_2)$$

$$\therefore P(A_1 \cup A_2) \le P(A_1) + P(A_2) \quad [\because P(A_1 \cap A_2) \ge 0] \quad \ldots (1)$$

Thus the theorem is true for $n = 2$.

Let the theorem holds for $n = m$

$$\therefore P\left(\bigcup_{i=1}^{m} A_i\right) \le \sum_{i=1}^{m} P(A_i) \quad \ldots (2)$$

Now $P\left(\left(\bigcup_{i=1}^{m} A_i\right) \cup A_{m+1}\right) = P\left(\bigcup_{i=1}^{m} A_i\right) + P(A_{m+1}) - P\left(\left(\bigcup_{i=1}^{m} A_i\right) \cap A_{m+1}\right)$

or, $P\left(\bigcup_{i=1}^{m+1} A_i\right) \le P\left(\bigcup_{i=1}^{m} A_i\right) + P(A_{m+1})$, by Axiom I,

$\le \sum_{i=1}^{m} P(A_i) + P(A_{m+1})$, since $P\left(\left(\bigcup_{i=1}^{m} A_i\right) \cap A_{m+1}\right) \ge 0$

$$=\sum_{i=1}^{m+1} P(A_i)$$

$$\therefore\ P\left(\bigcup_{i=1}^{m+1} A_i\right) \le \sum_{i=1}^{m+1} P(A_i).$$

Thus the theorem holds for $n=m+1$ whenever it is hold for $n=m$. But the theorem holds for $n=1,2$. Hence by induction, the theorem is true for any positive integral values of n.

(v) For any events $A_1, A_2, \cdots, A_n$,

$$P(A_1 \cap A_2 \cap \cdots \cap A_n) \ge 1-\sum_{i=1}^{n} P(\overline{A}_i)$$

Proof. By Boole's inequality we have

$$P(\overline{A}_1+\overline{A}_2+\cdots+\overline{A}_n) \le P(\overline{A}_1)+P(\overline{A}_2)+\cdots+P(\overline{A}_n)$$

or, $P\left(\overline{A_1 \cap A_2 \cap \cdots \cap A_n}\right) \le \sum_{i=1}^{n} P(\overline{A}_i)$, by De Morgan's law.

or, $1-P(A_1 \cap A_2 \cap \cdots \cap A_n) \le \sum_{i=1}^{n} P(\overline{A}_i)$

[$\because$ for any event A, $P(\overline{A})=1-P(A)$]

$$\therefore\ P(A_1 \cap A_2 \cap \cdots \cap A_n) \ge 1-\sum_{i=1}^{n} P(\overline{A}_i)$$

3.1.6. Techniques of Counting.

Fundamental Principle of Counting

If an operation can be performed in n_1 different ways and corresponding to each way, a second operation can be performed in n_2 different ways, then both the operations can be performed in $n_1 \times n_2$ different ways.

The rule can be extended to any finite number of operations.

Illustrations. (i) If a man has 5 shirts and 4 ties, then he has $5\times 4=20$ ways of choosing a shirt and a tie.

(ii) If there are 5, 7 and 10 men in three classes then a committee of three, one from each class, can be formed in $5\times 7\times 10=350$ ways.

Permutation. Any ordered arrangement of r symbols which can be made by taking $r(\leq n)$ symbols from a set of n symbols is called a permutation of n symbols taken r at a time. The number of such permutations is denoted by $P(n,r)$ or ${}^nP_r$.

Illustration. Consider a set of three letters *a, b, c*. Then *ab, ba, bc, cb, ca, ac* are permutations of the 3 letters taken 2 at a time and ${}^3P_2 = 6$

Theorems (i) $${}^nP_r = n(n-1)(n-2)\cdots(n-r+1) = \frac{n!}{(n-r)!}$$

(ii) $${}^nP_n = n(n-1)\cdots 2\cdot 1 = n!$$

Proof: Left for students

Note : From (ii), we have the total no. of permutations of n symbols taken all at a time is $n!$.

Example 1. How many different numbers can be formed by using 5 out of the nine digits 1, 2, ..., 9 ?

We have 9 distinct digits and we have to find the number of permutation taken 5 at a time. So the required number is ${}^9P_5 = 9\times 8\times 7\times 6\times 5 = 15,120$.

Example 2. In how many ways can a party of n persons arrange themselves (i) in a row ? (ii) in a circle ?

(i) Arrangement of n persons in a row = permutations of n persons taken all at a time. So the required no. of ways of arrangement in a circle is ${}^nP_n = n!$.

(ii) To arrange n persons in a circle, one person must be fixed at any position on the circle. Then other $(n-1)$ persons arranged themselves in $(n-1)!$ ways. So the required no. of ways is $(n-1)!$.

Theorem 1. **(i)** The number of permutation of n objects of which n_1 are indistinguishable from each other is $\frac{n!}{n_1!}$

(ii) The number of permutations of n objects of which n_1 are in distinguishable, another type of n_2 are indistinguishable $\cdots$ n_k are indistinguishable is $\frac{n!}{n_1!n_2!\cdots n_k!}$.

Proof: Left for students.

Example : In how many ways can 12 students take Physics, Mathematics and Chemistry if 4 students are to take each subject ?

Here we have to find the number of ordered partitions of 12 students into three groups containing 4 students. So this number is

$$\frac{12!}{4!4!\ 4!} = 34,650 .$$

Combinations. Any selection of r symbols i.e. any subset of r symbols which can be made from a set of n symbols is called a combination of n symbols taken r at a time. The number of such combinations is denoted by $C(n,r)$ or ${}^nC_r$ or $\binom{n}{r}$.

Illustration. Consider a set of three numbers 1, 2, 3.

Then 12, 23, 31 are combinations of three numbers taken two at a time and ${}^3C_2 = 3$.

Theorem 2. ${}^nC_r = \dfrac{n!}{r!(n-r)!} = \dfrac{{}^nP_r}{r!}$

Proof : Left for student.

Note. (1) ${}^nC_r = 0$ for integers $r > n$

${}^nC_0 = {}^nC_n = 1$

(2) ${}^nC_r = {}^nC_{n-r}$

Illustrative Example.

Ex. 1. *In how many ways 5 cards can be selected from an ordinary pack of 52 cards.*

The no. of ways of selecting 5 cards from 52 different cards is

$${}^{52}C_5 = \frac{52!}{5!\ 47!}.$$

Ex. 2. *In how many ways can a committee consisting of 5 men and 3 women be chosen from 8 men and 6 women ?*

The 5 men can be selected from 8 men in $\binom{8}{5}$ ways and 3 women can also be selected from 6 women in $\binom{6}{3}$ ways. Hence the required no. of committee is

$$\binom{8}{5}\times\binom{6}{3} = \frac{8!}{5!\ 3!}\times\frac{6!}{3!\ 3!} = \frac{8\cdot7\cdot6}{3\cdot2\cdot1}\times\frac{6\cdot5\cdot4}{3\cdot2\cdot1}$$

$$= 56\times20 = 1120.$$

Binomial Theorem and Binomial Coefficient.

The binomial theorem states that

$$(x+y)^n = \binom{n}{0}x^n + \binom{n}{1}x^{n-1}y + \binom{n}{2}x^{n-2}y^2 + \cdots$$

$$\cdots + \binom{n}{n-1}xy^{n-1} + \binom{n}{n}y^n$$

$$= \sum_{r=0}^{n} \binom{n}{r} x^{n-r} y^r \qquad \ldots (1)$$

where $\binom{n}{r}$ i.e. ${}^nC_r(r \le n)$ is called the Binomial Coefficient.

Note. Putting $x = y = 1$ in (1), we get $\sum_{r=0}^{n} \binom{n}{r} = 2^n$

So the total number of all possible combinations of n object is 2^n.

For example, the number of ways of invitation of relative, having 10 relatives $= {}^{10}C_1 + {}^{10}C_2 + \cdots + {}^{10}C_{10} = 2^{10}$.

Theorem 3. Pascal's Rule.

$$\binom{n+1}{r} = \binom{n}{r-1} + \binom{n}{r}, \quad 1 \le r \le n$$

Proof : Left as an exercise.

Theorem 4. $\sum_{r=0}^{k} \binom{m}{r}\binom{n}{k-r} = \binom{m+n}{k}$

Proof : By equating the coefficients of x^k on both sides of $(1+x)^{m+n} = (1+x)^m(1+x)^n$ we get the required result.

Note : (1) In the above theorem it is not necessary that m, n, are greater than k.

For example, for $m = 2, n = 3$ and $k = 4$,

we get, $$\sum_{r=0}^{k} \binom{m}{r}\binom{n}{k-r} = \binom{2}{0}\binom{3}{4} + \binom{2}{1}\binom{3}{3}$$

$$+\binom{2}{2}\binom{3}{2} + \binom{2}{3}\binom{3}{1} + \binom{2}{4}\binom{3}{0}$$

$$= 0 + 2\cdot 1 + 1\cdot 3 + 0 + 0 = 5 = \binom{5}{4} = \binom{m+n}{k}$$

(2) The above theorem can also be written as

$$\sum_{r=0}^{k}\binom{k}{r}\binom{m-k}{n-r}=\binom{m}{n}$$

(where k and n are positive integer such that $k \le n$)

(3) $\sum_{r=0}^{n}\binom{n}{r}^2=\binom{2n}{n}$

Ex. 3. *In how many ways can a teacher choose one or more students from five students.*

Since one or more students are chosen, so the total number of choices is

$$\binom{5}{1}+\binom{5}{2}+\binom{5}{3}+\binom{5}{4}+\binom{5}{5}=\sum_{r=0}^{5}\binom{5}{r}-\binom{5}{0}=2^5-1=31.$$

3.1.7. Conditional Probability.

We consider two events A and B connected with a random experiment E. Let us make the hypothesis that the event A has occured $n(A)$ times and B occurs simultaneously with A $n(A \cap B)$ times in the n repetations of experiment E. . Then ratio $\frac{n(A \cap B)}{n(A)}$ is called the conditional probability of B on the hypothesis that A has already occured and is denoted by $P(B/A)$.

Thus $P(B/A)=\frac{n(A \cap B)}{n(A)}$

$$=\frac{n(A \cap B)/n}{n(A)/n}=\frac{P(A \cap B)}{P(A)}, \text{ provided } P(A) \neq 0$$

Similarly the conditional probability of A on the hypothesis that B has already occured is $P(A/B)=\frac{P(A \cap B)}{P(B)}$, provided $P(B) \neq 0$.

Illustration : Let one card be drawn from a full pack. $A=$ 'spade' $B=$ 'king'

$\therefore$ Probability of King supposing Spade occurs $= P\left(B/A\right)$

$= \dfrac{n(A \cap B)}{n(A)} = \dfrac{1}{13}$

Multiplication Rule of Probability.

Thus we have

$$P(A \cap B) = P(A) \cdot P\left(B/A\right) = P(B) P\left(A/B\right) \quad \ldots (1)$$

if $P(A) \neq 0, P(B) \neq 0$

Theorem. (Generalization of the Multiplication Rule)

For n events $A_1, A_2, \cdots, A_n$

$$P(A_1 \cap A_2 \cap \cdots \cap A_n) = P(A_1) P\left(A_2/A_1\right) P\left(A_3/(A_1 \cap A_2)\right) \cdots P\left(A_n \Big/ \bigcap_{i=1}^{n-1} A_i\right)$$

provided $P\left(\bigcap A_i\right) \neq 0, \ i = 1, 2, \cdots, n-1$

Proof : Left as an exercise.

3.1.8. Stochastic Independence.

If for two events A and B, $P(A/B) = P(A)$ [i.e., the chance of occurrence of the event A is not affected by the occurrence of the event B], the event A is said to be independent of the event B.

The following theorem is the most important characterisation of being two events independent.

Theorem. Two events A and B are stochastically independent or statistically independent or independent if and only if $P(A \cap B) = P(A) \cdot P(B)$

Proof: $P(A/B) = P(A) \Leftrightarrow \dfrac{P(A \cap B)}{P(B)} = P(A) \Leftrightarrow P(A \cap B) = P(A) P(B)$

Hence the theorem.

Note : When $P(A \cap B) \neq P(A) \cdot P(B)$, the events A and B are said to be dependent.

Mutually Independent Events. The n events $A_1, A_2, \cdots, A_n$ are said to be mutually independent if the following conditions are satisfied

$$P(A_i \cap A_j) = P(A_i) P(A_j) \text{ for all } i \neq j; \ i, j = 1, 2, \cdots, n$$

$$P(A_i \cap A_j \cap A_k) = P(A_i) P(A_j) P(A_k) \text{ for all } i \neq j \neq k$$

... ...

... ...

$$P(A_1 \cap A_2 \cap \cdots \cap A_n) = P(A_1)P(A_2)\cdots P(A_n)$$

Pairwise Independent Events : The n events $A_1, A_2, \cdots, A_n$ are said to be pairwise independent if

$$P(A_i \cap A_j) = P(A_i)\,P(A_j) \text{ for all } i \neq j\,;\; i, j = 1, 2, \cdots, n$$

Illustrations. Let one card be drawn from a full pack. $A =$ 'spade', $B =$ 'King'. Then $P(B/A) = \dfrac{n(B \cap A)}{n(A)} = \dfrac{1}{13}$.

Again $P(B) = \dfrac{n(B)}{n(S)} = \dfrac{4}{52} = \dfrac{1}{13}$.

Thus $P(B/A) = P(B)$. So in this experiment 'king' and 'spade' are independent.

3.1.9. Baye's Theorem. If $A_1, A_2, \cdots, A_n$ be a given set of n pairwise mutually exclusive and exhaustive events then for any event A where $P(A) \neq 0$

(i) $P(A) = P(A_1)P(A/A_1) + P(A_2)P(A/A_2) + \cdots + P(A_n)P(A/A_n)$

$$= \sum_{i=1}^{n} P(A_i)P(A/A_i)$$

(ii) $P(A_i/A) = \dfrac{P(A_i)P(A/A_i)}{P(A)}$, for $i = 1, 2, \cdots, n$ [*W.B.U.Tech.2008*]

Proof: Beyond the scope of the book

Illustration. There are three urns. First urn contains 3 red, 4 black balls ; second urn contains 6 black, 2 red balls ; third urn contains 3 black balls and 5 red balls. One urn is chosen and then a ball is drawn from the urn. Let $A_1 =$ '1st urn is chosen' $A_2 =$ '2nd urn is chosen' $A_3 =$ 3rd urn is chosen ; $A =$ The ball is red.

According to Baye's theorem

$$P(A) = P(A_1)P(A/A_1) + P(A_2)P(A/A_2) + P(A_3)P(A/A_3)$$

$$= \frac{1}{3}\cdot\frac{3}{7} + \frac{1}{3}\cdot\frac{2}{8} + \frac{1}{3}\cdot\frac{5}{8} = \frac{1}{7} + \frac{1}{12} + \frac{5}{24} = \frac{73}{168}.$$

On the other hand the probability that the 3rd urn was chosen

supposing that the ball is red $= P(A_3/A) = \dfrac{P(A_3)P(A/A_3)}{P(A)}$

$$= \frac{\frac{1}{3}\cdot\frac{5}{8}}{\frac{73}{168}} = \frac{5}{24} \times \frac{168}{73} = \frac{35}{73}.$$

3.1.10. Illustrative Examples.

Ex. 1. *What is the chance that a leap year selected at random will contain 53 wednesdays ?* [*W.B.U.Tech 2002*]

A leap year contains 366 days that is 52 full weeks and two days extra. The extra two days will be either (i) Sunday, Monday or (ii) Monday, Tuesday or (iii) Tuesday, Wednesday or (iv) Wednesday, Thursday or (v) Thursday, Friday or (vi) Friday, Saturday or (vii) Saturday, Sunday.

So a leap year will contain 53 Wednesdays if one of the two extra days is Wednesday. Therefore out of the above seven cases two are favourable.

Hence the required probability is $\dfrac{2}{7}$.

Ex. 2. *The integers x and y are chosen at random with replacement from nine natural numbers 1, 2, ..., 8, 9. Find the probability that* $(x^2 - y^2)$ *is divisible by 2.*

$x^2 - y^2 = (x-y)(x+y)$ will be divisible by 2 iff x, y are either both even or both odd. Now two even numbers can be chosen from $\{1, 2, \cdots, 9\}$ with replacement in 4×4 ways. Similarly both odd number can be selected in 5×5 ways. So the total no. of favourable cases is $16+25=41$.

Again two integers x, y can be chosen at random with replacement from $\{1, 2, \cdots, 9\}$ is $9\times9=81$.

Hence the required probability is $\dfrac{41}{81}$.

Ex. 3. *Given* $P(A)=\frac{1}{2}, P(B)=\frac{1}{3}, P(AB)=\frac{1}{4}$. ($AB$ means $A\cap B$)

(a) find the values of the following probabilities

$$P(\overline{A}), P(A\cup B), P(A/B), P(\overline{A}B), P(\overline{A}\,\overline{B}), P(\overline{A}\cup B)$$

(b) State whether the events A and B are

(i) mutually exclusive (ii) exhaustive (iii) equally likely (iv) independent.

(a) $P(\bar{A}) = 1 - P(A) = 1 - \frac{1}{2} = \frac{1}{2}$

$P(A \cup B) = P(A) + P(B) - P(AB) = \frac{1}{2} + \frac{1}{3} - \frac{1}{4} = \frac{7}{12}$

$$P(A/B) = \frac{P(AB)}{P(B)} = \frac{\frac{1}{4}}{\frac{1}{3}} = \frac{3}{4}$$

$P(\bar{A}B) = P\{(S - A)B\} = P(SB - AB)$

$= P(B - AB) = P(B) - P(AB) = \frac{1}{3} - \frac{1}{4} = \frac{1}{12}$

$P(\overline{AB}) = 1 - P(A \cup B) = 1 - \frac{7}{12} = \frac{5}{12}$

$P(\bar{A} \cup B) = P(\bar{A}) + P(B) - P(\bar{A}B) = \frac{1}{2} + \frac{1}{3} - \frac{1}{12} = \frac{3}{4}$.

(b) (i) No ; because $P(AB) = \frac{1}{4} \neq 0$, i.e., $A \cap B \neq \phi$

(ii) No ; because $P(A \cup B) = \frac{7}{12} \neq 1$, i.e., $A \cup B \neq S$

(iii) No; because $P(A) = \frac{1}{2}, P(B) = \frac{1}{3}$ i.e., $P(A) \neq P(B)$

(iv) No ; because $P(AB) \neq P(A) \cdot P(B)$, here

$P(AB) = \frac{1}{4}$ but $P(A) \cdot P(B) = \frac{1}{6}$.

Ex. 4. *In an examination 30% of the students failed in Physics, 25% in Mathematics and 12% in both Physics and Mathematics. A student is selected at random. Find the probability that (i) the student has failed in Physics, if it is known that he has failed in Mathematics.*

(ii) the student has failed at least one of the two subjects

(iii) the student has passed at least one of the two subjects.

(iv) the student has passed in Mathematics if he failed in Physics.

Let A and B denote the events " a student failed in Physics" and "a student failed in Mathematics" respectively. Then

$P(A) = 0 \cdot 30,\ P(B) = 0 \cdot 25,\ P(A \cap B) = 0 \cdot 12$

Now (i) probability that a student has failed in Physics if it is known that he has failed in Mathematics is

$$P(A/B) = \frac{P(A \cap B)}{P(B)} = \frac{0 \cdot 12}{0 \cdot 25} = 0 \cdot 48 .$$

(ii) the probability that a student has failed at least one of the subjects is

$$P(A \cup B) = P(A) + P(B) - P(A \cap B) = 0 \cdot 30 + 0 \cdot 25 - 0 \cdot 12 = 0 \cdot 43$$

(iii) $\bar{A}$ = The student passed in Mathematics, $\bar{B}$ = he passed in Physics.

Then the probability that the student has passed at least one of the subject is

$$P(\bar{A} \cup \bar{B}) = 1 - P(\overline{\bar{A} \cup \bar{B}}) = 1 - P(A \cap B) \text{ [by D' Morgans law]}$$

$$= 1 - 0 \cdot 12 = 0 \cdot 88$$

and (iv) the probability that the student has passed in Mathematics if he failed in Physics is

$$P(\bar{B}/A) = 1 - P(B/A) = 1 - \frac{P(AB)}{P(A)} = 1 - \frac{0 \cdot 12}{0 \cdot 30} = 0 \cdot 60 .$$

Ex. 5. *Two urns contain respectively 2 red, 5 black, 7 green and 1 red, 4 black, 9 green balls. One ball is drawn from each urn. Find the probability that both the balls are of the same colour.*

Let A_1, A_2, A_3 be the event that both drawn balls are red, black and green respectively.

Then the required event is $A_1 \cup A_2 \cup A_3$, where A_1, A_2, A_3 are pairwise exclusive.

$$\therefore P(A_1 \cup A_2 \cup A_3) = P(A_1) + P(A_2) + P(A_3)$$

Now, $P(A_1) = \frac{2}{14} \times \frac{1}{14} = \frac{2}{196}$

$$P(A_2) = \frac{5}{14} \times \frac{4}{14} = \frac{20}{196}$$

$$P(A_3) = \frac{7}{14} \times \frac{9}{14} = \frac{63}{196}$$

So the required probability is $\frac{2}{196} + \frac{20}{196} + \frac{63}{196} = \frac{85}{196}$.

Ex.6. *Two cards are drawn from a well-shuffled pack. Find the probability that at least one of them is diamond.*

Let A be the event that at least one of the drawn cards is diamond. Then $\bar{A}$ be the event that the drawn cards is not diamond.

$$\therefore P(\overline{A})=\frac{^{39}C_2}{^{52}C_2}=\frac{19}{34} \quad \therefore P(A)=1-P(\overline{A})=1-\frac{19}{34}=\frac{15}{34}$$

So, the required probability is $\frac{15}{34}$.

Ex.7. *A card is drawn at random from an ordinary deck of 52 playing cards. Find the probability that it is (i) an ace (ii) a heart (iii) a nine or a club (iv) neither a spade nor a ten.*

Let H, D, C, and S be the event that the drawn balls are hearts, diamonds, clubs and spades respectively. Also let us use the numbers 1, 2, 3, ... 10 for ace, two, three, .. ten respectively.

Then (i) P (an ace) $= P(1) = \frac{^{4}C_1}{^{52}C_1}=\frac{4}{52}=\frac{1}{13}$

(ii) P (a heart) $= P(H)=\frac{^{13}C_1}{^{52}C_1}=\frac{13}{52}=\frac{1}{4}$

(iii) P (a nine or a club)

$= P(9\cup C) = P(9)+P(C)-P(9\cup C)$

$=\frac{^{4}C_1}{^{52}C_1}+\frac{^{13}C_1}{^{52}C_1}-\frac{1}{52}=\frac{4}{52}+\frac{13}{52}-\frac{1}{52}=\frac{4}{13}$

(iv) P (neither a spade nor a ten)

$= P(S'\cap 10')=P\left\{(S\cup 10)'\right\}=1-P(S\cup 10)$

$=1-\{P(S)+P(10)-P(S\cap 10)\}$

$=1-\left(\frac{13}{52}+\frac{4}{52}-\frac{1}{52}\right)=1-\frac{4}{13}=\frac{9}{13}$.

Ex. 8. *Two dice are thrown n times in succession. What is the probability of obtaining double six at least once.Hence find the minimum number of throws so that the probability of obtaining double six at least once is less than $\frac{1}{2}$.*

Let A be the event that there is at least once double six in n throws of two dice in succession. Then $\overline{A}$ be the event that there is no double six in n throws of two dice. So

$$P(\overline{A})=\left(\frac{35}{36}\right)^n \quad \therefore P(A)=1-P(\overline{A})=1-\left(\frac{35}{36}\right)^n$$

So the required probability is $1-\left(\frac{35}{36}\right)^n$.

Again when $P(A)<\frac{1}{2}$ then $1-\left(\frac{35}{36}\right)^n<\frac{1}{2}$

or, $\left(\frac{35}{36}\right)^n > \frac{1}{2}$ or, $n\log\left(\frac{35}{36}\right) > -\log 2$

or, $n > \dfrac{\log 2}{\log 36 - \log 35} \simeq 24.6 \quad \therefore n \geq 25$

Hence the minimum number of throws is 25.

Ex. 9. *A can hit a target 4 times in 5 shots ; B 3 times in 4 shots ;C twice in 3 shots. They fire a target. What is the probability that at least two shots hit ?*

Let A_1, A_2, A_3 be the event that A, B, C hit the target respectively. Then

$$P(A_1) = \frac{4}{5} \qquad \therefore P(\overline{A}_1) = 1 - \frac{4}{5} = \frac{1}{5}$$

$$P(A_2) = \frac{3}{4} \qquad \therefore P(\overline{A}_2) = 1 - \frac{3}{4} = \frac{1}{4}$$

$$P(A_3) = \frac{2}{3} \qquad \therefore P(\overline{A}_3) = 1 - \frac{2}{3} = \frac{1}{3}$$

For at least two hits, we may have

(i) A, B, C all hit the target, the probability for which is

$$P(A_1\, A_2\, A_3) = P(A_1)P(A_2)P(A_3) = \frac{4}{5}\cdot\frac{3}{4}\cdot\frac{2}{3} = \frac{2}{5}$$

(ii) A, B hit the target and C misses it, the probability for which is

$$P(A_1\, A_2\, \overline{A}_3) = P(A_1)P(A_2)P(\overline{A}_3) = \frac{4}{5}\cdot\frac{3}{4}\cdot\frac{1}{3} = \frac{1}{5}$$

(iii) A, C hit the target and B misses it, the probability for which is

$$P(A_1\overline{A}_2A_3) = P(A_1)P(\overline{A}_2)P(A_3) = \frac{4}{5}\cdot\frac{1}{4}\cdot\frac{2}{3} = \frac{2}{15}.$$

(iv) B, C hit the target and A misses it, the probability for which is

$$P(\overline{A}_1A_2A_3) = P(\overline{A}_1)P(A_2)P(A_3) = \frac{1}{5}\cdot\frac{3}{4}\cdot\frac{2}{3} = \frac{1}{10}.$$

So the required prob is $\frac{2}{5} + \frac{1}{5} + \frac{2}{15} + \frac{1}{10} = \frac{5}{6}$.

Ex. 10. *Show that the probability of occurrence of only one of the events A and B is* $P(A) + P(B) - 2P(AB)$. [*W.B.U.Tech 2006*]

Let C be the event of occurence of only one of the events A and B. Then

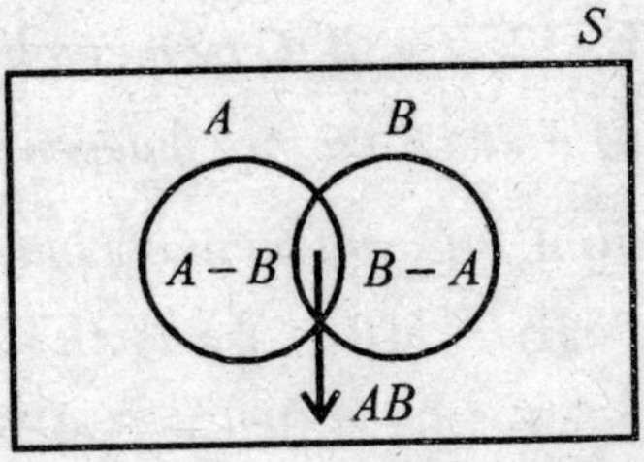

$C = (A \cup B) - (A \cap B)$

$= (A - B) \cup (B - A)$

$\therefore\ P(C) = P(A - B) + P(B - A)$ ... (1)

[$\because$ $A - B$ and $B - A$ are disjoint]

Now $A = (A - B) + AB$

$\therefore\ P(A) = P(A - B) + P(AB)$

$\therefore\ P(A - B) = P(A) - P(AB)$ ... (2)

Again $B = (B - A) + AB$

$\because\ P(B) = P(B - A) + P(AB)$

$\because\ P(B - A) = P(B) - P(AB)$ ... (3)

In virtue of (1), (2), and (3) we get

$P(C) = P(A) - P(AB) + P(B) - P(AB) = P(A) + P(B) - 2P(AB)$

Ex. 11. *If A and B are independent events, then show that the following pairs are independent :*

(i) $\overline{A}$ and $\overline{B}$ [*W.B.U.Tech 2002,2005,2007*]

(ii) A and $\overline{B}$ (iii) $\overline{A}$ and B.

Since A and B are independent, so

$P(A \cap B) = P(A)P(B)$ ... (1)

(i) Now $P(\overline{A} \cap \overline{B}) = P(\overline{A \cup B})$, by D' Morgan's law

$= 1 - P(A \cup B) = 1 - [P(A) + P(B) - P(A \cap B)]$

$= 1 - P(A) - P(B) + P(A \cap B) = 1 - P(A) - P(B) + P(A) \cdot P(B)$ by (1)

$= (1 - P(A))(1 - P(B)) = P(\overline{A})\, P(\overline{B})$.

$\therefore$ $\overline{A}$ and $\overline{B}$ are independent.

(ii) Again $A \cap B$ and $A \cap \overline{B}$ are mutually exclusive and

$(A \cap B) \cup (A \cap \overline{B}) = A$

$\therefore\ P(A) = P(A \cap B) + P(A \cap \overline{B})$

$\therefore P(A \cap \overline{B}) = P(A) - (A \cap B) = P(A) - P(A) \cdot P(B)$ by (1)

$= P(A)(1 - P(B)) = P(A)P(\overline{B})$

$\therefore$ A and $\overline{B}$ are independent.

(iii) Taking $B = (A \cap B) \cup (\overline{A} \cap B)$, we can prove the result as (ii)

Ex.12. *If A, B, C be mutually independent events, then prove that*

(i) A and $B+C$ are independent [*W.B.U.Tech 2003*]

(ii) $\overline{A},\overline{B},\overline{C}$ are mutually independent

(i) $P\{A(B+C)\}=P(AB+AC)=P(AB)+P(AC)-P(ABC)$

$=P(A)P(B)+P(A)P(C)-P(A)P(BC)$ [$\because$ A, B, C are mutually independent]

$=P(A)[P(B)+P(C)-P(BC)]=P(A)P(B+C)$

Hence A and $B+C$ are independent.

(ii) Now $\overline{A}$, $\overline{B}$ are independent as A, B are independent (by previous example). Similarly $\overline{B},\overline{C}$ and $\overline{C},\overline{A}$ are independent. Also since A and $B+C$ are independent so $\overline{A}$ and $\overline{B+C}$ i.e., $\overline{A}$ and $\overline{B}\overline{C}$ are independent $\left[\because\ \overline{B+C}=\overline{B}\overline{C}\right]$

$\therefore\ P\{\overline{A}(\overline{B}\overline{C})\}=P(\overline{A})\,P(\overline{B}\overline{C})=P(\overline{A})\,P(\overline{B})\,P(\overline{C})$

Hence $\overline{A},\overline{B},\overline{C}$ are mutually independent.

Ex.13. *The face cards are removed from a pack of 52 cards. Then 4 cards are drawn one by one from the remaining 40 cards. What is the probability that 4 cards belong to different suits and different denomination.*

4 cards can be drawn out of 40 cards one by one is $40\times39\times38\times37$ ways. So the total numbers of possible outcomes is $40\times39\times38\times37$.

Now the number of ways in which 4 cards belong to different suits and different denominators is ${}^{10}C_1\times{}^{9}C_1\times{}^{8}C_1\times{}^{7}C_1=10\times9\times8\times7$.

So the total number of favourable cases is $10\times9\times8\times7$

Hence the required prob. $=\dfrac{10\times9\times8\times7}{40\times39\times38\times37}=\dfrac{21}{9139}$.

Ex.14. *A and B throw alternatively with a pair of dice. A wins if he throws 8 before B throws 5 and B wins if he throws 5 before A throws 8. Find the probability that A wins.* [*W.B.U.Tech 2006*]

Here the total no. of event pts. is 36 in a trial of throwing two dice.

Let X, Y be the event that A throws 8 and B throws 5 with a pair of dice. Then

$X=\{(3,5),(5,3),(2,6),(6,2),(4,4)\}\quad\therefore\ n(X)=5$

$Y=\{(2,3),(3,2),(4,1),(1,4)\}\quad\therefore\ n(Y)=4$

$$\therefore\ P(X)=\frac{5}{36},\ P(Y)=\frac{4}{36}=\frac{1}{9}$$

$$\therefore\ P(\overline{X})=1-\frac{5}{36}=\frac{31}{36}$$

$$P(\overline{Y})=1-\frac{1}{9}=\frac{8}{9}.$$

So, the probability of A wins in the game is given by

$$P(X)+P(\overline{X}\,\overline{Y}\,X)+P(\overline{X}\,\overline{Y}\,\overline{X}\,\overline{Y}\,X)+\cdots$$

$$=P(X)+P(\overline{X})P(\overline{Y})\,P(X)+\cdots\ [\because\ \overline{X},\overline{Y},X \text{ are all independent }]$$

$$=\frac{5}{36}+\frac{31}{36}\cdot\frac{8}{9}\cdot\frac{5}{36}+\left(\frac{31}{36}\right)^2\left(\frac{8}{9}\right)^2\cdot\frac{5}{36}+\cdots$$

$$=\frac{\frac{5}{36}}{1-\frac{8}{9}\cdot\frac{31}{36}}=\frac{45}{76}.$$

Ex.15. *A pack of 2n cards, n of which are red and another n are black. It is divided into two equal parts and a card is drawn from each. Find the probability that the cards drawn are of the same colour.*

Let $2n$ cards be divided into such way that first part contains k red cards and $(n-k)$ black cards where $k=1,2,\cdots,n-1$. Then the 2nd part contains $(n-k)$ red cards and k black cards.

Then the probability for both the drawn cards are of black is $\frac{n-k}{n}\cdot\frac{k}{n}$ and that of red is $\frac{k}{n}\cdot\frac{n-k}{n}$.

So the probability for both the drawn cards are of the same colour is

$$\frac{n-k}{n}\cdot\frac{k}{n}+\frac{k}{n}\cdot\frac{n-k}{n}=2\frac{(n-k)k}{n^2},\ k=1,2,\cdots,n-1$$

Hence the required probability $=\sum_{k=1}^{n-1}\frac{2(n-k)k}{n^2}$

$$=\frac{2}{n^2}\left(n\sum_{k=1}^{n-1}k-\sum_{k=1}^{n-1}k^2\right)=\frac{2}{n^2}\left\{n\cdot\frac{(n-1)n}{2}-\frac{n(n-1)(2n-1)}{6}\right\}=\frac{n^2-1}{3n}.$$

Ex.16. *An urn contains a white and b black balls, from which k balls are drawn one by one and they laid aside without noticeing their colours. Then one more ball is drawn. Find the probability that it is white.*

Here the total number of cases of drawing $(k+1)$ balls from $(a+b)$ balls is $(a+b)(a+b-1)\cdots(a+b-k)$.

Since the last drawn ball shall be white, we choose one white ball from a white balls in a ways. Then k balls can be drawn from the rest $(a+b-1)$ balls is

$$(a+b-1)(a+b-2)\cdots(a+b-k) \text{ ways.}$$

So the total number of favourable cases in which the last drawn ball is white in $(k+1)$ drawn is $(a+b-1)(a+b-2)\cdots(a+b-k)a$.

Hence the required probability

$$=\frac{(a+b-1)(a+b-2)\cdots(a+b-k)\cdot a}{(a+b)(a+b-1)\cdots(a+b-k)}=\frac{a}{a+b}.$$

Ex.17. *15 new students are to be evenly distributed among 3 classes. Suppose that there are 3 whiz-kids among the fifteen. What is the probability that each class gets one whiz-kid and one class gets them all ?*

15 students can be evenly distributed among 3 classes in

$$^{15}C_5 \times {}^{10}C_5 \times {}^5C_5 \text{ ways } = \frac{(15)!}{(5!)^3} \text{ ways.}$$

So the total number of distributions is $\frac{(15)!}{(5!)^3}$.

(i) We can allot one whiz-kid to each of three classes in 3! ways. Then the other 12 students can be evenly distributed among 3 classes in

$$^{12}C_4 \times {}^8C_4 \times {}^4C_4 \text{ ways } = \frac{(12)!}{(4!)^3} \text{ ways.}$$

So the probability that each class gets one whiz-kid is $\dfrac{3!\dfrac{(12)!}{(4!)^3}}{\dfrac{(15)!}{(5!)^3}}=\dfrac{25}{91}$.

(ii) We can allot all the three whiz-kids to one class in 3 ways and the rest 12 students in $^{12}C_5 \times {}^7C_5 \times {}^2C_2 = \frac{(12)!}{(5!)^2\ 2!}$ ways.

So the prob. that one class gets all 3 whiz-kids is $\dfrac{3\times\dfrac{(12)!}{(5!)^2\ 2!}}{\dfrac{(15)!}{(5!)^3}}=\dfrac{6}{91}$.

Ex.18. *An urn contains 4 white and 6 black balls. Two balls are successively drawn from the urn without replacement of the first ball. If the first ball is seen to be white, what is the probability that the 2nd ball is also white ?*

Let A_1 be the event that the first drawn ball is white and A_2 be the event that the second drawn ball is white. Then $A_1 \cap A_2$ be the event that the both drawn ball is white.

$$\therefore \; P(A_1 \cap A_2) = \frac{4}{10} \times \frac{3}{9} = \frac{2}{15}$$

Also $P(A_1) = \frac{4}{10} = \frac{2}{5}$.

Then by definition of conditional probability,

$$P(A_2 / A_1) = \frac{P(A_1 \cap A_2)}{P(A_1)} = \frac{\frac{2}{15}}{\frac{2}{5}} = \frac{1}{3}$$

Hence the required probability $= \frac{1}{3}$.

Ex.19. *There are two identical urns containing respectively 4 white, 3 red balls and 3 white, 7 red balls. An urn is chosen at random and a ball is drawn from it. Find the probability that the ball is white. If the ball drawn is white, what is the probability that it is from the first urn ?*

Let A_1, A_2 be the event that the ball is drawn from the first and second urn respectively. Clearly the events A_1, A_2 are mutually exclusive and exhaustive events.

$$\therefore \; P(A_1) = P(A_2) = \frac{1}{2}$$

Also let A be the event that the drawn ball is white. Then we have

$$P(A / A_1) = \frac{4}{7}, \; P(A / A_2) = \frac{3}{10}$$

Now $P(A) = P(A_1) \cdot P(A / A_1) + P(A_2) \cdot P(A / A_2)$

$$= \frac{1}{2} \cdot \frac{4}{7} + \frac{1}{2} \cdot \frac{3}{10} = \frac{61}{140}.$$

So the probability that the drawn ball is white is $\frac{61}{140}$.

Now by Baye's theorem we have

$$P(A_1 / A) = \frac{P(A_1) P(A / A_1)}{P(A)} = \frac{\frac{1}{2} \cdot \frac{4}{7}}{\frac{61}{140}} = \frac{40}{61}.$$

So the prob. that the white ball is drawn from the first urn is $\frac{40}{61}$.

Ex.20. *Two urns contain respectively 5 white, 7 black balls and 4 white, 2 black balls. One of the urns is selected by the toss of a fair coin and then 2 balls are drawn without replacement from the selected urn. If both balls drawn are white, what is the probability that the first urn is selected ?*

Let A_1, A_2 be the event that the ball is drawn from the first and second urn respectively. Then as the urn is chosen by coin-tossing, so we have $P(A_1) = P(A_2) = \frac{1}{2}$.

Now let A be the event that the drawn two balls are white. Then

$$P(A/A_1) = \frac{{}^5C_2}{{}^{12}C_2} = \frac{\frac{5\cdot4}{2\cdot1}}{\frac{12\cdot11}{2\cdot1}} = \frac{5}{33}$$

$$P(A/A_2) = \frac{{}^4C_2}{{}^6C_2} = \frac{\frac{4\cdot3}{2\cdot1}}{\frac{6\cdot5}{2\cdot1}} = \frac{2}{5}$$

$\therefore$ by Baye's Theorem $P(A) = P(A_1)P(A/A_1) + P(A_2)P(A/A_2)$

$$= \frac{1}{2}\cdot\frac{5}{33} + \frac{1}{2}\cdot\frac{2}{5} = \frac{91}{330}$$

Again by Baye's theorem, the required probability

$$P(A_1/A) = \frac{P(A_1)P(A/A_1)}{P(A)} = \frac{\frac{1}{2}\cdot\frac{5}{33}}{\frac{91}{330}} = \frac{25}{91}.$$

Ex.21. *A speaks the truth 3 out of 4 times and B 7 times out of 10. They agree in their statement that from a bag contaning 6 balls of different colours a white ball has been drawn. Find the probability that the statement is true.*

Let A_1 and A_2 be the events that the joint statement of A and B is true and false respectively

Then $P(A_1) = \frac{1}{6}$, $P(A_2) = \frac{5}{6}$

Now let X be the event that both A and B agreed in their statement.

Then

$$P(X/A_1)=\frac{3}{4}\times\frac{7}{10}=\frac{21}{40}.$$

$$P(X/A_2)=\left(\frac{1}{4}\times\frac{1}{5}\right)\times\left(\frac{3}{10}\times\frac{1}{5}\right)=\frac{3}{1000}$$

$$\therefore\ P(X)=P(A_1)\cdot P(X/A_1)+P(A_2)\cdot P(X/A_2)$$

$$=\frac{1}{6}\times\frac{21}{40}+\frac{5}{6}\times\frac{3}{1000}=\frac{7}{80}+\frac{1}{400}=\frac{9}{100}.$$

Hence the probability of the statement being true is

$$P(A_1/X)=\frac{P(A_1)P(X/A_1)}{P(A)} \quad \text{(by Baye's theorem)}$$

$$=\frac{\frac{1}{6}\times\frac{21}{40}}{\frac{9}{100}}=\frac{35}{36}.$$

Ex.22. *Assuming that each child is as likely to be a boy as it is to be a girl, what is the conditional probability that in a family of two children both are boys, given that (i) the older child is a boy (ii) at least one of the children is a boy ?*

Let A_1 and A_2 be the event that the older child is a boy and the younger child is a boy

Then $P(A_1)=P(A_2)=\frac{1}{2}$

Also $A_1\cup A_2=$ at least one of the children is a boy

and $A_1\cap A_2=$ both children are boys. Since A_1, A_2 are independent

$$\therefore\ P(A_1\cap A_2)=P(A_1)P(A_2)=\frac{1}{2}\cdot\frac{1}{2}=\frac{1}{4}$$

$$P(A_1\cup A_2)=P(A_1)+P(A_2)-P(A_1\cap A_2)=\frac{1}{2}+\frac{1}{2}-\frac{1}{4}=\frac{3}{4}.$$

(i) Thus the probability that both children are boys given that the older is a boy is

$$P((A_1\cap A_2)/A_1)=\frac{P(A_1\cap A_2\cap A_1)}{P(A_1)}=\frac{P(A_1\cap A_2)}{P(A_1)}=\frac{1/4}{1/2}=\frac{1}{2}.$$

(ii) The probability that both the children are boys, given that at least one of them is a boy, is

$$P\{(A_1 \cap A_2)/(A_1 \cup A_2)\}$$

$$= \frac{P[(A_1 \cap A_2) \cap (A_1 \cup A_2)]}{P(A_1 \cup A_2)}$$

$$= \frac{P(A_1 \cap A_2)}{P(A_1 \cup A_2)} = \frac{1/4}{3/4} = \frac{1}{3}.$$

Ex.23. *In a bolt factory, machines A, B and C manufacture respectively 25%, 35% and 40% of the total of their output. 5%, 4% and 2% are defective bolts. A bolt is drawn at random from the product and is found to be defective.*

What are the probability that it was manufactured by machines A, B and C ? [*W.B.U.Tech 2003*]

Let X_1, X_2 and X_3 be the events that a bolt is manufactured by A, B and C respectively and X be the event that a bolt is defective. Then

$$P(X_1) = \frac{25}{100} = \frac{1}{4},\ P(X_2) = \frac{35}{100} = \frac{7}{20},\ P(X_3) = \frac{40}{100} = \frac{2}{5}$$

$$\therefore\ P(X/X_1) = \frac{5}{100} = \frac{1}{20}$$

$$P(X/X_2) = \frac{4}{100} = \frac{1}{25},\ P(X/X_3) = \frac{2}{100} = \frac{1}{50}$$

$$\therefore\ P(X) = P(X_1)P(X/X_1) + P(X_2)P(X/X_2) + P(X_3)P(X/X_3)$$

$$= \frac{1}{4} \times \frac{1}{20} + \frac{7}{20} \times \frac{1}{25} + \frac{2}{5} \times \frac{1}{50} = \frac{1}{80} + \frac{7}{500} + \frac{1}{125} = \frac{69}{2000}$$

Then by Baye's theorem, we have

$$P(X_1/X) = \frac{P(X_1)P(X/X_1)}{P(X)} = \frac{\frac{1}{4} \cdot \frac{1}{20}}{\frac{69}{2000}} = \frac{25}{69}$$

Similarly $P(X_2/X) = \frac{28}{69},\ P(X_3/X) = \frac{16}{69}$.

So, the required probability that defective bolt was manufactured by machines A, B, C are $\frac{25}{69}, \frac{28}{69}, \frac{16}{69}$ respectively.

Ex.24. *An urn contains 10 white and 3 black balls, while another urn 3 white and 5 black balls. Two balls are drawn from the first urn and put into the second arn and then a ball is drawn from the latter. What is the probability that it is a white ball ?*

Let A, B, C be the events that the drawn two balls from the first unr are both white, both black and one white and one black respectively. Then

$$P(A) = \frac{10_{c_2}}{13_{c_2}} = \frac{10\times 9}{13\times 12} = \frac{15}{26}$$

$$P(B) = \frac{3_{c_2}}{13_{c_2}} = \frac{3\times 2}{13\times 12} = \frac{1}{26}$$

$$P(C) = \frac{10_{c_1}\times 3_{c_1}}{13_{c_2}} = \frac{5}{13}$$

When two balls are transferred into the 2nd arn, it will contain either 5 white, 5 black balls. or, 3 white, 7 black balls or 4 white and 6 black balls according to the events A, B, C respectively.

Let W denote the event of drawing a white ball from the second urn.

Then $P(W/A) = \frac{5}{10} = \frac{1}{2}, P(W/B) = \frac{3}{10}$

$$P(W/C) = \frac{4}{10} = \frac{2}{5}.$$

So the required probability, $P(W) = P(A)\cdot P(W/A) + P(B)P(W/B)$

$$+P(C)\cdot P(W/C)$$

$$= \frac{15}{26}\cdot\frac{1}{2} + \frac{1}{26}\cdot\frac{3}{10} + \frac{5}{13}\cdot\frac{2}{5} = \frac{75+3+40}{260} = \frac{59}{130}.$$

Ex.25. *A student has to answer a multiple choice question with 5 alternatives. What is the probability that the student know the answer, given that he answered it correctly.*

Let B_1 and B_2 be the events that the student knew right answer and guesses the right answer respectively. Also let A be event that he getstheright answer. Again let p be the probability that he knew the corrcet answer.

$\therefore \quad P(B_1) = p, P(B_2) = 1 - p$

Also $P(A/B_1) = 1, P(A/B_2) = \frac{1}{5}$

$\therefore$ By Baye's theorem,

$$P(B_1/A) = \frac{P(B_1)P(A/B_1)}{P(B_1)P(A/B_1) + P(B_2)P(A/B_2)}$$

$$= \frac{p-1}{p \cdot 1 + (1-p)\frac{1}{5}}$$

$$= \frac{5p}{4p+1}.$$

EXERCISES

[I] Short Answer Questions

1. Twenty balls in an urn are numbered 1 through 20. A blind folded contestant draws five balls from the win, with the order of the draw recorded. What is the probability that the number 3 ball was selected ?

[Hints : The required probability $= \frac{19_{c_4}}{20_{c_5}} = \frac{1}{4}$.]

2. Prove that for any two events A, B with

$$B \subset A, P(A \cap \overline{B}) = P(A) - P(B).$$

[Hints : As $B \subset A$, so B and $A \cap \overline{B}$ are disjoint and $B \cup (A \cap \overline{B}) = A$

$\therefore \quad P(B) \cup P(A \cap \overline{B}) = P(A)$.]

3. If $p_1, p_2, \cdots p_n$ are the probabilities that n certain events happen, then find the probability that at least one of these events must happen.

Hints : The probabilities of their non-happening are

$1 - p_1, 1 - p_2, \cdots 1 - p_n$. So the probability of all of these failing is $(1-p_1)(1-p_2)\cdots(1-p_n)$.

Hence the chance is which at least one of these events must happen is

$$1 - (1-p_1)(1-p_2)\cdots(1-p_n).$$

4. A probability is mathematics in given to three students A, B, C whose chances of solving it are $\frac{1}{2}, \frac{1}{3}, \frac{1}{4}$ respectively. What is the probability that the problem will be solved ?

[Hints : The probability that the problem is not solved by any of A, B, C is $\left(1-\frac{1}{2}\right)\left(1-\frac{1}{3}\right)\left(1-\frac{1}{4}\right)=\frac{1}{4}$.

So the probability that the problem is solved by at least one of them in $1-\frac{1}{4}=\frac{3}{4}$.]

5. For two events A_1, A_2 let $P(A_1)=0.4, P(A_2)=p$ and $P(A_1 \cup A_2)=0.6$.

(i) Find p so that A_1 and A_2 are independent events.

(ii) For what value of p, the events A_1 and A_2 are mutually exclusive.

[Hints : $P(A_1 \cap A_2)=P(A_1)+P(A_2)-P(A_1 \cup A_2)$

$$= p-0.2.$$

(i) $P(A_1 \cap A_2)=P(A_1)\cdot P(A_2) \Rightarrow p \times 0.4 = p-0.2$

$$\Rightarrow p \times 0.6 = 0.2$$

$$\Rightarrow p=\frac{1}{3}.$$

(ii) $P(A_1 \cap A_2)=0 \Rightarrow p=0.2$.]

6. Two urns contain 4 white, 6 blue and 4 white, 5 blue balls respectively. One of the urns is selected at random and a ball is drawn from it. Find the probability that the drawn ball is white.

[Hints : $P(w)=P(A_1)\cdot P(w/A_1)+P(A_2)\cdot P(w/A_2)$

$$=\frac{1}{2}\cdot\frac{4}{10}+\frac{1}{2}\cdot\frac{4}{9}=\frac{19}{45}$$]

7. If $P(A)=\frac{2}{3}, P(A+B)=\frac{5}{6}$, find $P(\overline{A}\overline{B})$.

[Hints : $P(\overline{A}\overline{B})=1-P(A+B)=1-\frac{5}{6}=\frac{1}{6}$]

8. If $P(A/B)=1$, then prove that $P(ABC)=P(BC)$

[Hints : $P(A/B)=1 \Rightarrow P(AB)=P(B)$

$\therefore$ $P(ABC)=P(AB)\cdot P(C/AB)=P(B)\cdot P(C/B)$

$$=P(BC)$$]

9. If A and B are events with $P(A)=\frac{3}{8}, P(B)=\frac{5}{8}$ and $P(A\cup B)=\frac{3}{4}$, find $P(A/B), P(B/A)$. Are A and B independent ?

10. For any two event A_1, A_2 prove that $P(A_1)+P(A_2)\le 1+P(A_1A_2)$

11. If A,B are two events with $P(A)=0.4, P(B)=0.3$ and $P(A\cap B)=0.2$, find $P(A^c\cap B)$. [*W.B.U.Tech 2006 M 302*]

[Hints : $P(A\cup B)=0.4+0.3-0.2=0.5$

$\because\ P(A^c\cap B)=P(A\cap B)-P(A)=0.5-0.4=0.1$]

12. If two perfect coins are tossed simultaneously find the probability of getting at least one head.

13. Out of 120 tickets numbered consecutively from 1 to 120, one ticket is drawn at random. Find the probability of getting a multiple of 5.

14. If *A, B, C* are equally likely, mutually exclusive and exhaustive then find $P(A)$.

15. Find the probability of getting at least one 'Five' from 3 throws of a perfect die.

16. Two events *A* and *B* are such that $P(A)=0.4, P(A\cup B)=0.7$ and $P(B)=x$. For what values of *x* are *A* and *B* (i) mutually exclusive (ii) independent ?

17. Find the probability that there would be two tail if four unbiased coins are tossed.

18. Three balls are drawn at random from a box containing 6 red and 4 black balls. What is the probability that the two balls are red and one is black ?

19. Box *A* contains 3 black balls and 3 red, box *B* contains 6 black balls and 4 red. If a ball is randomly selected from each box, find the probability that the balls will be of same colour ?

20. 5 cards are drawn from a pack of 52 well-shuffled cards. Find the probability that 4 are aces and 1 is a king.

21. Five persons *X, Y, Z, W, S* speak at a seminar-lecture. What is the probability that *X* speaks immediately before *Y* ?

22. If 20 dates are named at random, what is the probability that 3 of them will be Sundays ?

23. The nine digits 1, 2, 3 , 4, 5, 6, 7, 8, 9 are arranged in random order to form a nine-digit number. Find the probability that 1, 2, and 3 appear as neighbours in the order mentioned.

24. Four dices are rolled. What is the chance of obtaining a sum of 18 ?

25. What is the probability of an odd sum when two dice are thrown ?

[Hints : The required Probability $=\frac{18}{6^2}=\frac{1}{2}$]

26. There are four persons in a room. Find the probability that (i) all of them have different birthdays (ii) at least 2 of them have the same birthday (iii) exactly 2 of them have the same birthday. (1 year = 365 days)

27. There are 20 people. Find the probability that among the 12 months in the year there are 4 months containing exactly 2 birthdays and 4 containing exactly 3 birthdays ?

28. If $P(A\cap B)=\frac{1}{3}, P(B)=\frac{3}{5}$ and $P(A)=\frac{1}{2}$ find $P(A\cup B), P(A^c\cap B^c)$, $P(A^c\cup B^c)$ and $P(A\cap B^c)$.

29. What is the chance that a non-leap year should have 53 sundays ?

(Hint. Same as Ex-1)

30. X and Y stand in a line at random with 10 other people. What is the prob. that there are 3 people between X and Y.

31. Prove that the probability of obtaining six at least once in 4 throws with a die is slightly greater than $\frac{1}{2}$.

[Hints : The required Probability $=1-\left(\frac{5}{6}\right)^9 \simeq 0.52>\frac{1}{2}$

32. What is the probability that a bridge hand will contain at least one ace ?

[Hints : $1-\frac{{}^{48}C_{13}}{{}^{52}C_{13}} \approx 0.696$]

33. Prove that for any two events A, B

(i) $P(A+\overline{B})=1-P(B)+P(AB)$

(ii) $P(A\overline{B})=P(A)-P(AB)$

(iii) $P(A\cap B)\le P(A)+P(B)$

(iv) $P(A\cup B)\ge P(A)$

(vi) $P(A/B)\le P(A)/P(B)$

(vii) $P(\overline{A}/B)=1-P(A/B)$

(viii) $P(A)/P(B)=P(A/B)/P(B/A)$.

34. If the two events A and B are independent and $P(B)=\frac{3}{5}, P(A)=\frac{2}{3}$, find $P(A \cup B), P(\bar{A}/B)$ and $P(\bar{A} \cap B)$.

35. Prove that if A and B are mutually exclusive events and $P(A) \neq 0,\ P(B) \neq 0$, then A and B are not independent.

36. Given $P(A \cup B)=\frac{2}{3},\ P(A \cap \bar{B})=\frac{1}{3}$, find $P(B)$ and $P(A)$ if $P(A/B)=\frac{1}{6}$.

37. If $P(A)>P(B)$, show that $P(A/B)>P(B/A)$.

38. For any three events A, B, C, prove that

(i) $P((A+B)/C)=P(A/C)+P(B/C)-P(AB/C)$

(ii) $P(A\bar{B}/C)+P(AB/C)=P(A/C)$

[Hints : (i) $P(AC+BC)=P(AC)+P(BC)-P(ABC)$

$$\therefore \frac{P(AC+BC)}{P(C)}=\frac{P(AC)}{P(C)}+\frac{P(BC)}{P(C)}-\frac{P(ABC)}{P(C)}$$

$$\therefore P[(A+B)/C]=P(A/C)+P(B/C)-P(AB/C)$$

(ii) $$P(A\bar{B}/C)+P(AB/C)=\frac{P(A\bar{B}C)}{P(C)}+\frac{P(ABC)}{P(C)}$$

$$=\frac{P(A\bar{B}C)+P(ABC)}{P(C)}=\frac{P(AC)}{P(C)}=P(A/C)\ ].$$

ANSWERS

1. $\frac{1}{4}$ 7. $\frac{1}{6}$ 9. $\frac{2}{5}, \frac{2}{3}$, No 12. $\frac{3}{4}$ 13. $\frac{1}{5}$ 14. $\frac{1}{3}$

15. $1-\left(\frac{5}{6}\right)^3$ 16. (i) 0.3, (ii) 0.5 17. $\frac{3}{8}$ 18. $\frac{1}{2}$

19. $\frac{1}{2}$ 20. $4/{}^{52}C_5$ 21. $\frac{1}{5}$ 22. $\frac{1140\times 6^{17}}{7^{20}}$ 23. $\frac{1}{72}$

24. 5/81 25. $\frac{1}{2}$ 26. (i) 0.984, (ii) 0.016, (iii) $\frac{6\times 364\times 363}{(365)^3}$

27. 1.0604×10^{-3} 28. $\frac{23}{30}, \frac{7}{30}, \frac{2}{3}, \frac{1}{6}$ 29. $\frac{1}{7}$ 30. 4/33

32. 0·696 34. 13/15, 1/3, 1/5 36. 1/3, $\frac{7}{18}$

[II] LONG ANSWER QUESTIONS

1. For a certain binary communication channel, probability that a trasmitted '0' is received as a '0' is 0.95 and the probability that a transmitted '1' is received as '1' is 0.90. If the probability that a '0' is transmitted is 0.4, find the probability that

(i) a '1' is received.

(ii) a '1' was transmitted given that a '1' was received.

[Hints : $A_1 =$ the event of transmitted '1'

$\overline{A}_1 =$ the event of transmitted '0'

$A_2 =$ the event of receiving '1'

$\overline{A}_2 =$ the event of receiving '0'

$$\therefore \quad (i) \ P(A_2) = P(A_1)\cdot P(A_2 / A_1) + P(\overline{A}_1)P(A_2 / \overline{A}_1)$$

$$= 0.6 \times 0.9 + 0.4 \times 0.5$$

$$= 0.56 \cdot$$

$$\text{(ii)} \ P(A_1 / A_2) = \frac{P(A_1)P(A_2 / A_1)}{P(A_2)} = \frac{0.6 \times 0.9}{0.56} = \frac{27}{28}$$]

2. A missile hits target with probability 0.3. How many missiles should be fired so that there is at least an 80% probability of hitting a target ?

[Hints : The probability of no hit $= 1 - 0.3 = 0.7$

$$\therefore \quad 1 - (0.7)^n > 0.8 \Rightarrow (0.7)^n < 0.2$$

$$\Rightarrow n > \frac{\log 0.2}{\log 0.7} = 4 \cdot 512$$

3. Two fair dice had two of their sides painted red, two painted black,one painted yellow and the other painted white. When this pair of dice are rolled, what is the probability that both land on the same colour ?

4. From an urn containing a white and b black balls, balls are successively drawn without replacement until only those of the same colour are left. Prove that the probability that the balls left are white is $\frac{a}{a+b}$.

5. Out of 100 tickets consecutively numbered, three are drawn at random. Find the chance that the numbers on them are in G.P.

[Hints:Total no of cases $= {}^{100}C_3$.

No of Favourable cases.

Common ratio	lowest no drawn	Total no of cases
2	1,2,3,···,25	25
3	1,2,3,···,11	11
4	1,2,3,4,5,6	6
5	1,2,3,4	4
6	1,2	2
7	1,2	2
8	1	1
9	1	1
10	1	1
		Total 53

Required probability $= \frac{53}{{}^{10}C_3}$.]

6. A bag contains 5 white and 4 black balls . If 3 balls are drawn at random, what are the probabilities of the following :

(i) 2 of them are white

(ii) at most one of them is white

(iii) at lest two are white. [*W.B.U.Tech 2007*]

[Hints : (i) $n(A) = {}^5C_2 \times {}^4C_1$ (ii) $n(A) = {}^5C_1 \times {}^4C_2 + {}^4C_3$

(iii) ${}^5C_2 \times {}^4C_1 + {}^5C_3$]

7. The probability that a contractor will get a plumbing contract is 23, and the probability that he will not get an electric contract is 59. If the probability of getting at least one contract is 4/5, what is the probability that he will get both the contracts?

8. From the numbers 1, 2, 3, 4, ... 999, 1000 one number is drawn at random. Find the probability that the number is multiple of (i) 12 and 18 (ii) 12 or 18

9. A box contains 5 defective and 10 non-defective lamps. 8 lamps are drawn at random in succession without replacement. What is the probability that the 8th lamp is the 5th defective?

[*W.B.U.Tech 2005,2007,2008*]

[Hints : $\frac{{}^5C_4 \times {}^{10}C_3 \times 7!}{15 \times 14 \times 13 \times 12 \times 11 \times 10 \times 9 \times 8}$]

10. The probability that Ashok can solve a problem in Business Statistics is 4/5, that Amal can solve it is 2/3, and that Abdul can solve it is 3/7. If all of them try independently, find the probability that the problem is solved.

11. Two sets of candidates are competing for the position of the Board of Directors of a company. The probabilities that the first and second sets will win are 0.6 and 0.4 respectively. If the first set wins, the probability of introducing a new product is 0.8, and the corresponding probability if the second set wins is 0.3. What is the probability that the new product will be introduced?

12. An urn contains 2 white and 2 black balls. Balls are drawn successively at random without replacement. What is the probability that a black ball appears (i) for the first time in the 3rd drawing (ii) for the 2nd time in the 4th drawing?

13. A packet of 10 electronic components is known to include 3 defectives. If 4 components are randomly chosen and tested, what is the probability of finding among them not more than one defective ?

14. A candidate is selected for interview for 3 posts. For the first post there are 3 candidates, for the second post there are 4, and for the third there are 2. What is the chance of getting at least one post ?

15. A bag contains 8 red balls and 5 white balls. Two successive draws of 3 balls are made without replacement. Find the probability that the first drawing will give 3 white balls and the second 3 red balls.

16. The probability that a teacher will give a surprise test during any class meeting is 1/5. If a student is absent on two days, what is the probability that he will miss at least one test ?

17. In a bridge game, North and South have 9 spades between them. Find the probability that either East or West has no spades. (There are only 13 spades in a pack of 52 cards and each player has 13 cards. The players are designated by the positions they occupy, viz. North, South, East, West.)

18. In a hand of bridge, what is the probability that you have 5 spades and your partner has the remaining 8.

19. A factory produces blades among which 20% are defective. If 5 blades are drawn at random from a day's production, find the probability that there will be (i) exactly 2 defectives (ii) not less than 2 defectives.

20. Three groups of children contain respectively 3 girls and 1 boy, 2 girls and 2 boys, 1 girl and 3 boys. One child is selected at random from each group. Show that the chance that the three selected children consisting of 1 girl and 2 boys is $\frac{13}{32}$.

[*W.B.U Tech 2005,2007*]

[Hint : $n(s)=4\times4\times4=64$

$A=$ "Two boy, One girl" $=\{$ 1girl from 1st group, 1boy from 2nd group, 1boy from 3rd group $\}\cup\{$ 1boy from 1st group, one girl from 2nd group , one boy from 3rd group $\}\cup\{$ 1 boy from 1st group, 1 boy from 2nd group, 1 girl from 3rd group $\}$

$\therefore \quad n(A)=(3\times2\times3)+(1\times2\times3)+(1\times2\times1)=26$

$\therefore \quad P(A)=\frac{26}{24}=\frac{13}{32}$]

21. A box contains 7 white and 5 black balls. If 3 balls are drawn simultaneously at random, what is the probability that they are not all of the same colour ? Calculate the probability of the same event for the case where the balls are drawn in succession with replacement between drawings.

22. An urn contains 3 white and 5 black balls. One ball is drawn and its colour is unnoted, kept aside and then another ball is drawn. What is the prob. that it (i) black (ii) white ?

[Hints : Let W_1, B_1 be the event that 1st drawn ball is white and black. Also let W_2, B_2 the event that 2nd drawn ball is white and black.

$P(W_1)=\frac{3}{8}, P(B_1)=\frac{5}{8}, \; P(W_2/W_1)=\frac{2}{7}, P(W_2/B_1)=\frac{3}{7}$

$\therefore\; P(W_2)=P(W_1)\cdot P(W_2/W_1)+P(B_1)\cdot P(W_2/W_1)=\frac{3}{8} \quad \therefore\; P(B_2)=\frac{5}{8}$.]

23. Each of the two cabinets has 3 drawers. Cabinet I contains a gold coin in each drawer, and cabinet II contains a gold coin in one of its drawers and a silver coin in the other. A cabinet is randomly selected, one of its drawers is opened, and a gold coin is found. Find the probability that there is a gold coin in the other drawer.

24. There are three identical boxes, each provides with two drawers. In the first, each drawer contains a gold coin, in the third, each drawer contains a silver coin and in the second, one drawer contains a gold and the other a silver coin. A box is selected at random and one of the drawers is opened. If a gold coin is found, what is the prob. that the box chosen is the second one ?

[Hints : $P(A_1) = P(A_2) = P(A_3) = \frac{1}{3}$

$P(A/A_1) = \frac{2}{2} = 1,\; P(A/A_2) = \frac{1}{2},,\; P(A/A_3) = 0$

$\therefore\; P(A) = 1 \cdot \frac{1}{3} + \frac{1}{2} \cdot \frac{1}{3} + \frac{1}{3} \cdot 0 = \frac{1}{2} \qquad \therefore\; P(A_2/A) = \frac{1}{3}$]

25. There are three identical urns containing white and black balls. The first urn contains 3 white, 4 black balls, the 2nd urn contains 4 white 5 black balls and the 3rd urn contains 2 white and 3 black balls. An urn is chosen at random and a ball is drawn from it. If the drawn ball is white, what is the prob. that the 2nd urn is chosen.[Hint. Same as Illust. Ex. 19] *[W.B.U.Tech 2007]*

26. There are two identical urns containing respectively 5 white, 4 red balls and 4 white, 9 red ball. An urn is chosen at random and 3 balls are drawn from it of two are white and one is red. What is the prob. that it is drawn from the first urn ?

[Hints : Same as Illust. 19]

27. Urn A contains 2 white, 4 red balls ; Urn *B* contains 1 white and 1 red bal. A ball is randomly chosen from urn *A* and put into urn *B* ,and a ball is then randomly selected from Urn *B*. Find (i) the probability that the ball selected from Urn *B* is white (ii) the conditional probability that the transferred ball was white, given that a white ball is selected from Urn *B*.

28. Each of the *n* urns contains *a* white and *b* black balls. One ball is tranformed from the first urn to the 2nd urn, then one ball from the latter to the 3rd and so on. Finally one ball is taken from the last urn. What is the prob. of its being white?

[Hints : Let P_k be the prob of drawing a white ball from the kth urn $(k \le n)$. Then

$$P_{k+1} = \frac{a+1}{a+b+1} P_k + \frac{a}{a+b+1}(1 - P_k) = \frac{P_k + a}{a+b+1}$$

Now, $P_1 = \frac{a}{a+b}$ $\therefore$ $P_2 = \frac{a}{a+b}$ and have $P_n = \frac{a}{a+b}$].

29. A box contains 3 types of disposable lights. The probability that type I light gives over 100 hours of use is 0.7, with the corresponding probabilities for type II and type III lights being 0.4 and 0.3, respectively. Let 20% of the lights in the box are type I, 30% are type II, and 50% are type III.

(i) Find the probability that a randomly chosen light gives more than 100 hours of use ?

(ii) Given the light lasted over 100 hours, what is the conditional probability that it was a type II light ?

30. 90% of all babies survive delivery. However, 15% of all births involve Cesarean-Section. When a Cesarean- Section is performed the baby survives 96% of all cases. If a randomly chosen pregnant woman does not have a Cesarean-Section find the probability that her baby survives.

31. A total of 46% of the voters in Kolkata classify themselves as supporter of party X, whereas 30% classify themselves as party Y and 24% as party Z. In CMC election 35% of party-X-supporter, 62% of party-Y, and 58% of the party-Z supporters voted. A voter is chosen at random. Given that this voter voted in the CMC-election, find the probability that the voter is (i) a supporter of X party (ii) a supporter of the party Y and (iii) a supporter of Z-party (iv) what fraction of voters participated in that election ?

32. An insurance company divides people in two classes : those who are accident prone and those who are not. An accident prone person has an accident at some time within a fixed one year period with probability 0.4 whereas this probability is 0.2 for a non-accident prone person. If we assume that 30% of the population is accident prone, what is the probability that a new policy holder has an accident within a year of buying a policy ?

33. Two cards are drawn successively from the pack without replacing the first. Find the prob. that the 2nd card is also spade, if the first card is spade.

[Hints. Let A, B be the event that the 1st and 2nd drawn card is spade.

$\therefore\ P(AB)=\dfrac{13\times 12}{52\times 51},\ P(A)=\dfrac{13}{52}\ P(B/A)=\dfrac{P(AB)}{P(A)}=\dfrac{12}{51}=\dfrac{4}{17}$.]

34. In a certain class 25% of the students failed in Mathematics, 15% failed in Chemistry, 10% failed in both Mathematics and Chemistry. A student is selected at random.

(i) If he failed in Mathematics, what is the prob. that he failed in Chemistry?

(ii) If he failed in Chemistry, what is the prob. that he failed in Mathematics?

(iii) What is the prob. that he failed both in Mathematics and Chemistry ? [*W.B.U.Tech 2002*]

(iv) What is the prob. that he failed in Mathematics or Chemistry.

35. Let 5% of men and 0.25% of women are colourblind. A colourblind person is chosen at random. Find the probability that the person being male. Assume that there are equal number of males and females. What is the population consisted of twice as many males as female ?

36. Section *A, B* and *C* have 50, 75 and 100 workers respectively, and , 50, 60 and 70 percent of these are men. Resignations are equally likely among all workers, irrespective of sex. One worker resigns, and this is a man. Find the probability that the person works in section *C*.

37. Three machines *X, Y, Z* produce respectively 60%, 30% and 10% of the total number of items of a factory. Of this output 2% , 3% and 4% are defective. An item is selected at random and is found defective. Find the prob. that the item was produced by machine *Z*.

38. Two persons *A* and *B* throw alternatively with a pair of dice. A wins if he throws 6 before *B* throws 7 and *B* wins if he throws 7 before A throws 6. If A begins, find his prob. of winning.

[*W.B. U.Tech 2002*]

[Hints. Same as Ex. 14]

39. A player rolls two dices. If the sum of the dice is either a 2, 3 or 12 the player loses ; if the sum is either a 7 or an 11 he wins. If the outcome is anything else, the player continues to roll the dice until he rolls either the initial outcome or a 7. If the 7 comes first, the player loses ; whereas if the initial outcome reappears before the 7, he wins. Find the probability of a player winning the game.

[Hint : $A =$ "the sum is 2 or 3 or 12"

$B =$ "the sum is 7 or, 11"

$C =$ 'the sum is anything-else'

$D =$ 'the sum is 7'

Think that the two dice are thrown infinity.

"The player wins" $= \{($B, any event, any event, $\ldots)\} + \{(C, C,$ any event, any event, $\ldots)\} + \{(C, \overline{C+D}, C,$ any event, any event, $\ldots)\} + \ldots$ upto ∞.

$\therefore$ Probabity that the player wins

$= P\{B,$ any event, any event, $\ldots)\} + P\{C, C,$ any event, any event, $\ldots)\} + P\{(C, \overline{C+D}, C,$ any event, any event, $\ldots)\} + P\{(C, \overline{C+D}, \overline{C+D}, C,$ any event, any event, $\ldots)\}$

$$=\left[P(B)\times 1\times 1\times\cdots\right]+\left[P(C)P(C)\cdot 1\cdot 1\cdots\right]+\left[P(C)\cdot P\left(\overline{C+D}\right),P(C)\cdot 1\cdot 1\cdots\right]$$

$$+\left[P(C)P\left(\overline{C+D}\right)P\left(\overline{C+D}\right)P(C)\cdot 1\cdot 1\cdots\right]+\cdots$$

$$=P(B)+\left\{P(C)^2\right\}+\left\{P(C)\right\}^2P\left(\overline{C+D}\right)+\left\{P(C)\right\}^2\left\{P\left(\overline{C+D}\right)\right\}^2$$

$$+\left\{P(C)\right\}^2\left\{P\left(\overline{C+D}\right)\right\}^3+\left\{P(C)\right\}^2\left\{P\left(\overline{C+D}\right)\right\}^4+\cdots\text{upto }\infty.$$

$$=P(B)+\left\{P(C)\right\}^2\left[1+P\left(\overline{C+D}\right)+\left\{P\left(\overline{C+D}\right)\right\}^2+\left\{P\left(\overline{C+D}\right)\right\}^3+\cdots\infty\right]$$

Now, $C=$ 'the sum is 4 or 5 or 6 or 8 or 9 or 10'

$$\therefore\quad n(C)=24.\quad\therefore\quad P(C)=\frac{n(C)}{n}=\frac{24}{6\times 6}=\frac{2}{3}$$

and $D=$ 'the sum is 7' $\therefore\ n(D)=6\quad\therefore\quad P(D)=\dfrac{6}{6\times 6}=\dfrac{1}{6}$

$$\therefore\quad P(C+D)=P(C)+P(D)=\frac{2}{3}+\frac{1}{6}=\frac{5}{6}$$

$$\therefore\quad P\left(\overline{C+D}\right)=1-\frac{5}{6}=\frac{1}{6}.$$

Now, $B=\{(1,6),(2,5),(3,4),(4,3),(5,2),(6,1),(5,6),(6,5)\}$

$$\therefore\quad n(B)=8\quad\therefore\quad P(B)=\frac{8}{36}=\frac{2}{9}.$$

$\therefore$ the probability that the player wins

$$=\frac{2}{9}+\left(\frac{2}{3}\right)^2\left\{1+\left(\frac{1}{6}\right)+\left(\frac{1}{6}\right)^2+\left(\frac{1}{6}\right)^3+\cdots\right\}$$

$$=\frac{2}{9}+\left(\frac{2}{3}\right)^2\left(1-\frac{1}{6}\right)^{-1}=\frac{2}{9}+\frac{4}{9}\times\left(\frac{5}{6}\right)^{-1}=\frac{2}{9}+\frac{4}{9}\times\frac{6}{5}=\frac{2}{9}+\frac{8}{15}$$

$$=\frac{34}{45}.$$

40. A box contains 10 pairs of shoes. If 8 shoes are randomly selected, what is the probability that there is (i) no complete pair (ii) exactly one complete pair?

[Hint (i) $\frac{^{10}C_8 \times 2^8}{^{20}C_8}$, (ii) $\frac{^{10}C_1 \times {}^9C_6 \times 2^6}{^{20}C_8}$. [*W.B.U.Tech 2005*]

41. A customer visiting a garment-shop purchases a suit with probability 0.22, a shirt with probability 0.30 and a tie with probability 0.28. The customer purchases both a suit and a shirt with probability 0.11, both a suit and a tie with probability 0.14, and both a shirt and a tie with probability 0.10. A customer purchases all three items with probability 0.06. What is the probability that a customer purchases (i) none of these items (ii) exactly one of these items.

42. Probability of recording temperature of Delhi 70° F is 0.3 and that of Mumbai is 0.4. The probability of recording the maximum of temperatures in Delhi and Mumbai as 70° F is 0.2. Find the probability that the minimum of the two city temperatures is 70° F.

43. Urn I contains 7 black and 5 white balls. Urn II contains 12 black and 3 white balls. We toss a fair coin. If head comes then a ball from Urn I is drawn,whereas if tail comes a ball from Urn II is drawn. Suppose that a white ball is drawn. What is the probability that the coin shows tails ?

44. Three machine men *A*, *B*, and *C*, produces a special kind of electronic-toy, with respective probabilities 0.02, 0.03 and 0.05 it fails to be established in market. In the factory where they work, A produces 50% of all toys, *B* 30% and *C* 20 %. What proportion of "non-establishment" is caused by *A* ?

45. There are 3 coins in an urn. One is a fair coin and another is a two headed coin; and the third is a biased coin that gives heads 75% times among all throws. When one of the three coins is selected at random and tossed, it gives head. Find the probability that it is the two-headed coin.

46. An Engineering system consisting of 4 components is such that the system functions if and only if at least 2 components work. Suppose that all components work independently of each other. If the *i*-th component works with probability $p_i (i = 1, 2, 3, 4)$, compute the probability that the system functions.

[Hint : Let the event, $A_i = i$ th component works $\therefore$ $P(A_i) = p_i$.

Now "The System does not function"

$$= (\bar{A}_1 \cap \bar{A}_2 \cap \bar{A}_3 \cap \bar{A}_4) \cup [A_1 \cap \bar{A}_2 \cap \bar{A}_3 \cap \bar{A}_4]$$

$$+ [\bar{A}_1 \cap A_2 \cap A_3 \cap \bar{A}_4] + [\bar{A}_1 \cap \bar{A}_2 \cap A_3 \cap \bar{A}_4] + [\bar{A}_1 \cap \bar{A}_2 \cap \bar{A}_3 \cap A_4]$$

$\therefore$ Probability that the system does not function

$$= P(\bar{A}_1 \cap \bar{A}_2 \cap \bar{A}_3 \cap \bar{A}_4) + \{P(A_1 \cap \bar{A}_2 \cap \bar{A}_3 \cap \bar{A}_4) + P(\bar{A}_1 \cap A_2 \cap \bar{A}_3 \cap \bar{A}_4)$$

$$+ P(\bar{A}_1 \cap \bar{A}_2 \cap A_3 \cap \bar{A}_4) + P(\bar{A}_1 \cap \bar{A}_2 \cap \bar{A}_3 \cap A_4)\}$$

$$= (1-p_1)(1-p_2)(1-p_3)(1-p_4) + \{p_1(1-p_2)(1-p_3)(1-p_4)$$

$$+(1-p_1)p_2(1-p_3)(1-p_4)$$

$$+(1-p_1)(1-p_2)p_3(1-p_4) + (1-p_1)(1-p_2)(1-p_3)p_4\}$$

Next see, Probability that the system function

$= 1-$ Probability that the system does not function.]

47. The probability that a construction job will be finished in time is 17/20; the probability that there will be no strike is 3/4 ; and the probability that the construction job will be finished in time, assuming that there will be no strike, is 14/15. Find the probability that (i) the construction job will be finished in time and there will be no strike. (ii) there will be strike or the job will not be finished in time.

48. One shot is fired from each of three guns. Let A, B, C denote the events that the target is hit by the first, second and the third gun respectively. Assuming that A, B, C are mutually independent events and that $P(A) = 0.5,\ P(B) = 0.6, P(C) = 0.8$ find the probability that at least one hit is registered.

49. There are three men aged 60, 65 and 70 years. The probability to live 5 years more is 0.8 for a 60 year old, 0.6 for a 65 year old, and 0.3 for a 70 year old person. Find the probability that at least two of the three persons will remain alive 5 years hence.

[Hint :

$(0.8)(0.6)(1-0.3)+(0.8)(1-0.6)(0.3)+(1-0.8)(0.6)(0.3)+(0.8)(0.6)(0.3)=612$]

50. A forest contains 20 lions, of which 5 are captured tagged and then released. A few months later 4 of the 20 lions are captured. What is the probability that 2 of these 4 have been tagged ? What assumption are you making ?

51. A teacher gives his class a set of ten problems with the information that the final examination consists of a random selection of 5 of them. If a student comes out how to do 7 of the problems what is the probability that he or she answers correctly ? (i) all 5 problems (ii) at least 4 of the problems ?

ANSWERS

1. (i) 0.56 (ii) $\frac{27}{28}$ **2.** 5 **3.** $\frac{5}{18}$ **5.** 53/161700

6. (i) $\frac{10}{21}$ (ii) $\frac{17}{42}$ (iii) $\frac{25}{42}$ **7.** 14/45 **8.** 0.027, 0.111

9. $\frac{5}{429}$ **10.** 101/105 **11.** 0.6 **12.** $\frac{1}{6}, \frac{1}{2}$ **13.** 2/3 **14.** 3/4

15. 7/429 **16.** 9/25 **17.** 11/115 **18.** 2.6084×10^{-6}

19. (i) 128/625, (ii) 821/3125 **20.** $\frac{13}{32}$ **21.** 35/44, 35/48

22. (i) $\frac{5}{8}$, (ii) $\frac{3}{8}$. **23.** $\frac{3}{4}$ **24.** $\frac{1}{3}$ **25.** $\frac{140}{401}$

26. 1430/1997 **27.** (i) $\frac{4}{9}$ (ii) $\frac{1}{2}$ **28.** $\frac{a}{a+b}$

29. (i) 0.41 (ii) 12/41 **30.** 0.9835

31. (i) 0.331 (ii) 0.383 (iii) 0.286 (iv) 48.62 percent.

32. 0.26 **33.** 4/17 **34.** (i) $\frac{2}{5}$, (ii) $\frac{2}{3}$, (iii) $\frac{1}{10}$, (iv) $\frac{3}{10}$.

35. $\frac{20}{21}; \frac{40}{41}$ **36.** $\frac{1}{2}$ **37.** 4/25 **38.** $\frac{30}{61}$ **39.** 34/45

40. .09145; .4268 **41.** (i) 0.49 (ii) 0.28 **42.** 0.5 **43.** 12/37

44. 34.48 **45.** 4/9 **46.** $\left[1-4\sum p_1p_2+2\sum p_1p_2p_3-3p_1p_2p_3p_4\right]$

47. 7/10, 3/10 **48.** 0.96 **50.** $\frac{70}{323}$ **51.** .0833 ; 0.5

[III] MULTIPLE CHOICE QUESTIONS

1. The probability that the sum 8 appear in a single toss of pair of a fair dice is

(a) $\frac{31}{36}$ (b) $\frac{5}{36}$

(c) $\frac{5}{6}$ (d) $\frac{1}{6}$.

2. Of 6 girls in a class, 3 have blue eyes. If 2 of the girls are chosen at random, the probability that both have blue eyes is

(a) $\frac{1}{5}$ (b) $\frac{4}{5}$

(c) $\frac{1}{10}$ (d) none.

3. If A and B are two independent events then

(a) A, B are exhaustive events

(b) $\overline{A}, \overline{B}$ are independent events.

(c) A, B are mutually exclusive events.

(d) None.

4. A die is tossed. If the number is odd, then the probability that it is prime, is

(a) $\frac{1}{2}$ (b) $\frac{2}{3}$

(c) $\frac{1}{3}$ (d) none.

5. If A and B be events with $P(A)=\frac{1}{3}, P(B)=\frac{1}{4}$, and $P(A \cup B)=\frac{1}{2}$ then $P(B/A)=$

(a) $\frac{3}{4}$ (b) $\frac{4}{3}$

(c) $\frac{1}{4}$ (d) $\frac{1}{3}$

6. If A and B be events with $P(A)=\frac{1}{2}, P(B)=\frac{1}{3}$ and $P(A\cup B)=\frac{2}{3}$ then $P(A\cap B)=$

(a) $\frac{1}{6}$ (b) $\frac{5}{6}$

(c) $\frac{1}{3}$ (d) none.

7. If X, Y are two events with $P(X)=\frac{5}{6}, P(Y)=\frac{2}{3}$, and $P(X\cap Y)=\frac{1}{4}$, then $P(X\cap \overline{Y})=$

(a) $\frac{5}{12}$ (b) $\frac{7}{12}$

(c) $\frac{1}{6}$ (d) none.

[Hints : $P(X\cap\overline{Y})=P(X)-P(X\cap Y)$

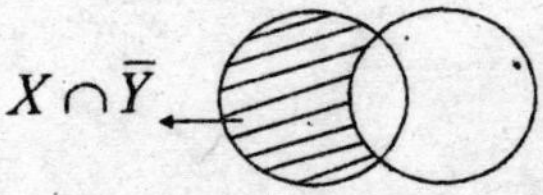

]

8. Two events are said to be independent if

(a) $P(A\cup B)=P(A)\cdot P(B)$ (b) $P(A/B)=P(A)$

(c) $P(A\cap B)\neq P(A)\cdot P(B)$ (d) $P(A/B)=P(B)$.

9. If A_1 and A_2 be two mutually exclusive i.c. disjoint events, then $P(A_1\cup A_2)$ is equal to [*W.B.U.Tech 2006, M 302*]

(a) $P(A_1)+P(A_2)-P(A_1\cap A_2)$ (b) $P(A_1)\cdot P(A_2)$

(c) $P(A_1)+P(A_2)$ (d) $P(A_1)-P(A_2)$.

10. Let A,B be events with, $P(A\cup B)=\frac{7}{8}, P(A\cap B)=\frac{1}{4}$ and $P(\overline{A})=\frac{5}{8}$. Then $P(B)=$

(a) $\frac{3}{8}$ (b) $\frac{3}{4}$

(c) $\frac{1}{2}$ (d) none.

11. A coin is tossed successively three times. The probability of getting exactly one head is

(a) $\frac{1}{8}$ (b) $\frac{1}{2}$

(c) $\frac{3}{8}$ (d) none.

12. If $P(A)=0.2, P(B)=0.4, P(A\cup B)=0.6$, then the events A, B are

(a) independent (b) mutually exhaustive

(c) mutually exclusive (d) none.

13. A and B are events with $P(A)=\frac{1}{3}, P(B)=\frac{2}{5}$ and $P(AB)=\frac{2}{15}$. Are A and B independent ?

(a) Yes (b) No.

14. If $P(A_1)+P(A_2)+P(A_3)=1$, then the events A_1, A_2, A_3 are

(a) Pairwise independent.

(b) mutually exclusive

(c) mutually exhaustive

(d) mutually exclusive and exhaustine.

15. If two events A and B are independent, then

(a) $P(A\cup B)=P(A)P(B)$ (b) $P(\overline{A}\cup B)=P(\overline{A})+P(B)$

(c) $P(\overline{A}\cap\overline{B})=P(\overline{A})P(\overline{B})$ (d) $P(\overline{A}\cup\overline{B})=P(\overline{A})+P(\overline{B})$.

16. If $P(A)=\frac{1}{3}, P(B)=\frac{2}{5}$, then $P(\overline{A}\overline{B})=?$

(a) $\frac{2}{15}$ (b) $\frac{1}{5}$

(c) $\frac{2}{5}$ (d) $\frac{4}{15}$.

17. When two events B_1, A_2 are not mutually exclusive, then

(a) $P(A+B)\le P(A)+P(B)$ (b) $P(AB)=P(A)+P(B)$

(c) $P(A+B)=P(A)+P(B)$ (d) $P(A+B)>P(A)+P(B)$.

18. If $P(B)=P(C)=1$. then both the events B,C are

(a) mutually exhustive (b) independent

(c) certain (d) mutually exclusive.

19. For any two events A_1, A_2 where $A_1 \subset A_2$, we have

(a) $P(A_2 - A_1) = P(A_2) - P(A_1)$

(b) $P(A_2 + A_1) = P(A_2) + P(A_1)$

(c) $P(A_2 - A_1) \leq P(A_2) - P(A_1)$

(d) $P(A_2 + A_1) \leq P(A_2) + P(A_1)$. [W.B.U.Tech.2008]

20. If $P(A+B)=\frac{2}{7}$, then $P(\overline{A}\overline{B})=?$

(a) $\frac{1}{7}$ (b) $\frac{2}{7}$

(c) $\frac{5}{7}$ (d) none.

[Hints : $P(\overline{A}\overline{B}) = 1 - P(A+B)$]

21. The chance that a leap year selected at random will contain 53 wednesdays is [W.B.U.Tech.2006, 2007]

(a) $\frac{2}{7}$ (b) 0

(c) 1 (d) $\frac{5}{6}$.

22. If $P(A)=\frac{1}{2}, P(B)=\frac{1}{3}, P(AB)=\frac{1}{4}$, then the value of $P(A \cup B)$ is [W.B.U.Tech 2007]

(a) $\frac{6}{7}$ (b) $\frac{3}{7}$

(c) 1 (d) $\frac{7}{12}$.

23. A coin is tossed. Event $\{H\},\{T\}$ are [W.B.U.Tech 2007]

(a) mutually exclusive (b) independent events

(c) dependent (d) both (a) and (c).

24. The condition for independence of two events A and B is

(a) $P(A \cap B) = P(A)P(B)$ (b) $P(A+B) = P(A) \cdot P(B)$

(c) $P(A-B) = P(A) \cdot P(B)$. (d) $P(A \cap B) = P(A)P(B / A)$

[*W.B.U.Tech. 2006*]

25. The probability of obtaining an even number in the throw of a fair die is

(a) $\frac{1}{2}$ (b) $\frac{1}{3}$

(c) $\frac{1}{4}$ (d) none of these.

26. 3 balls are drawn from a bag containing 6 red and 5 white balls. The probability of getting the balls all red is

(a) $\frac{6}{11}$ (b) $\frac{3}{7}$

(c) $\frac{1}{2}$ (d) none of these.

27. In a single cast with two dice the chance that one die turns up 3 and the other 4 is

(a) $\frac{1}{12}$ (b) $\frac{1}{18}$

(c) $\frac{1}{36}$ (d) $\frac{1}{3}$.

28. A bag contains 5 brown and 4 white balls. A man pulls out two balls from the bag. The probability that they are of the same colour is

(a) $\frac{1}{5}$ (b) $\frac{3}{7}$

(c) $\frac{4}{9}$ (d) none of these.

29. Three of six vertices of a regular hexagon are chosen at random. The probability that the triangle with three vertices is equilateral is

(a) $\frac{1}{6}$ (b) $\frac{1}{3}$

(c) $\frac{1}{7}$ (d) $\frac{1}{10}$.

30. An unbiased coin is tossed repeatedly. If 'tail' appears on first four tosses, then the probability of 'head' appearing on the fifth toss is

(a) $\frac{1}{2}$ (b) $\frac{1}{4}$

(c) $\frac{1}{3}$ (d) $\frac{2}{9}$.

31. Seven white balls and three black balls are randomly placed in a row. The probability that no two black balls are placed side by side is

(a) $\frac{3}{4}$ (b) $\frac{7}{15}$

(c) $\frac{2}{9}$ (d) $\frac{1}{4}$.

32. Five boys and three girls are seated at random in a row. The probability that no boy sits between two girls is

(a) $\frac{3}{28}$ (b) $\frac{1}{28}$

(c) $\frac{3}{25}$ (d) $\frac{4}{19}$.

33. Three different numbers are selected at random from the set $\{1,2,3,4,5,6,7,8,9,10\}$. The probability that the product of two of the numbers is equal to the third is

(a) $\frac{3}{5}$ (b) $\frac{3}{24}$

(c) $\frac{1}{38}$ (d) $\frac{1}{40}$.

34. The probability of a student getting first class, second class and third class at an examination are $\frac{1}{10}, \frac{2}{5}$ and $\frac{1}{5}$ respectively. The probability that he fails is

(a) $\frac{1}{10}$ (b) $\frac{4}{9}$

(c) $\frac{3}{10}$ (d) $\frac{7}{11}$.

35. The probability that Pradyut passes is 0.9 and Subrata passes is 0.8. The probability that at least one of them passes is

(a) 0.98 (b) 0.97

(c) 0.9 (d) none of these.

36. A fair die is thrown. The probability that either an odd number or a number greater then 4 will turn up is

(a) $\frac{2}{5}$ (b) $\frac{3}{7}$

(c) $\frac{2}{7}$ (d) $\frac{2}{3}$.

37. Three bombs are dropped to destroy a bridge. The probabilities of hitting the bridge by these bomb are 0.5, 0.2 and 0.1 respectively. The probability that the bridge is hit is

(a) 0.36 (b) 0.64

(c) 0.80 (d) 0.90.

38. The probability of an event A occuring is 0.5, and of B occuring is 0.3. If A and B are mutually exclusive events, then the probability of neither A nor B occuring is

(a) 0.2 (b) 0.1

(c) 0.8 (d) none of these.

39. If $P(B)=\frac{3}{4}, P(\overline{A}\cap B\cap\overline{C})=\frac{1}{3}, P(A\cap B\cap\overline{C})=\frac{1}{3}$ then $P(B\cap C)=$

(a) $\frac{3}{5}$ (b) $\frac{1}{14}$

(c) $\frac{1}{12}$ (d) $\frac{3}{7}$.

40. The probability that at least one of the events A and B occurs is 0.6. If A and B occur simultaneously with probability 0.2, then $P(\overline{A})+P(\overline{B})=$

(a) 1.5 (b) 2.4

(c) 1.2 (d) 0.3.

41. Two numbers are chosen from the set $\{1,2,3,4,5,6\}$ one after another without replacement. The probability that the smaller value of the two is less than 4 is

(a) $\frac{4}{5}$ (b) $\frac{1}{5}$

(c) $\frac{3}{5}$ (d) $\frac{7}{11}$.

Answers

1.b	2.a	3.b	4.b	5.c	6.a	7.b	8.b
9.c	10. b	11.c	12.c	13.a	14.d	15.c	16.c
17. a	18.c	19.a	20.c	21.a	21. a	22.d	23.d
24.a	25.a	26.d	27.b	28.c	29.d	30.a	31.b
32.a	33.d	34.c	35.a	36.d	37.b	38. a	39.c
40.c	41.a						

3.2 PROBABILITY DISTRIBUTION AND EXPECTATION

3.2.1. Random Variable.

Definition. Let S be a given sample space. Then a real valued function X defined on S is called a random or a stochastic variable or sometimes a variate. Thus for every point s of a sample space S, we have a unique real value of X i.e $X(s)$

The range of the function X i.e. the set of all values assumed by X is called the ***Spectrum*** of the random variable.

Discrete Random Variable. A random variable X is said to be discrete if the spectrum of X is finite or countably infinite i.e an infinite sequence of distinct values.

Continuous Random Variable. A random variable X is said to be continuous if it can assume every value in an interval .

Event Described by Random Variable. The set of all sample points s for which $X(s) \in A$, a given set of real numbers, is an event and is denoted by $(X \in A)$. In particular, the event $(X = a)$ is the set of all sample points corresponding to which X takes the value a.

Let [a, b] be a given closed interval. Then the set of all sample points s for which $a \le X(s) \le b$ is an event and is denoted by $(a \le X \le b)$. Similarly the events $(a < X \le b)$, $(a \le X < b)$ and $(a < X < b)$ are defined. Also the event $(-\infty < X \le x)$ is abbreviated as $(X \le x)$ where x is a real number. Further the events $(-\infty < X < \infty)$ and $(-\infty < X < -\infty)$ denote respectively the certain event S and the impossible event ϕ.

Thus we see an event can be described by a random variable.

Illustration. (i) Let us consider the random experiment of tossing two (unbiased) coins. Then the sample space S contains 4 sample points.

i.e., $S = \{HH, HT, TH, TT\}$.

Let the random variable X be such that X (an outcome) = "the number of heads". Then X is a function over S defined by

$$X(HH) = 2, \quad X(HT) = X(TH) = 1, \quad X(TT) = 0.$$

Thus the spectrum of X is $\{0, 1, 2\}$ which is a finite set. Hence X is a discrete random variable here. Here the event $(X = 1) = \{TH, HT\} =$ 'one head', the event $(-1 < X \le 0) = \{TT\} =$ 'Two tails'

(ii) Let the random variable X denote the weights (in kg) of a group of individuals. Then X can assume every value in an interval say (30, 100), suppasing there is no individual having weight less than 30 and greater than 100. Hence X is a continuous random variable. Here the event $(42 < X \le 50) =$ the group of individuals whose weight lie between 42 and 50, including 50 ; the event $(X = 70) =$ The group of individuals whose weight is 70 kg.

3.2.2. Probability Mass Function and Discrete Distribution

Let X be a discrete random variable which assumes the values $x_0, x_1, x_2, \cdots, x_n, \cdots$ Let $P(X = x_i) = f(x_i) = f_i$. So, the value of f_i depends on x_i i.e. i. This function f_i is called *Probability mass function* **(p.m.f)** of the random variable X. A particular value of f_i is called a probability mass.

The set of ordered pairs (x_i, f_i) is called the discrete probability distribution of the random variable X

Discrete distribution is presented in the following way :

X	:	x_0	x_1	x_2	...	...
f_i	:	f_0	f_1	f_2	...	...

Illustration. For the random experiment of tossing two coins as given in Illustration (i) of art 2.1 we see X assumes the values 0, 1 and 2. Moreover, $P(X = 0) = \frac{1}{4}, P(X = 1) = \frac{1}{2}, P(X = 2) = \frac{1}{4}$.

So, the distribution of the number of heads is given by

X	:	0	1	2
f_i	:	$\frac{1}{4}$	$\frac{1}{2}$	$\frac{1}{4}$

Fundamental Properties of pmf.

If

X	:	x_0	x_1	x_2	x_3	...	...
f_i	:	f_0	f_1	f_2	f_3	...	...

is a discrete distribution of X, then the pmf has following two properties :

(i) $f_i \geq 0$ (ii) $\sum_i f_i = 1$,

Proof: (i) $f_i = P(X = x_i) \geq 0$ since a probability is always ≥ 0

(ii) $\sum_i f_i = \sum P(X = x_i) = P(S) = 1$ as $S = \{x_0, x_1, x_2, \cdots \cdots\}$ = event space.

3.2.3. Distribution Function or Cumulative Distribution Function.

The distribution function (d.f) of a random variable X (discrete or continuous) is given by

$$F(x) = P(-\infty < X \leq x), \quad -\infty < x < \infty$$

Thus, if $x_i \leq x < x_{i+1}$, then

$$F(x) = P(X = x_0) + P(X = x_1) + \cdots + P(X = x_i) = \sum_{\alpha=0}^{i} f_\alpha .$$

Illustration. In the discrete distribution

X : 0 1 2

f_i : $\frac{1}{4}$ $\frac{1}{2}$ $\frac{1}{4}$, the d.f is

$$F(x) = 0, \quad x < 0$$

$$= \frac{1}{4}, \quad 0 \leq x < 1$$

$$= \frac{1}{4} + \frac{1}{2}, \quad 1 \leq x < 2$$

$$= \frac{1}{4} + \frac{1}{2} + \frac{1}{4}, 2 \leq x .$$

Properties of Distribution Function.

(i) The distribution function $F(x)$ is a monotonic non-decreasing function.

(ii) $F(-\infty) = 0$ and $F(\infty) = 1$ and hence $0 \leq F(x) \leq 1$.

(iii) $F(x)$ is continuous on the right at all points and has a jump discontinuity on the left at $x = a$, the height of jump being equal to $P(X = a)$ i.e., $\lim_{x \to a+} F(x) = F(a)$ and $F(a) - \lim_{x \to a-} F(x) = P(x = a)$

(iv) Suppose a and b are any real numbers such that $a < b$

Then $P(a < X \le b) = F(b) - F(a)$,

$P(a < X < b) = F(b) - F(a) - P(X = b)$

and $P(a \le X < b) = F(b) - F(a) - P(X = b) + P(X = a)$

Proof : Left as exercise.

Illustration. (i) Let X be a random variable denoting the number of points appearing in a toss of a die. The distribution of X is

X :	1	2	3	4	5	6
f_i :	$\frac{1}{6}$	$\frac{1}{6}$	$\frac{1}{6}$	$\frac{1}{6}$	$\frac{1}{6}$	$\frac{1}{6}$

Now, if $x < 1$, $F(x) = P(X \le x) = 0$

If $1 \le x < 2$, $F(x) = P(X \le x) = f_1 = \frac{1}{6}$

If $2 \le x < 3$, $F(x) = P(X \le x) = f_1 + f_2 = \frac{1}{6} + \frac{1}{6} = \frac{2}{6}$

and so on.

Thus the distribution function $F(x)$ is given by :

$$
\begin{aligned}
F(x) &= 0, \quad -\infty < x < 1 \\
&= \frac{1}{6}, \quad 1 \le x < 2 \\
&= \frac{2}{6}, \quad 2 \le x < 3 \\
&= \frac{3}{6}, \quad 3 \le x < 4 \\
&= \frac{4}{6}, \quad 4 \le x < 5 \\
&= \frac{5}{6}, \quad 5 \le x < 6 \\
&= 1, \quad 6 \le x < \infty
\end{aligned}
$$

The graph of the distribution $F(x)$ is as follow :

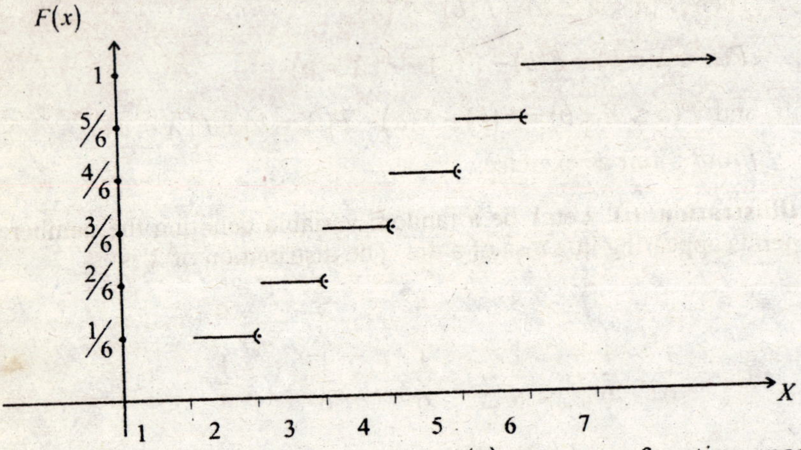

From the graph it is clear that $F(x)$ is a step function, non decreasing and is continuous on right at $x = 1, 2, 3, \ldots \ldots, 6$ and has a jump discontinuity on left at 1, 2, 3, ..., 6, the height of jump is $\frac{1}{6}$. Also, $F(-\infty) = 0,\ F(\infty) = 1$

(ii) Let three balls be drawn at random from a bag containing 5 white and 3 black balls ; X denotes the number of white balls drawn.

Then X can assume the values 0, 1, 2, 3.

Here, $f_0 = P(X = 0) =$ Probability of no white ball $= {}^3C_3 / {}^8C_3 = \frac{1}{56}$

$f_1 = P(X = 1) =$ Probability of one white ball

$$= {}^5C_1 \times {}^3C_2 / {}^8C_3 = \frac{15}{56}$$

$$f_2 = P(X = 2) = {}^5C_2 \times {}^3C_1 / {}^8C_3 = \frac{15}{28}$$

$$f_3 = P(X = 3) = {}^5C_3 / {}^8C_3 = \frac{5}{28}$$

Then the distribution of X is

X	:	0	1	2	3
f_i	:	$\frac{1}{56}$	$\frac{15}{56}$	$\frac{15}{28}$	$\frac{5}{28}$

From this we get the probability of the event like 'at most two white ball' $= P(X \le 2) = P(X=0) + P(X=1) + P(X=2)$

$$= f_0 + f_1 + f_2 = 1 - f_3 = 1 - \frac{5}{28} = \frac{23}{28}.$$

3.2.4. Probability Density Function (or density function) and Continuous Distribution.

Let X be a continuous random variable with distribution function $F(x) = P(-\infty < X \le x)$. Then a function $f(x)$ is said to be probability density function (*pdf*) of X if $f(x)$ is integrable on the interval $[a, x]$ for all a and if

$$F(x) = \int_{-\infty}^{x} f(t)dt$$

This holds for every real x.

Fundamental Properties of pdf.

If $f(x)$ is a *pdf* of a random variable X then

(i) $f(x) \ge 0$ (ii) $\int_{-\infty}^{\infty} f(x)dx = 1$

Proof: (i) From definition $F'(x) = f(x)$. Since $F(x)$ is increasing so $F'(x) \ge 0 \therefore f(x) \ge 0$

(ii) Since $F(\infty) = 1$ so, $\int_{-\infty}^{\infty} f(t)dt = F(\infty) = 1$.

Properties of pdf.

(i) As $F(x)$ is a continuous function, so we must have $P(X = a) = F(a) - \lim_{x \to a-} F(x) = 0$.

(ii) For a continuous distribution,

$$P(a \le X \le b) = P(a < X < b) = P(a < X \le b)$$

$$= P(a \le X < b) = \int_{a}^{b} f(t)dt = F(b) - F(a).$$

where f is the probability density function.

(iii) If $f(x)$ is continuous, then from (definition of pdf), we must have $f(x) = F'(x)$

(iv) In differential notation we have,

$$P(x < X \le x + dx) = F(x + dx) - F(x) = dF(x) = F'(x)dx = f(x)\,dx$$

which is known as the prob. differential of X.

Density Curve. The curve given by $y = f(x)$ ($f(x)$ is pdf) is called the probability density curve which gives the graphical representation of the corresponding continuous distribution.

Illustration. Consider a function $f(x)$ ($f(x)$ is pdf) which is defined as $f(x) = \frac{2}{x^3}$, $1 \le x < \infty$

$= 0$, elsewhere.

As $f(x) \ge 0$ everywhere and $\int_{-\infty}^{\infty} f(x)dx$

$$= \int_1^{\infty} \frac{2}{x^3} dx = \underset{P \to \infty}{Lt} \int_1^{P} \frac{2}{x^3} dx = \underset{P \to \infty}{Lt} \left(1 - \frac{1}{P^2}\right) = 1 - 0 = 1,$$

so this $f(x)$ is a probability density function of some random variable.

Now $F(x) = \int_{-\infty}^{x} f(x)dx = \int_{-\infty}^{x} 0 \cdot dx = 0$ when $-\infty < x < 1$

and $F(x) = \int_{-\infty}^{x} f(x)dx = \int_1^{x} \frac{2}{x^3} dx$ when $1 \le x < \infty$

$= 1 - \frac{1}{x^2}$.

So the distribution function of the above *pdf* is

$F(x) = 0, \quad -\infty < x < 1$

$= 1 - \frac{1}{x^2}, \quad 1 \le x < \infty$.

3.2.5. Expectation or Mean of a Random Variable

Expectation of a Discrete Random Variable

Let X be a discrete random variable whose distribution is

X	:	x_0	x_1	x_2	...	x_n	...
f_i	:	f_0	f_1	f_2	...	f_n	...

Then the mean or expectation or expected value of X, denoted by $E(X)$ or $m(X)$ or simply m is defined as

$E(X) = x_0 f_0 + x_1 f_1 + x_2 f_2 + \cdots \cdots = \sum_i x_i f_i$, provided the series is absolutely convergent if the above sum is an infinite series.

Expectation for Continuous Random Variable.

For a continuous random variable X with probability density function $f(x)$, the mean or expectation of X is defined as

$$E(X) = \int_{-\infty}^{\infty} x\, f(x)\, dx,$$

provided the infinite integral converges absolutely.

Similarly, the mean of a function $\psi(X)$ of the random variable X denoted by $E\{\psi(X)\}$ is defined as

$$E\{\psi(X)\} = \sum_i \psi(X_i) f_i, \qquad \text{for a discreate distribution}$$

$$= \int_{-\infty}^{\infty} \psi(x) f(x) dx, \qquad \text{for a continuous distribution.}$$

Illustration. (i) Suppose a die is rolled. Let X be the number of points on the die. Then its values are 1, 2, 3, 4, 5, 6.

$$\therefore\ P(X = i) = \frac{1}{6} \text{ for } i = 1, 2, 3, 4, 5, 6.$$

So the distribution of X is

X :	1	2	3	4	5	6
f_i :	$\frac{1}{6}$	$\frac{1}{6}$	$\frac{1}{6}$	$\frac{1}{6}$	$\frac{1}{6}$	$\frac{1}{6}$

Therefore its expectation,

$$E(X) = 1 \cdot \frac{1}{6} + 2 \cdot \frac{1}{6} + 3 \cdot \frac{1}{6} + \cdots + 6 \cdot \frac{1}{6} = \frac{7}{2}$$

and $E(X^2 + 1) = (1^2 + 1) \cdot \frac{1}{6} + (2^2 + 1) \cdot \frac{1}{6} + (3^2 + 1) \cdot \frac{1}{6} + (4^2 + 1) \cdot \frac{1}{6} + (5^2 + 1)$

$$\cdot \frac{1}{6} + (6^2 + 1) \cdot \frac{1}{6} = \frac{97}{6}$$

(ii) Let the p.d.f of a continuous random variable X is

$$f(x)=\frac{1}{2} \quad \text{in } -1<x<1$$
$$=0, \quad \text{elsewhere.}$$

Then the mean or expectation of X is

$$E(X)=\int_{-\infty}^{\infty} x\, f(x)\, dx=\int_{-1}^{1} x \frac{1}{2}\, dx=0$$

and also, $E(2X^3)=\int_{-\infty}^{\infty} 2x^3 f(x) dx=\int_{-1}^{1} 2x^3 \cdot \frac{1}{2} dx=\left[\frac{x^4}{4}\right]_{-1}^{1}=0$

Properties of Expectation

(i) $E(a)=a$, where a is a constant.

(ii) $E(aX)=aE(X)$, a being a constant.

(iii) $E(X \pm Y)=E(X) \pm E(Y)$, X, Y are two r.v

(iv) $E(XY)=E(X)E(Y)$ if the two r.v X and Y are independent.

Proof: Left as an exercise.

Remark. The mean has an important physical significance. In fact this represents the centre of mass of the probability distribution.

Illustration. A number is chosen at random from the set $\{1, 2, \cdots, 100\}$ and another number is chosen at random from the set $\{1, 2, 3, \cdots, 50\}$. We are to find the expected value of the product.

Let $X=$ first number and $Y=$ second number. So the distribution of X is

X	:	1	2	3	...	100
f_i	:	$\frac{1}{100}$	$\frac{1}{100}$	$\frac{1}{100}$	...	$\frac{1}{100}$

and the distribution of Y is

Y	:	1	2	3	...	50
f_i	:	$\frac{1}{50}$	$\frac{1}{50}$	$\frac{1}{50}$	...	$\frac{1}{50}$

So,

$$E(X)=\frac{1}{100}\cdot 1+\frac{1}{100}\cdot 2+\cdots+\frac{1}{100}\cdot 100=\frac{1}{100}(1+2+3+\cdots+100)$$

$$=\frac{1}{100}\cdot\frac{100(100+1)}{2}=\frac{101}{2}$$

and $E(Y)=\frac{1}{50}(1+2+3+\cdots+50)=\frac{1}{50}\cdot\frac{50(50+1)}{2}=\frac{51}{2}$

The expected value of the product

$=E(XY)=E(X)E(Y)$ [$\because$ the two numbers are drawn independently]

$=\frac{101}{2}\times\frac{51}{2}=\frac{5151}{4}$.

3.2.6. Variance and S.D.

The variance of a r.v X, denoted by $Var(X)$ is defined as $Var(X)=E\left((X-m)^2\right)$, where $m=E(X)$

The positive square root of $Var(X)$ is called the *standard deviation* of X and is denoted by $\sigma(X)$ or σ_x or simply σ. Thus $\sigma=+\sqrt{\text{var}(X)}$.

Remarks : (i) The variance describes how widely the probability masses are spread about the mean i.e it gives an inverse measure of concentration of the probability masses about the mean which is called the measure of dispersion.

(ii) As $Var(X)=0$ only when $X-m=0$ i.e, $X=m$, so in that case the whole mass is concentrated at the mean.

Theorem.

(i) $Var(X)=E(X^2)-m^2=E(X^2)-\{E(X)\}^2$

(ii) $Var(aX+b)=a^2Var(X)$

(iii) $Var(k)=0$ where k is constant.

(iv) $Var(X)=E\{X(X-1)\}-m(m-1)$ where m is mean of X.

Proof: (i) $Var(X)=E\left\{(X-m)^2\right\}=E(X^2-2mX+m^2)$

$$=E(X^2)-E(2mX)+E(m^2)$$

$$=E(X^2)-2mE(X)+m^2=E(X^2)-2m\cdot m+m^2=E(X^2)-m^2$$

(ii) and (iii) are left as exercise.

(iv) $(X-m)^2=X(X-1)-2mX+X+m^2$

$$\therefore\ E\left\{(X-m)^2\right\}=E\{X(X-1)\}-2mE(X)+E(X)+E(m^2)$$

$$=E\{X(X-1)\}-2m\cdot m+m+m^2$$

$$=E\{X(X-1)\}-m(m-1).$$

Note : In fact the result (i) and (iv) of the above theorem are used to evaluate variance and standard deviation.

Illustration. Consider the following distribution of a random variable X :-

$$f(x)=\frac{1}{2}x, \qquad 0 \le x \le 2$$
$$=0, \qquad \text{elsewhere.}$$

Then the expectation of X is

$$E(X)=\int_{-\infty}^{\infty} x\, f(x)dx=\int_{0}^{2} x\cdot\frac{1}{2}x\, dx=\left[\frac{x^3}{6}\right]_0^2=\frac{4}{3}.$$

Now, $E\left(X^2\right)=\int_{-\infty}^{\infty} x^2 f(x)dx=\int_{0}^{2} x^2\cdot\frac{1}{2}x\, dx=\left[\frac{x^4}{8}\right]_0^2=2$

$\therefore\ Var(X)=E(X^2)-\{E(X)\}^2=2-\frac{16}{9}=\frac{2}{9}$. So, s.d$=\sqrt{2}/3$.

Illustrative Examples.

Ex. 1. *Find the probability distribution (or probability function or p.m.f) of the number of heads when a fair coin is tossed repeatedly until the first tail appears.*

The sample space corresponding to the random experiment of tossing a fair coin is $S=\{T, HT, HHT, HHHT, \cdots\}$.

Let the random variable X denote "the number of heads in the experiment until the first tail appears".

Then the spectrum of X is $\{0,1,2,3,\cdots\}$

Now, $P(X=0)=P(T)=\frac{1}{2}$

$P(X=1)=P(HT)=P(H)\cdot P(T)$ [$\because$ the trials are independent]

$$=\frac{1}{2}\cdot\frac{1}{2}=\frac{1}{2^2}$$

$P(X=2)=P(HHT)=P(H)P(H)P(T)=\frac{1}{2^3}$ and so on.

Hence the probability distribution of X is

X	:	0	1	2	3	...	...
f_i	:	$\frac{1}{2}$	$\frac{1}{2^2}$	$\frac{1}{2^3}$	$\frac{1}{2^4}$	...	...

Ex. 2. *A random variable X has the following probability mass function* [*W.B.U.Tech 2007*]

X :	0	1	2	3	4	5	6	7
$P(X=k)=f(x)$:	0	k	$2k$	$2k$	$3k$	k^2	$2k^2$	$7k^2+k$

(i) Determine the constant k

(ii) Evaluate $P(X<6)$, $P(X\geq 6)$, $P(3<X\leq 6)$ *and* $P(3<X/X\leq 6)$

(iii) Find the minimum value of x so that $P(X\leq x)>\frac{1}{2}$

(iv) Obtain the distribution function $F(x)$. [*W.B.U.Tech 2004*]

(i) Since $f(x)$ is a p.m.f, $\sum_x f(x)=1$

$$\therefore \sum_{x=0}^{7} f(x)=1 \Rightarrow 0+k+2k+2k+3k+k^2+2k^2+7k^2+k=1$$

$$\Rightarrow 10k^2+9k-1=0 \Rightarrow (10k-1)(k+1)=0$$

$$\Rightarrow k=-1,\ \frac{1}{10}$$

$$\therefore k=\frac{1}{10}\quad [\because f(x)\geq 0,\ \forall\ x=0,1,2,\cdots 7 \text{ and so } k\neq -1]$$

(ii) $P(X<6)=1-P(X\geq 6)=1-\{P(X=6)+P(X=7)\}$

$$=1-\left\{2\left(\frac{1}{10}\right)^2+7\cdot\left(\frac{1}{10}\right)^2+\frac{1}{10}\right\}=\frac{81}{100}$$

$$\therefore\ P(X\geq 6)=1-P(X<6)=1-\frac{81}{100}=\frac{19}{100}$$

$$P(3<X\leq 6)=P(X=4)+P(X=5)+P(X=6)=\frac{33}{100}$$

$$P(3<X/X\leq 6)=\frac{P\{(3<X)\cap(X\leq 6)\}}{P(X\leq 6)}=\frac{P(3<X\leq 6)}{P(X\leq 6)}=\frac{33/100}{83/100}=\frac{33}{83}$$

(iii) Now $P(X\leq 1)=\frac{1}{10}<\frac{1}{2}$, $P(X\leq 2)=P(X=0)+P(X=1)+P(X=2)$

$$=\frac{1}{10}+2\cdot\frac{1}{10}=\frac{3}{10}<\frac{1}{2}$$

$$P(X\leq 3)=P(X=1)+P(X=2)+P(X=3)$$

$$=\frac{1}{10}+2\cdot\frac{1}{10}+2\cdot\frac{1}{10}=\frac{1}{2}$$

$$P(X \le 4) = P(X=1) + P(X=2) + P(X=3) + P(X=4)$$

$$= \frac{8}{10} = \frac{4}{5} > \frac{1}{2}.$$

Thus the minimum value of x so that $P(X \le x) > \frac{1}{2}$ is 4.

(iv) The distribution function $F(x)$ is given below

$$F(x) = \begin{cases} 0, & -\infty < x < 1 \\ \frac{1}{10} & 1 \le x < 2 \\ \frac{1}{10} + 2\cdot\frac{1}{10} = \frac{3}{10} & 2 \le x < 3 \\ \frac{3}{10} + 2\cdot\frac{1}{10} = \frac{1}{2} & 3 \le x < 4 \\ \frac{1}{2} + 3\cdot\frac{1}{10} = \frac{4}{5} & 4 \le x < 5 \\ \frac{4}{5} + \left(\frac{1}{10}\right)^2 = \frac{81}{100} & 5 \le x < 6 \\ \frac{81}{100} + 2\cdot\left(\frac{1}{10}\right)^2 = \frac{83}{100} & 6 \le x < 7 \\ \frac{83}{100} + 7\left(\frac{1}{10}\right)^2 + \frac{1}{10} = 1 & 7 \le x < \infty \end{cases}$$

Ex. 3. *Let* $F(x) = 0, \quad -\infty < x < 0$

$= \frac{1}{5}, \quad 0 \le x < 1$

$= \frac{3}{5}, \quad 1 \le x < 3$

$= 1, \quad 3 \le x < \infty$

Show that $F(x)$ *is a possible distribution function. Determine the spectrum and the probability mass of the distribution.* [W.B.U.Tech 2003]

Clearly $F(x)$ is monotonic non-decreasing function and is continuous on the right at all points.

Also, $F(-\infty) = 0$ and $F(\infty) = 1$. Hence $F(x)$ is a possible distribution function.

Again $F(x)$ is step function and step points are 0, 1, 3. So the spectrum is $\{0, 1, 3\}$.

Now, $P(X=0)=F(0)-\lim_{x\to 0-} F(x)=\frac{1}{5}-0=\frac{1}{5}$

$P(X=1)=F(1)-\lim_{x\to 1-} F(x)=\frac{3}{5}-\frac{1}{5}=\frac{2}{5}$

$P(X=3)=F(3)-\lim_{x\to 3-} F(x)=1-\frac{3}{5}=\frac{2}{5}$.

So the required prob. mass of the distribution is

X	:	0	1	3
f_i	:	$\frac{1}{5}$	$\frac{2}{5}$	$\frac{2}{5}$

Ex. 4 *The distribution function $F(x)$ of a variate X is defined as follows*

$$F(x)=A,\quad -\infty<x<-1$$
$$=B,\quad -1\le x<0$$
$$=C,\quad 0\le x<2$$
$$=D,\quad 2\le x<\infty$$

where A,B,C, D are constants Determine the values of A, B, C, D, given that $P(X=0)=\frac{1}{6}$ and $P(X>1)=\frac{2}{3}$. [W.B.U.Tech 2004]

We have $F(-\infty)=0$

$\therefore \lim_{x\to-\infty} F(x)=0$ i.e., $\lim_{x\to-\infty}(A)=0 \quad \therefore A=0$

Again $F(\infty)=1 \quad \therefore \lim_{x\to\infty} F(x)=1$

i.e., $\lim_{x\to\infty}(D)=1 \quad \therefore D=1$

Now, $\frac{1}{6}=P(X=0)=F(0)-\lim_{x\to 0-} F(x)\ [\because P(X=a)=F(a)-\lim_{x\to a-} F(x)]$

$=C-\lim_{x\to 0-} F(x) \quad =C-B$

$\therefore C-B=\frac{1}{6}$ ... (1)

Again $P(-\infty<X<\infty)=P(-\infty<X\le 1)+P(1<X<\infty)$

$\therefore 1=P(-\infty<X\le 1)+P(X>1) \quad =P(-\infty<X\le 1)+\frac{2}{3}$

$\therefore P(-\infty < X \le 1) = 1 - \frac{2}{3}$

i.e., $F(1) = \frac{1}{3}$ $\qquad [\because F(x) = P(-\infty < X \le x)]$

$\therefore \ C = \frac{1}{3}$

$\therefore$ From (1) $B = C - \frac{1}{6} = \frac{1}{3} - \frac{1}{6} = \frac{1}{6}$

$\therefore A = 0, \ B = \frac{1}{6}, \ C = \frac{1}{3}, \ D = 1$

Ex. 5. *The probability density function of a random variable X is $f(x) = k\ (x-1)(2-x)$ for $1 \le x \le 2$. Determine (i) the value of the constant k ; (ii) the distribution function $F(x)$ (iii) $P\left(\frac{5}{4} \le X \le \frac{3}{2}\right)$.*

[*W.B.U.Tech 2007*]

(i) Since $f(x)$ is a p.d.f,

we have, $\int_{-\infty}^{\infty} f(x)dx = 1$

i.e., $k\int_1^2 (x-1)(2-x)dx = 1$ $\qquad$ i.e., $k\int_1^2 (3x - x^2 - 2)dx = 1$

i.e., $k\left[\frac{3x^2}{2} - \frac{x^3}{3} - 2x\right]_1^2 = 1$

i.e., $\frac{k}{6} = 1$ $\qquad \therefore k = 6$. $\qquad \therefore$ the complete p.d.f of X is therefore

$f(x) = 6(x-1)(2-x)$, for $1 \le x \le 2$

$= 0,$ $\qquad$ elsewhere.

(ii) The distribution function $F(x)$ is defined by

$F(x) = \int_{-\infty}^{x} f(x)\,dx$ $\therefore$ $F(x) = 0$, for $x < 1$

Also $F(x) = \int_{-\infty}^{1} f(x)\,dx + \int_1^x f(x)dx$, for $1 \le x \le 2$

$$= \int_{-\infty}^{1} 0 \cdot dx + \int_1^x 6(x-1)(2-x)\,dx$$

$$= 6\left[\frac{3x^2}{2} - \frac{x^3}{3} - 2x\right]_1^x = 5 - 12x + 9x^2 - 2x^3.$$

Again $F(x)=\int_{-\infty}^{1} 0\cdot dx+\int_{1}^{2} 6(x-1)(2-x)\,dx+\int_{2}^{x} 0\cdot dx$, for $x>2$

$$=0+6\left[\frac{3x^2}{2}-\frac{x^3}{3}-2x\right]_1^2+0=1.$$

Therefore the distribution function is given by :

$$F(x)=0, \quad \text{for } x<1$$
$$=5-12x+9x^2-2x^3, \quad \text{for } 1\le x\le 2$$
$$=1, \quad \text{for } x>2$$

(iii) Using d.f :

$$P\left(\frac{5}{4}\le X\le\frac{3}{2}\right)=F\left(\frac{3}{2}\right)-F\left(\frac{5}{4}\right)$$

$$=\left\{5-12\cdot\frac{3}{2}+9\cdot\left(\frac{3}{2}\right)^2-2\left(\frac{3}{2}\right)^3\right\}-\left\{5-12\cdot\frac{5}{4}+9\cdot\left(\frac{5}{4}\right)^2-2\cdot\left(\frac{5}{4}\right)^3\right\}$$

$$=\frac{1}{2}-\frac{5}{32}=\frac{11}{32}$$

Alternative by (Using p.d.f) :

$$P\left(\frac{5}{4}\le X\le\frac{3}{2}\right)=\int_{5/4}^{3/2} f(x)\,dx=\int_{5/4}^{3/2} 6(x-1)(2-x)dx=\frac{11}{32}.$$

Ex. 6. *A random variable X has the density function*

$$f(x)=\frac{a}{x^2+1},\ -\infty<x<\infty.$$

Find (i) a (ii) the probability that X^2 lies between $\frac{1}{3}$ and 1 (iii) the distribution function of X.

(i) As $f(x)$ is a p.d.f, so we must have $\int_{-\infty}^{\infty} f(x)\,dx=1$

i.e., $\int_{-\infty}^{\infty}\frac{a}{1+x^2}dx=1$

i.e., $a\left[\tan^{-1}x\right]_{-\infty}^{\infty}=1$ i.e., $a\left(\frac{\pi}{2}+\frac{\pi}{2}\right)=1$ $\therefore a=\frac{1}{\pi}$

(ii) As $\frac{1}{3} \le X^2 \le 1 \Rightarrow \left(-1 \le X \le -\frac{1}{\sqrt{3}}\right) \cup \left(\frac{1}{\sqrt{3}} \le X \le 1\right)$

$$P\left(\frac{1}{3} \le X^2 \le 1\right) = P\left(-1 \le X \le -\frac{1}{\sqrt{3}}\right) + P\left(\frac{1}{\sqrt{3}} \le X \le 1\right)$$

$$= \frac{1}{\pi} \int_{-1}^{-\frac{1}{\sqrt{3}}} \frac{dx}{1+x^2} + \frac{1}{\pi} \int_{\frac{1}{\sqrt{3}}}^{1} \frac{dx}{1+x^2}$$

$$= \frac{1}{\pi}\left[\tan^{-1} x\right]_{-1}^{-\frac{1}{\sqrt{3}}} + \frac{1}{\pi}\left[\tan^{-1} x\right]_{\frac{1}{\sqrt{3}}}^{1}$$

$$= \frac{1}{\pi}\left(-\frac{\pi}{6} + \frac{\pi}{4}\right) + \frac{1}{\pi}\left(\frac{\pi}{4} - \frac{\pi}{6}\right) = \frac{1}{6}.$$

(iii) The distribution function of X is given by

$$F(x) = \frac{1}{\pi} \int_{-\infty}^{x} \frac{dx}{1+x^2} = \frac{1}{\pi}\left[\tan^{-1} x\right]_{-\infty}^{x} = \frac{1}{\pi}\left(\tan^{-1} x + \frac{\pi}{2}\right).$$

Ex. 7. *A random variable X is exponentially distributed with p.d.f*

$$f(x) = \frac{1}{40} e^{-\frac{x}{40}}, \quad x > 0$$

$$= 0, \qquad x \le 0$$

Find (i) $P(X \le 20)$ *(ii)* $P(32 \le X \le 48)$ *(iii)* $P(X \ge 25)$

(i) $P(X \le 20) = \int_{0}^{20} \frac{1}{40} e^{-\frac{x}{40}} dx = \left[-e^{-\frac{x}{40}}\right]_{0}^{20} = 1 - e^{-\frac{1}{2}}$

(ii) $P(32 \le X \le 48) = \int_{32}^{48} \frac{1}{40} e^{-\frac{x}{40}} dx = \left[-e^{-\frac{x}{40}}\right]_{32}^{48} = e^{-\frac{4}{5}} - e^{-\frac{6}{5}}$

(iii) $P(X \ge 25) = 1 - P(X < 25) = 1 - \int_{0}^{25} \frac{1}{40} e^{-\frac{x}{40}} dx$

$$= 1 - \left[-e^{-\frac{x}{40}}\right]_{0}^{25} = 1 - \left(1 - e^{-\frac{5}{8}}\right) = e^{-\frac{5}{8}}.$$

Ex. 8. *A special un-biased die with* $n+1$ *faces is rolled. Its faces are marked by the number* $0, \frac{1}{n}, \frac{2}{n}, \cdots, \frac{n-1}{n}, \frac{n}{n}$. *If X denotes the number shown then find (i) the expectation of X (ii) standard deviation of X (iii)* $E\left(X-\frac{1}{2}\right)^3$.

Note that probability of each face is $\frac{1}{n+1}$. So, the distribution of X is

X	:	0	$\frac{1}{n}$	$\frac{2}{n}$	....	$\frac{n-1}{n}$	$\frac{n}{n}$
f_i	:	$\frac{1}{n+1}$	$\frac{1}{n+1}$	$\frac{1}{n+1}$	....	$\frac{1}{n+1}$	$\frac{1}{n+1}$

obviously X assumes values $x_i = \frac{i}{n}$.

(i) $$E(X) = 0 \times \frac{1}{n+1} + \frac{1}{n} \times \frac{1}{n+1} + \frac{2}{n} \times \frac{1}{n+1} + \cdots + \frac{n-1}{n} \times \frac{1}{n+1} + \frac{n}{n} \times \frac{1}{n+1}$$

$$= \sum_{i=0}^{n} \frac{i}{n} \cdot \frac{1}{n+1} = \frac{1}{n(n+1)} \sum_{i=0}^{n} i = \frac{1}{n(n+1)} \frac{n(n+1)}{2} = \frac{1}{2}$$

(ii) $$E(X^2) = \sum_{i=0}^{n} x_i^2 f_i = \sum_{i=0}^{n} \left(\frac{i}{n}\right)^2 \times \frac{1}{n+1} = \frac{1}{n^2(n+1)} \sum_{i=0}^{n} i^2$$

$$= \frac{1}{n^2(n+1)} \cdot \frac{n(n+1)(2n+1)}{6} = \frac{2n+1}{6n}$$

$$\therefore Var(X) = E(X^2) - \{E(X)\}^2 = \frac{2n+1}{6n} - \frac{1}{4} = \frac{n+2}{12n}$$

$$\therefore \sigma_X = \sqrt{\frac{n+2}{12n}}$$

(iii) $$E\left(X-\frac{1}{2}\right)^3 = E(X^3) - \frac{3}{2}E(X^2) + \frac{3}{4}E(X) - \frac{1}{8}$$

Now, $$E(X^3) = \sum_{i=0}^{n} x_i^3 f_i = \sum_{i=0}^{n} \left(\frac{i}{n}\right)^3 \cdot \frac{1}{n+1} = \frac{1}{n^3(n+1)} \sum_{i=0}^{n} i^3$$

$$= \frac{1}{n^3(n+1)}\left(1^3 + 2^3 + \cdots + n^3\right) = \frac{1}{n^3(n+1)} \left\{\frac{n(n+1)}{2}\right\}^2 = \frac{n+1}{4n}$$

Hence $E\left(X-\frac{1}{2}\right)^3=0$.

Ex. 9. *If t is a positive real number and n is a discrete r.v assuming the values 0,1, 2, ... ∞ with p.m.f* $P(X=i)=e^{-t}\left(1-e^{-t}\right)^{i-1}$. *Find the mean and* $E(3X+2)$.

The p.m.f. $f_i=e^{-t}\left(1-e^{-t}\right)^{i-1}$

$\therefore$ the mean, $E(X)=\sum_{i=0}^{\infty} i\ f_i=\sum_{i=0}^{\infty} ie^{-t}\left(1-e^{-t}\right)^{i-1}$

$=e^{-t}\sum_{i=0}^{\infty} i\left(1-e^{-t}\right)^{i-1}=e^{-t}\sum_{i=0}^{\infty} iz^{i-1}$ [put $z=1-e^{-t}$]

$=e^{-t}\left(1+2z+3z^2+\cdots\text{up to }\infty\right)=e^{-t}(1-z)^{-2}$ [$\because z<1$]

$=e^{-t}\left\{1-\left(1-e^{-t}\right)\right\}^{-2}=e^{t}$

Now $E(3X+2)=3E(X)+2=3e^{t}+2$

Ex. 10. *If a person gains or loses an amount equal to the number appearing when a balanced die is rolled once according to whether the number is even or odd, how much money can he expect from the game in the long run ?*

Let $X=$ amount of gain or loss. Then X may assume the values $-1, 2, -3, 4, -5$ and 6.

Now, $P(X=-1)=$ Prob. of 'face 1' $=\frac{1}{6}$

$P(X=2)=$ Prob. of 'face 2' $=\frac{1}{6}$

$P(X=-3)=$ Prob. of 'face 3' $=\frac{1}{6}$

and so on.

Therefore the distribution of X is

X	:	-1	2	-3	4	-5	6
f_i	:	$\frac{1}{6}$	$\frac{1}{6}$	$\frac{1}{6}$	$\frac{1}{6}$	$\frac{1}{6}$	$\frac{1}{6}$

$\therefore$ the amount of expected money $=E(X)$

$=-1\times\frac{1}{6}+2\times\frac{1}{6}+(-3)\times\frac{1}{6}+4\times\frac{1}{6}+(-5)\times\frac{1}{6}+6\times\frac{1}{6}$

$$=\frac{1}{6}(-1+2-3+4-5+6)=\frac{1}{6}\times 3=\frac{1}{2}.$$

Ex. 11. *If a person gets Rs. $(2x+5)$ where x denotes the number appearing when a balanced die is rolled once, then how much money can he expect in the long run per game ?*

By problem $x=$ number appearing on the die.

$\therefore$ the distribution of x is

x	:	1	2	3	4	5	6
f_i	:	$\frac{1}{6}$	$\frac{1}{6}$	$\frac{1}{6}$	$\frac{1}{6}$	$\frac{1}{6}$	$\frac{1}{6}$

Now, $E(x)=1\cdot\frac{1}{6}+2\cdot\frac{1}{6}+3\cdot\frac{1}{6}+4\cdot\frac{1}{6}+5\cdot\frac{1}{6}+6\cdot\frac{1}{6}=\frac{21}{6}=\frac{7}{2}$

Expected amount of money $=E(2x+5)=E(2x)+E(5)$

$$=2E(x)+5=2\times\frac{7}{2}+5=12.$$

Ex. 12. *If the random variable X has p.d.f $f(x)=\frac{1}{4}, -2\le x\le 2$, find (i) $P(X<1)$ (ii) $P\left(|X-1|\ge\frac{1}{2}\right)$*

The p.d.f of X is given by

$f(x)=\frac{1}{4}$ in $-2\le x\le 2$

$=0$ elsewhere

(i) $P(X<1)=\int_{-\infty}^{1} f(x)dx=\int_{-\infty}^{-2} f(x)dx+\int_{-2}^{1} f(x)dx$

$$=\int_{-\infty}^{-2} 0\,dx+\int_{-2}^{1}\frac{1}{4}\,dx=\frac{1}{4}(1+2)=\frac{3}{4}$$

(ii) Now $|X-1|>\frac{1}{2}\Rightarrow\left(X-1\ge\frac{1}{2}\right)\cup\left(X-1\le-\frac{1}{2}\right)$

$$\Rightarrow\left(\frac{3}{2}\le X<\infty\right)\cup\left(-\infty<X\le\frac{1}{2}\right)$$

$$\therefore P\left(|X-1|>\frac{1}{2}\right)=P\left(\frac{3}{2}\le X<\infty\right)+P\left(-\infty<X\le\frac{1}{2}\right) = \int_{\frac{3}{2}}^{\infty} f(x)dx + \int_{-\infty}^{\frac{1}{2}} f(x)dx$$

$$=\int_{\frac{3}{2}}^{2}\frac{1}{4}dx+0+0+\int_{-2}^{\frac{1}{2}}\frac{1}{4}dx$$

$$=\frac{1}{4}\left(2-\frac{3}{2}\right)+\frac{1}{4}\left(\frac{1}{2}+2\right)=\frac{3}{4}.$$

Ex. 13. *Consider the p.d.f $f(x)=ae^{-b|x|}$ where x is a random variable whose allowable range of value are from $x=-\infty$ to ∞. Find (i) cumulative distribution function (ii) the relation between a and b (iii) $P(1\le X\le 2)$.*

The cumulative distribution function is given by

$$F(x)=\int_{-\infty}^{x} f(x)dx = \int_{-\infty}^{x} ae^{bx}dx \text{, for } -\infty<x<0$$

$$=\frac{a}{b}\left[e^{bx}\right]_{-\infty}^{x}=\frac{a}{b}e^{bx}$$

$$F(x)=\int_{-\infty}^{0} f(x)dx+\int_{0}^{x} f(x)dx \quad \text{for } 0\le x<\infty$$

$$=\int_{-\infty}^{0} ae^{bx}dx+\int_{0}^{x} ae^{-bx}dx \quad =\frac{a}{b}\left[e^{bx}\right]_{-\infty}^{0}+\frac{a}{b}\left[-e^{-bx}\right]_{0}^{x}$$

$$=\frac{a}{b}-\frac{a}{b}e^{-bx}+\frac{a}{b}=\frac{a}{b}\left(2-e^{-bx}\right)$$

Thus $F(x)=\frac{a}{b}e^{bx}, \qquad x<0$

$$=\frac{a}{b}\left(2-e^{-bx}\right), \quad x\ge 0$$

(ii) Since $F(\infty)=1$, so, $\frac{a}{b}(2-0)=1$

i.e., $\frac{2a}{b}=1 \qquad \therefore b=2a$

(iii) Now $P(1 \le X \le 2) = F(2) - F(1)$

$$= \frac{a}{b}\left(2 - e^{-2b} - 2 + e^{-b}\right) \quad = \frac{e^{-b}}{2}\left(1 - e^{-b}\right).$$

Ex.14. *Show that a function which is $|x|$ in $(-1, 1)$ and zero elsewhere is a possible p.d.f and find the corresponding distribution function.*

[*W.B.U.Tech 2006*]

Let us denote the given function by $f(x)$. Then

$$f(x) = |x| \text{ in } -1 < x < 1$$
$$= 0, \text{ elsewhere}$$

i.e.,
$$f(x) = -x, \quad -1 < x \le 0$$
$$= x, \quad 0 < x < 1$$
$$= 0, \quad \text{elsewhere}$$

Clearly we see that $f(x) \ge 0$ everywhere.

Also,
$$\int_{-\infty}^{\infty} f(x)dx = \int_{-\infty}^{-1} f(x)dx + \int_{-1}^{0} f(x)dx + \int_{0}^{1} f(x)dx + \int_{1}^{\infty} f(x)dx$$

$$= 0 + \int_{-1}^{0} (-x)dx + \int_{0}^{1} x\,dx + 0 \quad = \frac{1}{2} + \frac{1}{2} = 1$$

Hence $f(x)$ is a possible prob. density function.

Now the distribution function $F(x)$ is given by

$$F(x) = \int_{-\infty}^{x} f(x)dx = \int_{-\infty}^{x} 0 \cdot dx, \text{ for } -\infty < x \le -1$$

$$F(x) = \int_{-\infty}^{-1} f(x)dx + \int_{-1}^{x} f(x)dx, \text{ for } -1 < x \le 0$$

$$= \int_{-\infty}^{-1} 0 \cdot dx + \int_{-1}^{x} (-x)dx = \frac{1}{2}\left(1 - x^2\right)$$

$$F(x) = \int_{-\infty}^{-1} f(x)dx + \int_{-1}^{0} f(x)dx + \int_{0}^{x} f(x)dx, \text{ for } 0 < x \le 1$$

$$= 0 + \int_{-1}^{0} (-x)dx + \int_{0}^{x} x\,dx = \frac{1}{2}\left(1 + x^2\right), \text{ for } 0 < x \le 1.$$

$$F(x)=\int_{-\infty}^{-1} f(x)dx+\int_{-1}^{0} f(x)\,dx+\int_{0}^{1} f(x)dx+\int_{1}^{x} f(x),\ \text{for}\ 1<x<\infty$$

$$=0+\int_{-1}^{0}(-x)dx+\int_{0}^{1} x\,dx+0\ =\frac{1}{2}+\frac{1}{2}=1.$$

So the distribution function is given by

$$F(x)=\begin{cases}0, & -\infty<x\le -1\\ \frac{1}{2}\left(1-x^2\right), & -1<x\le 0\\ \frac{1}{2}\left(1+x^2\right), & 0<x\le 1\\ 1 & 1<x<\infty\end{cases}$$

Ex. 15. *X is a continuous random variable having p.d.f*

$$f(x)=\frac{4x}{5},\quad 0<x\le 1$$
$$=\frac{2}{5}(3-x),\quad 1<x\le 2$$
$$=0,\quad \text{elsewhere.}$$

Find the mean value of X.

Mean of $X=E(X)=\int_{-\infty}^{\infty} x\,f(x)dx$

$$=\int_{-\infty}^{0} x\times 0\,dx+\int_{0}^{1} x\cdot\frac{4x}{5}dx+\int_{1}^{2} x\frac{2}{5}(3-x)dx+\int_{2}^{\infty} x\cdot 0\,dx$$

$$=\frac{4}{5}\int_{0}^{1} x^2dx+\frac{2}{5}\int_{1}^{2} x(3-x)dx=\frac{17}{15}$$

Ex. 16. *Find the mean, variance of a continuous random variable having p.d.f.*

$$f(x)=1-|1-x|,\quad 0<x<2$$
$$=0,\quad \text{elsewhere}$$

Find also $E\left(X^3\right)$

The mean $=E(X)=\int_{-\infty}^{\infty} x\,f(x)dx \quad =\int_{-\infty}^{0} x\cdot 0dx+\int_{0}^{2} x\{1-|1-x|\}dx+\int_{2}^{\infty} x\cdot 0\,dx$

$$=\int_{0}^{2} x\{1-|1-x|\}dx=\int_{0}^{1} x\{1-(1-x)\}dx+\int_{1}^{2} x\{1+(1-x)\}dx$$

$$= \int_0^1 x^2 dx + \int_1^2 \left(2x - x^2\right) dx = 1$$

$$E\left(X^2\right) = \int_{-\infty}^{\infty} x^2 f(x) dx = \int_0^2 x^2 \{1 - |1 - x|\} dx$$

$$= \int_0^1 x^2 \{1 - (1 - x)\} dx + \int_1^2 x^2 \{1 + (1 - x)\} dx = \int_0^1 x^3 dx + \int_1^2 \left(2x^2 - x^3\right) dx = \frac{7}{6}$$

So, $Var(X) = E\left(X^2\right) - \{E(X)\}^2 = \frac{7}{6} - (1)^2 = \frac{1}{6}$

$$E\left(X^3\right) = \int_{-\infty}^{\infty} x^3 f(x) dx = \int_{-\infty}^{0} x^3 \cdot 0\, dx + \int_0^2 x^3 \{1 - |1 - x|\} dx$$

$$+ \int_2^{\infty} x^3 \cdot 0 dx = \int_0^2 x^3 \{1 - |1 - x|\} dx$$

$$= \int_0^1 x^3 \{1 - (1 - x)\} dx + \int_1^2 x^3 \{1 + (1 - x)\} dx = \frac{3}{2}$$

Ex. 17. *A variable X has the density function*

$$f(x) = \frac{x}{2}, \qquad 0 \le x \le 1$$

$$= \frac{1}{2}, \qquad 1 < x \le 2$$

$$= \frac{1}{2}(3 - x), \qquad 2 < x \le 3$$

Find the mean and variance of X. [*W.B.U.Tech 2002*]

$$\text{Mean} = E(X) = \int_0^3 x f(x) dx = \int_0^1 x \cdot \frac{x}{2} dx + \int_1^2 x \cdot \frac{1}{2} dx + \int_2^3 x \cdot \frac{1}{2}(3 - x) dx$$

$$= \frac{1}{6}(1 - 0) + \frac{1}{4}(4 - 1) + \frac{1}{2}\left(\frac{27}{2} - 9 - 6 + \frac{8}{3}\right) = \frac{3}{2}$$

$$E\left(X^2\right) = \int_0^3 x^2 f(x) dx = \int_0^1 x^2 \cdot \frac{x}{2} dx + \int_1^2 x^2 \cdot \frac{1}{2} dx + \int_2^3 x^2 \frac{1}{2}(3 - x) dx$$

$$= \frac{8}{3}$$

$$\therefore\ Var(X) = E\left(X^2\right) - \{E(X)\}^2 = \frac{8}{3} - \left(\frac{3}{2}\right)^2 = \frac{5}{12}$$

Ex. 18. *The demand for a new product of a company is assumed to be a random variable with p.d.f*

$$f(x) = \frac{x}{\lambda^2} e^{-\frac{x^2}{2\lambda^2}}, \qquad x \geq 0$$

$$= 0, \qquad x < 0$$

Find the mean, variance of this random variable and also find the probability that it will exceed λ.

$$\text{Mean} = E(X) = \int_0^\infty x \cdot \frac{x}{\lambda^2} e^{-\frac{x^2}{2\lambda^2}} dx$$

$$= \lambda\sqrt{2} \int_0^\infty u^{\frac{3}{2}-1} e^{-u} du, \text{ by putting } \frac{x^2}{2\lambda^2} = u$$

$$= \lambda\sqrt{2}\, \Gamma\left(\frac{3}{2}\right) = \lambda\sqrt{2} \frac{1}{2} \cdot \Gamma\left(\frac{1}{2}\right) = \lambda\sqrt{2} \cdot \frac{1}{2}\sqrt{\pi} = \lambda\sqrt{\frac{\pi}{2}}$$

(using the property $\Gamma(n+1) = n\Gamma(n)$)

$$\text{Now, } E(X^2) = \int_0^\infty x^2 \cdot \frac{x}{\lambda^2} e^{-\frac{x^2}{2\lambda^2}} dx$$

$$= 2\lambda^2 \int_0^\infty u^{2-1} e^{-u} du, \text{ by putting } \frac{x^2}{2\lambda^2} = u$$

$$= 2\lambda^2 \Gamma(2) = 2\lambda^2 \qquad [\because \Gamma(2) = 1!]$$

$$\because Var(X) = E(X^2) - \{E(X)\}^2 = 2 \cdot \lambda^2 - \lambda^2 \frac{\pi}{2} = (4-\pi)\frac{\lambda^2}{2}$$

$$\text{Now, } P(X > \lambda) = 1 - P(X \leq \lambda) \quad = 1 - \int_0^\lambda \frac{x}{\lambda^2} e^{-\frac{x^2}{2\lambda^2}} dx$$

$$= 1 - \int_0^{\frac{1}{2}} e^{-u} du, \text{ by putting } \frac{x^2}{2\lambda^2} = u$$

$$= 1 - \left(1 - e^{-\frac{1}{2}}\right) = e^{-\frac{1}{2}} = \frac{1}{\sqrt{e}}.$$

Ex. 19. *A random variable X has probability density*

$$f(x) = e^{-x}, \quad x \geq 0$$

$$= 0, \text{ otherwise}$$

Find (a) $E(X)$ *(b)* $E(X^2)$ *(c)* $E\{(X-1)^2\}$ *(d)* $E\left(e^{\frac{2X}{3}}\right)$.

(a) $E(X) = \int_{-\infty}^{\infty} x\, f(x)dx = \int_0^{\infty} xe^{-x}\, dx = \Gamma(2) = 1$

$$\left[\because \text{the Gamma Function } \Gamma(n) = \int_0^{\infty} e^{-x} x^{n-1} dx\right]$$

(b) $E(X^2) = \int_0^{\infty} x^2 e^{-x} dx = \Gamma(3) = 2! = 2$

(c) $E\{(X-1)^2\} = \int_0^{\infty} (x-1)^2 e^{-x} dx$

$$= \int_0^{\infty} x^2 e^{-x} dx - 2\int_0^{\infty} x\, e^{-x} dx + \int_0^{\infty} e^{-x} = \Gamma(3) - 2\,\Gamma(2) + \Gamma(1) = 2! - 2\cdot 1 + 1 = 1$$

(d) $E\left(e^{\frac{2X}{3}}\right) = \int_0^{\infty} e^{\frac{2x}{3}} \cdot e^{-x} dx = \int_0^{\infty} e^{-\frac{x}{3}} dx$

$$= \underset{U\to\infty}{Lt} \int_0^{U} e^{-\frac{x}{3}} dx = \underset{U\to\infty}{Lt} \left(-3e^{-U/3} + 3\right) = 0 + 3 = 3.$$

Ex. 20. *The distribution function of a random variable X is*

$$F(X) = cx^3, \quad 0 \le x < 3$$
$$= 1, \quad x \ge 3$$
$$= 0, \quad x < 0$$

If $P(X = 3) = 0$, find (i) the constant c (ii) the density function (iii) $P(X > 1)$ (iv) $P(1 < X \le 2)$ (v) $P(3X + 2 < 8)$

(i) The density function $f(x)$ is given by $f(x) = F'(x)$

$$\therefore f(x) = 3cx^2, \quad 0 \le x < 3$$
$$= 0, \quad \text{elsewhere}$$

Also we have $\int_{-\infty}^{\infty} f(x)dx = 1 \quad \therefore \int_0^3 3cx^2 dx = 1$

i.e., $27c = 1 \quad \therefore c = \frac{1}{27}$.

(ii) The required density function is given by

$$f(x) = \frac{x^2}{9}, \quad 0 \le x < 3$$
$$= 0, \quad \text{elsewhere}$$

(iii) $P(X>1)=\int_1^{\infty} f(x)\,dx=\frac{1}{9}\int_1^3 x^2 dx=\frac{26}{27}$

(iv) $P(1<X\le 2)=\int_1^2 f(x)\,dx=\frac{1}{9}\int_1^2 x^2 dx=\frac{2}{27}$

(v) $P(3X+2<8)=P(3X<6)=P(X<2)$

$$=\int_{-\infty}^{2} f(x)dx=\int_{-\infty}^{0} 0\cdot dx+\int_0^2 \frac{x^2}{9}dx=\frac{8}{27}.$$

Ex. 21. *Verify that the following is a distribution function*

$$F(x)=0, \qquad x<-a$$
$$=\frac{1}{2}\left(\frac{x}{a}+1\right), \qquad -a\le x\le a$$
$$=1, \qquad x>a$$

In order that $F(x)$ may be a distribution function we should have

(i) $F(-\infty)=0$ (ii) $F(\infty)=1$ (iii) $\frac{d\,F(x)}{dx}=f(x)\ge 0$, $-a\le x\le a$

(iv) $\int_{-\infty}^{\infty} f(x)\,dx=1$.

As $F(x)=0$ for $x<-a$ and $F(x)=1$ for $x>a$, so the conditions (i) and (ii) are satisfied.

Now $\frac{d\,F(x)}{dx}=\frac{1}{2a}\ge 0$ for $-a\le x\le a$

Again $\int_{-\infty}^{\infty} f(x)\,dx=\int_{-\infty}^{-a} 0\cdot dx+\int_{-a}^{a}\frac{1}{2a}dx+\int_a^{\infty} 0\cdot dx=\left[\frac{x}{2a}\right]_{-a}^{a}=1$

Hence $F(x)$ is a distribution function.

Ex. 22. *A continuous cumulative distribution function $F(x)$ is defined as follow:*

$$F(x)=0, \qquad x\le 1$$
$$=\frac{1}{16}(x-1)^4, \qquad 1<x\le 3$$
$$=1, \qquad x>3$$

Find the probability density function $f(x)$. Also find the mean of X.

The p.d.f $f(x)$ is given by $F'(x)=f(x)$

i.e., $f(x)=\frac{1}{4}(x-1)^3$, $1\le x\le 3$

$=0$, elsewhere

$\therefore$ mean of $X=\int_1^3 x\cdot\frac{1}{4}(x-1)^3 dx=\frac{1}{4}\int_1^3\left(x^4-3x^3+3x^2-x\right)dx$

$$=\frac{1}{4}\left[\frac{x^5}{5}-\frac{3x^4}{4}+x^3-\frac{x^2}{2}\right]_1^3=2\cdot 65.$$

Ex. 23. *Find the value of the constant k such that*

$f(x)=kx(1-x), 0<x\le 1$

$=0$, else where.

is a possible density function and compute $P\left(X>\frac{1}{2}\right)$. *Also find* $E(X)$.

Since $f(x)$ is a p.d.f, so

$$\int_{-\infty}^{\infty} f(x)dx=1$$

i.e., $k\int_0^1 x(1-x)dx=1$

i.e., $k\left[\frac{x^2}{2}-\frac{x^3}{3}\right]_0^1=1$

i.e., $k\left(\frac{1}{2}-\frac{1}{3}\right)=1$

$\therefore$ $k=6$.

$\therefore$ $f(x)=6x(1-x), 0<x\le 1$

$=0$, else where.

$\therefore$ $P\left(X>\frac{1}{2}\right)$

$=1-P\left(X\le\frac{1}{2}\right)$

$$=1-\int_0^{\frac{1}{2}} f(x)dx=1-6\int_0^{\frac{1}{2}} x(1-x)dx$$

$$= 1 - 6\left[\frac{x^2}{2} - \frac{x^3}{3}\right]_0^{\frac{1}{2}} = 1 - \frac{1}{2} = \frac{1}{2}.$$

Now, $E(x) = 6\int_0^1 x^2(1-x)dx$

$$= 6\left[\frac{x^3}{3} - \frac{x^4}{4}\right]_0^1$$

$$= 6\left(\frac{1}{3} - \frac{1}{4}\right) = \frac{1}{2}.$$

EXERCISES

[I] SHORT ANSWER QUESTIONS

1. Find mean and variance of the following distribution :

x_i	:	-1	0	1	2	3
f_i	:	0.3	0.1	0.1	0.3	0.2

[Hints : mean $= (-1)\times 0.3 + 0\times 0.1 + 1\times 0.1 + 2\times 0.3 + 3\times 0.2 = 1.0$

$$\text{Var} = (-1-1)^2 \times 0.3 + (0-1)^2 \times 0.1 + (1-1)^2 \times 0.1 + (2-1)^2 \times 0.3 + (3-1)^2 \times 0.2 = 2.4$$]

2. Find mean and S.D of the following distribution :

$$f(x) = \begin{cases} \frac{1}{4}e^{-\frac{x}{4}}, x > 0 \\ = 0 \quad , \text{else where.} \end{cases}$$

[Hints : mean $= \frac{1}{4}\int_0^\infty xe^{-\frac{x}{4}}dx = 4$

$$\sigma = \frac{1}{4}\int_0^\infty (x-4)^2 e^{-\frac{x}{4}}dx = 80$$]

3. Let $f(x) = \begin{cases} x & 0 < x \le 1 \\ 2 - x & 1 \le x \le 2 \\ 0 & \text{else where} \end{cases}$

Is the following function is a p.d.f. of random variable X ?

[Hints : Yes, as $f(x) \geq 0$ and $\int_{-\infty}^{\infty} f(x)dx = 1$]

4. Is the funtion $f(x)$ defined as

$$f(x) = \begin{cases} 0 & ,x < 2 \\ \frac{1}{18}(3+2x), & 2 \leq x \leq 4 \\ 0 & ,x > 4 \end{cases}$$

a density function ?

5. A random variable X has the p.d.f

$$f(x) = Axe^{-\lambda^2 x^2}, x > 0$$

Find A

[Hints : Use the result $\int_{-\infty}^{\infty} f(x)dx = 1$].

6. For what values of λ will the function

$$f(x) = \lambda x, x = 1, 2, 3, \cdots n$$

be a probability mass function of a random variable.

[Hints : Use the result

$\sum_i f_i = 1$ gives

$\lambda(1+2+\cdots+n) = 1 \Rightarrow \lambda \cdot \frac{n(n+1)}{2} = 1$].

7. If the random variable X has mean 'm' and S.D σ show that

$$E\left(\frac{X-m}{\sigma}\right)^2 = 1, E\left(\frac{X-m}{\sigma}\right) = 0.$$

8. Find the probability distribution of X, the number of 'sixes' in two tosses of a die.

9. Two cards are drawn successively without replacement from a well-shuffled deck of 52 cards. Find the probability distribution of the number of aces.

10. Five balls are drawn from a box containing 4 black and 6 white balls. Find the probability distribution of the number of black balls drawn without replacement.

11. A number is chosen at random from each of the two sets 0, 1, 2, 3 and 0, 1, 2, 3. Find the prob. distribution of the random variable denoting the sum of the numbers chosen.

12. Verify that the function $f(x)=\frac{1}{\lambda}e^{-\frac{x}{\lambda}},\quad 0<x<\infty$

$=0,\quad$ elsewhere

is a possible probability density function. Find $P(2<X<6)$.

13. (a) If a random variable X has the density function

$f(x)=\frac{1}{6},\ -3<x<3$

$=0,\ $ elsewhere,

obtain (i) $P(X<1)$, (ii) $P(|X|>1)$, (iii) $P(2X+3>5)$

(b) A random varaible X has the following pdf

$f(x)=k,\quad -2<x<2$

$=0,\ $ otherwise.

(i) Determine the constant k (ii) what is the value of $P(|x|>1)$?

[*W.B.U.Tech 2004*]

(c) Obtain $E(X)$, Var(X) of X whose *pdf* is given by

$f(x)=2x, 0\le x\le 1.$

$=0,$ otherwise.

Hence find $Var(3-5X)$.

14. Is the following a p.d.f ?

$f(x)=2x,\quad 0<x\le 1$

$=4-2x,\quad 1<x\le 2$

$=0,\quad$ elsewhere.

15. If the random variable X has the p.d.f

$f(x)=\frac{1}{4}, -2\le x\le 2$

$=0,\ $ elsewhere. Obtain $P(2x+3>5)$

16. A random variable X has probability function

$f(x)=\frac{1}{2^x}$, for $x=1,2,3,\cdots$ up to ∞

Find the mean,

17. An urn contains 7 white and 3 red balls. Two balls are drawn together at random from the urn. Find the mathematical expectation of the number of white balls drawn.

18. Three coins whose faces are marked 1 and 2, are tossed. What is the expectation of the total value of numbers on their faces ? [Hints : Use the formulae $E(X+Y+Z)=E(X)+E(Y)+E(Z)$]

19. An unbiased coin is tossed until a head or five tails come. Find the expected number of tosses of the coin.

20. Show that the function $f(x)$ given by

$$f(x)=k(x-9)(10-x), \qquad 9\le x\le 10$$
$$=0, \qquad \text{elsewhere,}$$

is a pdf for a suitable value of the constant k. Find mean of the distribution.

21. Find the mean and s.d of the continuous r.v having p.d.f

$$f(x)=\frac{1}{2}-kx, 0<x<4$$
$$=0, \qquad \text{elsewhere}$$

22. Show that the function $f(x)=\dfrac{x^n}{n!}e^{-x},\ x>0$, n being a positive integer, is a pdf. Find also its mean and variance.

23. The probability density of a continuous distribution is given by $f(x)=\frac{3}{4}x(2-x), 0<x<2$. Compute mean and variance.

[*W.B.U.Tech 2004,2007,2008*]

24. For a random variable X, $\text{var}(X)=1$, find $\text{var}(2X+3)$.

[*W.B.U.Tech 2004*]

25. Let $f(x)=ke^{-\alpha x}(1-e^{-\alpha x}), x\ge 0$. Find k such that $f(x)$ is a density function. Find also the corresponding distribution function.

[*W.B.U.Tech 2004*]

Answers

1. 1.0, 2.9 **2.** 4.80 **3.** yes **4.** Yes **5.** $2\lambda^2$ **6.** $\dfrac{2}{n(n+1)}$

8.

X	:	0	1	2
f_i	:	$\frac{25}{36}$	$\frac{10}{36}$	$\frac{1}{36}$

9.

X	:	0	1	2
f_i	:	$\frac{188}{221}$	$\frac{32}{221}$	$\frac{1}{221}$

10.

X	:	0	1	2	3	4
Prob.	:	$\frac{1}{42}$	$\frac{5}{21}$	$\frac{10}{21}$	$\frac{5}{21}$	$\frac{1}{42}$

11.

X	:	0	1	2	3	4	5	6
f_i	:	$\frac{1}{16}$	$\frac{1}{8}$	$\frac{3}{16}$	$\frac{1}{4}$	$\frac{3}{16}$	$\frac{1}{8}$	$\frac{1}{16}$

12. $e^{-2/\lambda}-e^{-6/\lambda}$ **13.** (a) (i) $\frac{2}{3}$ (ii) $\frac{2}{3}$ (iii) $\frac{1}{3}$ (c) $\frac{2}{3},\frac{1}{18},\frac{25}{18}$

14. not p.d.f. **15.** $\frac{1}{4}$, **16.** 2 **17.** 1.4 **18.** 4.5 **19.** $\frac{31}{16}$

20. $k=6$, mean $=9.5$ **21** $\frac{4}{3}$, $\frac{\sqrt{8}}{3}$, **22.** $n+1$, $n+1$

23. $1, \frac{1}{5}$, **24.** 4. **25.** $2\alpha, 1-2e^{-\alpha x}+e^{2\alpha x}$

[III] LONG ANSWER QUESTIONS

1. If $f(x)=kxe^{-x}, 0\le x<\infty$ be a continuous distribution, find k. Also Evaluate $P(2\le x\le 3)$ and $P(X\le 1)$.

2. A continuous random variable X has probability density function $f(x)$ as follows :

$$f(x)=kx, 0\le x<5$$
$$=k(10-x), 5\le x<10$$
$$=0, \text{ otherwise.}$$

Determine $k, P(X<6)$ and $E(X)$.

3. (a) The random variable X has the probability density function

$$f(x)=k, \text{ if } x=0$$
$$=2k \text{ if } x=1$$
$$=3k \text{ if } x=2$$
$$=0, \text{ elsewhere}$$

determine (i) value of k (ii) $P(X<2)$

(iii) The smallest value of k for whcih $P(X \le k) > \frac{1}{2}$

[W.B.U.Tech 2004]

(b) A random variable X has the following prob. distribution

$X = x_i$	:	0	1	2	3	4	5	6	7	8
f_i	:	k	$3k$	$5k$	$7k$	$9k$	$11k$	$13k$	$15k$	$17k$

(i) Determine the value of k

(ii) Find $P(X < 3), P(X \ge 3), P(2 \le X < 5)$

(iii) What is the smallest value of x for which $P(X \le x) > 0.5$?

4. A fair coin is tossed 3 times independently. Let X be the r.v whose value for any outcome is the number of heads obtained. Find the probability distribution of X and its distribution function also.

5. From a lot containing 12 items, 4 of which are defective, 5 are chosen at random. If X be the number of defectives found in the sample,write down

(i) the probability distribution of X

(ii) $P(X \le 1)$

(iii) $P(1 < X < 3)$

6. A random variable X has the following probability function :

$X = x_i$	:	-2	-1	0	1	2	3
f_i	:	0.15	m	0.25	$2m$	0.35	m

(i) Find the value of m, (ii) Obtain the distribution function $F(x)$.

7. A random variable X has the following probability function

$X = x$	:	-2	-1	0	1	2	3
$P(x)$	:	0.1	k	0.2	2k	0.3	3k

(i) Find k

(ii) Evaluate $P(X < 2), P(X \le 2), P(-2 < X < 2)$ [W.B.U.Tech 2006]

(iii) Determine the distribution function $F(x)$ of X.

[W.B.U.Tech 2005]

8. Verify whether the following function is a distribution function

$F(x)=0,\ -\infty<x<0$

$=\frac{1}{5}, 0\le x<1$

$=\frac{3}{5}, 1\le x<3$

$=1, 3\le x<\infty$

If so, find the spectrum and probability mass function.

9. The spectrum of the random variable X consists of the points $1,2,\cdots,n$ and $P(X=i)$ is proportional to $\frac{1}{i(i+1)}$. Determine the distribution function of X.

Compute $P(3<X\le n)$ and $P(X>5)$

[Hints : $\sum_{i=1}^{n}\frac{k}{i(i+1)}=1$

or, $k\sum_{i=1}^{n}\left(\frac{1}{i}-\frac{1}{i+1}\right)=1$ or, $\frac{kn}{n+1}=1$ or, $k=\frac{n+1}{n}$]

$$\therefore F(x)=\sum_{r=-\infty}^{i}P(X=x_r)=\sum_{r=1}^{i}\frac{k}{r(r+1)}=k\left(1-\frac{1}{i+1}\right)$$

$$=\frac{(n+1)i}{n(i+1)},\ i\le x<i+1$$

$$P(3<X\le n)=\sum_{4}^{n}f_i=k\sum_{4}^{n}\left(\frac{1}{i}-\frac{1}{i+1}\right)=\frac{n+1}{n}\left(\frac{1}{4}-\frac{1}{n+1}\right)=\frac{n-3}{4n}$$

$$P(X>5)=\sum_{i=6}^{n}f_i=\frac{n+1}{n}\left(\frac{1}{6}-\frac{1}{n+1}\right)=\frac{n-5}{6n}$$]

10. Find K for which the following gunction $f(x)$ will be a pmf of a discrete random variable $X:f(x)=K(1+x):x=2,3;\cdots n$. Find K, $P(X>2)$ and $P(X\ge 2/X\le 2)$. Also find the mean of the distribution. [$\frac{2}{(n-1)(n+4)}$; $\frac{(n-2)(n+5)}{(n-1)(n+4)}$; 1; $\frac{2(n^2+4n+6)}{3(n+4)}$]

11. The lifetime (in hours) of a certain machine is a continuous random variable X having range $0<x<\infty$ and p.d. function is

$f(x)=xe^{-kx},\quad 0<x<\infty$

$=0,\qquad$ otherwise

(a) Determine the constant k (b) Compute $P(20 < x < 30)$.

(c) Determine the distribution function.

12. A continuous r.v has the p.d.f.

$$f(x) = \frac{1}{2}, \quad 4 \le x \le 6$$
$$= 0, \quad \text{elsewhere.}$$

Find (i) $P(4 \le X \le 5)$ (ii) $P(X < 3.2)$ (iii) $P(X < 4.2)$ (iv) $P(3 \le X \le 4.5)$ (v) $P(X > 5.5)$ (vi) c.d.f (cumulative distribution function i.e., distribution function) (vii) k such that $P(X < k) = 0.7$

13. The p.d.f of a random variable X is

$$f(x) = cx^2, \quad 0 \le x \le 1$$
$$= 0, \quad \text{otherwise.}$$

Find (i) c (ii) $P\left(0 \le X \le \frac{1}{2}\right)$ (iii) $P(4X > 3)$ (iv) $P(1 < 4X < 3)$ (v) the c.d.f

14. $f(x) = \frac{1}{2} - kx,\ 0 \le x \le 4$ is the p.d.f of a r.v X. Determine (i) k (ii) the probability that X assumes the value between 2 and 3 (iii) the c.d.f (iv) $P(X \le 1)$ (v) $P(X \ge 2.5)$ (vi) $P(|X - 2| < 0.5)$

15. The length of the life of a tyre manufactured by a company follow a continuous distribution with density function $f(x) = \frac{k}{x^3}, 1000 \le x \le 1500$.

Find k, the probability that a tyre selected at random would function for at least 1200 hours.

16. The p.d.f of a r.v is $f(x) = k(x-1)(2-x), 1 \le x \le 2$. Find (i) k (ii) c.d.f (iii) the probability that X is less than $\frac{5}{4}$ (iv) the probability that $2X$ is greater than 3 (v) $P(5 < 4X < 6)$.

17. Show that

$$f(x) = x, \quad 0 \le x < 1$$
$$= k - x, \quad 1 \le x \le 2$$
$$= 0 \quad \text{elsewhere}$$

is a pdf of a random variable X for a suitable value of k which you are to determine. Then find the distribution function of the random variable X. Calculate the probability that the random

Variable lie between $\frac{1}{2}$ and $\frac{3}{2}$. [*W.B.U.Tech 2005, 2006*]

18. The pdf of a continuous r.v. is given by

$$f(x) = -cx + 3c, \quad 2 \le x \le 3$$
$$= c, \quad 1 < x < 2$$
$$= cx, \quad 0 \le x \le 1$$
$$= 0, \quad \text{elsewhere}$$

Find c and the distribution function.

19. Find the expectation, variance and standard deviation of each of the following distribution :

(i)

x_i	:	-5	-4	1	2
$f(x_i)$	:	$\frac{1}{4}$	$\frac{1}{8}$	$\frac{1}{2}$	$\frac{1}{8}$

(ii)

x_i	:	8	12	16	20	24
$f(x_i)$	:	$\frac{1}{8}$	$\frac{1}{6}$	$\frac{3}{8}$	$\frac{1}{4}$	$\frac{1}{12}$

20. A bag contains 5 white and 7 black balls. Find the expectation of a man who is allowed to draw two balls from the bag and who is to receive one rupee for each black ball and two rupees for each white ball drawn.

21. *A* and *B* play for a prize of Rs. 99. The prize is to be won by a player who first throws a '3' with one die. *A* first throws, and if he fails *B* throws, and if he fails *A* again throws, and so on. Find their respective expectation.

22. Two players A and *B* agree to play a game under the condition that *A* will get from *B* Rs. 3 if he wins and will pay *B* Rs. 3 if he losses. The probability of A's winning is *p*. Find the mean and variance of *A*'s gain.

[Hints : $X =$ *A*'s gain $= 3, -3$, $P(X=3) = p$, $P(X=-3) = 1-p$]

23. Evaluate the expectation of the number of sucess preceding the first failure in an infinite sequence of trails with probability of failure is *p*, where each trial gives two possible outcomes - success or failure.

[Hints : $X = 0, 1, 2, \cdots \cdots$ up to ∞ and $P(X=i) = (1-p)^i p$].

24. An unbiased die is tossed. Let *X* denote the double number appearing, *Y* denote 1 or 3 according as an odd or an even number appears. Find (a) the expectation and variance of (i) *X* (ii) *Y* (iii)

$X+Y$ (b) the distribution of XY. Hence its mean and standard deviation.

25. A biased coin is such that the probability of 'head' and 'tail' are respectively $\frac{3}{4}$ and $\frac{1}{4}$. This coin is tossed three times. Find the distribution, expectation and variance of the longest string of heads which occurs.

[Hints : If X = ' longest string of head which occurs' then for the out-come (H, T, H) X assumes value 2]

26. An unbiased coin in tossed four times. Find the variance of the number of times a head is followed immediately by a tail.

27. Two defective radio tubes were accidently placed in a box with 5 non-defective tubes. The tubes are tested one at a time until the second defective is found. What are the mean and variance of the number of radio tubes tested ?

[Hint : $X =$ No. of radio tubes tested. Then $P(X=i)=\frac{i-1}{21}$, $i=2,3,4,5,6,7$]

28. Find the mean and variance of the following distribution :

$$f_i=\frac{1}{1+\mu}\left(\frac{\mu}{1+\mu}\right)^i, \ \mu>0, \ x_i=i \ (i=0,1,2,\cdots).$$

29. The radius of a circle has distribution given by the pdf

$f(x)=1$, $1<x<2$

$=0$, elsewhere.

Find the mean and variance of the area of the circle.

[Hint:Mean $= E\left(\pi X^2\right)=\pi E\left(X^2\right)$, Variance $= Var\left(\pi X^2\right)$

$= E\left(\pi^2 X^4\right)-\left(\frac{7\pi}{3}\right)^2$].

30. A Continuous distribution of a variate X is defined by

$$f(x)=\frac{x}{2}, \quad 0<x\le 1$$

$$=\frac{1}{2}, \quad 1\le x\le 2$$

$$=\frac{1}{2}(3-x), \quad 2\le x\le 3$$

(i) Verify that the area under the curve is unity

(ii) Find mean and variance of X.

31. If the pdf of a r.v X is given by

$$f(x) = ce^{-(x^2+2x+3)}, -\infty < x < \infty$$

find the value of the constant c, the expectation and variance of the distribution.

32. (a) The demand for a new product of company is assumed to be a random variable with p.d.f.

$$f(x) = xe^{-\frac{x^2}{2}}, x \geq 0$$

$$= 0, \quad x < 0$$

Find the mean and variance of this random variable and find the probability that it will exceed 1.

(b) The length of the life of tyre manufactured by a company follows a continuous distribution given by the density function

$$f(x) = \frac{k}{x^3}, \quad 1000 \leq x \leq 1500$$

$$= 0, \quad \text{elsewhere}$$

Find k and find the probability that a randomly selected tyre would function for at least 1200 hours. *[W.B.U.Tech 2005]*

33. A continuous distribution is given by

$$f(x) = \frac{1}{x\sqrt{2\pi}} e^{-(\log x)^2/2}, \quad x > 0$$

$$= 0, \quad x < 0$$

Find the mean and standard deviation of the distribution.

[Assume $\int_0^\infty e^{-x^2} dx = \frac{\sqrt{\pi}}{2}$]

34. The pdf of a continuous r.v is given by

$$f(x) = Ke^{-b(x-a)}, a \leq x < \infty$$

$$= 0, \quad \text{elsewhere}$$

where K, a, b are constants. Prove that $K = b = \frac{1}{\sigma}$ and $a = m - \sigma$ where m is mean, σ is s.d. *[W.B.U.Tech 2003]*

35. The pdf of a continuous r.v. X is given by

$$f(x)=\frac{2(\beta+x)}{\beta(\alpha+\beta)}, \quad -\beta\le x<0$$
$$=\frac{2\alpha(\alpha-x)}{(\alpha+\beta)}, \quad 0\le x\le\alpha$$

Find the expectation and variance of X. Given $\alpha>\beta>0$.

[*W.B.U.Tech 2003*]

36. The pdf of a continuous random variable is given by

$$f(x)=kx+\frac{1}{2}, \quad -1<x<1$$
$$=0, \quad \text{elsewhere}$$

where k is constant. Find k for which variance of X is maximum.

[Hint : $Var(X)=\frac{4}{9}\left(\frac{3}{4}-k^2\right)$]

ANSWERS

1. $1, \frac{3e-4}{e^3}, \frac{e-2}{e}$ 2. $\frac{1}{25}, \frac{17}{25}, 5$

3. (a) (i) $k=\frac{1}{6}$ (ii) $\frac{1}{2}$ (b) (i) $k=\frac{1}{81}$, (ii) $\frac{1}{9}, \frac{8}{9}, \frac{7}{27}$, (iii) 6

4.

X	:	0	1	2	3
f_i	:	$\frac{1}{8}$	$\frac{3}{8}$	$\frac{3}{8}$	$\frac{1}{8}$

$$F(x)=0, \quad x<0$$
$$=\frac{1}{8}, \quad 0\le x<1$$
$$=\frac{1}{2}, \quad 1\le x<2$$
$$=\frac{7}{8}, \quad 2\le x<3$$
$$=1, \quad x\ge 3$$

5. (i) $X: \; 0 \quad 1 \quad 2 \quad 3 \quad 4$

$f_i: \; f_0 \quad f_1 \quad f_2 \quad f_3 \quad f_4$ where $f_i = \binom{4}{i}\binom{8}{5-i}\Big/\binom{12}{5}$

(ii) $P(X \le 1) = \left\{\binom{8}{5} + \binom{4}{1}\binom{8}{4}\right\}\Big/\binom{12}{5}$ (iii) $\binom{4}{2}\binom{8}{3}\Big/\binom{12}{5}$

6. (i) $m = \frac{1}{16}$,

(ii)
$$
\begin{aligned}
F(x) &= 0, & x &< -2 \\
&= 3/20, & -2 \le x &< -1 \\
&= 17/80, & -1 \le x &< 0 \\
&= 37/80, & 0 \le x &< 1 \\
&= 47/80, & 1 \le x &< 2 \\
&= 75/80, & 2 \le x &< 3 \\
&= 1, & x &\ge 3
\end{aligned}
$$

7. (i) $K = \frac{1}{15}$

(ii) 0.5, 0.8, 0.4

(iii)
$$
\begin{aligned}
F(x) &= 0, & x &< -2 \\
&= \frac{1}{10}, & -2 \le x &< -1 \\
&= \frac{1}{6}, & -1 \le x &< 0 \\
&= \frac{11}{30}, & 0 \le x &< 1 \\
&= \frac{1}{2}, & 1 \le x &< 2 \\
&= \frac{4}{5}, & 2 \le x &< 3 \\
&= 1, & 3 &\le x.
\end{aligned}
$$

9. $\{0,1,3\}, P(X=0)=\frac{1}{5}, P(X=1)=\frac{2}{5}, P(X=3)=\frac{2}{5}$

10. $\frac{2}{(n-1)(n+9)}, \frac{(n-2)(n+5)}{(n-1)(n+4)}; 1; \frac{2(n^2+4x+6)}{3(n+4)}$

11. (a) $k=1$, (b) $\frac{21}{e^{20}}-\frac{31}{e^{30}}$, (c) $1-(1+x)e^{-x}$

12. (i) 0.5 (ii) 0 (iii) 0.1 (iv) 0.25

(v) 0.25 (vi) $\frac{x-4}{2}$ (vii) 5.4

13. (i) 3, (ii) $\frac{1}{8}$, (iii) 37/64, (iv) 13/32, (v) x^3

14. (i) $\frac{1}{8}$ (ii) $\frac{3}{16}$ (iii) $\frac{x(8-x)}{16}$ (iv) $\frac{7}{16}$ (v) $\frac{9}{64}$ (vi) $\frac{1}{4}$

15. 36×10^5 ; $\frac{9}{20}$.

16. (i) 6 (ii) $5+9x^2-12x-2x^3$ (iii) $\frac{5}{32}$ (iv) $\frac{1}{2}$. (v) $\frac{11}{32}$.

17. $k=2,\ F(x)=0, -\infty<x<0$

$$=\frac{x^2}{2},\ 0\le x<1$$

$$=2x-\frac{x^2}{2}-1,\ 1\le x\le 2$$

$$=1,\ x>2.$$

18. $c=\frac{1}{2}, F(x)=0, x<0; \frac{x^2}{4}, 0\le x\le 1; \frac{2x-1}{4}, 1<x\le 2$;

$\frac{6x-x^2-5}{4}, 2<x\le 3$; $1, x>3$

19. (i) $-1, 8.25, 2.9$, (ii) 16, 20, $2\sqrt{5}$ **20.** Rs. 2.83

21. Rs. 54, 45 **22.** $3(2p-1), 36p(1-p)$ **23.** $\frac{1-p}{p}$

24. (a) (i) 7, 11.7 (ii) 2, 1 (iii) 9, 14.7,

(b)

XY :	2	6	10	12	24	36
f_i :	$\frac{1}{6}$	$\frac{1}{6}$	$\frac{1}{6}$	$\frac{1}{6}$	$\frac{1}{6}$	$\frac{1}{6}$

$E(XY) = 15$, $\sigma_{xy} = 11.6$

25.

X :	0	1	2	3
f_i :	$\frac{1}{64}$	$\frac{9}{64}$	$\frac{27}{64}$	$\frac{27}{64}$

Mean $= 25$; Variance $= 0.5625$.

26. $\frac{5}{16}$ **27.** $\frac{16}{3}, \frac{20}{9}$ **28.** $\mu, \mu(\mu+1)$ **29.** $\frac{7\pi}{3}, \frac{34\pi^2}{45}$

30. (ii) 1.5, 0.42 **31.** $e^2/\sqrt{\pi}, -1, \frac{1}{2}$

32. (a) $\sqrt{\frac{\pi}{2}}, \frac{4-\pi}{2}, \frac{1}{\sqrt{e}}$ (b) 0.45 **33.** $\sqrt{e}, \sqrt{e^2-e}$

35. $\frac{1}{3}(\alpha-\beta), \frac{1}{18}(\alpha^2+\alpha\beta+\beta^2)$ **36.** 0

[III] MULTIPLE CHOICE QUESTIONS

1. If $f(x)$ is a p.d.f of a random variable X, then

(a) $\int_0^\infty f(x)dx = 1$ (b) $\int_{-\infty}^\infty f(x)dx = 1$

(c) $\int_{-\infty}^x f(x)dx = 1$ (d) none.

2. The probability $P(a \le x \le b)$ is defined by (where $F(x)$ is the distribution function of the random variable X)

(a) $F(b) - F(a)$ (b) $F(b) + F(a)$

(c) $F(a) - F(b)$ (d) $F(a)F(b)$. [*W.B.U.Tech 2006*]

3. The distribution function $F(x)$ of a random variable X is given by (where $-\infty < x < \infty$)

(a) $P(-\infty < X < x)$ (b) $P(-\infty < X \leq x)$

(c) $P(-\infty < X < \infty)$ (d) none.

4. A random variable X has the following p.m.f :

x_i	1	2	3
f_i	$\frac{1}{2}$	$\frac{1}{3}$	k

Then the value of k is

(a) 1 (b) 0

(c) $\frac{5}{6}$ (d) $\frac{1}{6}$.

5. The p.d.f. of *a* random variable X is

$f(x) = k(2x-1), 0 \leq x \leq 2$.

The value of the constant k is

(a) 1 (b) $\frac{1}{2}$

(c) $\frac{1}{3}$ (d) $\frac{2}{3}$.

6. The variance of *a* random variable X is

(a) $E(X)^2$ (b) $[E(X)]^2$

(c) $E(X^2)-[E(X)]^2$ (d) $E(X^2)-E(X)$

[*W.B.U.Tech.2006*]

7. The expectation of the following distribution :

x_i	0	1	2	3
f_i	$\frac{1}{4}$	$\frac{1}{8}$	$\frac{1}{2}$	$\frac{1}{8}$

is

(a) $\frac{1}{2}$ (b) $\frac{5}{2}$

(c) $\frac{3}{2}$ (d) 1.

8. A random variable X has the following p.d.f.

$$f(x) = e^{-x}, x \geq 0$$
$$= 0, \text{ otherwise.}$$

Then the value of $E(X)$ is

(a) 0 (b) 1

(c) 2 (d) -1 [*W.B.U. Tech.2007*]

9. The distribution function of a random variable X is given by :

$$F(x) = 0, \quad \text{for } x < 2$$
$$= 3 - x^2, \quad \text{for } 2 \leq x \leq 5$$
$$= 1, \quad \text{for } x > 5.$$

Then $P(3 < X \leq 4) = ?$

(a) -7 (b) 7

(c) -13 (d) 6

10. A random variable X has the following p.d.f :

$$f(x) = k, -2 < x < 2$$
$$= 0, \text{ otherwise.}$$

Then the value of the constant k is

(a) $\frac{1}{8}$ (b) $\frac{1}{2}$

(c) $\frac{1}{4}$ (d) $\frac{1}{12}$ [*W.B.U. Tech. 2007*]

11. The distribution $F(x)$ of a variate X is defind as follows :

$$F(x) = 0, -\infty < x < 1$$
$$= \frac{2}{3}, 1 \leq x \leq 5$$
$$= \lambda, 5 \leq x < \infty.$$

Then the constant λ is equal to

(a) 1 (b) -1

(c) $-\frac{2}{3}$ (d) $\frac{1}{3}$.

12. Let X be a continuous random variable with distribution

$$f(x)=\begin{cases}\frac{1}{8}, & 0\le x\le 8\\ 0, & \text{else where}\end{cases}$$

Then $P(X\le 6)=$

(a) $\frac{1}{4}$ (b) $\frac{3}{4}$

(c) $\frac{7}{8}$ (d) $\frac{1}{2}$.

13. If the mean of a distribution is 5, then the value of $E(2X-9)$ is

(a) 10 (b) 1

(c) –4 (d) –9.

14. When the variance of a random variable is $\frac{2}{3}$, the Var $(3X+5)=$

(a) 8 (b) 2

(c) 6 (d) 11

15. The following distribution is a p.m.f of a random variable :

x_i	–1	0	1	2
f_i	0.3	0.4	0.6	0.1

The statement is

(a) True (b) False.

16. Choose the correct alternatives :

If X and Y are independent random variables, then

(a) $E(XY)=E(X)+E(Y)$

(b) $E(XY)=E(X)-E(Y)$

(c) $E(XY)=E(X)E(Y)$

(d) $E(XY)=E(X)/E(Y)$. [*W.B.U. Tech 2006 M 302*]

17. The following distribution is a p.d.f of a random variable X:

$f(x)=|x|, -1\le x\le 1$

$=0$, else where.

The statement is

(a) True (b) False.

18. If $f(x)=1-\frac{1}{2}x,\ 0\le x\le 2$ is a p.d.f of a random variable X, then $P(|X-1|<2)=$

(a) –1 (b) 1

(c) 0 (d) $\frac{1}{3}$.

19. The distribution of a random variable X is

$$\begin{aligned} F(x) &= 0, \quad x<0 \\ &= x^2+3, \quad 0\le x<1 \\ &= 1, \quad x\ge 1. \end{aligned}$$

Then the corresponding p.d.f of x is

(a) $\frac{x^3}{3}+3x,\ 0\le x<1$

0, else where.

(b) $2x, 0\le x<1$

0, else where.

(c) $2x, 0<x\le 1$

0, else where.

(d) $2x, -<x<\infty$.

20. The distribution of a random variable X is given by

$P(X=-1)=\frac{1}{8}, P(X=0)=\frac{3}{4}, P(X=1)=\frac{1}{8}$. Then S. D of X is

(a) $\frac{1}{2}$ (b) $-\frac{1}{2}$

(c) 0 (d) 1

21. If X assumes the values 3,5,6,9 and

$P(X=3)=.24, P(X=5)=0.1, P(X=6)=0.12$ and

$P(X=9)=0.55$. Then this makes distribution.

(a) Yes (b) No.

22. Which one of the followings is not true for a distribution function $\phi(x)$

(a) $\phi(\infty)=1$

(b) $\lim_{x\to-\infty}\phi(x)=0$

(c) $\phi(x)$ is stricly increasing function.

(d) $0\le\phi(x)\le1$.

23. If X is a continuous random variable and $F(x)$ is its distribution function then find which one of the followings is not true

(a) $P(X=2)=0$

(b) $P(2\le X\le3)=F(3)-F(2)$

(c) $P(2<X\le3)=F(3)-F(2)$

(d) $\frac{d}{dn}\{\text{its } pdf\}=F(x)$.

24. If X is discrete random variable then

(a) $E(|X|)\le|E(X)|$ (b) $E(|X|)\ge|E(X)|$

(c) $E(|X|)=|E(X)|$ (d) none of these.

[W.B.U.Tech.2008]

25. If $P(X=-1)=\frac{1}{4},P(X=0)=\frac{1}{2},P(X=1)=\frac{1}{4}$ and

$P(Y=-2)=\frac{1}{3},P(Y=10)=\frac{1}{3},P(Y=4)=\frac{1}{3}$ then $E(X+Y)=$

(a) 10 (b) 4

(c) 6 (d) $\frac{5}{6}$

26. If $P(X=-1)=\frac{1}{4},P(X=0)=\frac{1}{2},P(X=1)=\frac{1}{4}$ and

$P(Y=-2)=\frac{1}{3},P(Y=10)=\frac{1}{3},P(Y=4)=\frac{1}{3}$ and if X,Y are independent then $E(XY)=$

(a) 10 (b) 4

(c) 0 (d) 1

27. Four coins are tossed. Expectation of number of heads is

(a) 1 (b) 2

(c) 3 (d) 4.

28. For two random variables X and Y, $E(XY)=E(X)E(Y)$ hold if X and Y are

(a) independent (b) uncorrelated

(b) continuous variate (d) none of these.

29. If X is a random variable then standard deviation of $3X+10$ is

(a) $3\sigma_x$ (b) $9\sigma_x$

(c) $3\sigma_x+10$ (d) $\pm 3\sigma_x$

30. If X is a random variable and its s.d $\sigma_x=0$ then X is

(a) discrete (b) continuous

(c) constant (d) 0.

31. If the mean of X is 3 and Var $(X)=E(X^2)-E(X)-K$ then $K=$

(a) 6 (b) 5

(c) 36 (d) -6

32. If $f(x)=\frac{1}{2}x,\ 0\le x\le 2$ and $f(x)=0$ elsewhere then the expectation of X^2 is

(a) 4 (b) 2

(c) 6 (d) 1

33. The pdf of a random variable is

$$f(x)=\frac{3}{4}x(2-x), 0\le x\le 2$$

$$=0 \qquad , \text{other wise.}$$

then $F(1)$ (where $F(x)$ is the distribution function) is

(a) $\frac{1}{3}$ (b) $\frac{1}{4}$

(c) $\frac{1}{2}$ (d) 1

34. If X is a random variable which assumes negative value only and if $E(|X|)=2$ then $E(X)=$

(a) 2 (b) 0

(c) –2 (d) none of these.

35. A discrete distribution of a random variable is

X	:	1	2	3	4	5
f	:	$\frac{1}{15}$	$\frac{2}{15}$	$\frac{3}{15}$	$\frac{4}{15}$	$\frac{5}{15}$.

Then $P(x>2/x\le 4)=$

(a) $\frac{7}{5}$ (b) $\frac{5}{7}$

(c) $\frac{1}{5}$ (d) $\frac{7}{10}$

36. Value of c for which the function $f(x)=cx$, for $x=0,1,2,3,4,5$

$=0$, otherwise

becomes a pmf is

(a) $\frac{7}{15}$ (b) $\frac{1}{15}$

(c) $\frac{2}{15}$ (d) $\frac{4}{15}$

37. For the discrete distribution

X	:	1	2	3	4	5
f_i	:	$\frac{1}{15}$	$\frac{2}{15}$	$\frac{3}{15}$	$\frac{4}{15}$	$\frac{5}{15}$

$P(X<a)=\frac{1}{5}$. Then $a=$

(a) 1 (b) 2

(c) 3 (d) 4

38. For the pmf $f(x)$ defined by $f(0)=\frac{1}{7}, f(1)=\frac{2}{7}, f(2)=\frac{1}{7}$ $f(3)=\frac{2}{7}, f(4)=\frac{1}{7}$ the value of $P(X \geq K)=\frac{6}{7}$. Then $K=$

(a) 1
(b) 2
(c) 3
(d) 4

39. For the pmf $f(x)$ defined by $f(0)=\frac{1}{8}, f(1)=\frac{3}{8}, f(2)=\frac{3}{8}$, $f(3)=\frac{1}{8}$ the value of the distribution function at 2 is

(a) $\frac{2}{5}$
(b) $\frac{3}{7}$
(c) $\frac{1}{7}$
(d) $\frac{7}{8}$

40. The variance of the discrete distribution

x	:	0	1	2
$p(x)$	:	$\frac{1}{4}$	$\frac{1}{2}$	$\frac{1}{4}$ is

(a) 0.6
(b) 0.7
(c) 0.2
(d) 0.5

41. Throwing two unbiased coins simultaneously Diku bets Buku that he will receive Rs 4 from Buku if he gets 2 heads and he will give Rs 4 to Buku, otherwise. If $P(x)$ be the pmf of Diku's gain then P(Diku's Rs 4 loss)=

(a) $\frac{1}{4}$
(b) $\frac{4}{7}$
(c) $\frac{3}{4}$
(d) $\frac{1}{5}$

42. For a random variable X, $E\{(X-1)^2\}=10$ and $E\{(X-2)^2\}=6$. Then $E(X)=$

(a) $\frac{5}{2}$
(b) 6
(c) $\frac{9}{2}$
(d) $\frac{7}{2}$

43. For a random variable $X, E\{(X-2)^2\} = 6, E\{(X-1)^2\} = 10$. Then $\sigma_x =$

(a) $\sqrt{3}$ (b) $\sqrt{2}$

(c) 0 (d) 2

44. Let X be a discrete random variable assuming values $1,2,3,4,\cdots$ and suppose that $E(X)$ exists. Then $E(X) = \sum_{n=1}^{\infty} P(X \geq n)$

(a) True (b) False.

45. The probability distribution of the number of TV set sold in a day by a salesman is

X	:	0	1	2	3	4	5
$P(X)$	:	0.1	0.2	0.3	0.2	0.1	0.1.

The average number of TV set sold in a day is

(a) 2 (b) 2.3

(c) 2.6 (d) 3.1.

46. If X and Y are both negative random variables, mutually independent, $E(XY) = 6$ and $E(|X|) = 2$. Then $E(Y) =$

(a) -1 (b) -2

(c) -3 (d) -4

47. If $P(X = i) = p$ and $P(Y = j) = q, (i, j = 1,2)$ and X and Y are two mutually independent random variables, $E(XY) =$

(a) $\frac{1}{4}$ (b) $\frac{7}{4}$

(c) $\frac{9}{4}$ (d) none of these.

48. A random variable X has the pdf

$$f(x) = \frac{1}{2}, -2 < x < 2$$
$$= 0, \text{ otherwise.}$$

Then $P(|X| > 1) =$

(a) $\frac{1}{2}$ (b) $\frac{1}{4}$

(c) 3 (d) none of these.

49. The pdf of a continuous random variable X is

$$f(x) = 6x(1-x), 0 \le x \le 4$$
$$= 0, \text{ otherwise.}$$

Then the mean of X is

(a) $\frac{1}{3}$ (b) $\frac{1}{2}$

(c) $\frac{1}{7}$ (d) $\frac{1}{9}$.

50. If X is a random variable which assumes values $1, 2, 3, 4, \cdots$ with respective probability given by $P(X = K) = q^{K-1}p, p + q = 1$. Then Mean of X is

(a) $\frac{1}{q}$ (b) $\frac{1}{p+1}$

(c) $\frac{1}{p}$ (d) $\frac{p}{q}$.

ANSWERS

1.b	2.a	3.b	4.d	5.b	6.c	7.c	8.b
9.a	10.c	11.a	12.b	13.b	14.c	15.b	16.c
17.a	18.b	19.b	20.a	21.b	22.c	23.d	24.b
25.b	26.c	27.b	28.a	29.a	30.c	31.a	32.b
33.c	34.c	35.d	36.b	37.c	38.a	39.d	40.d
41.c	42.d	43.b	44.a	45.b	46.c	47.c	48.a
49.b	50.c						

3.3 SPECIAL TYPE OF DISTRIBUTION

3.3.1. Introduction :

A random variable is said to have a special type of distribution if its pmf / pdf gets a special form.

Seven such type of distributions are discussed in this chapter. Among these the two discrete distributions Binomial and Poisson are very well known and useful in everyfield of life. The continuous distribution'Normal distribution' has also a wide range of application in sociology, market research and specially in industry. Definition of each of the distribution is first given and their properties and field of fitness are illustrated in this chapter.

3.3.2. Binomial Distribution.

A discrete random variable X is said to have a binomial distribution with parameters $p(0 < p < 1)$ and n (a positive integer) if its distribution is given by

X	:	0	1	2	...	...	n
f_i	:	f_0	f_1	f_2	...	...	f_n

where the pmf $f_i = P(X = i) = \binom{n}{i} p^i (1-p)^{n-i}$, $i = 0, 1, 2, \cdots, n$

For instant, one probability mass $f_1 = \binom{n}{1} p(1-p)^{n-1} = np(1-p)^{n-1}$ etc.

Note : (1) The pmf f_i satisfy the two fundamental properties $f_i \geq 0$ and $\sum_{i=0}^{n} f_i = 1$, which can be easily verified.

(2) When the random variable X has a binomial distribution with parameters n, p we write $X \sim b(n, p)$ and we say X is a binomial variate.

(3) The significance of the parameters n and p would be given in subsequent theorem.

Cases where Binomial Distribution fits.

Let A be an event of a random experiment E. We call "the event A" = 'success' and "the event Not A" = 'failure'. Let p = probability of 'success' in a single trial of E. E be repeated, independently, n times. Let X = number of succss in n trials. Then X may assume the values 0, 1, 2, ... n. For example, the event $(X=3)$ means " 3 success in n trials". It can be shown that $P(X=i)=\binom{n}{i}p^i(1-p)^{n-i}$

Thus the distribution of X becomes

X	:	0	1	2	3	...	...	n
f_i	:	f_0	f_1	f_2	f_3	...	...	f_n

where $f_i=P(X=i)=\binom{n}{i}p^i(1-p)^{n-i}$ which is the pmf of Binomial distribution.

Thus 'No. of Success' in n trials is a Binomial variate with parameter $p=$ probability of success in a single trial and $n=$ number of trial.

Illustration. The efficiency of a fighter-plane is such that the probability of a bomb hitting a target is $2/5$. The fighter is assigned to completely destroy a camp of enemy-side. The plane carries 6 bombs, i.e., 6 bombs can be aimed at the camp. Here throwing a bomb is the experiment. It can be repeated 6 times; ' A bomb hits the camp' $=$ Success and $X=$ number of success in 6 trials. Then X has the Binomial distribution,

X	:	0	1	2	3	4	5	6
f_i	:	f_0	f_1	f_2	f_3	f_4	f_5	f_6

where $f_i=\binom{6}{i}\left(\frac{2}{5}\right)^i\left(1-\frac{2}{5}\right)^{6-i}=\binom{6}{i}\left(\frac{2}{5}\right)^i\left(\frac{3}{5}\right)^{6-i}$

If it is known that at least four direct hits are necessary to destroy the camp then the probability of complete destruction of the camp

$$=P(X\geq 4)=P(X=4)+P(X=5)+P(X=6)=f_4+f_5+f_6$$

$$=\binom{6}{4}\left(\frac{2}{5}\right)^4\left(\frac{3}{5}\right)^2+\binom{6}{5}\left(\frac{2}{5}\right)^5\left(\frac{3}{5}\right)+\binom{6}{6}\left(\frac{2}{5}\right)^6=\frac{112}{625}$$

Theorem. If X has Binomial Distribution with parameter n and p then (i) its mean is np (ii) its variance is npq where $q=1-p$.

[*W.B.U.Tech 2005*]

Proof. Here $X: 0\ 1\ 2 \cdots n$

and its pmf is $f_i=P(X=i)=\binom{n}{i}p^i(1-p)^{n-i}$

(i) Mean $=E(X)=\sum_{i=1}^{n} if_i=\sum_{i=1}^{n} i\binom{n}{i}p^i(1-p)^{n-i}$

$$= np\sum_{i=1}^{n}\binom{n-1}{i-1}p^{i-1}(1-p)^{n-i} = np\sum_{r=0}^{n-1}\binom{n-1}{r}p^{r}(1-p)^{n-1-r}$$

replacing $i-1$ by r

$$= np(p+1-p)^{n-1} = np$$

(ii) Now

$$E\{X(X-1)\} = \sum_{i=0}^{n} i(i-1)\binom{n}{i}p^{i}(1-p)^{n-i}$$

$$= \sum_{i=0}^{n} i(i-1)\frac{n(n-1)}{i(i-1)}\binom{n-2}{i-2}p^{i}(1-p)^{n-i}$$

$$= n(n-1)p^{2}\sum_{i=2}^{n}\binom{n-2}{i-2}p^{i-2}(1-p)^{n-i}$$

$$= n(n-1)p^{2}\sum_{r=0}^{n-2}\binom{n-2}{r}p^{r}(1-p)^{n-2-r}, \text{ replacing } i-2 \text{ by } r$$

$$= n(n-1)p^{2}(p+1-p)^{n-2} = n(n-1)p^{2}.$$

Now by a previous result on variance (See Art 2.6) we have
$Var(X) = E\{X(X-1)\} - m(m-1)$, where $m =$ mean
$= n(n-1)p^2 - np(np-1) = np(1-p) = npq$ where $q = 1-p$
$\therefore$ standard deviation $\sigma = \sqrt{npq}$.

Illustration. An unbiased die is tossed four times. Let 'multiple of three' be success; otherwise it is failure. Here $p =$ probability of sucess in a single trial $= \frac{2}{6} = \frac{1}{3}$. Let $X =$ number of 'multiple of three' appeared among these four trial. Then, as we discussed before, X has Binomial distribution with parameter $n = 4$ and $p = \frac{1}{3}$. The expected number of 'multiple of 3' = mean of $X = 4\times\frac{1}{3} = \frac{4}{3}$. The standard deviation of $X = \sqrt{4\times\frac{1}{3}\left(1-\frac{1}{3}\right)} = \sqrt{4\times\frac{1}{3}\times\frac{2}{3}} = \frac{2\sqrt{2}}{3}$.

Illustrative Examples

Ex. 1. *The mean and s.d of a binomial distribution are respectively 4 and* $\sqrt{\frac{8}{3}}$*.Find the values of n and p.Hence evaluate* $P(X = 0)$.

[*W.B.U.Tech 2006*]

We know the mean and s.d of a binomial variate are respectively np and $\sqrt{np(1-p)}$.

$\therefore np = 4$ or, $np(1-p) = \frac{8}{3}$

$\therefore 4(1-p) = \frac{8}{3} \Rightarrow p = \frac{1}{3}$

$\therefore n = 4 \times 3 = 12$

$$P(X=0) = f_0 = {}^{12}C_0\left(\frac{1}{3}\right)^0\left(1-\frac{1}{3}\right)^{12-0} = \left(\frac{2}{3}\right)^{12}.$$

Ex. 2. *Comment on the statement "a binomial variate has mean 4 and s.d 3".*

Here, $np = 4$ and $\sqrt{np(1-p)} = 3$ i.e. $np(1-p) = 9$

$\therefore 4(1-p) = 9 \quad \therefore 1-p = \frac{9}{4} \therefore p = -\frac{5}{4}$ which is not possible since $0 < p < 1$. So the statement is false.

Ex. 3. *If the mean of a binomial distribution is 3 and the variance is $\frac{3}{2}$, find the probability of obtaining at most 3 success.* [*W.B.U.Tech 2007*]

Let X be the r.v corresponding to the number of success. Then the *pmf* of X is $f_i = P(X=i) = {}^nC_i p^i (1-p)^{n-i}$, $i = 0, 1, 2, \cdots, n$

$\therefore$ mean $= np = 3$, Variance $= np(1-p) = \frac{3}{2}$

or, $\frac{np(1-p)}{np} = \frac{3}{2} \times \frac{1}{3}$ $\qquad \therefore p = \frac{1}{2}$ and so $n = 6$

Now probability of at most 3 success $= P(X \le 3)$

$= P(X=0) + P(X=1) + P(X=2) + P(X=3)$

$$= {}^6C_0\left(\frac{1}{2}\right)^0\left(\frac{1}{2}\right)^6 + {}^6C_1\left(\frac{1}{2}\right)^1\left(\frac{1}{2}\right)^5 + {}^6C_2\left(\frac{1}{2}\right)^2\left(\frac{1}{2}\right)^4 + {}^6C_3\left(\frac{1}{2}\right)^3\left(\frac{1}{2}\right)^3 = \frac{21}{32}.$$

Ex. 4. *The probability that a pen manufactured by a company will be defective is $\frac{1}{10}$. If 12 such pens are manufactured, find the prob. that (i) exactly two will be defecive (ii) none will be defective (iii) at least two will be defective*

Let 'defective pen' = success $\therefore$ prob of success in a single trial $p = \frac{1}{10}$.

The experiment is repeated 12 times.

Let the random variable X = number of defective pens. Then X is a binomial variate where the parameters $n=12$ and $p=\frac{1}{10}$.

$\therefore$ Its pmf is $f_i = P(X=i)=\binom{12}{i}\left(\frac{1}{10}\right)^i\left(1-\frac{1}{10}\right)^{12-i}$

$\therefore$ (i) The required probability $= P(X=2)=\binom{12}{2}\left(\frac{1}{10}\right)^2\left(1-\frac{1}{10}\right)^{12-2}$

$$=\frac{66\times 9^{10}}{10^{12}}=0.2301.$$

(ii) required probality $P(X=0)=\binom{12}{0}\left(\frac{1}{10}\right)^0\left(1-\frac{1}{10}\right)^{12-0}=\left(\frac{9}{10}\right)^{12}=0.2833$.

(iii) $P(X=1)=\binom{12}{1}\left(\frac{1}{10}\right)^1\left(1-\frac{1}{10}\right)^{12-1}=12\times\frac{9^{11}}{10^{12}}=.3755$.

$\therefore$ $P(X\geq 2)=1-P(X<2)=1-[P(X=0)+P(X=1)]=0.3412$.

Ex. 5. *A defective die is thrown ten times independently. The probability that an even number will appear 5 times is twice the probability that an even number will appear 4 times. What is the probability that odd face appear in each of the ten throws.*

Let "even face" = "success".

and X = number of even face among 10 trials.

$\therefore X$ has Binomial distribution with parameter $n=10$ and p. So the pmf is given by $P(X=i)=f_i={}^{10}C_i p^i(1-p)^{10-i}$.

By problem $P(X=5)=2P(X=4)$ or, $f_5=2\times f_4$

or, ${}^{10}C_5\times p^5(1-p)^{10-5}=2\times{}^{10}C_4 p^4(1-p)^{10-4}$

or, ${}^{10}C_5 p=2\times{}^{10}C_4(1-p)$

or, $252p=2\times 210(1-p)$ or, $3p=5(1-p)$ or $p=\frac{5}{8}$

Required probability $=P(X=0)=f_0={}^{10}C_0 p^0(1-p)^{10-0}$

$$=\left(1-\frac{5}{8}\right)^{10}=\left(\frac{3}{8}\right)^{10}.$$

Ex. 6. *The overall percentage of failures in a certain examination is 40. What is the probability that out of a group of 6 candidates at least 4 passed the examination?*

Now the overall percentage of success in a certain exam is 60.

Consider the experiment that one student is drawn and seen whether he is passed. Probability that he is a passed student is $p=\frac{60}{100}=0.6$. The experiment be repeated 6 times. Let $X=$ number of passed-student in 6 such trials. So X is a Binomial variate with parameter $n=6,\ p=0.6$.

So the distribution of X is

X	:	0	1	2	3	4	5	6
$P(X=i)=f_i$	:	f_0	f_1	f_2	f_3	f_4	f_5	f_6,

where $f_i={}^6C_i p^i(1-p)^{6-i}={}^6C_i(0.6)^i(1-0.6)^{6-i}={}^6C_i(0.6)^i(0.4)^{6-i}$

Required probability $=P(X\geq 4)=f_4+f_5+f_6$

$={}^6C_4(0.6)^4(0.4)^{6-4}+{}^6C_5(0.6)^5(0.4)^{6-5}+{}^6C_6(0.6)^6(0.4)^{6-6}=0.54432$.

Ex. 7. *A family has 6 children. Find the probability that (i) 3 boys and 3 girls (ii) fewer boys than girls.*

Probability of any particular child being a boy is $\frac{1}{2}$.

Let 'boy' = success. The experiment of noticeing whether the child is boy or girl is repeated 6 times. $X=$ No. of boy i.e. No. of success

$\therefore\ X\sim b(n,p)$ where $n=6, p=\frac{1}{2}$. So the pmf of X is

$$f_i=P(X=i)={}^6C_i\left(\frac{1}{2}\right)^i\left(\frac{1}{2}\right)^{6-i}={}^6C_i\left(\frac{1}{2}\right)^6.$$

(i) Now Probability of " 3 boys and 3 girls"

$$=P(X=3)={}^6C_3\left(\frac{1}{2}\right)^6=5/16$$

(ii) Probability of "fewer boys than girls'

$$=P(X\leq 2)=P(X=0)+P(X=1)+P(X=2)$$

$$={}^6C_0\left(\frac{1}{2}\right)^6+{}^6C_1\left(\frac{1}{2}\right)^6+{}^6C_2\left(\frac{1}{2}\right)^6=\frac{11}{32}.$$

Ex. 8. *A die is tossed thrice. A success is "getting 1 or 6" on a toss. Find the mean and variance of the number of success.*

Let X denote the number of successs. Clearly X can take the values 0, 1, 2 or 3

and X follows binomial distribution with

$$p = \text{Probability of success } = \frac{2}{6} = \frac{1}{3}$$

$$\text{and } q = 1 - \frac{1}{3} = \frac{2}{3}$$

$$\text{and pmf } P(X = i) = f_i = {}^3C_i \; p^i \; q^{3-i}$$

$$P(X = 0) = q \cdot q \cdot q = \frac{2}{3} \times \frac{2}{3} \times \frac{2}{3} = \frac{8}{27}$$

$$P(X = 1)$$

$$= {}^3C_1 \times p \times q \times q = 3 \times \frac{1}{3} \times \frac{2}{3} \times \frac{2}{3} = \frac{4}{9}$$

$$P(X = 2) = {}^3C_2 \times p \times p \times q = 3 \times \frac{1}{3} \times \frac{1}{3} \times \frac{2}{3} = \frac{2}{9}$$

$$P(X = 3) = p \times p \times p = \frac{1}{3} \times \frac{1}{3} \times \frac{1}{3} = \frac{1}{27}$$

Therefore the prob. distribution of X is

x_i	:	0	1	2	3
$P(X = x_i)$	:	$\frac{8}{27}$	$\frac{4}{9}$	$\frac{2}{9}$	$\frac{1}{27}$

$$\therefore \text{ Mean } = E(X) = \sum_i x_i P(X = x_i)$$

$$= 0 \times \frac{8}{27} + 1 \times \frac{4}{9} + 2 \times \frac{2}{9} + 3 \times \frac{1}{27} = 1$$

$$E(X^2) = \sum_i x_i^2 P(X = x_i)$$

$$= 0^2 \times \frac{8}{27} + 1^2 \times \frac{4}{9} + 2^2 \times \frac{2}{9} + 3^2 \times \frac{1}{27} = \frac{5}{3}$$

$$\therefore \text{ Variance } = E(X^2) - \{E(X)\}^2 = \frac{5}{3} - (1)^2 = \frac{2}{3}.$$

Ex. 9. *Find the probability distribution of the number of boys in a family with 3 children, assuming equal probabilities for boys and girls. Graph the distributions. Also find the distribution function $F(x)$ for the random variable X.*

Let E be the experiment of picking a child in the family. The event 'boy' = success. We have p = probability of boy $= \frac{1}{2}$ by hypothesis. E is repeated 3 times. Let X denote the number of boys. Then X can assume the values 0, 1, 2, 3. X has the Binomial distribution.

X	:	0	1	2	3
f_i	:	f_0	f_1	f_2	f_3

where $f_i = \binom{3}{i}\left(\frac{1}{2}\right)^i\left(\frac{1}{2}\right)^{3-i} = \binom{3}{i}\left(\frac{1}{2}\right)^3 = \binom{3}{i}\frac{1}{8}$

Therefore, the probability distribution of boys is

X	:	0	1	2	3
f_i	:	$\frac{1}{8}$	$\frac{3}{8}$	$\frac{3}{8}$	$\frac{1}{8}$

The graph of the distribution is given below :

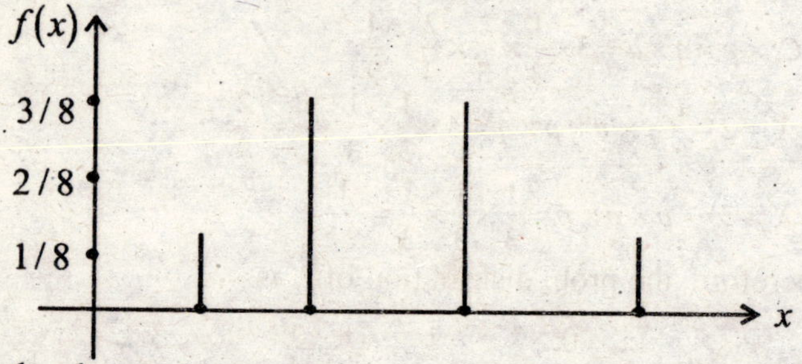

Using the above table we obtain the distribution function $F(x)$ as

$$F(x) = 0, \quad -\infty < x < 0$$
$$= \frac{1}{8}, \quad 0 \le x < 1$$
$$= \frac{1}{8} + \frac{3}{8}, \quad 1 \le x < 2$$
$$= \frac{1}{2} + \frac{3}{8}, \quad 2 \le x < 3$$
$$= \frac{7}{8} + \frac{1}{8}, \quad 3 \le x < \infty$$

i.e., $F(x) = 0, \quad -\infty < x < 0$

$$= \frac{1}{8}, \quad 0 \le x < 1$$

$$= \frac{1}{2}, \quad 1 \le x < 2$$
$$= \frac{7}{8}, \quad 2 \le x < 3$$
$$= 1, \quad 3 \le x < \infty.$$

Ex. 10. *Suppose that half the population of a town are consumers of rice. 100 investigators are appointed to find out its truth. Each investigator interviews 10 individuals. How many investigator do you expect to report that three or less of the people interviewed are consumers of rice ?*

Consider the experiment "One investigator is drawn and seen whether he reports that three or less of the people interviewed are consumers of rice" ≡ whether it is success. Let probability of success $= p$

Now the investigator draws one individual and sees whether he is consumer of rice. Let q be the probability of this event $= \frac{1}{2}$. The investigator repeats this experiment 10 times.

Let $Y =$ Number of such individual among ten. So Y has binomial distribution with parameter $n = 10,\ q = \frac{1}{2}$

Therefore $p = P(Y \le 3)$

$$= P(Y=0) + P(Y=1) + P(Y=2) + P(Y=3)$$

$$= {}^{10}C_0\left(\frac{1}{2}\right)^0\left(1-\frac{1}{2}\right)^{10-0} + {}^{10}C_1\left(\frac{1}{2}\right)^1\left(1-\frac{1}{2}\right)^{10-1}$$

$$+ {}^{10}C_2\left(\frac{1}{2}\right)^2\left(1-\frac{1}{2}\right)^{10-2} + {}^{10}C_3\left(\frac{1}{2}\right)^3\left(1-\frac{1}{2}\right)^{10-3}$$

$$= \left(\frac{1}{2}\right)^{10} + 10 \times \frac{1}{2}\left(\frac{1}{2}\right)^9 + 45\left(\frac{1}{2}\right)^2\left(\frac{1}{2}\right)^8 + 120\left(\frac{1}{2}\right)^3\left(\frac{1}{2}\right).$$

$$= \left(\frac{1}{2}\right)^{10}\{1 + 10 + 45 + 120\} = 176 \times \left(\frac{1}{2}\right)^{10}.$$

Now, X has bonomial distribution with parameter

$n = 100,\ p = 176 \times \left(\frac{1}{2}\right)^{10}$. Therefore required expectation $= E(X)$

$$= 100 \times 176 \times \left(\frac{1}{2}\right)^{10} = \frac{17600}{2^{10}} = \frac{17600}{1024} \simeq 17$$

∴ 17 investigators are expected to report so.

Ex. 11. *If X be a binomially distributed with* $E(X)=2$ *and* $\text{var}(X)=\frac{4}{3}$, *find the distribution of X.*

We have $E(X)=2, \text{var}(X)=\frac{4}{3}$.

$\therefore \quad np=2 \qquad \ldots (1)$

$np(1-p)=\frac{4}{3} \qquad \ldots (2)$

Solving (1), (2), we get

$$p=\frac{1}{3}, n=6, \quad q=1-p=1-\frac{1}{3}=\frac{2}{3}.$$

$$\therefore \quad f_0=P(X=0)={}^6C_0\left(\frac{1}{3}\right)^0\left(\frac{2}{3}\right)^6=\frac{64}{729}$$

$$f_1=P(X=1)={}^6C_1\left(\frac{1}{3}\right)^1\left(\frac{2}{3}\right)^5=\frac{64}{249}$$

$$f_2=P(X=2)={}^6C_2\left(\frac{1}{3}\right)^2\left(\frac{2}{3}\right)^4=\frac{80}{243}.$$

Similarly $f_3=P(X=3)=\frac{160}{729}, f_4=P(X=4)=\frac{20}{243}$

$$f_5=P(X=5)=\frac{4}{243}, f_6=P(X=6)=\frac{1}{729}.$$

Thus the required distribution of X is

X	:	0	1	2	3	4	5	6
f_i	:	$\frac{64}{729}$	$\frac{64}{243}$	$\frac{80}{243}$	$\frac{160}{729}$	$\frac{20}{243}$	$\frac{4}{729}$	$\frac{1}{729}$

Ex. 12. *The probability of a man hitting a target is* $\frac{1}{3}$

(a) If he fires 5 times, what is the probability of his hitting the target at least twice?

(b) *How many times must he fire so that the probability of his hitting the target at least once is more than 90%.*

Here $p=$ probability of hitting $=\frac{1}{3}$.

$\therefore$ $q=$ probability of no hit $=\frac{2}{3}$.

(a) Let X be the number of hits. Here $n=5$

$$\therefore \quad P(X\geq 2)=1-P(X<2)$$

$$=1-P(X=0)-P(X=1)$$

$$=1-{}^5C_0\left(\frac{1}{3}\right)^0\left(\frac{2}{3}\right)^5-{}^5C_1\left(\frac{1}{3}\right)^1\left(\frac{2}{3}\right)^4$$

$$=\frac{131}{243}.$$

(b) Let n be the smallest number of fires so that the probability of hitting the target of least once is more than 90%.

$\therefore$ By the condition

$$P(X\geq 1)>\frac{90}{100}$$

or, $1-P(X=0)>0.9$

or, $1-\left(\frac{2}{3}\right)^n>0.9$

or, $\left(\frac{2}{3}\right)^n<0.1$

or, $n\log\frac{2}{3}<\log 0.1$

or, $n>\dfrac{\log 0.1}{\log\frac{2}{3}}$

$\therefore$ $n>5\cdot 679$

$\therefore$ $n=6$.

Thus he must fire 6 times.

3.3.3. Poisson Distribution.

A discrete random variable X is said to have a poisson distribution with parameter $\mu(>0)$ if its distribution is given by

X	: 0	1	2	3	...	...
f_i	: f_0	f_1	f_2	f_3	...	...

where the pmf, $f_i = P(X=i) = \dfrac{e^{-\mu}\mu^i}{i!}$

e.g one probability mass, $f_2 = \dfrac{e^{-\mu}\mu^2}{2!} = \dfrac{e^{-\mu}\mu^2}{2}$ etc.

Note : (1) Since the parameter $\mu > 0$, $f_i = \dfrac{e^{-\mu}\mu^i}{i!} \geq 0$

Again $\displaystyle\sum_{i=0}^{\infty} f_i = \sum_{i=0}^{\infty} \frac{e^{-\mu}\mu^i}{i!} = e^{-\mu}\sum_{\mu=0}^{\infty}\frac{\mu^i}{i!} = e^{-\mu}e^{\mu} = 1$

Thus the pmf f_i satisfies the two fundamental properties of pmf.

(2) When the random variable X has a poisson distribution with parameter $\mu(>0)$, we write $X \sim P(\mu)$ and we say X is a poisson variate.

(3) Actually poisson distribution is a limiting case of Binomial distribution when n is very large and p is very small so that $\mu = np$ is of finite magnitude.

(4) The significance of the parameter μ is given in next theorem.

Cases where Poisson Distribution fits.

Let us consider a sequence of changes. If the random variable $X(t)$ denotes the number of changes during the interval $(0, t)$, then $X(t)$ assumes the values 0, 1, 2, 3, ...

It can be shown that $P(X(t)=i) = f_i = e^{-\lambda t}\dfrac{(\lambda t)^i}{i!}$, $i = 0, 1, 2, \cdots \cdots$

where $\lambda =$ number of changes per unit time.

Thus the distribution of $X(t)$ becomes

$X(t)$	:	0	1	2	3	...	...
f_i	:	f_0	f_1	f_2	f_3	...	...

where $f_i = e^{-\lambda t}\dfrac{(\lambda t)^i}{i!}$ which is the pmf of poisson distribution.

Thus 'changes in an interval' is a poisson variate with parameter $\mu = \lambda t =$ average changes in the interval.

Note. (1) The interval $(0, t)$ may not be a time-interval. See the following illustration

(2) We use the notation $X(t)$ in place of X because it depends on t.

Illustration. **(i)** Let a huge metal sheet be produced by a machine in a factory. Defects are noticed in the sheet. The machine is such that the average number of defects per unit area is 3. A piece of '10 unit area of the sheet is purchased by a company. Let $X =$ Number of defects in this piece. Then X may assume values 0, 1, 2, 3 ... up to ∞. Here 'change' means 'defect' and the interval $(0, t)$ stands for '10 unit area of the sheet' or $(0, 10)$. Then X (or, $X(10)$) has the poisson distribution

X	:	0	1	2	3	...	...
f_i	:	f_0	f_1	f_2	f_3	...	...

where $f_i = e^{-\mu} \dfrac{\mu^i}{i!}$ with the parameter $\mu = 3 \times 10 = 30 =$ average number of defects per 10 unit area of the sheet.

For example, $f_0 = P(X = 0) = e^{-30} \dfrac{30^0}{0!} = e^{-30}$

i.e., probability no defects in the piece $= e^{-30}$

In other words $100e^{-30}\%$ such pieces will be free from defects.

(ii) The number of deaths in a state in one year is a poisson variate

(iii) The number of radio active atoms decaying in time t follows the poisson distribution with parameter $\mu = t \times$ average number of decayed radioactive atoms per unit time.

Theorem. If X is a poisson variate with parameter μ then

(i) Mean of X is μ

(ii) Variance of X is μ *[W.B.U.Tech 2006,2007]*

Proof. The values assumed by X are 0, 1, 2, ..., ... with probability $P(X = i) = f_i = \dfrac{e^{-\mu} \mu^i}{i!}$.

(i) Mean $= E(X) = \sum_{i=0}^{\infty} i\, f_i = \sum_{i=0}^{\infty} i \frac{e^{-\mu}\mu^i}{i!}$

$$= e^{-\mu} \sum_{i=0}^{\infty} \frac{\mu^i}{(i-1)!} = e^{-\mu}\left(\frac{\mu}{0!} + \frac{\mu^2}{1!} + \frac{\mu^3}{2!} + \frac{\mu^4}{3!} + \cdots \text{ up to } \infty\right)$$

$$= \mu e^{-\mu}\left(1 + \mu + \frac{\mu^2}{2!} + \frac{\mu^3}{3!} + \cdots \text{ up to } \infty\right) = \mu e^{-\mu} \cdot e^{\mu} = \mu$$

(ii) Now, $E\{X(X-1)\} = \sum_{i=0}^{\infty} i(i-1) e^{-\mu} \frac{\mu^i}{i!} \quad = e^{-\mu}\mu^2 \sum_{i=2}^{\infty} \frac{\mu^{i-2}}{(i-2)!}$

$$= e^{-\mu}\mu^2\left(1 + \frac{\mu}{1!} + \frac{\mu^2}{2!} + \frac{\mu^3}{3!} + \cdots \cdots \text{ up to } \infty\right)$$

$$= e^{-\mu}\mu^2 e^{\mu} = \mu^2$$

So, by an earlier result, (See the Theorem of Art 2.6) the variance, $Var(X) = E\{X(X-1)\} - m(m-1)$, where $m =$ mean

$$= \mu^2 - \mu(\mu - 1) = \mu$$

Note. In light of the above theorem we considered the parameter $\mu = \lambda t$ in our previous illustration.

3.3.4. Binomial Approximation to Poisson Distribution.

The range of applications of Poisson Distribution becomes more wider as it is used as an approximation for a Binomial distribution.In case of a Binomial distribution when n becomes large, p is small enough so that np is a moderate fixed value,the Binomial variate becomes approximately equal to a Poisson variate.This is given by the following theorem.

Theorem. Let the random variable X follows Binomial distribution with pmf , $f_i = \binom{n}{i} p^i (1-p)^{n-i}$, $i = 0, 1, 2, \cdots, n$ where n and p are parameters.If $n \to \infty$ and $p \to 0$ such that $np = \mu$,a fixed quantity then

$\lim_{n\to\infty} \binom{n}{i} p^i (1-p)^{n-i} = \frac{e^{-\mu}\mu^i}{i!}$ for a fixed i, *i.e.* f_i of Binomial distribution $\simeq f_i$ of Poisson distribution.

Proof. Beyond the scope of the book.

Illustration. Let a box contains 200 fuses.Experience tells that 2% of such fuses are defective.Let us consider the experiment of drawing a fuse and testing whether this is defective or not. Let $X =$ number of defective fuse. $\therefore$ X has Binomial distribution with parameter

$n=200,\ p=\dfrac{2}{100}=.02$. The pmf of X is $f_i=\binom{n}{i}p^i(1-p)^{n-i}=\binom{200}{i}$ $(.02)^i(1-.02)^{200-i}$.

Here we see n is so large and p is small so we can write

$$\binom{200}{i}(.02)^i(1-.02)^{200-i}\simeq\frac{e^{-200\times.02}}{i!}=\frac{e^{-4}}{i!}$$

Now $P(X\le3)=f_0+f_1+f_2+f_3\simeq\dfrac{e^{-4}}{0!}+\dfrac{e^{-4}}{1!}+\dfrac{e^{-4}}{2!}+\dfrac{e^{-4}}{3!}$.

Illustrative Examples.

Ex. 1. *For a poisson variate if* $P(X=2)=P(X=1)$ *,find* $P(X=1\text{ or }0)$. *Find also mean of X.*

Let m be the parameter of the poisson variate.

$$\therefore P(X=i)=f_i=\frac{e^{-m}m^i}{i!}.$$

Now, $f_2=f_1$, or, $\dfrac{e^{-m}\cdot m^2}{2!}=\dfrac{e^{-m}\cdot m^1}{1!}$ $\therefore m=2$

$\therefore\ P(X=1\ or\ 0)=f_1+f_0=e^{-m}(1+m)=e^{-2}(1+2)=3e^{-2}$

Mean of $X=m=2$.

Ex. 2. *A car-hire firm has two cars which it hires out day by day. The number of demands for a car on each day is distributed as a poisson distribution with average number of demand per day 1.5. Calculate the proportion of days on which neither car is used and the proportion of days on which some demand is refused* $(e^{-1.5}=0.2231)$.

[W.B.U.Tech,2003, 2006,2007]

Let X be the random variable denoting the number of demands for a car on any day. Then X is poisson distributed with parameter $\mu=1.5$. So its pmf $P(X=i)=f_i=\dfrac{e^{-\mu}\mu^i}{i!}$ where $\mu=1.5$.

$\therefore$ Proportion of days on which neither car is used

= Prob. of there being no demand for the car

$=P(X=0)=\dfrac{\mu^0e^{-\mu}}{0!}=e^{-1.5}=0.2231$.

Proportion of days on which some demand is refused

= Prob. for the number of demands to be more than two

$=P(X>2)=1-P(X\le2)$

$=1-\{P(X=0)+P(X=1)+P(X=2)\}$

$$=1-\left\{e^{-\mu}+\frac{\mu e^{-\mu}}{1!}+\frac{\mu^2 e^{-\mu}}{2!}\right\}=1-e^{-1.5}\left(1+1.5+\frac{(1.5)^2}{2}\right)=0.19126.$$

Ex. 3. *A radio active source emits on the average 2.5 particles per second. Calculate the prob. that 2 or more particles will be emitted in an interval of 4 seconds.*

Here $\lambda =$ number of changes (which is particle emitted) per unit time on an average $=2.5$.

Let X be the random variable denoting the number of particles emitted in the given interval. Then X is poisson distributed with parameter $\mu =$ average number of particle in 4 seconds.

So the p.m.f, $f_i = P(X=i)=\mu^i e^{-\mu}/i! = 10^i e^{-10}/i!$

So the required Prob. $= P(X\geq 2)=1-P(X<2)$

$$=1-\{P(X=0)+P(X=1)\} \quad =1-\{e^{-10}+10e^{-10}\}=1-11e^{-10}.$$

Ex. 4. *In a certain factory turning razor blades, there is a small chance, 1/500 for any blade to be defective. The blades are in packets of 10. Use poisson distribution to calculate the approximate number of packets containing (i) no defective (ii) one defective (iii) two defective blades respectively in one consignment of 10, 000 packets. (Given* $e^{-.02}=.9802$*).* [*W.B.U.Tech 2004*]

On an average there are 1 defective blade per 500 blades. So the average number of defective blades in a packet of 10 is $10\times\frac{1}{500}=\frac{1}{50}=0.02$.

Let $X=$ number of defective blades in a packet. X follows poisson distribution with parameter $\mu=0.02$. So the pmf is

$$P(X=i)=f_i=\frac{e^{-\mu}\mu^i}{i!}.$$

(i) Now probability that one packet contains no defective blade

$$=P(X=0)=f_0=\frac{e^{-\mu}\mu^0}{0!}=e^{-\mu}=e^{-0.02}=0.9802.$$

$\therefore$ Number of packets in the consignment containing no defective blades $=0.9802\times 10{,}000=9802$

(ii) Probability that one packet contains one defective blade

$$=P(X=1)=f_1=\frac{e^{-\mu}\mu}{1!}=e^{-0.02}\times 0.02 \quad =0.9802\times 0.02=0.019604$$

$\therefore$ Number of blades in the consignment

$= 0.019604 \times 10{,}000 = 196.04 \simeq 196$

(iii) $P(X=2) = \dfrac{e^{-\mu}\mu^2}{2!} = e^{-.02} \times \dfrac{(.02)^2}{2} = 0.9802 \times .0002 = 0.00019604$

$\therefore$ Required number $= 0.00019604 \times 10{,}000 = 1.9604 \simeq 2$.

Ex. 5. *If a random variable has a poisson distribution such that $P(1) = P(2)$, find (i) mean of the distribution (ii) standard derivation (iii) $P(X=4)$* [W.B.U.Tech 2007]

Let X be a poisson variate. Then the p.m.f of X is

$$f_i = P(X=i) = e^{-\mu}\frac{\mu^i}{i!}, \quad i = 0, 1, 2, \cdots$$

As $P(1) = P(2)$, so $e^{-\mu}\mu = e^{-\mu}\dfrac{\mu^2}{2!}$ $\quad \therefore \mu = 2$

(i) So the mean of the distribution is 2

(ii) Now $Var(X) = \mu = 2$ $\quad \therefore$ standard derivation $= \sqrt{2}$.

(iii) $P(X=4) = e^{-2}\dfrac{2^4}{4!} = \dfrac{2}{3}e^2$.

Ex. 6. *Find the probability that at most 5 defective fuses will be found in a box of 200 fuses if experience shows that 2 percent of such fuses are defective.*

Let X denote the number of defective fuses in the box. Then clearly X has a binomial distribution with parameters $n = 200$, $p = \dfrac{2}{100} = \dfrac{1}{50}$

$\therefore \mu = np = 200 \times \dfrac{1}{50} = 4$

Using an approximation, by the poisson distribution, we have

$$P(X \le 5) = \sum_{i=0}^{5} \frac{e^{-4}4^i}{i!} = e^{-4}\left(1 + \frac{4}{1!} + \frac{4^2}{2!} + \cdots + \frac{4^5}{5!}\right) = 0.785$$

Ex. 7. *Six coins are tossed 6400 times. Using the poisson distribution find the approximate probability of getting six heads 8 times.*

Let X denote the number of six heads in the toss of six coin. Then X is a binomial variate with parameter $n = 6400$, $p = \left(\dfrac{1}{2}\right)^6 = \dfrac{1}{64}$. Here n is so large and p is small, but $\mu = np = 6400 \times \dfrac{1}{64} = 100$

So using the poisson approximation, we have,

$$P(X=8) = {}^nC_8 p^8 (1-p)^{n-8} \simeq e^{-100}\frac{(100)^8}{8!}$$

Ex. 8. *2% of the items made by a machine are defective. Find the probability that 3 or more items are defective in a sample of 100 items. (Given $e^{-1} = 0.368$, $e^{-2} = 0.135$, $e^{-3} = 0.498$)*

Consider the experiment – one item is drawn and found whether it is defective (success). Let this experiment be repeated 100 times. $X =$ number of defective items. Then X follows Binomial distribution with parameter $n = 100$, $p = 2\% = \frac{2}{100} = .02$. $\therefore$ the pmf, $f_i = {}^nC_i p^i (1-p)^{n-i}$.

Now since n is large and p is small and $np = 100 \times .02 = 2$ so Binomial pmf is approximately equal to Poisson pmf with parameter np.

$$\therefore f_i \simeq \frac{e^{-2} \cdot 2^i}{i!}$$

Required probability $= P(X \geq 3) = 1 - P(X < 3)$

$$= 1 - \{f_0 + f_1 + f_2\} = 1 - e^{-2}\left(1 + 2 + \frac{2^2}{2!}\right) = 1 - 0.135 \times 5 = 0.325 .$$

3.3.5. Uniform or Rectangular Distribution.

A random variable X is said to have a uniform distribution on the interval $[a, b], -\infty < a < b < \infty$, if its probability density function is given by

$$f(x) = \frac{1}{b-a}, \quad a \leq x \leq b$$
$$= 0, \quad \text{otherwise}$$

where a, b are two parameters of the distribution. Clearly $f(x) \geq 0$, for $a < x < b$ and $\int_{-\infty}^{\infty} f(x)dx = \int_a^b f(x)dx = \int_a^b \frac{dx}{b-a} = 1$. So the fundamental properties of the density function is satisfied. If X has a uniform distribution on [a, b], then we write $X \sim U[a, b]$.

The density curve is shown in the following fig.

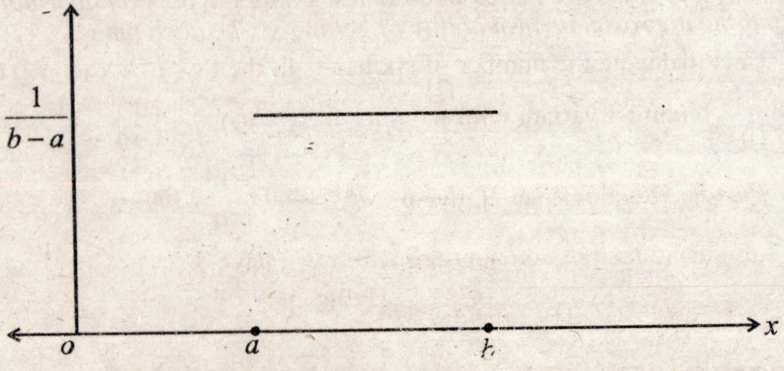

The distribution function $F(x)$ may be easily calculated by

$$F(x) = \int_{-\infty}^{x} f(x)dx$$

and is given by

$$F(x) = 0, \quad -\infty < x < a$$
$$= \frac{x-a}{b-a}, \quad a \le x < b$$
$$= 1, \quad b \le x < \infty$$

The graph of the distribution function is given below.

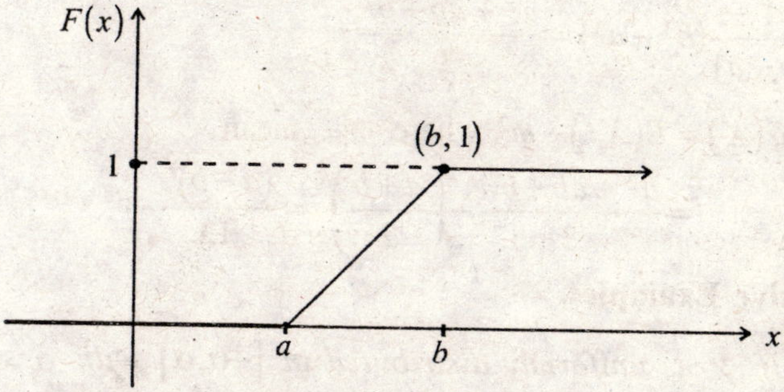

A Case where Uniform Distribution fits.

Let [a. b] be an interval. A point P is taken at random in the interval [a, b]. Let $OP = x$.

Let X be a random variable which assumes the values x. Then X would have uniform distribution with pdf

$$f(x) = \frac{1}{b-a}, \ a < x < b$$
$$= 0, \quad \text{elsewhere}$$

Theorem. If a continuous random variable X has uniform distribution with parameter a and b then

(i) the mean is $\frac{1}{2}(a+b)$

(ii) the variance is $\frac{(a-b)^2}{12}$

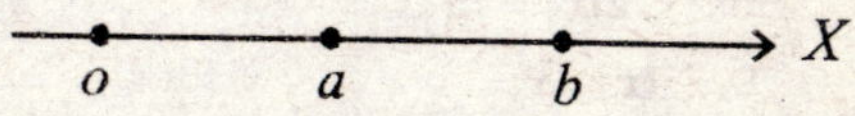

Proof. The p.d.f of X is given by

$$f(x) = \frac{1}{b-a}, \quad a \le x \le b$$
$$= 0, \quad \text{elsewhere}$$

(i) the mean $= E(X) = \int_{-\infty}^{\infty} xf(x)dx$

$$= \int_{-\infty}^{a} x \cdot 0dx + \int_{a}^{b} x \cdot \frac{1}{b-a}dx + \int_{b}^{\infty} x \cdot 0dx$$

$$= \frac{1}{b-a}\frac{1}{2}\left[x^2\right]_a^b = \frac{1}{2}(a+b)$$

(ii) Now, $E(X^2) = \int_a^b x^2 \frac{1}{b-a}dx = \frac{1}{3(b-a)}\left[x^3\right]_a^b$

$$= \frac{1}{3(b-a)}\left(b^3 - a^3\right) = \frac{a^2+ab+b^2}{3}$$

$\therefore Var(X) = E(X^2) - m^2$ where $m =$ mean

$$= \frac{a^2+ab+b^2}{3} - \left(\frac{a+b}{2}\right)^2 = \frac{(a-b)^2}{12}.$$

Illustrative Examples.

Ex. 1. *If X is uniformly distributed in $[-\alpha, \alpha]$ with $\alpha > 0$, then determine α such that $P(X > 1) = \frac{1}{3}$.*

Here p.d.f of X is $f(x) = \frac{1}{2\alpha}$ is $-\alpha < x < \alpha$

$= 0$ elsewhere.

when $\alpha < 1$,

$P(X > 1)$ should be zero, as X lies outside the given interval $[-\alpha, \alpha]$. So α must be greater than 1.

$\therefore P(X > 1)$

$$= \int_1^{\alpha} f(x)dx = \frac{\alpha - 1}{2\alpha}$$

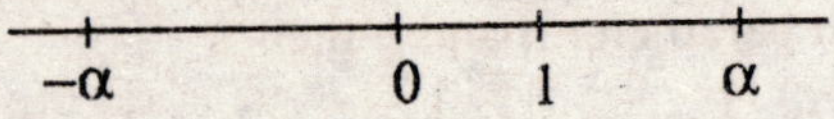

$\therefore \frac{\alpha-1}{2\alpha} = \frac{1}{3}$.

$\therefore \alpha = 3$.

Ex. 2. *If X is uniformly distributed in $-2 \le x \le 2$, find $P\left(|X-1| \ge \frac{1}{2}\right)$.*

Here p.d.f of X is

$f(x) = \frac{1}{4}$ is $-2 \le x \le 2$

$= 0$, elsewhere.

$$\therefore \quad P\left(|X-1| \geq \frac{1}{2}\right)$$

$$= P\left(-2 \leq X \leq \frac{1}{2}\right) + P\left(\frac{3}{2} \leq X \leq 2\right)$$

$$= \int_{-2}^{\frac{1}{2}} f(x)dx + \int_{\frac{3}{2}}^{2} f(x)dx$$

$$= \frac{\frac{1}{2}+2}{4} + \frac{2-\frac{3}{2}}{4}$$

$$= \frac{3}{4}.$$

Ex. 3. *If X is uniformly distributed over* [1, 2] *find U so that* $P\left(X > U + \overline{X}\right) = \frac{1}{6}$. [*W.B.U.Tech 2006,2008*]

Since X is uniformly distributed over [1, 2] therefore its pdf,

$$f(x) = \frac{1}{2-1}, \quad 1 < x < 2$$
$$= 0, \quad \text{elsewhere}$$

i.e., $f(x) = 1, \quad 1 < x < 2$

$$= 0, \quad \text{elsewhere}$$

$$\therefore \ \overline{X} = E(X) = \int_{-\infty}^{\infty} xf(x)dx = \int_{1}^{2} x \cdot 1\, dx = \left[\frac{x^2}{2}\right]_1^2 = \frac{3}{2}$$

Now, $P\left(X > U + \overline{X}\right) = \frac{1}{6}$ or, $P\left(U + \frac{3}{2} < X < \infty\right) = \frac{1}{6}$

or, $\int_{U+\frac{3}{2}}^{2} dx = \frac{1}{6}$ or, $2 - U - \frac{3}{2} = \frac{1}{6}$ or, $U = \frac{1}{3}$

3.3.6. Exponential Distribution.

Two parameters exponential distribution. A random variable X is said to have a two-parameter-exponential distribution if its probability density function is given by

$$f(x) = \frac{1}{b} e^{-\frac{x-a}{b}}, \quad x \geq a$$
$$= 0, \quad \text{elsewhere}$$

where $a, b(b > a)$ are two parameters of the distribution.

Note. (1) Clearly $f(x) \geq 0$ for all x and

$$\int_{-\infty}^{\infty} f(x)dx = \int_{a}^{\infty} \frac{1}{b} e^{-\frac{x-a}{b}} dx = 1.$$

So the two fundamental properties of pdf are satisfied

(2) If X has an exponential distribution with parameters a and b we write $X \sim E[a,b]$.

(3) The density curve is shown in the following figure

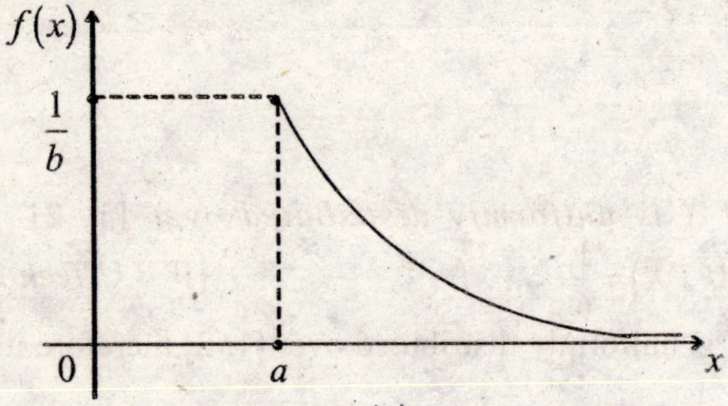

(4) The distribution function $F(x)$ is given by

$F(x) = 0, \qquad -\infty < x < a$

$= 1 - e^{-\frac{x-a}{b}}, \quad a \leq x < \infty$

(5) The graph of the distribution function $F(x)$ is given in the following figure

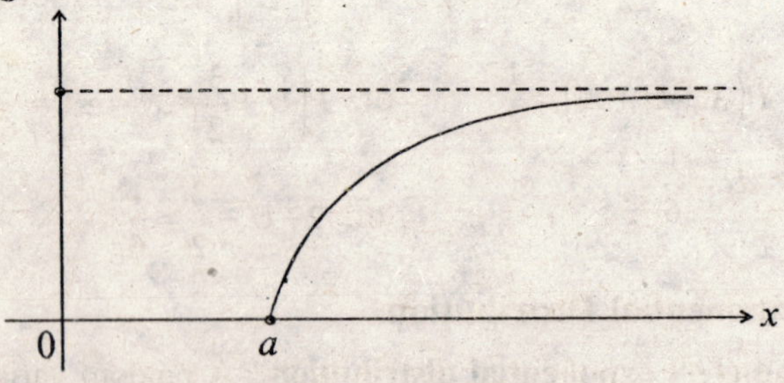

One Parameter Exponential Distribution

We say that X has one parameter exponential distribution if its pdf is

$f(x) = \lambda e^{-\lambda x}, \qquad x \geq 0$

$= 0, \qquad$ elsewhere

where λ is the only parameter.

Note. (1) By the term X has exponential distribution' we mean X has one-parameter exponential distribution.

(2) Clearly the pdf $f(x)$ satisfies the two fundamental properties of pdf.

(3) If X has exponential distribution with parameter λ we write $X \sim E(0, \lambda)$.

(4) This distribution is obtained from the previous by putting $a = 0,\ b = \frac{1}{\lambda}$.

Theorem. If a continuous random variable X has exponential distribution with parameter λ then

(i) the mean is $\frac{1}{\lambda}$ (ii) the variance is $\frac{1}{\lambda^2}$

Proof. The p.d.f. of this distribution be

$$f(x) = \lambda e^{-\lambda x}, \quad x \geq 0$$
$$= 0, \quad \text{elsewhere}$$

$\therefore$ mean, $m = E(x) = \int_0^\infty x\lambda e^{-\lambda x}dx = \frac{1}{\lambda}\int_0^\infty ue^{-u}du$, by putting $\lambda x = u$

$= \frac{1}{\lambda}\Gamma(2)$ $[\because \Gamma(n) = \int_0^\infty e^{-x}x^{n-1}dx$ from definition of Gamma function]

$= \frac{1}{\lambda}$ $[\because \Gamma(n+1) = n!$ when n is a +ve integer]

Now, $E(X^2) = \int_0^\infty x^2\lambda e^{-\lambda x}dx = \frac{1}{\lambda^2}\int_0^\infty u^2 e^{-u}du$ by putting $\lambda x = u$

$= \frac{1}{\lambda^2}\Gamma(3) = \frac{2!}{\lambda^2} = \frac{2}{\lambda^2}$

$\therefore Var(X) = E(X^2) - m^2$ [where m is mean] $= \frac{2}{\lambda^2} - \frac{1}{\lambda^2} = \frac{1}{\lambda^2}$.

Illustrative Example.

Ex. 1. *Suppose that during rainy season, on a tropical island, the length of shower has an exponential distribution with average length of shower* $\frac{1}{2}$ *mins. What is the probability that a shower will last more than three minutes ? If a shower has already lasted for 2 minutes, what is the probability that it will last for at least one more minute ?*

Let $X =$ length of shower in minute.

By problem X has exponential distribution with parameter λ where $\frac{1}{\lambda}=\frac{1}{2}$, i.e., $\lambda=2$

$\therefore$ its p.d.f, $f(x)=2e^{-2x}$, $x>0$

Now, Probability that a shower lasts more than three minutes

$$=P(X>3)=\int_3^{\infty}2e^{-2x}dx=2\lim_{X\to\infty}\int_3^{X}e^{-2x}dx$$

$$=2\lim_{X\to\infty}\left[\frac{e^{-2x}}{-2}\right]_3^X=-\lim_{X\to\infty}\left\{\frac{1}{e^{2X}}-\frac{1}{e^6}\right\}=\frac{1}{e^6}$$

Probability that a shower lasts more than two minutes

$$=P(X>2)=\int_2^{\infty}2e^{-2x}dx=\frac{1}{e^4}.$$

Now the required probability $=P(X\geq 3/X\geq 2)$

$$=\frac{P((X\geq 3)\cap(X\geq 2))}{P(X\geq 2)}=\frac{P(X\geq 3)}{P(X\geq 2)}=\frac{\frac{1}{e^6}}{\frac{1}{e^4}}=\frac{1}{e^2}.$$

3.3.7. Normal Distribution.

A continuous random variable X is said to have a normal distribution if its probability density function is given by

$$f(x)=\frac{1}{\sigma\sqrt{2\pi}}e^{-\frac{(x-\mu)^2}{2\sigma^2}},\ -\infty<x<\infty$$

where μ and $\sigma>0$ are the two parameter of the distribution.

Note. (1) In this case we say X is a normal variate with parameters μ and σ and we denote it by $X\sim N(\mu,\sigma)$

(2) Clearly $f(x)\geq 0$ for all x.

Moreover, $\int_{-\infty}^{\infty}f(x)dx=\frac{1}{\sigma\sqrt{2\pi}}\int_{-\infty}^{\infty}e^{-\frac{(x-\mu)^2}{2\sigma^2}}dx$

$$=\frac{1}{\sqrt{\pi}}\int_{-\infty}^{\infty}e^{-z^2}dz\text{, by putting } z=\frac{x-\mu}{\sigma\sqrt{2}}$$

$$= \frac{2}{\sqrt{\pi}} \int_0^\infty e^{-z^2} dz = \frac{1}{\sqrt{\pi}} \int_0^\infty e^{-u} u^{-\frac{1}{2}} du \text{, by putting } z^2 = u$$

$$= \frac{1}{\sqrt{\pi}} \Gamma\left(\frac{1}{2}\right) = \frac{1}{\sqrt{\pi}} \cdot \sqrt{\pi} = 1.$$

So the two necessary conditions for the probability density function are satisfied.

(3) The significance of the parameters μ and σ are given in the next theorem.

(4) The distribution function $F(x)$ is given by

$$F(x) = \int_{-\infty}^{x} f(u) du = \frac{1}{\sigma\sqrt{2\pi}} \int_{-\infty}^{x} e^{-\frac{(u-\mu)^2}{2\sigma^2}} du$$

Normal Curve

The graph of the pdf of a normal variate is called normal curve.

The curve is shown in the following figure for different values of σ.

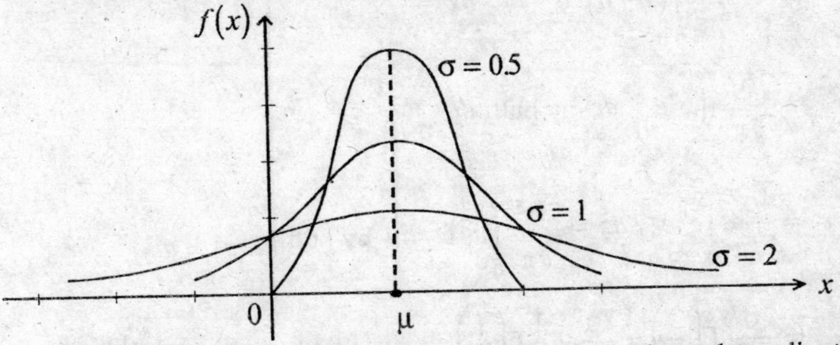

The normal curve is bell shaped and symmetric about the ordinate $x = \mu$. For small values of σ, the curve has a small peak and as σ increases, the normal curve tends to be flatter.

Theorem 1. If X has normal distribution with parameter μ and σ then (i) the mean of X is μ (ii) the s.d of X is σ.

Proof. Here p.d.f of X is given by

$$f(x) = \frac{1}{\sigma\sqrt{2\pi}} e^{-\frac{(x-\mu)^2}{2\sigma^2}}, \quad -\infty < x < \infty$$

(i) Mean $= E(X) = \frac{1}{\sigma\sqrt{2\pi}} \int_{-\infty}^{\infty} x e^{-\frac{(x-\mu)^2}{2\sigma^2}} dx$

$$=\frac{1}{\sigma\sqrt{2\pi}}\int_{-\infty}^{\infty}(x-\mu)e^{-\frac{(x-\mu)^2}{2\sigma^2}}dx+\frac{\mu}{\sigma\sqrt{2\pi}}\int_{-\infty}^{\infty}e^{-\frac{(x-\mu)^2}{2\sigma^2}}dx$$

$$=\frac{1}{\sqrt{\pi}}\int_{-\infty}^{\infty}ue^{-u^2}du+\frac{\mu}{\sqrt{\pi}}\int_{-\infty}^{\infty}e^{-u^2}du\text{, by putting } u=\frac{x-\mu}{\sigma\sqrt{2}}$$

$$=0+\frac{\mu}{\sqrt{\pi}}2\cdot\int_{0}^{\infty}e^{-u^2}du$$ [$\because$ the integrand of the 1st integral is odd and that of the 2nd integral is even]

$$=\frac{\mu}{\sqrt{\pi}}\int_{0}^{\infty}z^{\frac{1}{2}-1}e^{-z}dz\text{, by putting } u^2=z$$

$$=\frac{\mu}{\sqrt{\pi}}\Gamma\left(\frac{1}{2}\right)=\frac{\mu}{\sqrt{\pi}}\sqrt{\pi}=\mu \qquad \left[\because\Gamma\left(\frac{1}{2}\right)=\sqrt{\pi}\right]$$

(ii) Now, $Var(X)=E\{(X-\mu)^2\}$

$$=\frac{1}{\sigma\sqrt{2\pi}}\int_{-\infty}^{\infty}(x-\mu)^2e^{-\frac{(x-\mu)^2}{2\sigma^2}}dx$$

$$=\frac{2\sigma^2}{\sqrt{\pi}}\int_{-\infty}^{\infty}z^2e^{-z^2}dz\text{ by putting }\frac{x-\mu}{\sigma\sqrt{2}}=z$$

$$=\frac{4\sigma^2}{\sqrt{\pi}}\int_{0}^{\infty}z^2e^{-z^2}dz=\frac{2\sigma^2}{\sqrt{\pi}}\int_{0}^{\infty}u^{\frac{1}{2}}e^{-u}du\text{ by putting } z^2=u$$

$$=\frac{2\sigma^2}{\sqrt{\pi}}\Gamma\left(\frac{3}{2}\right)=\frac{2\sigma^2}{\sqrt{\pi}}\cdot\frac{1}{2}\Gamma\left(\frac{1}{2}\right) \qquad [\because\Gamma(n+1)=n\Gamma(n)]$$

$$=\frac{\sigma^2}{\sqrt{\pi}}\cdot\sqrt{\pi}=\sigma^2$$

$\therefore$ the s.d of X is σ.

A Case where Normal Distribution Fits.

Let a thermal plant supplies electric power in a certain city. Let $X=$ amount of electric power (in watt) supplied by the plant in a day. Clearly X varies from day to day. It can be assumed that X is a continuous variate. It can be shown that X has normal distribution with parameter $\mu=$ average power supply per day and $\sigma=$ standard deviation of all the values assumed by X.

Standard Normal Distribution.

The normal distribution with mean 0 and standard deviation 1 is called standard normal distribution. The random variable having standard normal distribution is called standard normal variate or $X \sim N(0,1)$.

Thus the p.d.f of the standard normal distribution is given by

$$\phi(x) = \frac{1}{\sqrt{2\pi}} e^{-\frac{x^2}{2}}, \quad -\infty < x < \infty .$$

and the corresponding distribution function is given by

$$\Phi(x) = \frac{1}{\sqrt{2\pi}} \int_{-\infty}^{x} e^{-\frac{u^2}{2}} du$$

Standard Normal Curve.

The graph of the p.d.f of a standard normal distribution is called standard normal curve. This is shown in the following figure.

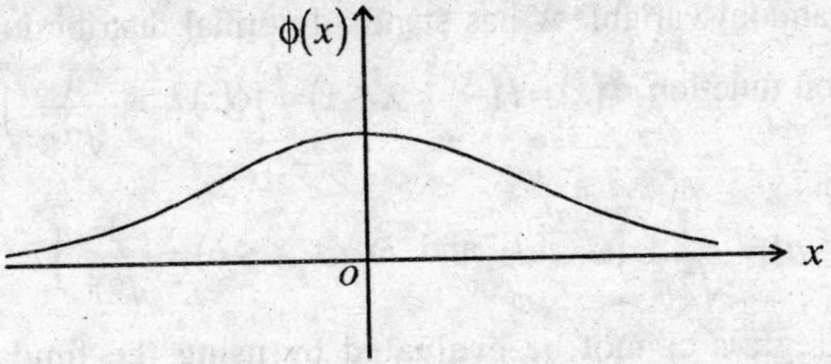

It is symmetric about y axis, bell shaped. It has maximum value at $x = 0$. Since $\int_{-\infty}^{\infty} \phi(x) dx = 1$ so area under this curve is 1. X axis is its asymptote.

Standard Normal Distribution Curve.

The graph of the distribution function $\Phi(x)$ is shown in the following figure.

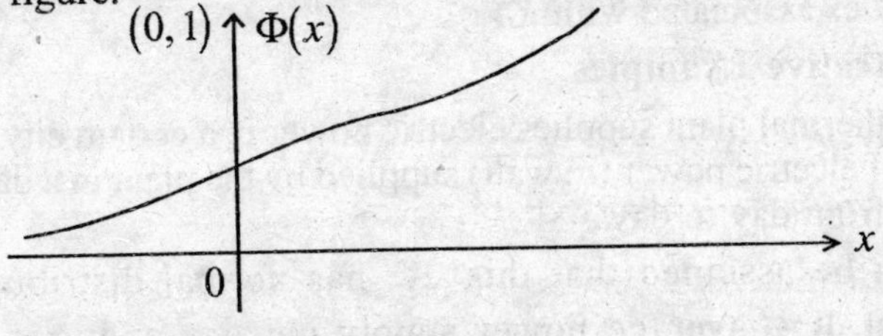

Theorem 2. (An Important Result)

If the continuous random variable X has normal distribution with parameter μ and σ then $Z = \frac{X - \mu}{\sigma}$ has standard normal distribution.

Proof. That Z has normal distribution is shown in the next chapter. Yet it is observed that the mean of Z,

$$\bar{Z} = E(Z) = E\left(\frac{X-\mu}{\sigma}\right) = \frac{1}{\sigma}\{E(X) - E(\mu)\}$$

$$= \frac{1}{\sigma}\{\mu - \mu\} = 0$$

$$Var(Z) = Var\left(\frac{X-\mu}{\sigma}\right) = \frac{1}{\sigma^2}Var(X-\mu)\text{, from the property of variance}$$

$$= \frac{1}{\sigma^2}\{Var(X) - Var(\mu)\}\text{, from the property of variance}$$

$$= \frac{1}{\sigma^2}\{\sigma^2 - 0\} = 1.$$

Tabulation of the Standard Normal Distribution.

Let the random variable X has standard normal distribution. Then its distribution function $\Phi(x) = P(-\infty < X \le x) = \int_{-\infty}^{x} \phi(z)dz = \frac{1}{\sqrt{2\pi}}\int_{-\infty}^{x} e^{-\frac{z^2}{2}} dz$

$$\therefore\ P(X \le a) = \frac{1}{\sqrt{2\pi}}\int_{-\infty}^{a} e^{-\frac{z^2}{2}} dz \text{ and } P(a \le X \le b) = \frac{1}{\sqrt{2\pi}}\int_{a}^{b} e^{-\frac{z^2}{2}} dz$$

These integrals cannot be evaluated by using the fundamental theorem of integral calculus. However, method of numerical integral can be used to evaluate the integral.

The values of the probabilities for different values of a have been tabulated in the Table - 1 at the end of the book. From this we can find $\Phi(x)$ for different values of x.

Since $P(a \le x \le b) = \Phi(b) - \Phi(a)$, we have if X has distribution $N(0,1)$ then from the tabulated value of Φ we can evaluate the probabilities associated with X.

An Illustrative Examples

Let a thermal plant supplies electric power in a certain city. Let $X =$ amount of electric power (in watt) supplied by the plant in a day. Clearly X varies from day to day.

It can be assumed that this X has normal distribution with parameter $\mu =$ average power supply per day and $\sigma =$ standard deviation of all the values assumed by X. Now if it is known that these $m = 300$ and $\sigma = 10$ then by the previous theorem $Z = \frac{X-300}{10}$ is a standard normal variate.

Suppose we are asked to find the number of days on which power supplied will lie between 280 to 310 MW.

Then we find $P(280 < X < 310)$

$$= P\left(\frac{280-300}{10} < \frac{X-300}{10} < \frac{310-300}{10}\right) = P(-2 < Z < 1)$$

$$= \int_{-2}^{1} \phi(z)dz \text{ [where } \phi \text{ is the pdf of standard normal distribution]}$$

= area enclosed by standard normal curve, *X axis*, the ordinates $z = -2$ and $z = 1$ (shown by shade in the figure)

$= 0.4772 + 0.3413$ (obtained from the tabulated value)

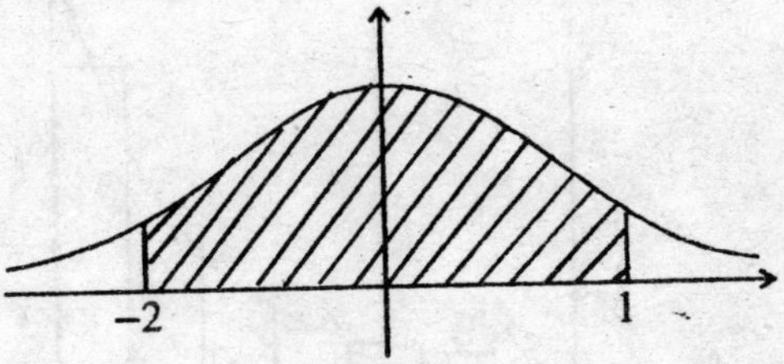

$= 0.8185$. So the probability of having power supply between 280 and 310 MW in a day is 0.8185. Thus 81.85% day will receive a power supply between 280 and 310 MW.

3.3.8. Binomial Approximation to Normal Distribution.

Theorem. Let the random variable X follows Binomial distribution with pmf

$$f_i = \binom{n}{i} p^i (1-p)^{n-i}, \quad i = 0, 1, 2, \cdots, n$$

where n and p are parameters. If $n \to \infty$ and p is not very small then the distribution of the r.v $Z = \frac{X - np}{\sqrt{np(1-p)}}$ approches to the standard normal distribution.

Proof. Beyond the scope.

Note. (1) In light of above theorem we understand X is approximately a normal variate with mean np and s.d. $\sqrt{npq}$

(2) The variable $Z = \frac{X - np}{\sqrt{np(1-p)}}$ is called standardized Binomial variate.

Illustration. Let an unbiased coin is tossed 12 times. Let X = number of heads appeared. Then X has binomial distribution with parameter $n=12$ and $p=\frac{1}{2}=0.5$. So the probability masses are

$$f_0={}^{12}C_0(0.5)^0(0.5)^{12-0}=0.000244$$

$$f_1={}^{12}C_1(0.5)^1(0.5)^{12-1}=0.00292$$

$$f_2={}^{12}C_2(0.5)^2(0.5)^{12-2}=0.01611$$

and so on.

In the following diagram the heights of the rectangles are these.

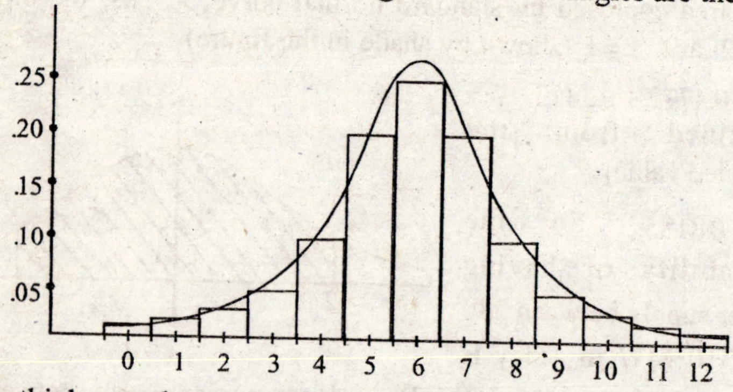

We think n as large. So X is approximately normal variate with parameter $\mu=np=12\times.5=6$ and $\sigma=\sqrt{np(1-p)}=\sqrt{12\times.5\times.5}=1.73$

i.e., $Z=\frac{X-6}{1.73}$ is approximately a standard normal variate. In the above figure the curve is $(6, 1.73)$ normal curve. This approximately fits the height of the rectangle.

(i) Now, suppose we seek $P(4\le X\le 7)$. But if we find this from the normal distribution then we must find $P(3.5\le X\le 7.5)$ as indicated in the above diagram. Now when $X=3.5$, $Z=\frac{3.5-6}{1.73}=-1.46$, when $X=7.5, Z=\frac{7.5-6}{1.73}=0.87$

Then $P(4\le X\le 7)\simeq P(-1.46\le X\le 0.87)$

$$=\int_{-1.46}^{0.87}\phi(t)dt$$

$$=\int_{-1.46}^{0}\phi(t)dt+\int_{0}^{0.87}\phi(t)dt$$

$$= 0.4279 + 0.3078$$

(obtained from the statistical table-I)

$$= 0.7357 \cdot$$

(ii) Now, suppose we seek $P(X=8)$.

We calculate this with the help of approximate normal distribution then we find $P(7.5 \le X \le 8.5) = P(0.87 \le Z \le 1.44)$ which would be obtained from statistical table.

Illustrative Examples.

Ex. 1. *If X is normally distributed with zero mean and unit variance find the expectation of* X^2. *[W.B.U.Tech, 2002, 2007]*

By problem $E(X)=0,\ Var(X)=1$ or, $E(X^2)-\{E(X)\}^2=1$

or, $E(X^2)-0^2=1$ $\therefore\ E(X^2)=1$

Ex. 2. *If X is normally distributed with mean 3 and s.d. 2, find c such that* $P(X>c)=2\,P(X\le c)$.

Given $\int_{-\infty}^{.43}\phi(t)dt=\frac{1}{3}$. *[W.B.U.Tech 2007]*

Let Z be the standard normal variate. Then $Z=\frac{X-3}{2}$

$$\therefore\ P(X>c)=P\left(\frac{X-3}{2}>\frac{c-3}{2}\right)=P\left(Z>\frac{c-3}{2}\right)$$

$$=1-P\left(Z\le\frac{c-3}{2}\right)=1-\int_{-\infty}^{\frac{c-3}{2}}\phi(t)dt$$

Also $P(X\le c)=P\left(\frac{X-3}{2}\le\frac{c-3}{2}\right)=P\left(Z\le\frac{c-3}{2}\right)=\int_{-\infty}^{\frac{c-3}{2}}\phi(t)dt$

As $P(X>c)=2P(X\le c)$, we have

$$1-\int_{-\infty}^{\frac{c-3}{2}}\phi(t)dt=2\times\int_{-\infty}^{\frac{c-3}{2}}\phi(t)dt \qquad \text{or, } \int_{-\infty}^{\frac{c-3}{2}}\phi(t)dt=\frac{1}{3}$$

or, $\frac{c-3}{2}=.43$ (from the given data)

$\therefore\ c=3.86$.

Ex. 3. *The length of bolts produced by a machine is normally distributed with mean 4 and and s.d. 0.5. A bolt is defective if its length does not lie in the interval* $(3.8,\ 4.3)$. *Find the percentage of defective bolts produced by the machine.*

$$\left[\frac{1}{\sqrt{2\pi}}\int_{-\infty}^{0.6} e^{-\frac{t^2}{2}}dt = 0.7257,\ \frac{1}{\sqrt{2\pi}}\int_{-\infty}^{0.4} e^{-\frac{t^2}{2}}dt = 0.6554\right]$$ [W.B.U.Tech 2004]

Let $X =$ length of bolt. By problem X has normal distribution with mean $\mu = 4$ and s.d $\sigma = 0.5$. First we shall find the probability $P(3.8 < X < 4.3)$.

Now, $Z = \dfrac{X-4}{0.5}$ has standard normal distribution. Now when $X = 3.8,\ Z = \dfrac{3.8-4}{0.5} = -0.4$; when $X = 4.3,\ Z = \dfrac{4.3-4}{0.5} = 0.6$

$\therefore\ P(3.8 < X < 4.3) = P(-0.4 < Z < 0.6)$

= Area under standard normal curve enclosed between the two ordinate $Z = -.04$ and $Z = 0.6$ (shaded part in figure) $= \int_{-0.4}^{0.6} \phi(t)dt$

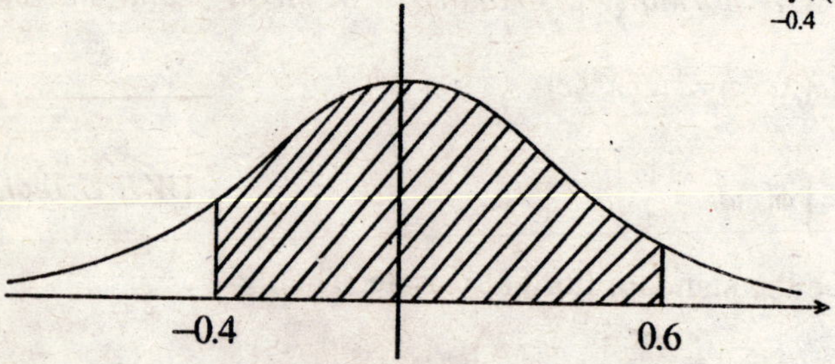

$$= \int_{-\infty}^{0.6} \phi(t)dt - \int_{-\infty}^{-0.4} \phi(t)dt = 0.7257 - \int_{0.4}^{\infty} \phi(t)dt \quad [\because \phi \text{ curve is symmetric}]$$

$$= 0.7257 - \left(1 - \int_{-\infty}^{0.4} \phi(t)dt\right) = 0.7257 - (1 - 0.6554) = 0.3811$$

$\therefore$ Probability that the length of the bolt lies between 3.8 and 4.3 is 0.3811.

$\therefore$ Probability that the length of the bolt does not lie between 3.8 and $4.3 = 1 - 0.3811 = 0.6189$.

$\therefore$ Probability that the bolt is defective $= 0.6189$

$\therefore$ Percentage of defective bolts produced $= 0.6189 \times 100 = 61.89 \simeq 62$.

Ex. 4. *If the weekly wage of 10,000 workers in a factory follows normal distribution with mean and s.d Rs. 70 and Rs. 5 respectively, find the expected number of workers whose weekly wages are (i) between Rs. 66 and Rs. 72 (ii) less than Rs. 66 and (iii) more than Rs. 72* [W.B.U.Tech 2006]

[*Given that* $\frac{1}{\sqrt{2\pi}}\int_0^z e^{-t^2/2}dt = 0.1554$ *and* 0.2881 *according as* $z = 0.4$ *and* $z = 0.8$]

Let $X =$ wage of a worker. X has normal distribution with $\mu = 70,\ \sigma = 5$.

$\therefore\ Z = \frac{X-70}{5}$ has standard normal distribution.

(i) When $X = 66,\ Z = \frac{66-70}{5} = -0.8$; when $X = 72,\ Z = \frac{72-70}{5} = 0.4$

$\therefore\ P(66 < X < 72) = P(-0.8 < Z < 0.4)$

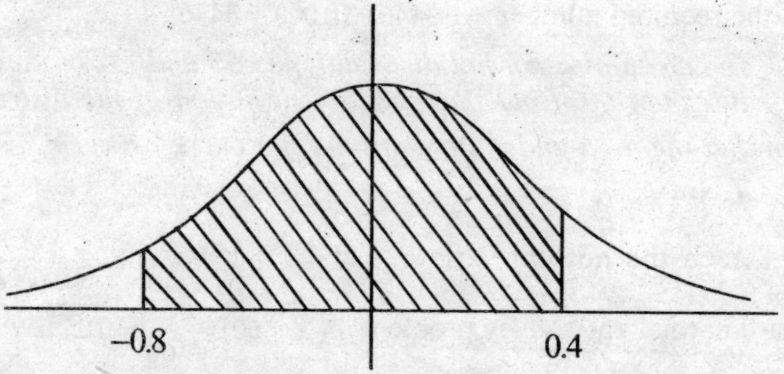

$=$ area under st. normal curve enclosed between the two ordinates $Z = -0.8$ and $Z = 0.4$ (shown by shade in the figure)

$$= \int_{-0.8}^{0.4}\phi(t)dt = \int_{-0.8}^{0}\phi(t)dt + \int_{0}^{0.4}\phi(t)dt$$

$$= \int_{0}^{0.8}\phi(t)dt + \int_{0}^{0.4}\phi(t)dt \text{, since } \phi \text{ curve is symmetric about } Y \text{ axis.}$$

$= 0.2881 + 0.1554 = 0.4435$

$\therefore$ Probability that the wage of a worker lies between Rs 66 and Rs. 72 is 0.4435.

$\therefore$ the number of worker whose wage lie between Rs. 66 and Rs. 72 $= 0.4435 \times 10{,}000 = 4435$.

(ii) When $X = 66,\ Z = -0.8$

So, $P(X < 66) = P(Z < -0.8)$

$=$ area under the st. normal curve enclosed on left side of the ordinate $Z = -0.8$

$$= \int_{-\infty}^{-0.8}\phi(t)dt = \int_{0.8}^{\infty}\phi(t)dt \text{ (by symmetry)}$$

$$= 0.5 - \int_{0}^{0.8}\phi(t)dt = 0.5 - 0.2881 = 0.2119$$

$\therefore$ the expected number of workers $= 0.2119 \times 10{,}000 = 2119$

(iii) When $X = 72,\ Z = 0.4$

$$\therefore\ P(X > 72) = P(Z > 0.4) = \int_{0.4}^{\infty}\phi(t)dt$$

$$= 0.5 - \int_{0}^{0.4}\phi(t)dt = 0.5 - 0.1554 = 0.3446$$

$\therefore$ the required number $= 0.3446 \times 10{,}000 = 3446$.

Ex. 5. *The mean of a normal distribution is 50 and 5% of the values are greater than 60. Find the standard deviation of the distribution (Given that the area under standard normal curve between $z = 0$ and $z = 1.64$ is 0.45)*

Let X be the normal variate. Let its s.d. be σ $\therefore$ $Z = \dfrac{X-50}{\sigma}$ is standard normal variate. By problem $P(X > 60) = \dfrac{5}{100} = 0.05$.

When $X = 60,\ Z = \dfrac{60-50}{\sigma} = \dfrac{10}{\sigma}$

$\therefore$ from above, $P\left(Z > \dfrac{10}{\sigma}\right) = 0.05$ ... (1)

From the supplied data we have $P(0 < Z < 1.64) = .45$

or, $P(Z > 1.64) = 0.5 - 0.45 = 0.05$ ... (2)

Comparing (1) and (2) we get $\dfrac{10}{\sigma} = 1.64$

or, $\sigma = \dfrac{10}{1.64} = 6.097 \simeq 6.1$ $\therefore$ the s.d is 6.1.

Ex. 6. *In a normal distribution, 31% of the items are under 45 and 8% are above 64. Find the mean and standard deviation. [Given $P(0 < Z < 1.405) = 0.42$, $P(-0.496 < Z < 0) = 0.19$]* [*W.B.U.Tech 2003*]

Let X be the normal variate with mean μ and s.d σ. So, $Z = \dfrac{X-\mu}{\sigma}$ is standard normal variate. Now, by problem, $P(X < 45) = \dfrac{31}{100}$ and

$P(X>64)=\frac{8}{100}$. When $X=45, Z=\frac{45-\mu}{\sigma}$; when $X=64$, $Z=\frac{64-\mu}{\sigma}$.

So from above $P\left(Z<\frac{45-\mu}{\sigma}\right)=0.31$ and $P\left(Z>\frac{64-\mu}{\sigma}\right)=0.08$.

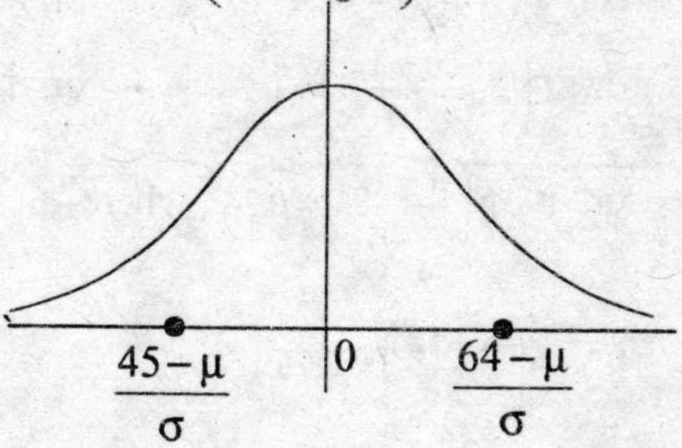

Since $0.31<0.5$ so $\frac{45-\mu}{\sigma}$ is on negative side.

From above

$P\left(\frac{45-\mu}{\sigma}<z<0\right)=0.5-0.31$

$=0.19$

Comparing this with the second given data we have $\frac{45-\mu}{\sigma}=-1.496$

or, $45-\mu=-1.496\sigma$ or, $\mu-0.496\sigma=45$ ... (1)

Since $0.08<0.5$ so $\frac{64-\mu}{\sigma}$ is lies on +ve side.

$\therefore P\left(0<z<\frac{64-\mu}{\sigma}\right)=0.5-0.08=0.42$

Comparing this with the first given data we get $\frac{64-\mu}{\sigma}=1.405$

or, $64-\mu=1.405\sigma$

or, $\mu+1.405\sigma=64$ ... (2)

Solving (1) and (2) $\sigma=9.995$ and $\mu=49.958$.

$\therefore$ Mean $=49.958$ and s.d. $=9.995$.

Ex. 7. *If X is normally distributed with mean 12 and s.d. 4, find the probability of (i)* $X\geq 20$ *(ii)* $0\leq X\leq 12$ *and (iii) also find a such that* $P(X>a)=0.24$. [Use table]

Let Z be the standard normal variate. Then

$Z=\frac{X-\mu}{\sigma}=\frac{X-12}{4}$ $\therefore X=4Z+12$

$\therefore$ (i) $P(X\geq 20)=P(4Z+12\geq 20)=P(Z\geq 2)=1-P(Z<2)=1-\Phi(2)$

$=1-0.9772=0.0228$.

(ii) $P(0 \le X \le 12) = P(0 \le 4Z + 12 \le 12) = P(-3 \le Z \le 0)$

$= P(0 \le Z \le 3)$ (due to symmetry)

$= \Phi(3) - \Phi(0) = 0.9986 - 0.5 = 0.4986$ (from table)

(iii) $P(X > a) = 0.24$ or, $P(4Z + 12 > a) = 0.24$

or, $P\left(Z > \frac{a-12}{4}\right) = 0.24$ or, $1 - P\left(Z \le \frac{a-12}{4}\right) = 0.24$

or, $1 - \Phi\left(\frac{a-12}{4}\right) = 0.24$, where Φ is c.d.f of st. normal variate

or, $\Phi\left(\frac{a-12}{4}\right) = 0.76$

or, $\frac{a-12}{4} = 0.71$, from table-I

$\therefore\ a = 14.84$

Ex. 8. *If a random variable X follows a normal distribution such that* $P(9.6 \le X \le 13.8) = 0.7008$, *and* $P(X \ge 9.6) = 0.8159$ *where* $\frac{1}{\sqrt{2\pi}} \int_{-\infty}^{0.9} e^{-t^2/2} dt = 0.8159$, $\frac{1}{\sqrt{2\pi}} \int_{-\infty}^{1.2} e^{-t^2/2} dt = 0.8849$, *find mean and variance of X.*

Let $E(X) = \mu$, $Var(X) = \sigma^2$. Now, if Φ is c.d.f of st normal variate

$$\Phi(x) = P\left(\frac{X-\mu}{\sigma} \le x\right) = \frac{1}{\sqrt{2\pi}} \int_{-\infty}^{x} e^{-u^2/2} du$$

$\therefore\ \Phi(0.9) = .8159,\ \Phi(1.2) = 0.8849$

$\therefore\ \Phi(-0.9) = 1 - \Phi(0.9) = 0.1841$

Now, $P\left(-0.9 \le \frac{X-\mu}{\sigma} \le 1.2\right) = \Phi(1.2) - \Phi(-0.9)$

$= 0.8849 - 0.1841 = 0.7008$

$= P(9.6 \le X \le 13.8)$

$= P\left(\frac{9.6-\mu}{\sigma} \le \frac{X-\mu}{\sigma} \le \frac{13.8-\mu}{\sigma}\right)$

$\therefore\ \frac{13.8-\mu}{\sigma} = 1.2,\ \frac{9.6-\mu}{\sigma} = -0.9$.

Solving we get $\mu = 11.4$, $\sigma = 2$

$\therefore$ mean $= 11.4$, $\text{var}(X) = 4$.

Ex. 9. *A fair cain is tassed 400 times. Using normal approximation to binomial distribution find the probability of obtaining (i) exactly 200 heads (ii) between 190 and 210 heads, both inclusive. Given that the area under standard normal curve between Z=0 and Z=0.05 is 0.0199 and between Z=0 and Z=1.05 is 0.3531.* [*W.B.U.Tech 2007*]

Let the random variable X denotes the number of heads in 400 tosses. Then clearly X has a binomial distribution with parameter $n = 400$, $p = \frac{1}{2}$. Since n is large we suppose X is an approximately normal variate with parameter $\mu = np = 400 \times \frac{1}{2} = 200$

and $\sigma = \sqrt{np(1-p)} = \sqrt{400 \times \frac{1}{2} \times \frac{1}{2}} = 10$

i.e. $Z = \frac{X - 200}{10}$ is approximate standard normal variate.

(i) Now, $X = 199 \cdot 5 \Rightarrow X = \frac{199 \cdot 5 - 200}{10} = -0 \cdot 05$ and $X = 200 \cdot 5 \Rightarrow Z = 0 \cdot 05$.

Then using the normal approximation, we have the required probability, $P(X = 200) \approx P(199 \cdot 5 \leq X \leq 200 \cdot 5) = P(-0 \cdot 05 \leq Z \leq 0 \cdot 05)$

$$= 2P(0 \leq Z \leq 0 \cdot 05) = 2 \times 0 \cdot 0199 = 0 \cdot 0398$$

(ii) Now, $X = 189.5 \Rightarrow Z = -1.05$ and $X = 210 \cdot 5 \Rightarrow Z = 1.05$.

Then using the normal approximation, we have the required probability,

$P(190 \leq X \leq 210)$

$\simeq P(189.5 \leq X \leq 210.5)$ (in terms of normal approximation)

$= P(-1.05 \leq Z \leq 1.05) = 2P(0 \leq Z \leq 1.05) = 2 \times 0.3531 = 0.7062$.

Ex. 10. *Among 10,000 random digits, find the probability that the digit 3 appears at most 950 times.*

Let X denotes the number of times the digit 3 appears. Then X is a $b(n, p)$ variate where $n = 10{,}000$, $p = \frac{1}{10}$

$\therefore np = 1000$, $np(1-p) = 10000 \times \frac{1}{10} \times \frac{9}{10} = 900$

$\because n$ is large we find $P(X \le 950)$ by approximating with normal distribution. X is approximately $\left(np, \sqrt{np(1-p)}\right) = (1000, 30)$ normal variate i.e., $Z = \dfrac{X-1000}{30}$ is approximately standard normal variate. Since discrete variate is approximated by continuous variate so we find $P(X \le 950.5)$. Now $X = 950.5 \Rightarrow Z = -1.65$

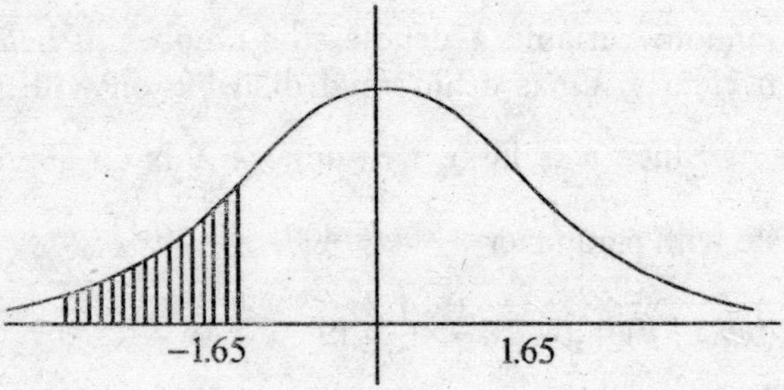

$\therefore$ the required probability $\simeq P(X \le 950.5) = P(Z \le -1.65)$

$= 0.5 - P(-1.65 \le Z < 0) = 0.5 - P(0 \le Z \le 1.65)$

$= 0.5 - \int_0^{1.65} \phi(t)\,dt = 0.5 - 0.4505$ (From statistical table)

$= 0.0495$.

EXERCISES

[I] SHORT ANSWER QUESTIONS

1. If a conference room cannot be reserved for more than 4 hours, find the probability that a given conference lasts more than three hours.

   [Hints : Here $f(x) = \dfrac{1}{4}$

   $$P(X \ge 3) = \int_3^4 f(x)\,dx = \frac{1}{4}\,]$$

2. Suppose 2% of the items made by a factory are defective. Find the probability p that there are 3 defective items in a sample of 100 items.

   [Hints : Here $n = 100, p = 0.02$

   $\therefore \; \mu = np = 2$

   $\therefore \; P = P(X = 3) = \dfrac{e^{-2}2^3}{3!} = 0.180\,]$

3. Given that for a standared normal variable z. $P(0<z<0.8)=.2881$. Find out $P(|z|\geq 0.8)$.

[Hints : $P(|z|\geq 0.8)=2P(z\geq 0.8)$, due to symmetry

$$=2[0.5-P(0<z<0.8)]$$
$$=2[0.5-0.2881]$$
$$=0.4238\,]$$

4. X is distributed as $N(50,4)$ find $P(41\leq X\leq 56)$.

[Hints : For $x=41, z_1=\dfrac{51-50}{4}=-2.25$

$$x=56, z_2=\frac{56-60}{4}=1.5$$

$$\therefore\quad P(41\leq X\leq 56)=P(-2.25\leq z\leq 1.5)$$
$$=P(-2.25\leq z\leq 0)+P(0<z\leq 1.5)$$
$$=P(0\leq z\leq 2.25)+P(0<z\leq 1.5)$$
$$=0.4878+0.4332=0.921\,]$$

5. If for a poisson variate X, $E(X^2)=20$, find $E(X)$ and $\text{var}(X)$.

[Hints : $E(X^2)=\mu^2+\mu$

$$\therefore\quad \mu^2+\mu=20$$
$$\therefore\quad \mu=<4$$
$$\therefore\quad E(X)=\mu=4$$
$$\text{and } \text{var}(X)=\mu=4\,]$$

6. For a Binomial distribution, the mean is 3 and the variance is 2. Find the values of n and p. Hence find the probability that X (the variable value) is 5.

7. A discrete random variable X has the mean 6 and variance 2. Assuming the distribution is binomial, find the probability that $5\leq X\leq 7$. [*W.B.U.Tech, 2002*]

8. If 20% of the articles produced by a machine are defective, determine the probability that out of the 4 articles chosen at random less than 2 articles will be defective. [*W.B.U.Tech 2002*]

9. Show that the Binomial distribution is symmetric when $p=\dfrac{1}{2}$.

10. If the probability of a defective bolt is $\frac{1}{10}$, find (i) mean (ii) variance for the distribution of defective bolts in a total of 400.

11. If the sum of mean and variance of a Binomial distribution is 4.8 for five trials, find the distribution.

12. In a shooting competition, the probability of a man hitting a target is $\frac{1}{5}$. If he fires 5 times, what is the probability of hitting the target at least twice.

13. Let the probability of a patient recovering from a certain disease is 0.75. Find the distribution of the number of recoveries among 4 patients. Hence find mean and s.d.

14. Prove that the probability of obtaining double six at least once in 24 throws with two dice is slightly less than $\frac{1}{2}$.

15. Find the minimum number of times a die has to be thrown such that the prob. of 'no. six' is less than $\frac{1}{2}$.

[Hints : The prob. of 'no. six' in n throws of a die $= \left(\frac{5}{6}\right)^n$

$\therefore \left(\frac{5}{6}\right)^n < \frac{1}{2} \quad \therefore n > \frac{-\log 2}{\log 5 - \log 6} = 3.81$]

16. A and B play a game in which A's chance of winning is $\frac{2}{9}$. In a series of 8 games, what is the chance that A will win at least 6 games ?

17. Find the probability that at most 5 defective pen will be found in a box of 200 pen if experience shows that 2% of such pen are defective.

18. Five balls are drawn with replacement from a box containing 5 white and 4 black balls. Find the probability of getting 3 white balls.

[Hint. $X =$ No. of white balls, $X \sim b(n, p)$ with $n = 5, p = \frac{5}{9}; P(X = 3) = ?$]

19. Two dice are thrown n times in succession. What is the probability of obtaining double six at least once ?

20. If probability of success is 0.1, how many trials are neccessary in order that probability of at least one success is $> \frac{1}{2}$.

21. The mean of the poisson distribution is μ. Then what is its standard deviation? [*W.B.U.Tech 2004*]

22. If a r.v. follows poisson distribution such that $P(1) = P(2)$, find (i) mean of the distribution (ii) $P(4)$ [*W.B.U. Tech 2003, 2004*]

23. If X is a poisson variate such that $P(X=1)=0.2$ and $P(X=2)=0.2$, find $P(X=0)$ [*W.B.U.Tech, 2002*]

24. Show that the poisson distribution is a probability distribution.

25. Show that for a poisson variate standard deviation $=\sqrt{mean}$.

26. Let X be a chance variate having a normal distribution with mean 30 and s.d 5. Find the probability that (i) $26 \le X \le 40$ (ii) $|X-30|>5$

ANSWERS

1. $\frac{1}{4}$ **2.** 0.180 **3.** 0.4238 **4.** 0.921 **5.** 4, 4 **6.** $9, \frac{1}{3}, 224/2187$

8. 0.8192 **10.** (i) 40, (ii) 36 **11.** $B\left(5, \frac{4}{5}\right)$ **12.** $\frac{821}{3125}$

13.

X :	0	1	2	3	4
f_i :	$\frac{1}{256}$	$\frac{3}{64}$	$\frac{27}{128}$	$\frac{27}{64}$	$\frac{31}{256}$

Mean $=3$, s.d $=\frac{\sqrt{3}}{2}$

16. 1024/2187 **17.** 0.785 **18.** 10/21 **19.** $1-\left(\frac{35}{36}\right)^n$ **20.** 69

21. $\sqrt{\mu}$ **22.** (i) 2 **23.** e^{-2} **26.** (i) 0.7654 (ii) 0.3174

[II] LONG ANSWER QUESTIONS

1. Assume that 50% of all engineering students are good in mathematics. Determine the probabilities that among 18 engineering students (a) exactly 10 (b) at least 10 (c) at most 8 (d) at least 2 and at most 9 are good in maths.

2. On an average 20 red blood cells are found in a fixed volume of blood for a normal person. Determine the probability that the blood sample of a normal person will contain less than 15 red cells.

[Hints : Here $\mu = 20$

$$\therefore \quad P(X < 15) = \sum_{i=0}^{14} \frac{c^{-20} 20^i}{i!} = 0.105 \,]$$

3. The probability that screw manufactured by a machine to be defective is $\frac{1}{50}$. A lot of 6 screws are taken at random. Find the probability that (i) there are exactly 2 defective screws in the lot (ii) no defective screw and (iii) at most 2 defective screws.

4. The probability of a missile hitting a target is $\frac{1}{4}$.

(i) If 7 such missiles are sent, what is the probability of hitting the target at least twice

(ii) How many missiles must be sent so that the probability of hitting the target at least once is greater that $\frac{2}{3}$?

5. In a bombing attack there is a 50% chance that a bomb strikes the target. Two direct hits are required to destroy a bridge completely. How many bombs must be dropped to give a 99% chance or better to completely destroy the target.

6. (a) Amal has probability $\frac{2}{3}$ of winning a game. If he plays 4 such games, find the probability that Amal wins (i) exactly 2 games (ii) at least 1 game (iii) more than half of the game.

(b) A and B play a game in which their chances of winining are in the ratio 3 : 2. Find A's chances of winning at least three games out of the five games played. *[W.B.U.Tech 2004]*

7. The probability is 0.02 that a bicycle produced by a factory is defective. An consignment of 10000 bicycles is sent to a country. Find the expected number of defective bicycles and standard deviation.

8. It is seen that a cricket player becomes out within 10 runs in 3 out of 10 innings. If he plays 4 innings, what is the probability that he will becomes (i) out twice (ii) out at least once within 10 runs. *[W.B.U.Tech 2007]*

[Hints : $p = \frac{3}{10}$ (the probability of success in a trial)

Let $X =$ No. of times of becoming out within 10 runs when he plays 4 innings.

$=$ No. Of success in 4 trials.

X follows Binomial distribution with $p = \frac{3}{10}, n = 4$.

$$\therefore \quad f_i = P(X = i) = {}^4C_i \left(\frac{3}{10}\right)^i \left(1 - \frac{3}{10}\right)^{4-i} = {}^4C_i \left(\frac{3}{10}\right)^i \left(\frac{7}{10}\right)^{4-i}$$

$$= {}^4C_i \frac{3^i \times 7^{4-i}}{10^4}$$

$$\therefore \text{(i) } P(X=2) = {}^4C_2 \cdot \frac{3^2 \times 7^{4-2}}{10^4} = \cdot 2646$$

$$\text{(ii) } P(X \geq 1) = 1 - P(X=0) = 1 - f_0 = 1 - {}^4C_0 \cdot \frac{3^4 \times 7^{4-0}}{10^4}$$

$$= 1 - \frac{7^4}{10^4} = 1 - \frac{2401}{10000} = \cdot 7599 \text{]}$$

9. Show that the neccessary and sufficient condition for two given numbers a and b to be respectively the mean and variance of some binomial distribution are that $a > b > 0$ and $\frac{a^2}{a-b}$ is an integer. Show further that when these conditions are satisfied the binomial distribution is uniquely determined.

10. Determine the expected number of boys in a family with 8 children, assuming the sex distribution to be equally probable. What is the probability that the expected number of boys does occur ?

11. In sampling a large number of parts manufactured by a machine, the mean number of defectives in a sample of 20 is 2. Out of 1000 such samples, how man would be expected to contain at least 3 defective parts.

12. Let X be a binomially distributed random variable with parameters n and p. For what values of p is $\text{var}(X)$ maximum, if you assume that n is fixed.

13. X is poisson variate with parameter 3. Find the probability that X assumes the values (i) 3, 2, 1, 0 (ii) less than 3 (iii) at least 2.

[Given $e^{-3} = 0.0498$]

14. A bank receives on an average 2.5 customers per hour. Find the probability that in a certain hour the bank receives (i) no customer (ii) exactly 4 customers. Assume that the number of customers received in an hour is poissonly distributed. ($e^{-2.5} = .0821$)

15. For a poisson variate with mean 2, find the probabilities

$P(X=1), P(X \leq 1), P(X<1), P(X>1), P(1 \leq X \leq 3)$ [Given $e^{-2} = .1353$]

16. (i) The average number of defective spot per sheet of a metal sheet factory is 2. Assuming poisson distribution, what is the probability that a particular sheet is free from spot ? If the factory supplies 1000 such sheets, how many sheet contain more than 2 spots ?

(ii) Suppose that the number of telephone calls an opereto[r] receives from 11.00 am to 11.05 a.m. follows a poisso[n] distribution with $m=3$ (i) Find the probability that the operato[r] will receive no calls in that time interval tomorrow (ii) Find the probability that in the next 3 days the operator will receive a total of 1 call in that time interval. $(e=2.718)$

17. If 5% of the electric bulbs manufactured by a company are defective, use poisson distribution to find the probability that in a sample of 100 bulbs (i) none is defective (ii) 5 bulbs will be defective (given $e^{-5}=.007$)

18. Find the probability that at most 5 defective bolts will be found in a box of 200 bolts, if it is known that 2% of such bolts are expected to be defective. (You may take the distribution to be poisson). Given $e^{-4}=0.0183$.

19. The manufacturer of a telephone instrument knows that 3% of his product is defective. He sold the instruments in cartoon of 100 and gurantees that not more then 3 in any cartoon will be defective. Find the probability that a cartoon fails to meet the gurantee ? (Given $e^{3}=20.1$)

20. In a certain factory blades are manufactured in packets of 10. There is a 0.2% probability for any blade to be defective. Using poisson distribution calculate approximately the number of packets containing two defective blades in a consignment of 20,000 packets. (given that $e^{-0.02}=0.9802$)

21. Suppose 220 misprints are distributed at random throughout a book of 200 pages. Find the probability that a given page contains (i) no misprints (ii) one misprint (iii) 2 or more misprints.

22. A source of liquid contains bacteria with the average number of bacteria per c.c equal to 3. Ten 1 c.c. test tubes are filled with the liquid. Assuming the poisson distribution is applicable calculate the probability (i) that all 10 test tubes show growth, that is, contains at least 1 bacterium each (ii) that exactly 7 test tubes show growth [Given $e^{-3}=0\cdot04975$]

23. A large number of observation on a given solution, which contained bacteria, were made taking samples of 1 c.c each noting down the number of bacteria present in each sample. Assuming the poisson distribution and given that 10% samples contained no bacteria, find the average number of bacteria per c.c. [$\log_e 10=2.3026$].

24. 100 litres of water are supposed to be polluted with 10^6 bacteria. Find the probability that a sample of 1 c.c of the same water is free from bacteria.

25. An office switch board receives phone calls at the rate of 2 in every 5 minutes on the average. What is the probability of getting exactly 4 calls in 15 minutes ?

26. In turning out certain toys in a manufacturing process in a factory, the average number of defectives is 10%. What is the probability of getting exactly 3 defectives in a sample of 10 toys chosen at random, by using Poisson approximation to the Binomial distribution (Take $e = 2.72$)

(Hint : the method is adopted in an Illustrative Ex. page 89;

27. (i) An aeroplane runs with the help of its 1000 components. Failure of a component is independent of other. Chance of failure of a component is 0.1%. The plane takes off from one airport and reached safely its destination if every component of its does not fail. Find the probability of reaching safe. (Hint. Use method of approximation of Binomial variate by poisson)

(ii) If 5% of the books bound at a certain bindery have defective bindings, find the probability that 2 of 100 books bound by this bindery will have defective bindings using (a) the binomial distribution (b) the possion approximation [$e^{-5} = 0.007$].

(iii) Suppose 1% of the items made by a machine are defective. Find by Poisson aprox to Binomial the probability that 3 or more items are defective in a sample of 100 items (Given $e^{-1} = .632$)

(iv) In a lottery with 10,000 tickets there are 100 prizes. A man buys 100 tickets. Apply poisson approximation to binomial distribution to find the approximate probability of his winning at least one ticket $\left(\frac{1}{e} = 0.368\right)$.

28. The distribution of height of men is normally distributed with mean 64.5" and s.d 4.5". Among 10,000 men find the number of men whose height is (a) less than 69" but greater than 55.5" (b) less than 55.5" and (c) more than 73.5"

[Given $\int_0^1 \phi(t)dt = .0398$, $\int_0^2 \phi(t)dt = 0.4772$]

29. A sample of 100 dry battery cells tested to find the length of life produced the following results : $\bar{x} = 12$ hours, $\sigma = 3$ hours. Assuming that the data are normally distributed, what percentage of battery cells to have life (i) more than 15 hours (ii) less than 6 hours (iii) between 10 and 14 hours ? Given

$\int_0^z \phi(t)dt = 0.4938, 0.4772, 0.3413$ and 0.2487 according as

$z = 2.5, 2, 1$ and 0.67.

30. The weight of students in a college is normally distributed with $m = 40kg,\ \sigma = 5kg$. Find the percentage of the students that have weight

(i) greater than 40 kg.

(ii) greater than 50 kg.

(iii) between 38 kg and 52 kg

Given $\frac{1}{\sqrt{2\pi}}\int_{-\infty}^{2} e^{-\frac{t^2}{2}}dt = 0.9772$, $\frac{1}{\sqrt{2\pi}}\int_{-\infty}^{0.4} e^{-\frac{t^2}{2}}dt = 0.6554$,

$\frac{1}{\sqrt{2\pi}}\int_{-\infty}^{2.4} e^{-\frac{t^2}{2}}dt = 0.9918$ [*W.B.U.Tech,2003*]

31. The life of tyre manufactured by *ABC* company is normally distributed with mean 34,000 miles and standard deviation 4000 miles. Find (i) the probability that such a tyre lasts over 40,000 miles (ii) the probability that it lasts between 30,000 and 35, 000 miles? (iii) given that it has survived 30,000 miles what is the probabiltiy that it survives another 10,000 miles ?

You may use the appropriate statistical table.

32. A random variable X follows normal distribution with mean 100 and variance 25. Find $P(|X - 100| \le 5)$ given that

$$\frac{1}{\sqrt{2\pi}}\int_{-\infty}^{1} e^{-\frac{x^2}{2}}dx = 0.8413$$

33. The mean and s.d of the I.Q of a group of 500 children is 90 and 20 respectively. Assuming that the I.Q is normally distributed, find the number of children with $I.Q \ge 100$. Given that

$$\frac{1}{\sqrt{2\pi}}\int_{0.5}^{\infty} e^{-\frac{x^2}{2}}dx = 0.308$$

34. If X has distribution $N(m, \sigma)$, find m and σ. If $P(X < 89) = 0.90$, $P(X < 94) = 0.95$. [Given $\int_0^{1.28} (x)dx = 0.4$ and $\int_0^{1.645} (x)dx = 0.45$]

35. Suppose that 10% of the population for a normal distribution $N(m, \sigma)$ is below 60 and 5% is above 90. What are the values of m and σ?

36. Assuming that the height distribution of a group of men is normally, find the mean and standard deviation, if 84% of the men have heights less than 65.2 inches and 68% have height lying between 65.2 and 62.8 inches.

37. (a) As a result of test on electric light bulbs, it was found that the lifetime of a particular make was distributed normally with an average life of 1000 burning hours and standard deviation of 200 hours. Out of 10,000 bulbs produced by the company how many bulbs are expected to fail (i) in the first 800 burning hours (ii) between 800 and 1200 burning hours ?

[Given $\Phi(1) = 0.84134$] [Hint : (i) $P(X < 800) = ?$ $\Phi(1)$ means $\int_{-\infty}^{1} \phi(t)dt$]

(b) The marks obtained by 1000 students in a final examinations are found to be approximately normal distributed with mean 70 and standard deviation 5. Estimate the number of students whose marks will be between 60 and 75 both inclusive. Given the area under the normal curve $\phi(z) = \frac{1}{\sqrt{2\pi}} \exp\left(-\frac{z^2}{2}\right)$ between $z = 0$ and $z = 2$ is 0.4772 and between $z = 0$ and $z = 1$ is 0.3413.

38. The mean of the inner diameters (in inch) of a sample of 200 tubes produced by a machine is 0.502 and the s.d is 0.005. The purpose for which these tubes are intended allows a maximum tolerance in the diameter of 0.496 to 0.508 (i.e. otherwise the tubes are considered defective). What percentage of the tubes produced by the machine is defective, if the diameters are found to be normally distributed ? (Area under the standard normal curve between $z = 0$ and $z = 1.2$ is 0.3849)

39. If skulls are classified as *A, B, C* according as the length-breadth index is under 75, between 75 and 80, over 80, find approximately (assuming that the distribution is normal) the mean and standard deviation of a series in which *A* are 58%, *B* are 38% and *C* are 4%, being given that if $f(t) = \frac{1}{\sqrt{2\pi}} \int_0^t \exp\left(-\frac{1}{2}x^2\right) dx$, then $f(0.20) = 0.08$ and $f(1.75) = 0.46$.

40. Screws are manufactured to be 3 c.m. in lengths. Yet they are acceptable if their length lie between 2.99 cm and 3.01 cm. It is noted that 5% screws are rejected over-size and 5% are rejected under-size. The lengths of the screws are nomally distributed. Find the standard deviation of the distribution. Hence find the proportion of 'would be rejects' if the permissible limits were widened to 2.983 cm and 3.015 cm. Use table for necessary data.

(Hint : $\frac{3 - 2.99}{\sigma} = 1.645$)

41. The average height of soldiers is 68.22 inches with s.d $\sqrt{10.8}$ inch. How many soldiers in a reginment of 1000 would you expect to be over 6 feet tall, given that the area under the S.N.C between the ordinates $z=0$ and $z=0.35$ is 0.1368 and between $z=0$ and $z=1.15$ is 0.3749)

42. In firing a target assume that the horizontal distance of a shot from the centre is normally distributed with standard deviation 2 feet. In 200 shots, how many would be expected to miss the target, if it is 10 feet wide and sufficiently high ? Use the appropriate stat. table for relevant data.

43. The marks obtained by candidates in Mathematics in B.Tech examination are normally distributed. If 12.5 percent of the candidates score 60% or more marks, 39% obtain less than 30 marks ; find the mean number of marks obtained by the candidates. You may use the table given below :

Z :	0.27	0.28	0.29	1.14	1.15	1.16
area :	0.1064	0.1103	0.1141	0.3729	0.3749	0.3770

(under S.N.C between 0 and z)

44. (a) A fair coin is tossed 10 times. Find the probability of obtaining between 4 and 7 heads inclusive by using (i) the binomial distribution (ii) the normal approximation. (See stat. table foı relevant data)

(b) 100 unbiased coins are tossed. Using normal approximation to Binomial distribution calculate the probability of getting

(i) exactly 40 heads, (ii) 55 heads or more. Given $\Phi(2.1)=0.9821$, $\Phi(1.9)=0.9713$, $\Phi(0.9)=0.8159$. [*W.B.U. Tech 2005*]

45. Among 625 random digits, find the probability that the digit 7 appears (i) between 50 and 60 times inclusive (ii) between 60 and 70 times inclusive. (See stat. table for relevant data)

46. A pair of dice is rolled 900 times and X denote the number of times a total of 9 occurs. Find $P(80 \le X \le 120)$

47. The ideal size of a class of a college is 150 students. The college, experienced from past, knows that only 30% of the admitted students will actually attend. The college uses a policy of approving the applications of 450 students. Find the probability that more than 150 students attend the class. Area under the standard normal curve enclosed between the ordinates $z=0$ and $z=1.59$ is 0.441.

ANSWERS

1. (a) 0.167 (b) 0.4073 (c) 0.4073 (d) 0.5920 **2.** 0.105

3. (i) $\binom{6}{2}\left(\frac{1}{50}\right)^2\left(\frac{49}{50}\right)^4$ (ii) $\left(\frac{49}{50}\right)^6$ (iii) $\left(\frac{49}{50}\right)^8+\binom{6}{1}\left(\frac{1}{50}\right)\left(\frac{49}{50}\right)^5$

4. (i) $\frac{4547}{8192}$ (ii) 4 **5.** 11

6. (a) (i) $8/27$ (ii) $80/81$ (iii) $16/27$ (b) 0.68256

7. 200, 14 **8.** (i) 0.2646 (ii) 0.7599

10. 4, 0.27 **11.** 323 **12.** $\frac{1}{2}$

13. (i) .2241, .2241, .1494, .0498 (ii) .4233 (iii) .8008

14. (i) 0.0821 (ii) 0.1336

15. 0.2706, 0.4059, 0.1353, 0.5941, 0.7216

16. (i) $e^{-2}; 1000(1-5e^{-2})$ (ii) $e^{-3}, 9e^{-9}$

17. (i) 0.007, (ii) 0.182 **18.** 0.78 **19.** $0.35(=1-13e^{-3})$ **20.** 4

21. (i) 0.333 (ii) 0.366 (iii) 0.301 **22.** (i) 0.60003 (ii) 0.01036

23. $\log_e 10 = 2.3026$ **24.** e^{-10} **25.** 0.1339 **26.** 0.061

27. (i) e^{-1} (ii) (a) 0.081 (b) 0.084 (iii) 0.080 (iv) 0.632

28. (a) 8,200 (b) 200 (c) 200

29. (i) 15.87%, (ii) 2.28%, (iii) 49.74%

30. (i) 50%, (ii) 2.28%, (iii) 64.72%

31. (i) 0.0668 (ii) 0.44 (iii) 0.079

32. 0.6826 **34.** 71.465, 13.6 **35.** 73.1, 10.2 **36.** 64.0, 1.2

37. (a) (i) 1587 (ii) 6827 (b) 818.5 **38.** 23

39. mean $= 74.4$ s.d. $= 3.2$ **40.** s.d $= 0.806, 1.24\%$ **41.** 125

42. $\frac{5}{2} \simeq 3$ **43.** 36 **44.** (a) 0.7734, 0.7718 (b) (i) .0108, (ii) .1841

45. 0.3518, 0.5131 **46.** 0.96 **47.** .0559

[III] MULTIPLE CHOICE QUESTIONS

1. The mean of Binomial distribution $B(n,p)$ (where n,p are the no. of trials and probability of success) is

(a) $\frac{n}{p}$ (b) 0

(c) np (d) 1 [*W.B.U.Tech 2007*]

2. If X has Binomial distribution with parameter n and p, then its variance is

(a) np (b) np^2

(c) $np(1-p)$ (d) $np(1+p)$

3. The mean and variance of a Binomial distribution with parameter $6, \frac{2}{3}$ are

(a) $6, \frac{4}{3}$ (b) $4, \frac{4}{3}$

(c) $0, \frac{2}{3}$ (d) $9, \frac{4}{3}$.

4. The mean of the Binomial distribution $B\left(10, \frac{2}{5}\right)$ is

(a) 4 (b) 6

(c) 5 (d) 0. [*W.B.U. Tech. 2007*]

5. The statement "a Binomial variate has mean 7 and S.D 5" is

(a) False (b) True.

6. If X has binomial distribution with parameter 4 and $\frac{1}{3}$, then $P(X=1) = ?$

(a) $\frac{4}{3}$ (b) $\frac{16}{27}$

(c) $\frac{32}{81}$ (d) $\frac{16}{9}$.

7. The mean and standard deviation of a Binomial distribution are respectively 4 and $\sqrt{\frac{8}{3}}$. The values of n and p are (where n and p are the parameters of the distribution) [*W.B.U.Tech 2006*]

(a) $11, \frac{3}{4}$
(b) $12, \frac{2}{7}$
(c) $12, \frac{1}{3}$
(d) $11, \frac{4}{3}$

8. If X has a poisson distribution with parameter μ, then mean of X is

(a) μ
(b) μ^2
(c) $\mu(\mu-1)$
(d) 1.

9. The distribution for which mean and variance are equal is

(a) Poisson
(b) Normal
(c) Binomial
(d) Exponential.

[*W.B.U.Tech 2006*]

10. If a random variable X has a poisson distribution with mean 0.4, then $P(X \le 1)$ is

(a) $e^{-0.4}$
(b) $1.4e^{-0.4}$
(c) $1.48e^{-0.4}$
(d) none.

11. A random variable has a poisson distribution such that $P(1) = P(2)$. Then the S.D of X is [*W.B.U.Tech 2007*]

(a) 0
(b) 2
(c) $\sqrt{2}$
(d) -2.

12. The Poisson distribution is a limiting case of Binomial distribution when

(a) n is very large and p is vary small
(b) n is very small and p is very large
(c) n, p both are very small
(d) n, p both are very large.

13. The mean of a uniform distribution with parameters a, b is

(a) $b-a$
(b) $b+a$
(c) $\frac{a+b}{2}$
(d) $\frac{b-a}{2}$ [*W.B.U.Tech. 2006*]

14. The variance of a uniform distribution with parameter a,b is

(a) $\frac{(a-b)^2}{12}$ (b) $\frac{(a+b)^2}{12}$

(c) $\frac{(a-b)^2}{2}$ (d) $\frac{(a+b)^2}{2}$

15. The p.d.f of a random variable X is

$$f(x)=\frac{1}{6},\ -3<x<3$$

$$=0,\ \text{elsewhere.}$$

The mean of the distribution is

(a) 0 (b) 3

(c) –3 (d) $\frac{1}{3}$.

16. If X is normally distributed with zero mean and unit variance, then the expectation of X^2 is [*W.B.U.Tech. 2007*]

(a) 1 (b) 2

(c) 8 (d) 20.

17. If X is a normal variate with parameter μ and σ *i.e.* $X \sim N(\mu,\sigma)$ then variance of X is

(a) σ (b) σ^2

(c) $\sqrt{\sigma}$ (d) $\frac{1}{\sigma}$.

18. The mean and S.D of a standard normal dsitribution are

(a) 1,0 (b) 0, 1

(c) 1, 1 (d) 0, 0.

19. The mean of Poisson variate is

(a) greates then its variance

(b) less than its variance

(c) equal to its variance

(d) twice its variance [*W.B.U.Tech. 2007 M 302*]

0. The mean and variance of a binomial distribution are 4 and 3 espectively. Then the parameters of the distribution are

(a) $16, \frac{1}{4}$ (b) $8, \frac{1}{2}$

(c) $12, \frac{1}{3}$ (d) none.

21. A random variable X has the uniform distribution on $(-5,5)$. Then $P(X=3)=?$

(a) $\frac{1}{3}$ (b) $\frac{3}{10}$

(c) $\frac{1}{10}$ (d) $\frac{2}{5}$.

22. If $F(x)=0, -\infty<x<4$

$= \frac{x-4}{2}, 4 \le x < 6$

$= 1, 6 \le x < \infty$

be the distribution function of a uniform random variable then value of its pdf on $[4,6]$ is

(a) $\frac{1}{3}$ (b) $\frac{1}{6}$

(c) $\frac{1}{4}$ (d) $\frac{1}{2}$.

23. If X is a poisson variate such that $P(X=1)=0.2$ and $P(X=2)=0.2$ find $P(X=0)$. [W.B.U.Tech 2002]

24. If $f(x)=\frac{1}{3\sqrt{2\pi}}e^{-\frac{(x-2)^2}{18}}, -\infty<x<\infty$ is the pdf of a normal variate then its mean and variance are

(a) 2, 3 (b) 2, 9

(c) 0, 9 (d) 1, 2.

25. Normal curve of a variate represents the

(a) distribution function

(b) skewness

(c) probability density function

(d) none of these.

26. If X is a normal variate with mean -2 and variance 25 then which one of the following is standard normal variate

(a) $\frac{X-2}{5}$ (b) $\frac{X+2}{25}$

(c) $\frac{X-25}{2}$ (d) $\frac{X+2}{5}$.

27. If $3z-x=5$ where z is a standard normal variate then

(a) x is a normal variate with mean 5, s.d. 3

(b) x is a normal variate with mean 3, s.d 5

(c) x is a normal variate with mean -5, s.d. 3

(d) none of these.

28. If X has Binomial distribution with parameter $n=500$ and $p=.001$ then which of the following variate is approximately standard normal variate?

(a) $\frac{X-0.5}{.4995}$ (b) $\frac{X-.001}{500}$

(c) $\frac{X-.001}{\sqrt{500}}$ (d) $\frac{X-0.5}{\sqrt{.4995}}$.

29. If X is a normal variate with mean 4 and s.d 0.5 then $P(3.8<X<4.3)=$

(a) $P(-0.4<z<0.6)$ (b) $P(0.4<z<0.6)$

(c) $P(0<z<0.6)$ (d) none of these.

30. If X is a normal variate with mean 70 and variance 25 and if z a standard normal variate than $X=66$ implies $z=$

(a) 0.8 (b) 0.6

(c) -0.8 (d) 0.

31. If X is a normal variate with mean 4 and s.d. 0.5 and z be standard normal variate then $z=0.6$ implies $X=$

(a) 4 (b) 5

(c) 1.6 (d) 4.3

32. X has normal distribution with s.d. 2 and z is standard normal variate. If $X = 9.6$ implies $z = -0.9$ then the mean of X is

(a) 11 (b) 11.4

(c) 13 (d) none of these.

ANSWERS

1.c	**2.c**	**3.b**	**4.a**	**5.a**	**6.c**	**7.c**	**8.a**
9.a	**10.b**	**11.c**	**12.a**	**13.c**	**14.a**	**15.a**	**16.a**
17.b	**18.b**	**19.c**	**20.a**	**21.c**	**22.d**	**23.d**	**24.b**
25.c	**26.d**	**27.c**	**28.d**	**29.a**	**30.c**	**31.d**	**32.b**

Module -IV

Partial Differential Equation (PDE)

Series Solution of Ordinary Differential Equation (ODE)

Partial Differential Equation (PDE)

4.1 Solution of PDE By Separation of Variables

4.1.1. Introduction.

In science and Technology often we face some problems involving partial differential equations (p.d.e) with constant co-efficients together with somé boundary conditions. These are known as Boundary value problem.

In this Chapter we shall discuss some special type p.d.e known as wave equation, heat equation, Laplace equation and their solution by using a Technique known as, 'separation of variables.'

4.1.2. Method of separation of variables :

Suppose the given p.d.e involves two independent variables x, y and one dependent variable u. In this method we asume the solution is of the form

$$u(x,y) = X(x)Y(y)$$

where X, Y are functions of x and y respectively. Substituting this into the given equation we shall obtain two ordinary differential equations one in each of the unknown functions X and Y. Then also the boundary conditions will be modified involving X and Y separately. After that solving two ordinary equations using boundary conditions we get two particular solutions and hence we obtain the most general solution of the given p.d.e equation. We illustrate the method by the following examples :

Ex. 1. *Solve by the method of separation of variable* $\frac{\partial u}{\partial x} = 4\frac{\partial u}{\partial y}$ *where* $u(0,y) = 8e^{-3y}$.

Here u is a function of x, y. So let

$$u = X(x)Y(y) \qquad \text{... (i)}$$

where X is a function of x only and Y is a function of y only. Then

$$\frac{\partial u}{\partial x} = X'Y, \frac{\partial u}{\partial y} = XY'.$$

Substituting in the given equation, we get

$$X'Y = 4XY'$$

i.e. $$\frac{X'}{X} = 4\frac{Y'}{Y} \qquad \text{... (ii)}$$

L.H.S of (ii) is a function of x alone and R.H.S of (ii) is a function of y alone. So they must be equal to a constant, say λ

$$\therefore \frac{X'}{X}=\lambda \text{ and } 4\frac{Y'}{Y}=\lambda$$

i.e. $\frac{dX}{X}=\lambda dx, \frac{dY}{Y}=\frac{\lambda}{4}dy$.

Integrating we get

$$\log X=\lambda x+\log c_1, \log Y=\frac{1}{4}\lambda y+\log c_2$$

$$\therefore X=c_1e^{\lambda x}, Y=c_2e^{\frac{1}{4}\lambda y}.$$

Thus from (i) $u=c_1c_2e^{\lambda x}\cdot e^{\frac{1}{4}\lambda y}=c_1c_2e^{\lambda x+\frac{1}{4}\lambda y}$.

Since $u(0,y)=8e^{-3y}$, so

$$8e^{-3y}=c_1c_2e^{\lambda\cdot 0+\frac{1}{4}\lambda\cdot y}$$

$$=c_1c_2e^{\frac{1}{4}\lambda\cdot y}$$

$\therefore c_1c_2=8, \frac{1}{4}\lambda=-3$ i.e. $\lambda=-12$.

Thus the required solution is

$$u=8e^{-12x+\frac{1}{4}(-12)y}$$

i.e. $u=8e^{-12x-3y}$.

Ex. 2. *Using the method of separation of variable, solve*

$$\frac{\partial^2 u}{\partial x^2}=\frac{\partial u}{\partial t}, u(0,t)=0, u(4,t)=0, \; u(x,0)=\sin 3x.$$

Let $u=X(x)T(t)$

$$\therefore \frac{\partial^2 u}{\partial x^2}=X''T, \frac{\partial u}{\partial t}=XT'.$$

Substituting in the given equation we get

$$X''T=XT'$$

$$\therefore \frac{X''}{X}=\frac{T'}{T} \qquad \ldots \qquad \text{(i)}$$

As L.H.S of (i) is a function of x alone and R.H.S of (i) is a function of t alone they must be equal to a constant, say $-\lambda^2$ (for fitting the boundary conditions).

$$\therefore \frac{X''}{X} = -\lambda^2, \frac{T'}{T} = -\lambda^2$$

i.e. $X'' + \lambda^2 X = 0, \frac{dT}{T} = -\lambda^2 dt$

whose solution are given by

$$X(x) = c_1 \cos\lambda x + c_2 \sin\lambda x \quad \dots \quad \text{(ii)}$$

$$T(t) = c_3 e^{-\lambda^2 t} \quad \dots \quad \text{(iii)}$$

We are given $u(0,t) = 0$

$\therefore X(0)T(t) = 0$

$$\therefore X(0) = 0 \quad \dots \quad \text{(iv)}$$

Also we have $u(4,t) = 0$ i.e. $X(4)T(t) = 0 \therefore X(4) = 0$ ... (v)

Using (iv), we get from (ii), $c_1 = 0$

$\therefore X(x) = c_2 \sin\lambda x$

Again using (v), we have $c_2 \sin 4\lambda = 0$.

$\therefore \sin 4\lambda = 0$ $[\because c_2 = 0 \Rightarrow \sin(x) = 0\ \forall x$ which impossible$]$

which implies λ is an integer

$\therefore u(x,t) = c_2 \sin\lambda x \cdot c_3 e^{-\lambda^2 t}$

$= c \sin\lambda x e^{-\lambda^2 t}$, where $c = c_2 c_3$.

$\therefore u(x,0) = c \sin\lambda x$.

As we are given $u(x,0) = \sin 3x$,

so $c \sin\lambda x = \sin 3x$.

$\Rightarrow c = 1, \lambda = 3$.

Thus the required solution is

$u(x,t) = \sin 3x e^{-9t}$.

4.1.3. One Dimensional Wave Equation.

Formulation. Consider an elastic string which is stretched to length l between two points O and A. Let the string be distorted and at a certain instant it is released and allowed to vibrate. Also let the time $t = 0$, at this certain instant and $u(x,t)$ be its deflection at any point x

and at any time t. Let OA be taken as the x-axis and a perpendicular line through O as u-axis.

We obtain the equation of motion of the string under the following assumptions :

(i) The entire motion takes place in the xu-plane.

(ii) The string is homogeneous i.e. mass of the string per unit length is constant, say ρ and perfectly elastic so that it does not offer resistance to bending.

(iii) The tension in the string is so large that the forces due to weight of the string can be neglected.

(iv) The string makes small transverse vibration so that higher power of u and $\dfrac{\partial u}{\partial x}$ are small.

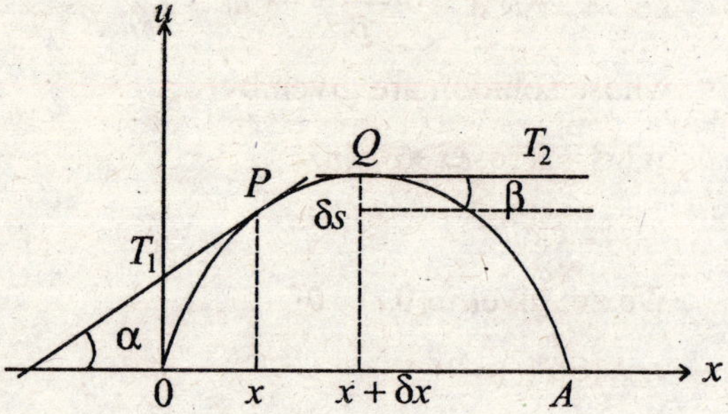

Consider the motion of small portion PQ of the string of length δs. As the string does not offer resistance to bending by assumption, so the tensions T_1 and T_2 at P and Q will be tangential to the curve. Also since there is no motion in the horizontal direction, we have $T_1 \cos\alpha = T_2 \cos\beta = T$ (constant), say. ... (1)

Now the equation of motion of PQ in the vertical direction is

$$\rho\delta s \frac{\partial^2 u}{\partial t^2} = T_2 \sin\beta - T_1 \sin\alpha$$

i.e. $$\frac{\rho\delta s}{T}\frac{\partial^2 u}{\partial t^2} = \frac{T_2 \sin\beta}{T} - \frac{T_1 \sin\alpha}{T}$$

$$= \frac{T_2 \sin\beta}{T_2 \cos\beta} - \frac{T_1 \sin\alpha}{T_1 \cos\alpha}, \text{ by (1)}$$

i.e. $$\frac{\partial^2 u}{\partial t^2} = \frac{T}{\rho\delta s}(\tan\beta - \tan\alpha). \qquad \text{... (2)}$$

Now $$\tan\alpha = \left(\frac{\partial u}{\partial x}\right)_P = \frac{\partial u}{\partial x}(x,t)$$

$$\tan\beta = \left(\frac{\partial u}{\partial x}\right)_Q = \frac{\partial u}{\partial x}(x+\delta x,t)$$

$$\therefore \frac{\partial^2 u}{\partial t^2} = \frac{T}{\rho} \frac{\frac{\partial u}{\partial x}(x+\delta x, t) - \frac{\partial u}{\partial x}(x,t)}{\delta x}$$

[$\because \delta s = \delta x$ to a first approximation]

$$= \frac{T}{\rho} \frac{\partial^2 u}{\partial x^2}, \text{ as } \delta x \to 0$$

$$\therefore \frac{\partial^2 u}{\partial t^2} = c^2 \frac{\partial^2 u}{\partial x^2} \text{ where } c^2 = T/\rho$$

which is the ***one dimensional wave equation***

Solution of Wave Equation, $\frac{\partial^2 u}{\partial t^2} = c^2 \frac{\partial^2 u}{\partial x^2}$. [*W.B.U.Tech 2003*]

Let $u(x,t) = X(x)T(t)$ ... (1)

be a solution of the wave equation

$$\frac{\partial^2 u}{\partial t^2} = c^2 \frac{\partial^2 u}{\partial x^2} \quad \text{... (2)}$$

where X is a function of x alone and T is a function of t alone. Then

$$\frac{\partial^2 u}{\partial t^2} = XT'', \frac{\partial^2 u}{\partial x^2} = X''T.$$

$\therefore$ From (2), we get

$$XT'' = c^2 X'' T$$

i.e. $\frac{X''}{X} = \frac{1}{c^2}\frac{T''}{T}$ ... (3)

As L.H.S. of (3) is a function of x only and R.H.S is a function of t only, so they must be equal to a constant, say λ

$\therefore \frac{X''}{X} = \lambda$ i.e. $X'' - \lambda X = 0$ ... (4)

and $\frac{1}{c^2}\frac{T''}{T} = \lambda$ i.e. $T'' - \lambda c^2 T = 0$ ... (5)

Three cases arise according as λ is zero, $+ve$ or $-ve$.

Case (a). Let $\lambda = 0$.

Then solutions of (4) and (5) are

$X = a_1 x + a_2, T = a_3 t + a_4$.

Case (b). Let λ be $+ve$, say p^2. Then (4) and (5) becomes respectively

$X'' - p^2 X = 0$ and $T'' - p^2 c^2 T = 0$

Solving these differential equations we obtain

$X = b_1 e^{px} + b_2 e^{-px}, T = b_3 e^{pct} + b_4 e^{-pct}$.

Case (c). Let λ be $-ve$, say $-p^2$. Then (4) and (5) becomes respectively

$X'' + p^2 X = 0$ and $T'' + p^2 c^2 T = 0$.

whose solution are

$X = c_1 \cos px + c_2 \sin px, T = c_3 \cos pct + c_4 \sin pct$.

Thus the various possible solution of the wave equation are :

$$u(x,t) = (a_1 x + a_2)(a_3 t + a_4) \quad \ldots \quad (6)$$

$$u(x,t) = (b_1 e^{px} + b_2 e^{-px})(b_3 e^{pct} + b_4 e^{-pct}) \quad \ldots \quad (7)$$

$$u(x,t) = (c_1 \cos px + c_2 \sin px)(c_3 \cos pct + c_4 \sin pct) \quad \ldots \quad (8)$$

Of these solutions, we have to choose that solution which is consistent with the physical nature of the problems.

As we deal with problems on vibrations, $u(x,t)$ must be a periodic function. Hence the solution given by (8) is the only suitable solution.

Now applying boundary conditions i.e. $u(x,t) = 0$ at $x = 0$ and $x = l$ we get from (8)

$0 = c_1(c_3 \cos pct + c_4 \sin pct) = 0$ i.e. $c_1 = 0$

and $0 = (c_1 \cos pl + c_2 \sin pl)(c_3 \cos pct + c_4 \sin pct)$

i.e. $c_1 \cos pl + c_2 \sin pl = 0$

i.e. $c_2 \sin pl = 0 \, [\because c_1 = 0]$.

$\therefore \sin pl = 0$. [$\because c_2 = 0 \Rightarrow u(x,t) = 0 \forall x, t$ which is impossible]

$\therefore pl = n\pi, n = 1,2,3,\cdots$

$\therefore p = \dfrac{n\pi}{l}$.

$\therefore$ From (8), we get

$$u(x,t) = c_2 \sin\frac{n\pi x}{l}\left(c_3 \cos\frac{n\pi ct}{l} + c_4 \sin\frac{n\pi ct}{l}\right)$$

$$= \left(c_n \cos\frac{n\pi ct}{l} + d_n \sin\frac{n\pi ct}{l}\right)\sin\frac{n\pi x}{l} \quad \text{for } n = 1,2,3,\cdots$$

where $c_n = c_2 c_3, d_n = c_2 c_4$.

Thus for different values of n, we consider the more general form of the solution as

$$u(x,t) = \sum_{n=1}^{\infty} u_n(x,t) = \sum_{n=1}^{\infty}\left(c_n \cos\frac{n\pi ct}{l} + d_n \sin\frac{n\pi ct}{l}\right)\sin\frac{n\pi x}{l} \quad \ldots (9)$$

where c_n, d_n can be obtained using initial conditions.

Illustrative Examples.

Ex. 1. *Solve the wave equation* $\dfrac{\partial^2 u}{\partial t^2} = c^2 \dfrac{\partial^2 u}{\partial x^2}$ *given that* $u(0,t) = u(l,t) = 0, u(x,0) = f(x)$ *and* $\dfrac{\partial u}{\partial t}(x,0) = 0,\ 0 < x < l$.

We have from (9),

$$u(x,t) = \sum_{n=1}^{\infty}\left(c_n \cos\frac{n\pi ct}{l} + d_n \sin\frac{n\pi ct}{l}\right)\sin\frac{n\pi x}{l}$$

$$\therefore \frac{\partial u}{\partial t} = \sum_{n=1}^{\infty}\frac{n\pi c}{l}\left(-c_n \sin\frac{n\pi ct}{l} + d_n \cos\frac{n\pi ct}{l}\right)\sin\frac{n\pi x}{l}.$$

Now applying the initial conditions,

$u(x,0) = f(x), \dfrac{\partial u}{\partial t}(x,0) = 0$, we get

$$f(x) = \sum_{n=1}^{\infty} c_n \sin\frac{n\pi x}{l} \quad \ldots (10)$$

$$\text{and } 0 = \sum_{n=1}^{\infty} d_n \frac{n\pi c}{l}\sin\frac{n\pi x}{l} \quad \ldots (11)$$

Since (10) represents Fourier series of $f(x)$, we have

$$c_n = \frac{2}{l}\int_0^l f(x)\sin\frac{n\pi x}{l}dx \qquad \dots \quad (12)$$

Also from (11), we have.

$d_n = 0$ for all n.

Hence $u(x,t) = \sum_{n=1}^{\infty} c_n \cos\frac{n\pi ct}{l}\sin\frac{n\pi x}{l}$

where c_n is given by (12).

Ex. 2. *A tightly stretched string with fixed end point $x=0$ and $x=l$ is initially in a position given by $u = u_0 \sin^3\frac{\pi x}{l}$. If it is released from rest from this position, find the displacement $u(x,t)$.*

Here the boundary conditions are $u(0,t) = u(l,t) = 0$ and the initial conditions are $u(x,0) = u_0\sin^3\frac{\pi x}{l}$ and $\frac{\partial u(x,0)}{\partial t} = 0$ (The value of u, u_t etc. at time $t=0$ are known as initial condition.)

As discussed in Art 3.3, we have from (9),

$$u(x,t) = \sum_{n=1}^{\infty}\left(c_n\cos\frac{n\pi ct}{l} + d_n\sin\frac{n\pi ct}{l}\right)\sin\frac{n\pi x}{l}.$$

Thus using the initial conditions we have

$$c_n = \frac{2}{l}\int_0^l u_0\sin^3\frac{\pi x}{l}\sin\frac{n\pi x}{l}dx \qquad \dots \quad (13)$$

and $d_n = 0$ for all n (proceeding as Ex 1.)

$$\therefore u(x,t) = \sum_{n=1}^{\infty} c_n\cos\frac{n\pi ct}{l}\sin\frac{n\pi x}{l} \text{ where } c_n \text{ is given by (13)} \dots \quad (14)$$

Now $c_n = \frac{2u_0}{l}\int_0^l \frac{1}{4}\left(3\sin\frac{\pi x}{l} - \sin\frac{3\pi x}{l}\right)\sin\frac{n\pi x}{l}dx$

$= \frac{u_0}{2l}\int_0^{\pi}(3\sin z\cdot\sin nz - \sin 3z\cdot\sin nz)\frac{l}{\pi}dz$, putting $\frac{\pi x}{l} = z$

$= \frac{u_0}{2\pi}\int_0^{\pi}(3\sin z\cdot\sin nz - \sin 3z\sin nz)dz$

But $\int_0^{\pi} \sin px \sin qx dx = 0$, if $p \neq q$

$$= \frac{\pi}{2} \text{ if } p = q .$$

Hence $c_1 = \frac{u_0}{2\pi}\left(3 \cdot \frac{\pi}{2} - 0\right) = \frac{3u_0}{4}$

$c_2 = 0, c_3 = \frac{u_0}{2\pi}\left(0 - \frac{\pi}{2}\right) = -\frac{u_0}{4}$

and $c_n = 0 \ \forall n \geq 4$.

Putting these values of of c_n in (14), we get the required solution as

$$u(x,t) = \frac{3u_0}{4}\cos\frac{\pi ct}{l}\sin\frac{\pi x}{l} - \frac{u_0}{4}\cos\frac{3\pi ct}{l}\sin\frac{3\pi x}{l}$$

$$= \frac{u_0}{4}\left(3\cos\frac{\pi ct}{l}\sin\frac{\pi x}{l} - \cos\frac{3\pi ct}{l}\sin\frac{3\pi x}{l}\right).$$

Ex 3. *A tightly stretched string with fixed end points $x = 0$ and $x = l$ is initially at rest in its equilibrium position. If it is set vibrating by giving to each of its points a velocity $\lambda x(l - x)$, find the displacement of the string at any distance x from one end at any time t.*

Here the boundary conditions are

$u(0,t) = u(l,t) = 0$ and the initial conditions are $u(x,0) = 0$,

$\frac{\partial u}{\partial t}(x,0) = \lambda x(l - x)$.

Using the boundary conditions, we have from (8)

$$u(x,t) = \sum_{n=1}^{\infty}\left(c_n \cos\frac{n\pi ct}{l} + d_n \sin\frac{n\pi ct}{l}\right)\sin\frac{n\pi x}{l} \quad \ldots \quad (15)$$

Then using the initial condition $u(x,0) = 0$, we have

$$0 = \sum_{n=1}^{\infty} c_n \sin\frac{n\pi x}{l}$$

$\therefore c_n = 0 \ \forall n$.

$\therefore$ From (15), $u(x,t)=\sum_{n=1}^{\infty} d_n \sin\frac{n\pi ct}{l}\sin\frac{n\pi x}{l}$ ... (16)

$$\therefore \frac{\partial u}{\partial t}=\sum_{x=1}^{\infty}\frac{n\pi c}{l}d_n\cos\frac{n\pi ct}{l}\sin\frac{n\pi x}{l}.$$

Then using $\frac{\partial u}{\partial t}(x,0)=\lambda x(l-x)$

$$\lambda x(l-x)=\sum_{n=1}^{\infty}\frac{n\pi c}{l}d_n\sin\frac{n\pi x}{l}$$

which is the Half Range Fourier sine Series for $\lambda x(l-x), 0<x<l$

$\therefore$ the Fourier co-efficient $\frac{n\pi c}{l}d_n=\frac{2}{l}\int_0^l \lambda x(l-x)\sin\frac{n\pi x}{l}dx$

$$=\frac{2\lambda}{l}\int_0^l\left(lx-x^2\right)\sin\frac{n\pi x}{l}dx.$$

$$=\frac{2\lambda}{l}\left[-\left(lx-x^2\right)\frac{l}{n\pi}\frac{\cos n\pi x}{l}+(l-2x)\frac{l^2}{n^2\pi^2}\sin\frac{n\pi x}{l}-2\frac{l^3}{n^3\pi^3}\cos\frac{n\pi x}{l}\right]_0^l,$$

integrating by parts

$$=\frac{2\lambda}{l}\left[0+0-\frac{2l^3}{n^3\pi^3}(\cos n\pi-1)\right]$$

$$=\frac{4\lambda l^2}{n^3\pi^3}(1-\cos n\pi)=\frac{4\lambda l^2}{n^3\pi^3}\left[1-(-1)^n\right]$$

$=0$ when n is even

$=\frac{8\lambda l^2}{n^3\pi^3}$ when n is odd.

$\therefore d_n=\frac{8\lambda l^3}{cn^4\pi^4}$ when n is odd

$=0$ when n is even

$\therefore$ From (16) we get $u(x,t)=\frac{8\lambda l^3}{c\pi^4}\sum_{n=1,3,5,\ldots}\frac{1}{n^4}\sin\frac{n\pi ct}{l}\sin\frac{n\pi x}{l}$

which is the required displacement.

4.1.4. One Dimensional Heat Flow Equation.

Formulation. Consider the flow of heat by conduction in a uniform one-dimensional bar OA. Let OA be taken along the x-axis. Consider an elementary cross section $PQRS$ as shown in figure. Clearly the temperature u at any point P is a function of x and the time t. We now obtain the equation of heat flow under the following assumptions :

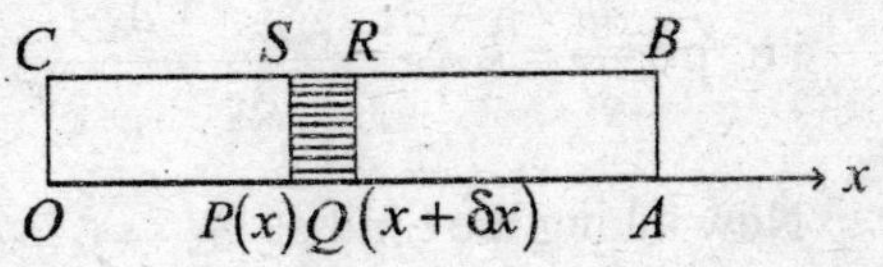

(i) The bar is homogeneous i.e. the mass per unit volume ρ is constant.

(ii) The sides of the bar are insulated and loss of heat from the sides by conduction or radiation is negligible.

(iii) The amount of heat crossing any section of the bar is $KA\left(\frac{\partial u}{\partial x}\right)\cdot\delta t$ where K is the thermal conductivity of the material of the bar, A, area of cross - section, $\frac{\partial u}{\partial x}$, the temperature gradient and δt is the time of heat flow.

Then the quantity of heat flowing across the section PS in time δt is $-KA\left(\frac{\partial u}{\partial x}\right)_P\cdot\delta t$ i.e. $-KA\left(\frac{\partial u}{\partial x}\right)_x\cdot\delta t$ and that across the section QR is $-KA\left(\frac{\partial u}{\partial x}\right)_Q\cdot\delta t$ i.e. $-KA\left(\frac{\partial u}{\partial x}\right)_{x+\delta x}\cdot\delta t$ (the negative sign is taken because u decreases as x increases).

Hence the amount of heat retained by elementary section $PQRS$ is

$$KA\delta t\left[\left(\frac{\partial u}{\partial x}\right)_{x+\delta x}-\left(\frac{\partial u}{\partial x}\right)_x\right] \quad \ldots \quad (1)$$

But the rate of increase of heat in $PQRS$ is $(\rho A\delta x)\sigma\frac{\partial u}{\partial t}$ ... (2)

where σ is specific heat of the bar.

$\therefore$ From (1) and (2)

$$(\rho A\delta x)\sigma\frac{\partial u}{\partial t}\cdot\delta t = KA\delta t\left[\left(\frac{\partial u}{\partial x}\right)_{x+\partial x}-\left(\frac{\partial u}{\partial x}\right)_x\right]$$

i.e. $$\rho\sigma\frac{\partial u}{\partial t} = k\frac{\left(\frac{\partial u}{\partial x}\right)_{x+\delta x} - \left(\frac{\partial u}{\partial x}\right)_{x}}{\delta x}.$$

Now taking the limit as $\delta x \to 0$, we have

$$\rho\sigma\frac{\partial u}{\partial t} = k\frac{\partial^2 u}{\partial x^2}$$

i.e. $$\frac{\partial u}{\partial t} = \frac{k}{\rho\sigma}\frac{\partial^2 u}{\partial x^2}$$

i.e. $$\frac{\partial u}{\partial t} = c^2\frac{\partial^2 u}{\partial x^2}$$

where $c^2 = \dfrac{k}{\rho\sigma}$ is known as diffusivity of the material of the bar.

This is the one dimensional heat equation.

Solution of Heat Equation.

Let $u(x,t) = X(x)T(t)$ ... (1)

be the solution of the heat equation

$$\frac{\partial u}{\partial t} = c^2\frac{\partial^2 u}{\partial x^2} \quad \text{... (2)}$$

where X is a function of x alone and T is a function of t alone. Then

$$\frac{\partial u}{\partial t} = XT', \frac{\partial^2 u}{\partial x^2} = X''T$$

$\therefore$ From (2), $XT' = c^2 X'' T$

i.e. $$\frac{X''}{X} = \frac{1}{c^2}\frac{T'}{T} \quad \text{... (3)}$$

As L. H.S of (3) is a function of x only and R.H.S is a function of t only, so they must be equal to a constsnt, say λ.

$$\therefore \frac{X''}{X} = \lambda, \frac{1}{c^2}\frac{T'}{T} = \lambda$$

i.e. $X'' - \lambda X = 0$ ... (4)

$T' - \lambda c^2 T = 0$ ... (5)

Three cases arise according as λ is zero, positive or negative.

Case (a). Let $\lambda = 0$.

Then solution of (4) and (5) are

$X = a_1 x + a_2, T = a_3$.

Case (b). Let λ be $+ve$, say p^2. Then (4) and (5) becomes

$X'' - p^2 X = 0$

$T' - p^2 c^2 T = 0$.

Solving these differential equations we get

$X = b_1 e^{px} + b_2 e^{-px}, T = b_3 e^{c^2 p^2 t}$.

Case (c). Let λ be $-ve$, say $-p^2$. Then (4) and (5) becomes

$X'' + p^2 X = 0,\ T' + p^2 c^2 T = 0$ whose solutions are

$X = c_1 \cos px + c_2 \sin px, T = c_3 e^{-c^2 p^2 t}$.

Thus the various possible solution of the heat equation are :

$$u(x,t) = (a_1 x + a_2) a_3 \quad \ldots \quad (6)$$

$$u(x,t) = \left(b_1 e^{px} + b_2 e^{-px}\right) b_3 e^{p^2 c^2 t} \quad \ldots \quad (7)$$

$$u(x,t) = (c_1 \cos px + c_2 \sin px) c_3 e^{-p^2 c^2 t} \quad \ldots \quad (8)$$

Of these three solutions, we have to choose that solution which is consistent with the physical nature of the problem. As we deal with the problems of heat flow where u decreases as t increases, the only suitable solution is given by (8).

Illustrative Examples.

Ex. 1. *The faces* $x = 0$ *and* $x = l$ *of an infinite slab are maintained at zero temperature. Given that the temperature* $u(x,t) = f(x)$ *at* $t = 0$. *Determine the temperature at a subsequent time t.*

Here the boundary conditions are $u(0,t) = u(l,t) = 0$ and the initial condition are $u(x,0) = f(x)$.

The temperature $u(x,t)$ satisfies the heat equation

$$\frac{\partial u}{\partial t} = c^2 \frac{\partial^2 u}{\partial x^2}$$

whose solution is given by (8) as.

$$u(x,t)=(c_1\cos px+c_2\sin px)c_3e^{-p^2c^2t}$$
$$=(a_1\cos px+a_2\sin px)e^{-p^2c^2t} \quad \dots \quad (1)$$

where $a_1=c_1c_3, a_2=c_2c_3$.

Then using the boundary conditions $u(0,t)=u(l,t)=0$, we get

$a_1e^{-p^2c^2t}=0$ i.e. $a_1=0$.

and $(a_1\cos pl+a_2\sin pl)e^{-p^2c^2t}=0$.

i.e. $a_2\sin pl\, e^{-p^2c^2t}=0$.

i.e. $\sin pl=0$ [$\because a_2\neq 0$, for $a_2=0\Rightarrow u=0\ \forall x,t$ which is not possible].

$\therefore pl=n\pi$,

$p=\dfrac{n\pi}{l}, n=1,2,3,\cdots$.

Hence (1) takes the form

$$u_n(xt)=b_n\sin\frac{n\pi x}{l}e^{-\frac{n^2\pi^2}{l^2}c^2t}, \quad n=1,2,\cdots\cdots$$

Now we consider the most general solution as

$$u(x,t)=\sum_{n=1}^{\infty}b_n\sin\frac{n\pi x}{l}e^{-\frac{n^2\pi^2}{l^2}c^2t} \quad \dots \quad (2)$$

and then using the initial cindition $u(x,0)=f(x)$, we get

$$f(x)=\sum_{n=1}^{\infty}b_n\sin\frac{n\pi x}{l} \quad \dots \quad (3)$$

which is the Fourier sine series.

$$\therefore b_n=\frac{2}{l}\int_0^l f(x)\sin\frac{n\pi x}{l}dx, \quad n=1,2,3,\cdots \quad \dots \quad (4)$$

Hence the required temperature $u(x,t)$ is given by (3) where b_n is given by (4).

Ex. 2. *A rod of length l with insulated sides, is initially at a uniform temperature u_0. Its ends are suddenly cooled at $0°c$ and are kept at that temperature. Find the temperature function $u(x,t)$.*

Proceeding as Ex. 1, we have

$$u(x,t)=\sum_{n=1}^{\infty}b_n\sin\frac{n\pi x}{l}e^{-\frac{n^2\pi^2}{l^2}c^2t} \quad \dots \quad (1)$$

where $b_n = \frac{2}{l}\int_0^l f(x)\sin\frac{n\pi x}{l}dx$.

Here $f(x) = u_0$

$$\therefore b_n = \frac{2}{l}\int_0^l b_0 \sin\frac{n\pi x}{l}dx$$

$$= \frac{2u_0}{l}\frac{l}{n\pi}\left[-\cos\frac{n\pi x}{l}\right]_0^l$$

$$= \frac{2u_0}{n\pi}\left[1-(-1)^n\right]$$

$= 0$ when n is even

$= \frac{4u_0}{n\pi}$ when n is odd i.e. $n = 1,3,5,\cdots$.

$$\therefore \text{From (1), } u(x,t) = \frac{4u_0}{\pi}\sum_{n=1,3,5,\cdots}\frac{1}{n}\sin\frac{n\pi x}{l}e^{-\frac{n^2\pi^2c^2}{l^2}t}$$

$$= \frac{4u_0}{\pi}\sum_{n=1}^{\infty}\frac{1}{2m-1}\sin\frac{(2m-1)\pi x}{l}e^{-\frac{(2m-1)^2\pi^2c^2}{l^2}t}$$

which is the required temperature functions.

Ex. 3. *Show that the solution of the equation* $\frac{\partial^2\theta}{\partial x^2} = \frac{\partial\theta}{\partial t}$ *satisfying the conditions (i)* $\theta \to 0$ *as* $t \to \infty$ *(ii)* $\theta \to 0$ *when* $x = \pm l$ *for all values of* $t > 0$ *(iii)* $\theta = x$ *when* $t = 0$, $-l < x < l$ *is*

$$\theta = \frac{2l}{\pi}\sum_{n=1}^{\infty}\frac{(-1)^{n-1}}{n}\sin\frac{n\pi x}{l}\exp\left(-\frac{n^2\pi^2 t}{l^2}\right).$$

Here the only suitable solution of the equation $\frac{\partial^2\theta}{\partial x^2} = \frac{\partial\theta}{\partial t}$ is

$$\theta(x,t) = (a_1\cos px + a_2\sin px)e^{-p^2t} \quad \cdots \quad (1)$$

Obviously (1) satisfies the given condition (i). Then using the condition $\theta \to 0$ when $x = \pm l$, we have

$$(a_1\cos pl + a_2\sin pl)e^{-p^2t} = 0$$

and $(a_1\cos pl - a_2\sin pl)e^{-p^2t} = 0$.

i.e. $a_1 \cos pl + a_2 \sin pl = 0$ and $a_1 \cos pl - a_2 \sin pl = 0$.

Solving we get $a_1 = 0$ and $a_2 \sin pl = 0$

i.e. $\sin pl = 0$ [$\therefore a_2 = 0 \Rightarrow \theta = 0 \forall x, t$ which is not possible]

$\therefore pl = n\pi \quad \therefore p = \frac{n\pi}{l}, n = 1, 2, 3, \cdots$

Hence (1) takes the form

$$\theta_n(x,t) = b_n \sin\frac{n\pi x}{l} e^{-\frac{n^2\pi^2}{l^2}t}, \ n = 1, 2, \cdots$$

Now we consider the most general solution as

$$\theta(x,t) = \sum_{n=1}^{\infty} b_n \sin\frac{n\pi x}{l} e^{-\frac{n^2\pi^2}{l^2}t} \qquad \cdots \qquad (2)$$

Then using the condition $\theta = x$ when $t = 0$, we get

$$x = \sum_{n=1}^{\infty} b_n \sin\frac{n\pi x}{l}$$

which is the Fourier sine series.

$$\therefore b_n = \frac{2}{l}\int_0^l x \sin\frac{n\pi x}{l} dx$$

$$= \frac{2}{l}\left[-x \cdot \frac{l}{n\pi}\cos\frac{n\pi x}{l} + \frac{l^2}{n^2\pi^2} + \sin\frac{n\pi x}{l}\right]_0^l$$

$$= \frac{2}{l}\left[-\frac{l^2}{n\pi}\cos n\pi + 0\right]$$

$$= -\frac{2l}{n\pi}(-1)^n = \frac{2l}{\pi}\frac{(-1)^{n-1}}{n}$$

Thus from (2), we get

$$\theta(x,t) = \frac{2l}{\pi}\sum_{n=1}^{\infty}\frac{(-1)^{n-1}}{n}\sin\frac{n\pi x}{l} e^{-\frac{n^2\pi^2}{l^2}t}.$$

4.1.5. Two Dimensional Heat Flow Equation.

Formulation. Consider the flow of heat by conduction in a plate in xy-plane. When temparature at any point is a function of x, y and time t but is independent of z, the flow is called two dimentional heat flow in xy-plane. Now consider a elementary rectangular portion $PQRS$ of the plate with sides $\delta x, \delta y$ and thickness α as shown in the adjacent fig. Then the quantity of heat entering across the sides PQ and PS in time δt is given by $-k\alpha\delta x\left(\frac{\partial u}{\partial y}\right)_y \delta t$ and $-k\alpha\delta y\left(\frac{\partial u}{\partial x}\right)_y \delta t$ respectively.

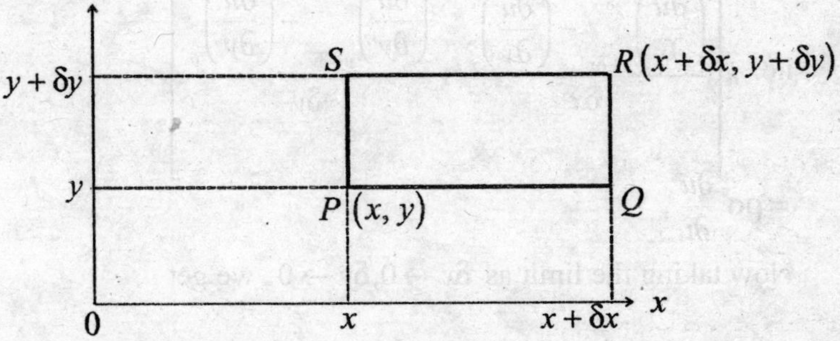

Also the quantity of heat flowing out across the sides SR and QR in time δt is given by

$$-k\alpha\delta x\left(\frac{\partial u}{\partial y}\right)_{y+\partial y} \delta t \text{ and } -k\alpha\delta y\left(\frac{\partial u}{\partial x}\right)_{x+\partial x} \delta t \text{ respectively.}$$

Hence the amount of heat gained by $PQRS$ in time δt is

$$-k\alpha\delta x\delta t\left(\frac{\partial u}{\partial y}\right)_y - k\alpha\delta y\delta t\left(\frac{\partial u}{\partial x}\right)_x + k\alpha\delta x\delta t\left(\frac{\partial u}{\partial x}\right)_{y+\delta y} + k\alpha\delta y\delta t\left(\frac{\partial u}{\partial x}\right)_{x+\delta x}$$

$$= k\alpha\delta x\delta y\delta t\left[\frac{\left(\frac{\partial u}{\partial x}\right)_{x+\delta x} - \left(\frac{\partial u}{\partial x}\right)_x}{\delta x} + \frac{\left(\frac{\partial u}{\partial y}\right)_{y+\delta y} - \left(\frac{\partial u}{\partial y}\right)_y}{\delta y}\right] \quad \ldots \text{(a)}$$

Again the rate of increase of heat in $PQRS$ is

$$(\rho\alpha\delta x\delta y)\sigma\frac{\partial u}{\partial t} \quad \ldots \text{(b)}$$

where σ is the specific heat and ρ in the density of the plate.

From (a) and (b),

$$k\alpha\delta x\delta y\delta t\left[\frac{\left(\frac{\partial u}{\partial x}\right)_{x+\delta x}-\left(\frac{\partial u}{\partial x}\right)_{x}}{\delta x}+\frac{\left(\frac{\partial u}{\partial y}\right)_{y+\delta y}-\left(\frac{\partial u}{\partial y}\right)_{y}}{\delta y}\right]$$

$$=\rho\alpha\delta x\delta y\sigma\frac{\partial u}{\partial t}\cdot\delta t$$

$$\text{i.e. } k\left[\frac{\left(\frac{\partial u}{\partial x}\right)_{x+\delta x}-\left(\frac{\partial u}{\partial x}\right)_{x}}{\delta x}+\frac{\left(\frac{\partial u}{\partial y}\right)_{y+\delta y}-\left(\frac{\partial u}{\partial y}\right)_{y}}{\delta y}\right]$$

$$=\rho\sigma\frac{\partial u}{\partial t}.$$

Now taking the limit as $\delta x \to 0, \delta y \to 0$, we get

$$k\left(\frac{\partial^2 u}{\partial x^2}+\frac{\partial^2 u}{\partial y^2}\right)=\rho\sigma\frac{\partial u}{\partial t}$$

$$\therefore \frac{\partial u}{\partial t}=c^2\left(\frac{\partial^2 u}{\partial x^2}+\frac{\partial^2 u}{\partial y^2}\right) \qquad \ldots (1)$$

where $c^2=\frac{k}{\rho\sigma}$ is the diffusivity of the material of the plate.

which is the **two dimensional heat equation** and gives the temperature distribution of the plate in the transient state.

Note. In the steady state, u is independent of t, so that $\frac{\partial u}{\partial t}=0$ and the above heat equation reduces to

$$\frac{\partial^2 u}{\partial x^2}+\frac{\partial^2 u}{\partial y^2}=0 \text{ (or, } \nabla^2 u=0) \qquad \ldots (2)$$

which is known as **Laplace's Equation** in two dimension.

Solution of Laplace's Equation. [*W. B.U. Tech 2002*]

Let $u(x,y) = X(x)Y(y)$ ... (1)

be the solution of the Laplace's equation

$$\nabla^2 u \equiv \frac{\partial^2 u}{\partial x^2} + \frac{\partial^2 u}{\partial y^2} = 0 \quad \text{... (2)}$$

$\therefore$ From (1), $\dfrac{\partial^2 u}{\partial x^2} = X''Y, \dfrac{\partial^2 u}{\partial y^2} = XY''$.

Substituting in (2), we have

$$X''Y + XY'' = 0$$

i.e. $\dfrac{X''}{X} = -\dfrac{Y''}{Y}$ ... (3)

Since x, y are independent variables, so (3) is true if each side is equal to a constant, say λ

$\therefore \dfrac{X''}{X} = \lambda$ i.e. $X'' - \lambda X = 0$ ... (4)

and $\dfrac{-Y''}{Y} = \lambda$ i.e. $Y'' + \lambda Y = 0$ ... (5)

We consider the following three cases according as λ is zero, $+ve$ or $-ve$:

Case (i). Let $\lambda = 0$.

Then solution of (4) and (5) are

$X = a_1 x + a_2, Y = a_3 y + a_4$.

Case (ii). Let λ be $+ve$, say p^2.

Then (4) and (5) become respectively

$X'' - p^2 X = 0$

$Y'' + p^2 Y = 0$.

Solutions of these differential equations are

$X = b_1 e^{px} + b_2 e^{-px}$

$Y = b_3 \cos py + b_4 \sin py$.

Case (iii). Let λ be $-ve$, say $-p^2$.

Then (4) and (5) become respectively

$X'' + p^2 X = 0, Y'' - p^2 Y = 0$.

Solutions of these equation are

$X = c_1 \cos px + c_2 \sin px, Y = c_3 e^{py} + c_4 e^{-py}$.

Thus the various possible solution of Laplace's equation (1) are

$$u(x,y) = (a_1 x + a_2)(a_3 y + a_4) \quad \text{...} \quad (6)$$

$$u(x,y) = (b_1 e^{px} + b_2 e^{-px})(b_3 \cos py + b_4 \sin py) \quad \text{...} \quad (7)$$

$$u(x,y) = (c_1 \cos px + c_2 \sin px)(c_3 e^{py} + c_4 e^{-py}) \quad \text{...} \quad (8)$$

Of these three solutions, we have to choose that solution which is consistent with the physical nature of the problem and the given boundary conditions.

Illustrative Examples.

Ex. 1. *Find the steady temperature distribution in a thin plate bounded by the lines $x = 0, x = l, y = 0$ and $y = \infty$ assuming that heat cannot escape from either surface; the sides $x = 0, x = l$ being kept at temperature zero. The lower edge $y = 0$ is kept at temperature $f(x)$ and the edge $y = \infty$ at temperature zero.*

The temperature $u(x,y)$ at any point $P(x,y)$ in steady state satisfies the Laplace's equation

$$\frac{\partial^2 u}{\partial x^2} + \frac{\partial^2 u}{\partial y^2} = 0 \quad \text{...} \quad (1)$$

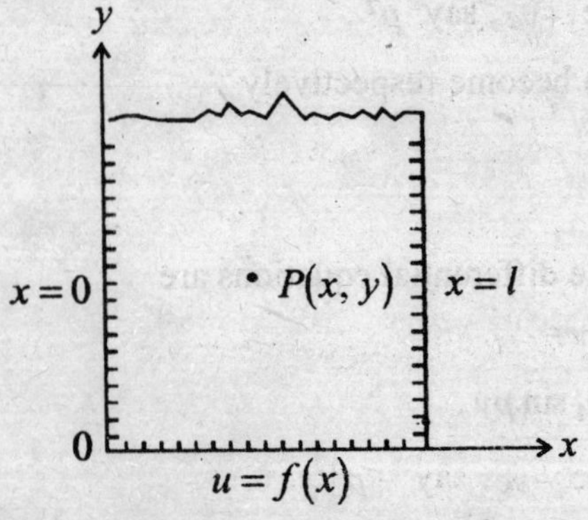

Here the boundary conditions are

$$u(0,y) = u(l,y) = 0 \qquad \cdots \qquad (2)$$

$$u(x,y) \to 0 \text{ as } y \to \infty \qquad \cdots \qquad (3)$$

and $u(x,0) = f(x)$ ... (4)

The only possible solution of (1) is of the form

$$u(x,y) = (c_1 \cos px + c_2 \sin px)(c_3 e^{py} + c_4 e^{-py}) \qquad \cdots \qquad (5)$$

Using the boundary condition (2), we get from (5),

$$c_1(c_3 e^{py} + c_4 e^{-py}) = 0$$

i.e. $c_1 = 0$

and $(c_1 \cos pl + c_2 \sin pl)(c_3 e^{py} + c_4 e^{-py}) = 0$.

i.e. $c_2 \sin pl(c_3 e^{py} + c_4 e^{-py}) = 0 (\because c_1 = 0)$

$\therefore \sin pl = 0 \quad [\because c_2 = 0 \Rightarrow u(x,y) = 0 \; \forall x, y$ which is impossible]

$$\therefore pl = n\pi$$

$$\therefore p = \frac{n\pi}{l}, n = 1,2,3,\cdots$$

Again condition (3) implies $c_3 = 0$.

Thus (5) takes the form

$$u_n(x,y) = b_n \sin\frac{n\pi x}{l} e^{-\frac{n\pi}{l}y}, n = 1,2,\cdots$$

Now we consider the most general solution of (1) as

$$u(x,y) = \sum_{n=1}^{\infty} b_n \sin\frac{n\pi x}{l} e^{-\frac{n\pi y}{l}} \qquad \cdots \qquad (6)$$

Then using the condition (4), we get

$$f(x) = \sum_{n=1}^{\infty} b_n \sin\frac{n\pi x}{l} \qquad \cdots \qquad (7)$$

which is the Fourier sine series and hence

$$b_n = \frac{2}{l}\int_0^l f(x) \sin\frac{n\pi x}{l} dx \qquad \cdots \qquad (8)$$

Thus the required temperature $u(x,y)$ is given by (6) where b_n is given by (8).

Ex. 2. *A long rectangular plate of width l unit with insulated surface has its temperature u equal to zero on both the long sides and one of the short sides so that* $u(0,y)=u(l,y)=u(x,\infty)=0$ *and* $u(x,0)=\lambda x$. *Show that the steady state temperature within the plate is*

$$u(x,y)=\frac{2l\lambda}{\pi}\sum_{n=1}^{\infty}\frac{(-1)^{n+1}}{n}e^{-\frac{n\pi y}{l}}\sin\frac{n\pi x}{l}.$$

Proceeding as Ex. 1 we get

$$u(x,y)=\sum_{n=1}^{\infty}b_n\sin\frac{n\pi x}{l}e^{-\frac{n\pi y}{l}} \quad \ldots (1)$$

where b_n is given by

$$b_n=\frac{2}{l}\int_0^l f(x)\sin\frac{n\pi x}{l}dx \quad \ldots (2)$$

Here $f(x)=\lambda x$.

$$\therefore b_n=\frac{2}{l}\int_0^l \lambda x\sin\frac{n\pi x}{l}dx$$

$$=\frac{2\lambda}{l}\left[\frac{-x\cos\frac{n\pi x}{l}}{\frac{n\pi}{l}}+\frac{\sin\frac{n\pi x}{l}}{\frac{n^2\pi^2}{l^2}}\right]_0^l$$

$$=-\frac{2\lambda}{l}\frac{l}{n\pi}l\cos n\pi$$

$$=-\frac{2\lambda l}{\pi}\frac{(-1)^n}{n}=\frac{2\lambda l}{\pi}\frac{(-1)^{n+1}}{n}.$$

Thus the required steady state temperature within the plate is given by (1) as

$$u(x,y)=\frac{2\lambda l}{\pi}\sum_{n=1}^{\infty}\frac{(-1)^{n+1}}{n}e^{-\frac{n\pi y}{l}}\sin\frac{n\pi x}{l}.$$

Ex. 3. *Solve* $\frac{\partial^2 u}{\partial x^2}+\frac{\partial^2 u}{\partial y^2}=0$, *subject to the conditions*

$$u(0,y)=u(l,y)=u(x,0)=0,\ u(x,a)=\sin\frac{n\pi}{l}x.$$

Here only possible solution of the equation $\frac{\partial^2 u}{\partial x^2}+\frac{\partial^2 u}{\partial y^2}=0$ is

$u(x,y)=(c_1\cos px+c_2\sin px)(c_3e^{py}+c_4e^{-py})$

which can be written in the form

$$u(x,y)=(c_1\cos px+c_2\sin px)(c_5\cosh py+c_6\sinh py) \quad \ldots \quad (1)$$

Using the condition $u(0,y)=u(l,y)=0$, we get from (1)

$c_1(c_5\cosh py+c_6\sinh py)=0$

i.e. $c_1=0$

and $(c_1\cos pl+c_2\sin pl)(c_5\cosh py+c_6\sinh py)=0$

i.e. $c_2\sin pl(c_5\cosh py+c_6\sinh py)=0$

i.e. $\sin pl=0$ [$\because c_2=0\Rightarrow u(x,y)=0\ \forall\ x,y$ which is impossible].

$\therefore pl=n\pi$

$\therefore p=\frac{n\pi}{l}, n=1,2,3,\cdots$

Again using the condition $u(x,0)=0$, we get from (1),

$(c_1\cos px+c_2\sin px)c_5=0$.

$\therefore c_5=0$.

Thus (1) takes the form

$u_n(x,y)=b_n\sin\frac{n\pi x}{l}\sinh\frac{n\pi y}{l}, n=1,2,\cdots$ where $b_n=c_2c_6$.

Now we consider the most general solution as

$$u(x,y)=\sum_{n=1}^{\infty}b_n\sin\frac{n\pi x}{l}\sinh\frac{n\pi y}{l} \quad (2)$$

Then using the condition $u(x,a)=\sin\frac{n\pi}{l}x$ we get

$$\sin\frac{n\pi}{l}x=\sum_{n=1}^{\infty}b_n\sin\frac{n\pi x}{l}\sinh\frac{n\pi a}{l}$$

which is the Fourier sine series

$$\therefore b_n\sinh\frac{n\pi a}{l}=\frac{2}{l}\int_0^l\sin^2\frac{n\pi}{l}x\,dx=\frac{1}{l}\int_0^l\left(1-\cos\frac{2n\pi x}{l}\right)dx$$

$$=\frac{1}{l}\left[x-\frac{l}{2n\pi}\sin\frac{2n\pi x}{l}\right]_0^l=1. \quad \therefore b_n=\frac{1}{\sinh\frac{n\pi a}{l}}.$$

Thus the required solution is

$$u(x,y)=\sum_{n=1}^{\infty}\frac{\sin\frac{n\pi x}{l}\sinh\frac{n\pi y}{l}}{\sinh\frac{n\pi a}{l}}.$$

EXERCISES

1. Solve the following equations by the method of separation of variables :

(i) $3\frac{\partial u}{\partial x}+2\frac{\partial u}{\partial y}=0$ where $u(x,0)=4e^{-x}$.

(ii) $4\frac{\partial u}{\partial x}+\frac{\partial u}{\partial y}=3u$, given $u=3e^{-y}-e^{-5y}$ when $x=0$.

(iii) $\frac{\partial u}{\partial x}=2\frac{\partial u}{\partial t}+u$ given $u(x,0)=6e^{-3x}$.

(iv) $\frac{\partial u}{\partial t}=\frac{\partial^2 u}{\partial x^2}, u(0,t)=0, u(4,t)=0, u(x,0)=6\sin\frac{\pi x}{2}+3\sin\pi x$.

2. Solve the equation $\frac{\partial^2 u}{\partial x^2}-2\frac{\partial u}{\partial x}+\frac{\partial u}{\partial y}=0$ by method of separation of variables.

3. Solve by method of separation of variables $\frac{\partial u}{\partial x}+\frac{\partial u}{\partial y}-2(x+y)u=0$.

4. (a) Solve $\frac{\partial^2 u}{\partial t^2}=Ax+\frac{\partial^2 u}{\partial x^2}, 0<x<1, t>0$; given that

$$u(0,t)=u(1,t)=u(x,0)=\frac{\partial u}{\partial t}(x,0)=0.$$

(b) Solve $\frac{\partial^2 u}{\partial x^2}=\frac{\partial u}{\partial y}+2u$ where $u=0, \frac{\partial u}{\partial x}=1+e^{-3y}$ when $x=0$ for all values of y.

5. Solve $\frac{\partial^2 z}{\partial x^2}+\frac{\partial^2 z}{\partial y^2}=0$ by method of separation of variables, given $z(x,\pi/2)=0$, $z(x,0)=4e^{-3x}$.

6. A string is stretched and fastened to two points. Motion is started by displacing the string in the form $y = a\sin\frac{\pi x}{l}$ from which it is released at time $t = 0$. Show that the displacement of any point at a distance x from one end at time t is given by $y(x,t) = a\sin\frac{\pi x}{l}\cos\frac{\pi ct}{l}$.

7. Solve the equation $\frac{\partial^2 u}{\partial t^2} = 4\frac{\partial^2 u}{\partial x^2}$ given $u(0,t) = u(5,t) = 0, u(x,0) = 0$, and $\left(\frac{\partial u}{\partial t}\right)_{t=0} = 5\sin\pi x$.

8. Solve $\frac{\partial^2 y}{\partial t^2} = a^2\frac{\partial^2 y}{\partial x^2}$ under the boundary conditions $y(0,t) = 0, t > 0; y(l,t) = 0, t > 0$ and the initial conditions $y(x,0) = \lambda(lx - x^2), 0 < x < l$ and $\left(\frac{\partial y}{\partial t}\right)_{t=0} = 0, 0 < x < l$, l being a suitable constant.

9. Solve $\frac{\partial^2 u}{\partial t^2} = c^2\frac{\partial^2 u}{\partial x^2}$ where $u(0,t) = u(l,t) = 0$

$u(x,0) = A\sin\pi x$ and $\left(\frac{\partial u}{\partial t}\right)_{t=0} = 0$.

10. A string is stretched between the fixed points $(0,0)$ and $(l,0)$ and released at rest from the initial deflection given by

$$f(x) = \begin{cases} \frac{2k}{l}x, & 0 < x < l/2 \\ \frac{2k}{l}(l - x), & \frac{l}{2} < x < l. \end{cases}$$

Find the deflection of the string at any time t.

11. Solve the equation $\frac{\partial u}{\partial t} = \frac{\partial^2 u}{\partial x^2}$ with boundary conditions $u(x,0) = 3\sin n\pi x$, $u(0,t) = u(1,t) = 0$ where $0 < x < 1, t > 0$.

12. Determine the solution of one-dimensional heat equation $\frac{\partial u}{\partial t} = c^2\frac{\partial^2 u}{\partial x^2}$ where the boundary conditions are $u(0,t) = 0, u(l,t) = 0 (t > 0)$ and the initial condition $u(x,0) = x$, l being the length of the bar.

13. Find the solution of $\frac{\partial u}{\partial t} = k\frac{\partial^2 u}{\partial x^2}$ having given that $u = u_0 \sin nt$ when $x = 0$ for all values of t and $u = 0$ when x is very large.

14. Show that the solution of $\frac{\partial u}{\partial t} = k\frac{\partial^2 u}{\partial x^2}$ subject to the conditions

(i) u is not infinite for $t \to \infty$

(ii) $\frac{\partial u}{\partial x} = 0$ for $x = 0$ and $x = l$

(iii) $u = lx - x^2$ for $t = 0$ between $x = 0$ and $x = l$ is

$$u = \frac{1}{6}l^2 - \frac{l^2}{\pi^2}\sum_{n=1}^{\infty}\frac{1}{n^2}\cos\frac{n\pi x}{l}e^{-\frac{4\pi^2 n^2 kt}{l}}.$$

15. Solve $\frac{\partial u}{\partial t} = c^2\frac{\partial^2 u}{\partial x^2}$ for $0 < x < \pi$, when $u_x(0,t) = u_x(\pi,t) = 0$ and $u(x,0) = \sin x$.

16. Solve the equation $\frac{\partial u}{\partial t} = c^2\frac{\partial^2 u}{\partial x^2}$, $0 < x < l$ given that $u_x(0,t) = 0, u_x(l,t) = 0, u(x,0) = x$.

17. Solve $\frac{\partial \theta}{\partial t} = k\frac{\partial^2 \theta}{\partial x^2}$ such that (i) θ is finite when $t \to \infty$ (ii) $\frac{\partial \theta}{\partial x} = 0$ when $x = 0$ and $\theta = 0$ when $x = l$ for all t (iii) $\theta = \theta_0$ when $t = 0$ for all values of x.

18. Find the temperature $u(x,t)$ in a bar of length l which is perfectly insulated laterally whose ends are kept at temperature zero and the initial temperature is $f(x) = \begin{cases} x, & 0 < x < l/2 \\ l - x, & l/2 < x < l. \end{cases}$

19. Solve the one dimensional heat equation $\frac{\partial^2 \theta}{\partial x^2} = \frac{1}{k}\frac{\partial \theta}{\partial t}$ in $0 \le x \le \pi, t \ge 0$ when (a) θ remains finite as $t \to \infty$ (b) $\theta = 0$ if $x = 0$ or π for all values of t (c) at $t = 0$, $\theta = \begin{cases} x, & 0 \le x \le \pi/2 \\ \pi - x, & \pi/2 \le x \le \pi. \end{cases}$

20. A bar with insulated sides is initially at temperature $0^{o}c$ throughout. The end $x=0$ is kept at $0^{o}c$ and heat is suddenly applied at the end $x=l$ so that $\frac{\partial\theta}{\partial x}=\theta_0$ for $x=l$ where θ_0 is a constant. Find the temperature function $\theta(x,t)$.

21. Solve $\frac{\partial^2 u}{\partial x^2}+\frac{\partial^2 u}{\partial y^2}=0$ for $0<x<\pi, 0<y<\pi$, given that

$$u(0,y)=u(\pi,y)=u(x,\pi)=0, u(x,0)=\sin^2 x.$$

22. Solve $\frac{\partial^2 u}{\partial x^2}+\frac{\partial^2 u}{\partial y^2}=0$ within the rectangle $0\le x\le a, 0\le y\le b$ given that $u(0,y)=u(a,y)=u(x,b)=0$, and $u(x,0)=x(a-x)$.

23. Evaluate the steady temperature in a rectangular plate of length a and width b, the side of which are kept at temperature zero, the lower end is kept at temperature $f(x)$ and the upper edge is kept insulated

24. An infinitely long plane uniform plate is bounded by two parallel edges and an end at right angles to them. The breadth is π; this end is maintained at a temperature u_0 at all points and other edges are at zero temperature. Determine the temperature at any point of the plate in the steady state.

25. Solve Laplace's equation $\frac{\partial^2 V}{\partial x^2}+\frac{\partial^2 V}{\partial y^2}=0$ subject to the following boundary conditions : $V=0$ when y is very large for all x; $V=0$ when $x=0$ and 1 for all y; and $V=x(1-x)$ when $y=0$ for $0<x<1$.

ANSWERS

1. (i) $u=3e^{-5x-3y}+2e^{-3x-2y}$. (ii) $u=3e^{x-y}-e^{2x-5y}$.

(iii) $u=6e^{-(3x+2t)}$. (iv) $u=6e^{-\frac{\pi^2 t}{4}}\sin\frac{1}{2}\pi x-3e^{-\pi^2 t}\sin\pi x$.

2. $u=\left[Ae^{\left(1+\sqrt{1+k}\right)x}+Be^{\left(1-\sqrt{1+k}\right)x}\right]e^{-ky}$.

3. $u=Ae^{\left[x^2+y^2+k(x-y)\right]}$.

4. (b) $=\frac{1}{\sqrt{2}}\sinh x\sqrt{2}+e^{-3y}\sin x$.

5. $z = 4e^{-3x}\cos 3y$.

7. $u = \dfrac{5}{2\pi}\sin \pi x \sin 2\pi t$.

8. $y = \dfrac{8\lambda l^2}{\pi^3}\sum_{n=1}^{\infty}\dfrac{1}{(2m-1)^2}\sin\dfrac{(2m-1)\pi x}{l}\cos\dfrac{(2m-1)\pi ct}{l}$.

9. $u(x,t) = A\cos(\pi ct)\sin \pi x$

10. $y(x,t) = \dfrac{8k}{\pi^2}\sum_{n=1}^{\infty}\dfrac{1}{n^2}\sin\dfrac{n\pi}{2}\cos\dfrac{n\pi ct}{l}\sin\dfrac{n\pi x}{l}$.

11. $u(x,t) = 3\sum_{n=1}^{\infty}\sin n\pi x\, e^{-n^2\pi^2 t}$.

12. $u(x,t) = \dfrac{-2l}{\pi}\sum_{n=1}^{\infty}\dfrac{\cos n\pi}{n}\sin\dfrac{n\pi x}{l}\times e^{-\frac{c^2n^2\pi^2 t}{l^2}}$.

13. $u = u_0 e^{-\sqrt{\frac{n}{2k}}x}\cdot\sin\left(nt - \sqrt{\dfrac{n}{2k}}x\right)$.

15. $u(x,t) = \dfrac{2}{\pi} - \dfrac{4}{\pi}\sum_{n=1}^{\infty}\dfrac{1}{4m^2-1}e^{-4m^2c^2t}\cos 2mx$

17. $\theta = \dfrac{4\theta_0}{\pi}\sum_{n=1}^{\infty}\dfrac{(-1)^{m-1}}{2m-1}e^{-\frac{(2m-1)^2\pi^2}{4l^2}kt}\cos\dfrac{(2m-1)\pi x}{2l}$

18. $u(x,t) = \dfrac{4l}{\pi^2}\sum_{n=1}^{\infty}\dfrac{1}{n}\sin\dfrac{n\pi}{2}\sin\dfrac{n\pi x}{l}e^{-\frac{n^2\pi^2 k}{l^2}}$

19. $\theta(x,t) = \dfrac{4}{\pi}\sum_{m=1}^{\infty}\dfrac{(-1)^m}{(2m+1)^2}\sin(2m+1)xe^{-(2m+1)^2 k\pi t}$

20. $\theta(x,t) = \theta_0 x + \dfrac{8\theta_0 l}{\pi^2}\sum_{m=1}^{\infty}\dfrac{(-1)^m}{(2m-1)^2}\sin\dfrac{(2m-1)\pi x}{l}e^{-\frac{(2m-1)^2c^2\pi^2 t}{4l^2}}$

21. $u(x,y) = -\dfrac{8}{\pi}\sum_{n=1,3,5,\ldots}\dfrac{\sin nx\sinh n(\pi - y)}{n(n^2-4)\sinh n\pi}$

22. $u(x,y)=\dfrac{8a^2}{\pi^3}\sum_{n=0}^{\infty}\dfrac{1}{(2n+1)^3}\sin\dfrac{(2n+1)\pi x}{a}\dfrac{\sinh\dfrac{(2n+1)\pi}{a}(b-y)}{\sin h\dfrac{(2n+1)\pi b}{a}}.$

23. $u(x,y)=\sum_{n=1}^{\infty}b_n\dfrac{\sin\dfrac{n\pi x}{a}\cosh\dfrac{n\pi}{a}(b-y)}{\cosh\dfrac{n\pi b}{a}}$ where $E_n=\dfrac{2}{a}\int_0^a f(x)\sin\dfrac{n\pi x}{a}dx$.

24. $u(x,y)=\dfrac{4u_0}{\pi}\sum_{m=1}^{\infty}\dfrac{1}{2m-1}e^{-(2m-1)y}\sin(2m-1)x$.

25. $V(x,y)=\dfrac{8}{\pi^3}\sum_{m=1}^{\infty}\dfrac{1}{(2m-1)^3}e^{-(2m-1)\pi y}\sin(2m-1)\pi x$.

4.2 Solution of PDF By Integral Transform

4.2.1. Introduction :

In various physical and geometrical problems functions of two or more independent variables are involved and so partial differential equations arise in connection with these. In this chapter we solve three types of important partial differential equation occuring in engineering field by using Laplace transform and Fourier transform.

4.2.2. One Dimensional Wave Equation

Solution of one dimensional wave equation by Laplace transform.

Ex. 1. Solve the wave equation

$$\frac{\partial^2 u}{\partial t^2}=c^2\frac{\partial^2 u}{\partial x^2}\quad (x>0, t>0)$$

where $u(x,0)=0,\ \frac{\partial u}{\partial t}(x,0)=0, x>0$

$$u(0,t)=F(t),\ u(\infty,t)=0, t\geq 0$$

Solution. Taking Laplace transform of both sides of the equation and then using the boundary conditions we have

$$L\left\{\frac{\partial^2 u}{\partial t^2}\right\}=c^2 L\left\{\frac{\partial^2 u}{\partial x^2}\right\}$$

or, $s^2\bar{u}(x,s)-s\,u(x,0)-u_t(x,0)=c^2\frac{d^2\bar{u}}{dx^2}$,where is $\bar{u}(x,s)$ in the Laplace transform of $u(x,t)$

or, $s^2\bar{u}(x,s)-s.0-0=c^2\frac{d^2\bar{u}}{dx^2}$

$$\therefore \frac{d^2\bar{u}}{dx^2}-\frac{s^2}{c^2}\bar{u}=0 \quad \cdots \quad (1)$$

Also $\bar{u}(0,s)=\int_0^\infty F(t)\ e^{-st}dt$

$$=\bar{F}(s) \quad \cdots \quad (2)$$

and $\bar{u}(s)=0 \quad \cdots \quad (3)$

Now, solution of (1) is given by

$$\bar{u} = c_1 e^{\frac{sx}{a}} + c_2 e^{-\frac{sx}{a}}$$

Using (2), (3) we have

$$c_1 = 0, c_2 = \bar{F}(s)$$

Hence $\bar{u} = \bar{F}(s)\, e^{-\frac{sx}{a}}$

$$\therefore u(x,t) = L^{-1}\left\{\bar{F}(s) e^{-\frac{sx}{a}}\right\}$$

$$= \begin{cases} F\left(t - \dfrac{x}{a}\right), & t > \dfrac{x}{a} \\ 0 & t < \dfrac{x}{a} \end{cases} \qquad \left[\because L^{-1}\{\bar{F}(s)\} = F(t)\right]$$

$\therefore u(x,t) = F\left(t - \dfrac{x}{a}\right) H\left(t - \dfrac{x}{a}\right)$ where H is the unit step Heaviside function.

Ex 2. A string is stretched between two fixed points (0, 0) and (l, 0). If it is displaced into the curve $y = b\sin\left(\frac{\pi x}{l}\right)$ and released from rest in that position at time $t = 0$, find its displacement at any time $t > 0$ and any point $0 < x < l$.

Solution. Here the displacement $y(x,t)$ of the string is governed by the p.d. equation

$$\frac{\partial^2 y}{\partial t^2} = c^2 \frac{\partial^2 y}{\partial x^2} \qquad \ldots (1)$$

with boundary conditions $y(0,t) = 0, y(l,t) = 0$

and initial condition $y(x,0) = b\sin\dfrac{\pi x}{l}$ and $y_t(x,0) = 0$

Taking the Laplace transforms of (1) and the boundary conditions, we have

$$L\left\{\frac{\partial^2 y}{\partial t^2}\right\} = c^2 L\left\{\frac{\partial^2 y}{\partial x^2}\right\}$$

or, $s^2\bar{y}(x,s)-s\,y(x,0)-y_t(x,0)=c^2\dfrac{d^2\bar{y}}{dx^2}$

or, $s^2\bar{y}-s.b\sin\dfrac{\pi x}{l}-0=c^2\dfrac{d^2\bar{y}}{dx^2}$, where $\bar{y}=L(y)$

$$\therefore \frac{d^2\bar{y}}{dx^2}-\frac{s^2}{c^2}\bar{y}=-\frac{bs}{c^2}\sin\frac{\pi x}{l} \qquad \ldots (2)$$

Also $\bar{y}(0,s)=0$ and $\bar{y}(l,s)=0$

Now solution of (2) is given by

$$\bar{y}=c_1e^{\frac{sx}{c}}+c_2e^{-\frac{sx}{c}}+\frac{bs\sin\frac{\pi x}{l}}{s^2+\frac{\pi^2c^2}{l^2}}$$

$\therefore \bar{y}(0,s)=0$ gives $c_1+c_2=0$

and $\bar{y}(l,s)=0$ gives $c_1e^{\frac{sl}{c}}+c_2e^{-\frac{sl}{c}}=0$

Solving we get

$$c_1=c_2=0$$

Hence $\bar{y}=\dfrac{bs\sin\frac{\pi x}{l}}{s^2+\frac{\pi^2c^2}{l^2}}$

$$\therefore y(x,t)=b\sin\frac{\pi x}{l}\,L^{-1}\left\{\frac{s}{s^2+(\frac{\pi l^2}{l})^2}\right\}$$

$$=b\sin\frac{\pi x}{l}\cos\frac{\pi ct}{l}$$

Ex 3. An infinite long string having one end $x=0$ is initially at rest on the x-axis. The end $x=0$ undergoes a periodic transverse displacement given by a $\sin nt, t>0$, find the displacement of any point on the string at $t>0$

Solution. Here the displacement of any point of the string is governed by the equation

$$\frac{\partial^2 u}{\partial t^2} = c^2 \frac{\partial^2 u}{\partial x^2}, x > 0, t > 0 \qquad \dots \qquad (1)$$

with the boundary and initial conditions

$$u(0,t) = a \sin nt, t > 0$$

$$u(x,0) = 0, u_t(x,0) = 0, x \geq 0$$

and the displacement is finite.

Taking the Laplace transform of (1) and the boundary conditions, we have

$$L\left\{\frac{\partial^2 u}{\partial t^2}\right\} = c^2 L\left\{\frac{\partial^2 u}{\partial x^2}\right\}$$

or, $s^2\bar{u}(x,s) - su(x,0) - u_t(x_1,0) = c^2 \dfrac{d^2\bar{u}}{dx^2}$

or, $s^2\bar{u} - s.0 - 0 = c^2 \dfrac{d^2\bar{u}}{dx^2}$

$$\therefore \quad \frac{d^2\bar{u}}{dx^2} - \frac{s^2}{c^2}\bar{u} = 0$$

Also $\bar{u}(0,s) = aL(\sin nt) = \dfrac{an}{s^2 + n^2}$

and $\bar{u}(x,s)$ is finite for $x \geq 0$

Now solution of (2) is given by

$$\bar{u}(x,s) = c_1 e^{\frac{sx}{c}} + c_2 e^{-\frac{sx}{c}}$$

Since $\bar{u}(x,s)$ is finite, so we must have $c_1 = 0$

$$\therefore \ \bar{u} = c_2 e^{-\frac{sx}{c}}$$

Now $\bar{u}(0,s) = \dfrac{an}{s^2 + n^2}$ gives

$$c_2 = \frac{an}{s^2 + n^2}$$

$$\therefore\ \bar{u} = \frac{an}{s^2+n^2} e^{-\frac{sx}{c}}$$

$$\therefore\ u(x,t) = anL^{-1}\left\{\frac{1}{s^2+n^2} e^{-\frac{sx}{c}}\right\}$$

$$= \begin{cases} a \sin x\left(t-\dfrac{x}{c}\right), & t > \dfrac{x}{c} \\ 0, & t > \dfrac{x}{c} \end{cases} \qquad \left[\because L^{-1}\left\{\frac{1}{s^2+n^2}\right\} = \frac{\sin nt}{n}\right.$$

Solution of one dimensional wave equation by Fourier transform.

Ex. 4. Consider the equation $\dfrac{\partial^2 u}{\partial t^2} = c^2 \dfrac{\partial^2 u}{\partial x^2}$ subject to the conditions $u(x,0) = f(x)$

$u_t(x,0) = g(x)$, $-\infty < x < \infty$, $0 < t < \infty$.

Taking Fourier Transform on both side of the partial differential equation

$$\frac{\partial^2 u}{\partial t^2} = c^2 \frac{\partial^2 u}{\partial x^2} \quad \text{we have } \mathcal{F}\left(\frac{\partial^2 u}{\partial t^2}\right) = \mathcal{F}\left(c^2 \frac{\partial^2 u}{\partial x^2}\right)$$

$$\text{or, } \int_{-\infty}^{\infty} \frac{\partial^2 u}{\partial t^2} e^{isx} dx = c^2\, \mathcal{F}\left(\frac{\partial^2 u}{\partial x^2}\right) = c^2(-is)^2\, \mathcal{F}(u)$$

$$\text{or, } \frac{d^2}{dt^2}\left(\int_{-\infty}^{\infty} u e^{isx} dx\right) = -c^2 s^2\, \mathcal{F}(u)$$

$$\text{or, } \frac{d^2 U(s)}{dt^2} + c^2 s^2 U = 0$$

[where $U = \mathcal{F}(u)$ is Fourier Transform of $u(x,t)$].

This is an ordinary second order homogeneous differential equation with constant co-efficient. Its auxilliary equation is $m^2 + c^2 s^2 = 0$ $\therefore m = \pm ics$. We suppose $s \neq 0$.

$\therefore$ its general solution is $U(s,t) = A\cos sct + B\sin sct$ ... (1)

where A and B are arbitrary constants. We are going to find these.

Taking Fourier transform on the condition $u(x,0)=f(x)$ we have

$$\mathcal{F}(u)=\mathcal{F}(f(x))$$

or, $U(s)=F(s)$ where $F(s)$ is Fourier transform of $f(x)$.

From (1) we get $U(s,0)=A\times 1+B\times 0=A$

so from above, $F(s)=A$ at $t=0$.

Again taking Fourier transform on the condition $u_t(x,0)=g(x)$ we get

$$\mathcal{F}\left(\frac{\partial u}{\partial t}\right)=\mathcal{F}(g(x))$$

or, $\int_{-\infty}^{\infty}\frac{\partial u}{\partial t}e^{isx}dx=G(s)$ where $G(s)$ is Fourier transform of $g(x)$.

or, $\frac{d}{dt}\left(\int_{-\infty}^{\infty}ue^{isx}dx\right)=G(s)$ or, $\frac{dU}{dt}=G(s)$

From (1) we get

$$\frac{\partial U}{\partial t}=-scA\sin sct+scB\cos sct$$

or, $U_t(s,t)=-scA\sin sct+scB\cos sct$

or, $U_t(s,0)=0+scB\times 1$

or, $G(s)=scB$ at $t=0$ (from above)

or, $B=\frac{G(s)}{sc}$

So, from (1) we get $U(s,t)=F(s)\cos sct+\frac{G(s)}{sc}\sin sct$

where, $U(s)$, $F(s)$, $G(s)$ are Fourier transforms of $u(x)$, $f(x)$ and $g(x)$ respectively.

$$\therefore\ u(x,t)=\mathcal{F}^{-1}\{F(s)\cos sct\}+\mathcal{F}^{-1}\left\{\frac{G(s)}{sc}\sin sct\right\} \quad \dots \quad \dots \quad (2)$$

By shifting property we get

$$\mathcal{F}\{f(x-ct)\}=e^{icts}F(s)$$

and $\mathcal{F}\{f(x+ct)\} = e^{-icts}F(s)$

$$\therefore\ \mathcal{F}\{f(x-ct)+f(x+ct)\} = \left(e^{icts}+e^{-icts}\right)F(s) = 2\cos cts\,F(s)$$

or, $\mathcal{F}\left\{\dfrac{f(x-ct)+f(x+ct)}{2}\right\} = \cos cts\,F(s)$

$$\therefore\ \mathcal{F}^{-1}(\cos cts\,F(s)) = \frac{f(x-ct)+f(x+ct)}{2}.$$

Now, since $s \neq 0$, $\mathcal{F}(f(x)) = \dfrac{2\sin sct}{s}$

where $f(x) = 1,\ |x| < ct$

$= 0,\ |x| > ct$

i.e. $\mathcal{F}\left(\dfrac{1}{2c}f(x)\right) = \dfrac{\sin sct}{sc}$

or, $\mathcal{F}^{-1}\left(\dfrac{\sin sct}{sc}\right) = \dfrac{1}{2c}f(x)$

By Convolution Theorem,

$$\mathcal{F}^{-1}\left\{\frac{G(s)}{sc}\sin sct\right\} = \mathcal{F}^{-1}\left\{\frac{\sin sct}{sc}G(s)\right\}.$$

$= \left\{\dfrac{1}{2c}f(x)\right\} * g(x)$ $[\mathcal{F}(g(x)) = G(s)]$

$= \displaystyle\int_{-\infty}^{\infty}\frac{1}{2c}f(u)g(x-u)du$ [by definition of convolution]

$= \dfrac{1}{2c}\displaystyle\int_{-ct}^{ct}1\cdot g(x-u)du = \dfrac{1}{2c}\int_{-ct}^{ct}g(z)dz$, putting $z = x-u$

$= \dfrac{1}{2c}\displaystyle\int_{-ct}^{ct}g(x)dx.$

So, from (2) we get $u(x,t) = \dfrac{f(x-ct)+f(x+ct)}{2} + \dfrac{1}{2c}\displaystyle\int_{-ct}^{ct}g(x)dx.$

Ex. 5. *Use the method of Fourier transform to determine the displacement $u(x,t)$ of an infinite string given that the string is initially at rest and that the initial displacement is $f(x)$, $-\infty < x < \infty$.*

We know, u, x, t are related by $\frac{\partial^2 u}{\partial t^2} = c^2 \frac{\partial^2 u}{\partial x^2}$; $u(x,0) = f(x)$, and the velocity $u_t(x,0) = 0$. So in the above article if we put $g(x) = 0$ we get the result which is

$$u(x,t) = \frac{f(x-ct) + f(x+ct)}{2} + \frac{1}{2c}\int_{-ct}^{ct} 0\,dx = \frac{f(x-ct) + f(x+ct)}{2}.$$

4.2.3. One Dimensional Heat Conduction Equation

Solution of one dimensional heat equation by Laplace transform

Ex 1. Find the solution of the equation

$$\frac{\partial u}{\partial t} = k\frac{\partial^2 u}{\partial x^2}$$

which tends to zero as $x \to \infty$ and which satisfies the conditions

$$u = f(t) \text{ when } x = 0, t > 0$$

and $\quad u = 0$ when $x > 0, t = 0$

Solution. Taking the Laplace transform of both sides of the given equation, we have

$$L\left\{\frac{\partial u}{\partial t}\right\} = k\,L\left\{\frac{\partial^2 u}{\partial x^2}\right\}$$

or, $s\,\bar{u}(x,s) - u(x,0) = k\,\frac{d^2\bar{u}}{dx^2}$

$\therefore \frac{d^2\bar{u}}{dx^2} - \frac{s}{k}\bar{u} = 0$

whose solution is

$$\bar{u} = c_1 e^{\sqrt{\frac{s}{k}}x} + c_2 e^{-\sqrt{\frac{s}{k}}x} \quad \ldots \quad (1)$$

Since $u \to 0$ as $x \to \infty$, so $\bar{u} \to 0$ as $x \to \infty$

$\therefore$ From (1), we must have $c_1 = 0$

$$\therefore \bar{u} = c_2 e^{-\sqrt{\frac{s}{k}}x} \quad \ldots \quad (2)$$

Again, when $x = 0$, $\bar{u} = L\{f(t)\} = \bar{f}(s)$, say

$\therefore$ From (2), we have

$$c_2 = \bar{f}(s)$$

Hence, $\bar{u} = \bar{f}(s)e^{-\sqrt{\frac{s}{k}}x}$

Applying Laplace inversion theorem, we have

$$u(x,t) = \frac{1}{2\pi i}\int_{v-i\infty}^{v+i\infty} e^{st}\,\bar{f}(s)e^{-\sqrt{\frac{s}{k}}x}\,ds$$

Ex 2. A sem-infinite solid $x > 0$ is initially at temperature zero. At time $t > 0$, a constant temperature $V_0 > 0$ is applied and maintained at the face $x = 0$. Find the temperature at any point of the solid at any time $t > 0$.

Solution : Here the temperature $u(x,t)$ at any point of the solid at any time $t > 0$ is governed by one dimensional heat equation

$$\frac{\partial u}{\partial t} = c^2\frac{\partial^2 u}{\partial x^2}, (x > 0, t > 0) \qquad (1)$$

with the boundary and initial conditions

$$u(0,t) = V_0,\ u(x,0) = 0$$

Taking the Laplace transform of both sides of the equation (1), we have

$$L\left(\frac{\partial u}{\partial t}\right) = c^2 L\left\{\frac{\partial^2 u}{\partial x^2}\right\}$$

or, $s\,\bar{u}(x,s) - u(x,0) = c^2\dfrac{d^2\bar{u}}{dx^2}$

$$\therefore \frac{d^2\bar{u}}{dx^2} - \frac{s}{c^2}\bar{u} = 0$$

whose solution is

$$\bar{u} = c_1 e^{\sqrt{s/c^2}\,x} + c_2 e^{-\sqrt{s/c^2}\,x} \qquad \ldots (2)$$

Since $u(x,t)$ is finite when $x \to \infty$, so $\bar{u}(x,s)$ is also finite when $x \to \infty$

$\therefore$ From (2), we must have $c_1 = 0$

$$\therefore \bar{u} = c_2 e^{-\sqrt{\frac{s}{c^2}}x} \quad \ldots \quad (3)$$

Also we have

$u(0,t) = V_0$

$$\therefore \bar{u}(0,s) = L(V_0) = V_0 L\{1\} = \frac{V_0}{s}$$

$\therefore$ From (3), $c_2 = \dfrac{V_0}{s}$.

Hence $\bar{u}(x,s) = \dfrac{V_0}{s} e^{-\sqrt{s/c^2}\,x}$

$$\therefore u(x,t) = V_0 L^{-1}\left\{\frac{1}{s} e^{-\sqrt{s/c^2}\,x}\right\}$$

$$= V_0 \, erfc\left(\frac{x}{2c\sqrt{t}}\right)$$

$$\left[\because L^{-1}\left\{\frac{e^{-\lambda\sqrt{s}}}{s}\right\} = erfc\left(\frac{\lambda}{2\sqrt{t}}\right)\right]$$

Ex 3. The initial temperature of a slab of homogeneous material bounded by the planes $x = 0$ and $x = l$ is u_0. Find the temperature in this solid after the $x = 0$ is insulated and the temperature of face $x = l$ is reduced to zero.

Solution. The temperature $u(x,t)$ in the solid is governed by the heat equation

$$\frac{\partial u}{\partial t} = c^2 \frac{\partial^2 u}{\partial x^2}, \; (0 < x < l, t > 0) \quad \ldots \quad (1)$$

with the boundary condition $u_x(0,t) = 0$, $u(l,t) = 0 (t > 0)$ and initial condition $u(x,0) = u_0 (0 < x < l)$

Taking the Laplace transform of the equation (1) and the boundary conditions, we have

$$L\left\{\frac{\partial u}{dt}\right\}=c^2 L\left\{\frac{\partial^2 u}{\partial x^2}\right\}$$

or, $s\,\bar{u}(x,s)-u(x,0)=c^2\dfrac{d^2\bar{u}}{dx^2}$

$$\therefore \frac{d^2\bar{u}}{dx^2}-\frac{s}{c^2}\bar{u}=-\frac{x_0}{c^2} \qquad \ldots \qquad (2)$$

The boundary conditions $u_x(0,t)=0, u(l,t)=0$ give

$$\frac{d\bar{u}}{dx}(0,s)=0 \qquad \ldots \qquad (3)$$

and $\bar{u}(l,s)=0$ ... (4)

The general solution of (2) is given by

$$\bar{u}(x,s)=c_1\cos h\, x\sqrt{\frac{s}{c^2}}+c_2\sin h\, x\sqrt{\frac{s}{c^2}}+\frac{x_0}{s}$$

Using (3) and (4), we get

$$c_2=0$$

$$c_1=-\frac{x_0}{s\cos h\, l\sqrt{p/c^2}}$$

Hence, $\bar{u}(x,s)=\dfrac{u_0}{s}-\dfrac{u_0}{s}\dfrac{\cos h\, x\sqrt{p/c^2}}{\cos h\, l\sqrt{p/c^2}}$

$$=\frac{u_0}{s}-\frac{u_0}{s}\frac{\cos h\, xq}{\cos h\, lq} \quad \text{where } q=\sqrt{s/c^2}$$

$$=\frac{u_0}{s}-\frac{u_0}{s}\frac{e^{xq}+e^{-xq}}{e^{lq}+e^{-lq}}$$

$$= \frac{u_0}{s} - \frac{u_0}{s} e^{-lq}\left(e^{xq} + e^{-xq}\right)\left(1 + e^{-2lq}\right)^{-1}$$

$$= \frac{u_0}{s} - \frac{u_0}{s} \left\{e^{-(x+l)q} + e^{-(-x+l)q}\right\}\sum_{n=0}^{\infty}(-1)^n e^{-2nlq}$$

$$= \frac{u_0}{s} - \frac{u_0}{s} \sum_{n=0}^{\infty}(-1)^n\left[e^{-q(ml+n)} + e^{-q(ml-n)}\right] \text{ where } m = 2n+1$$

$$\therefore u(n,t) = u_0 L^{-1}\left(\frac{1}{s}\right) - u_0 \sum_{n=0}^{\infty}(-1)^n\left[L^{-1}\left\{\frac{e^{-q(ml+n)}}{s}\right\}\right] + L^{-1}\left\{\frac{e^{-q(ml-n)}}{s}\right\}$$

$$= x_0 - u_0 \sum_{n=0}^{\infty}(-1)^n\left[L^{-1}\left\{\frac{e^{-\frac{\sqrt{s}(ml+x)}{\sqrt{c^2}}}}{s}\right\}\right] + L^{-1}\left\{\frac{e^{-\frac{\sqrt{s}(ml-x)}{\sqrt{c^2}}}}{s}\right\}$$

$$= u_0 - u_0 \sum_{n=0}^{\infty}(-1)^n\left[erfc\frac{ml+x}{2\sqrt{c^2 t}} + erfc\frac{ml-x}{2\sqrt{c^2 t}}\right]$$

$$\left[\because L^{-1}\left\{\frac{e^{-\alpha\sqrt{s}}}{s}\right\} = crfc\left(\frac{\alpha}{2\sqrt{t}}\right)\right]$$

$$\therefore u(x,t) = u_0 - u_0 \sum_{n=0}^{\infty}(-1)^n\left[crfc\frac{(2n+1)l+x}{2\sqrt{c^2 t}} + crfc\frac{(2n+1)l-x}{2\sqrt{c^2 t}}\right]$$

Solution of One Dimensional Wave Heat Equation by Fourier transform

Ex. 4. *Solve the heat wave equation*

$$k\frac{\partial^2 u}{\partial x^2} = \frac{\partial u}{\partial t}, \quad -\infty < x < \infty, \quad t > 0$$

when $u(x,t) = 0$ *at* $x = \pm\infty$

$$\frac{\partial u}{\partial x} = 0 \text{ at } x = \pm\infty$$

and $u(x,0) = f(x), \quad -\infty < x < \infty.$

Taking Fourier Transform on both side of the equation

$$k\frac{\partial^2 u}{\partial x^2}=\frac{\partial u}{\partial t}$$

we get $\int_{-\infty}^{\infty} k\frac{\partial^2 u}{\partial x^2}e^{isx}dx=\int_{-\infty}^{\infty}\frac{\partial u}{\partial t}e^{isx}dx$

or, $k(-is)^2\int_{-\infty}^{\infty} ue^{isx}dx=\frac{\partial}{\partial t}\int_{-\infty}^{\infty} ue^{isx}dx$.

[By property of 'derivative on Fourier Transform']

or, $-ks^2U=\frac{\partial U}{\partial t}$.

[where $U(s,t)$ is Fourier Transform of $u(x,t)$]

or, $-ks^2U(s,t)=\frac{\partial U}{\partial t}$

or, $\frac{\partial U}{\partial t}+ks^2U=0$ ... ... (1)

Now, taking Fourier transform on both side of the given condition $u(x,0)=f(x)$ we get

$$\int_{-\infty}^{\infty} u(x,0)e^{isx}dx=\int_{-\infty}^{\infty} f(x)e^{isx}dx$$

or, $U(s,0)=F(s)$ ... ... (2)

where $F(s)$ is Fourier Transform of $f(x)$.

Now we shall solve the differential equation (1). Its auxilliary equation is

$m+ks^2=0$ or, $m=-ks^2$

$\therefore$ the solution of (1) is $U=Ae^{-ks^2t}$ i.e. $U(s,t)=Ae^{-ks^2t}$

Putting $t=0$, $U(s,0)=A$ or, $F(s)=A$ by (2)

$\therefore$ the solution of (1) is $U(s,t)=F(s)e^{-ks^2t}$.

Since $U(s,t)$ is Fourier Transform of $u(x,t)$ so applying Fourier Integral Theorem we get

$$u(x,t)=\frac{1}{2\pi}\int_{-\infty}^{\infty}U(s,t)e^{-isx}ds=\frac{1}{2\pi}\int_{-\infty}^{\infty}F(s)e^{-ks^2t}e^{-isx}ds$$

$$\text{or, } u(x,t)=\frac{1}{2\pi}\int_{-\infty}^{\infty}F(s)e^{-ks^2t-isx}ds \quad \ldots \quad \ldots \quad (3)$$

This may be kept as the required solution. Here $F(s)$ is Fourier transform of $f(x)$ which would be known.

We proceed further to express (3) in terms of $f(x)$.

We know $\mathcal{F}\left(e^{-\frac{x^2}{2}}\right)=\sqrt{2\pi}\ e^{-\frac{s^2}{2}}$.

By change of scale property $\mathcal{F}\left(e^{-\frac{\left(\frac{x}{\sqrt{2kt}}\right)^2}{2}}\right)=\sqrt{2\pi}\sqrt{2kt}\,e^{-\frac{\left(\sqrt{2kt}s\right)^2}{2}}$

$$\text{or, } \mathcal{F}\left(e^{-\frac{x^2}{4kt}}\right)=2\sqrt{\pi kt}\ e^{-ks^2t} \quad \text{or, } \mathcal{F}\left(\frac{1}{2\sqrt{\pi kt}}e^{-\frac{x^2}{4kt}}\right)=e^{-ks^2t}$$

or, $\mathcal{F}(g(x))=G(s)$ say

From (3) we have

$$u(x,t)=\frac{1}{2\pi}\int_{-\infty}^{\infty}F(s)G(s)e^{-isx}ds$$

$$=\frac{1}{2\pi}\mathcal{F}^{-1}\{F(s)G(s)\}=(f*g) \text{ (convolution of } f \text{ and } g)$$

$$=\int_{-\infty}^{\infty}f(x)g(x-u)du=\int_{-\infty}^{\infty}f(u)\frac{1}{2\sqrt{\pi kt}}e^{-\frac{(x-u)^2}{4kt}}du$$

$$=\frac{1}{\sqrt{4\pi kt}}\int_{-\infty}^{\infty}f(u)e^{-\frac{(x-u)^2}{4kt}}du$$ which is the required solution.

Ex. 5. *Solve the heat wave equation*

$$k\frac{\partial^2 u}{\partial x^2}=\frac{\partial u}{\partial t},\ -\infty<x<\infty,\ 0<t<\infty$$

when $u(x,t)=0$ *at* $x=\pm\infty$ $\frac{\partial u}{\partial x}=0$ *at* $x=\pm\infty$ *and* $u(x,0)=1$, $-\infty<x<\infty$.

This example is nothing but the previous example with $f(x)=1$. So from previous example we get the solution as

$$u(x,t)=\frac{1}{\sqrt{4\pi kt}}\int_{-\infty}^{\infty}e^{-\frac{(x-u)^2}{4kt}}du \quad \left[\text{Put } z=\frac{x-u}{2\sqrt{kt}},\ \text{i.e. } dz=-\frac{du}{2\sqrt{kt}}\right]$$

$$=\frac{-2\sqrt{kt}}{\sqrt{4\pi kt}}\int_{\infty}^{-\infty}e^{-z^2}dz=\frac{1}{\sqrt{\pi}}\cdot 2\int_0^{\infty}e^{-z^2}dz=\frac{1}{\sqrt{\pi}}\cdot 2\cdot\frac{\sqrt{\pi}}{2}=1.$$

$\therefore u(x,t)$ is the required solution .

Ex. 6. *Determine the distribution of temperature* $u(x,t)$ *in the semi-infinite medium* $x\geq 0$ *when* $u(0,t)=0$ *and* $u(x,0)=f(x)$.

The heat flow is governed by the equation

$$k\frac{\partial^2 u}{\partial x^2}=\frac{\partial u}{\partial t}\quad (x>0,\ t>0) \quad \dots \quad \dots \quad (1)$$

Since the range of x is $x>0$, we take Fourier sine transform on (1) and get

$$\frac{\partial U_s}{\partial t}=k\left(su(0,t)-s^2U_s\right) \text{ or, } \frac{\partial U_s}{\partial t}+ks^2U_s=O \quad \dots \quad \dots \quad (2)$$

where U_s is Fourier sine transform of $u(x,t)$.

Again taking Fourier sine transform on the condition $u(x,0)=f(x)$ we get

$$U_s=F_s(s) \text{ at } t=0 \quad \dots \quad \dots \quad (3)$$

where F_s is Fourier sine transform of $f(x)$. Solving the differential equation (2) and using (3) we get $U_s=F_s(s)e^{-ks^2t}$.

Then by Fourier sine integral theorem we get

$$u(x,t)=\frac{2}{\pi}\int_0^{\infty}F_s(s)e^{-ks^2t}\sin xs\,ds \text{ which gives the required solution.}$$

4.2.4. Two Dimensional Laplace Equation

Solution of Laplace Equation

Ex. 1. *Solve the Laplace Equation*

$$\frac{\partial^2 u}{\partial x^2}+\frac{\partial^2 u}{\partial y^2}=0 \text{ inside the region } 0<x<\infty,\ 0<y<b \text{ such that}$$

$u=f(x),\ y=0,\ 0<x<\infty$

$u=0,\ y=b,\ 0<x<\infty$

$u=0,\ x=0,\ 0<y<b$

$$\left(\frac{\partial u}{\partial x}\right)_{x=\infty}=0\ (i.e.\ \frac{\partial u}{\partial x}\to 0 \text{ as } x\to\infty).$$

The Laplace equation is

$$\frac{\partial^2 u}{\partial x^2}+\frac{\partial^2 u}{\partial y^2}=0;\ 0<x<\infty,\ 0<y<b.$$

Taking Fourier Sine Transform on both side we get [$\because 0<x<\infty$, we take Four. Sin. Trans.], considering function of x,

$$\mathcal{F}_s\left(\frac{\partial^2 u}{\partial x^2}\right)+\mathcal{F}_s\left(\frac{\partial^2 u}{\partial y^2}\right)=\mathcal{F}_s(0)$$

$$\text{or, } \int_0^\infty \frac{\partial^2 u}{\partial x^2}\sin sx\,dx+\int_0^\infty \frac{\partial^2 u}{\partial y^2}\sin sx\,dx=0$$

$$\text{or, } \left[\sin sx\frac{\partial u}{\partial x}\right]_0^\infty - s\int_0^\infty \cos sx\frac{\partial u}{\partial x}dx+\frac{d^2}{dy^2}\left\{\int_0^\infty u(x,y)\sin sx\,dx\right\}=0$$

(Integrating by parts)

$$\text{or, } -s\int_0^\infty \cos sx\frac{\partial u}{\partial x}dx+\frac{d^2}{dy^2}U_s(s)=0 \text{ by given condition}$$

$$\text{or, } -s\left\{[\cos sx\,u]_0^\infty-\int_0^\infty -s\sin sx\,u\,dx\right\}+\frac{d^2}{dy^2}U_s(s)=0$$

$$\text{or, } -s\left\{s\int_0^\infty \sin sx\,u\,dx\right\}+\frac{d^2}{dy^2}U_s(s)=0$$

or, $-s^2U_s(s)+\frac{d^2}{dy^2}U_s(s)=0$

or, $\frac{d^2U_s}{dy^2}-s^2U_s=0.$

Its auxilliary equation is $m^2-s^2=0$ or, $m=\pm s$.

Since s may be positive or negative so we take the general solution as

$U_s = A\cosh sy + B\sinh sy$ ... ... (1)

(not in the form of $Ae^{sy}+Be^{-sy}$ or $A\cos sy+B\sin sy$).

Now taking Fourier Sine transform on the given condition

$u=f(x)$ at $y=0$, $0<x<\infty$

we have $U_s(s)=\int_0^\infty f(x)\sin sx\,dx$ at $y=0$, ... (2)

Again taking Fourier sine transform on the given condition

$u=0$, $y=b$, $0<x<\infty$

we get $U_s(s)=0$ at $y=b$ ... ... (3)

From (1) we have, when $y=0$,

$(U_s)_{y=0}=A\times1+B\times0$ $[\because \cos h0=1$ and $\sin h0=0]$

or, $A=\int_0^\infty f(x)\sin sx\,dx$ from (2).

From (1) we have, when $y=b$, $(U_s)_{y=b}=A\cos hsb+B\sin shb$

or, $0=A\cos hsb+B\sin shb$ from (3)

or, $B=-\frac{\cos hsb}{\sin hsb}A=-\frac{\cos hsb}{\sin hsb}\int_0^\infty f(x)\sin sx\,dx$.

Putting these values of A and B in (1) we get

$$U_s(s)=\cos hs\,y\int_0^\infty f(x)\sin sx\,dx \quad -\frac{\cos hsb}{\sin hsb}\sin hs\,y\int_0^\infty f(x)\sin sx\,dx$$

$$=\left\{\cosh sy-\frac{\cosh sb\cdot\sinh sy}{\sinh sb}\right\}\int_0^\infty f(x)\sin sx\,dx$$

$$= \frac{\cosh sy \sinh sb - \cosh sb \sinh sy}{\sinh sb} \int_0^\infty f(x) \sin sx \, dx$$

$$= \frac{\sinh(sb - sy)}{\sinh sb} \int_0^\infty f(x) \sin sx \, dx \quad [\because \sinh A \cosh B - \cosh A \sinh B = \sinh(A - B)]$$

or, $U_s(s) = \int_0^\infty f(x) \sin sx \frac{\sinh s(b-y)}{\sinh sb} dx$.

Now, by Fourier Sine Integral Theorem [i.e. taking inverse Fourier sine transform]

$U(x) = \frac{2}{\pi} \int_0^\infty U_s(s) \sin sx \, ds$, at the point of continuity

i.e. $U(x,y) = \frac{2}{\pi} \int_0^\infty \left\{ \int_0^\infty f(x) \sin sx \frac{\sinh s(b-y)}{\sinh sb} dx \right\} \sin sx \, ds$

$$= \frac{2}{\pi} \int_0^\infty \left\{ \int_0^\infty f(t) \sin st \frac{\sinh s(b-y)}{\sinh sb} dt \right\} \sin sx \, ds$$

or, $U(x,y) = \frac{2}{\pi} \int_0^\infty f(t) dt \int_0^\infty \frac{\sinh s(b-y)}{\sinh sb} \sin st \sin sx \, ds$

which is the required solution.

Ex. 2. *Write down the differential equation with boundary value of the following problem and hence find the steady temperature in the semi infinite solid $y > 0$ using cosine transform :*

The temperature in the semi solid $y > 0$ is steady when the temperature on the surface $y = 0$ is kept at unity over the strip $|x| < a$ and at zero outside the strip.

The result $\int_0^\infty \frac{e^{-sx} \sin rx}{x} dx = \tan^{-1} \frac{r}{s}$ where $r > 0$, $s > 0$ may be assumed.

As we have discused earlier the steady temperature $U(x,y)$ satisfies the two dimensional Laplace Equation

$$\frac{\partial^2 U}{\partial x^2} + \frac{\partial^2 U}{\partial y^2} = 0 \qquad 0 < y < \infty, \ -\infty < x < \infty \qquad \dots (1)$$

The boundary conditions are

$U(x,y)=1,\ y=0,\ -a<x<a$ (i.e. $|x|<a$) and $U(x,y)=0,\ y=0$ when, $|x|>a$

we assume $\left(\frac{\partial U}{\partial x}\right)_{x=\infty}=0$ and $\left(\frac{\partial U}{\partial x}\right)_{x=0}=0,\ U(x,\infty)=0$ ... (2)

Taking Fourier Cosine transform on both side of (1) we get

$\mathcal{F}_c\left(\frac{\partial^2 U}{\partial x^2}\right)+\mathcal{F}_c\left(\frac{\partial^2 U}{\partial y^2}\right)=0$ (considering as function of x)

or, $\int_0^\infty \frac{\partial^2 U}{\partial x^2}\cos sx\,dx+\int_0^\infty \frac{\partial^2 U}{\partial y^2}\cos sx\,dx=0$

or, $\left[\cos sx\frac{\partial U}{\partial x}\right]_0^\infty+\int_0^\infty s\sin sx\frac{\partial u}{\partial x}dx+\frac{d^2}{dy^2}\int_0^\infty U\cos sx\,dx=0$

or, $s\int_0^\infty \sin sx\frac{\partial U}{\partial x}dx+\frac{d^2U_c(s)}{dy^2}=0$

where $U_c(s)$ is Fourier Cosine transform of $U(x,y)$. Again integrating by part we get, from above,

$$s\left\{\left[\sin sxU(x,y)\right]_0^\infty-\int_0^\infty s\cos sxU(x,y)dx\right\}+\frac{d^2U_c}{dy^2}=0$$

or, $-s^2\int_0^\infty \cos sxU\,dx+\frac{d^2U_e}{dy^2}=0$ by asumption (2)

or, $-s^2U_c+\frac{d^2U_c}{dy^2}=0$ or, $\frac{d^2U_c}{dy^2}-s^2U_c=0$.

Its auxillary equation is $m^2-s^2=0\ \therefore m=\pm s$.

We consider, $s>0$.

$\therefore$ the general solution of the above equation is

$U_c(s)=Ae^{sy}+Be^{-sy}$ ... ... (3)

Now, $U_c(s)=\int_0^\infty U(x,y)\cos sx\,dx$

$\therefore$ at $y=\infty$, $U_c(s)=\int_0^\infty 0\cos sx\,dx$, by hypothesis

$=O$.

$\therefore$ from (3) we get $\dfrac{U_c(s)}{e^{sy}}=A+\dfrac{B}{e^{2sy}}$

Letting $y\to\infty$ we get, $O=A+B\times 0 \quad \therefore A=0$.

So, from (3), $U_c(s)=Be^{-sy}$ ... ... (4)

Again, when $y=0$,

$$U_c(s)=\int_0^a 1\cdot\cos sx\,dx+\int_a^\infty 0\cdot\cos sx\,dx=\left[\frac{\sin sx}{s}\right]_{x=0}^{a}=\frac{\sin sa}{s}$$

$\therefore$ from (4) at $y=0$, $\dfrac{\sin sa}{s}=B\cdot e^0=B$.

$\therefore U_c(s)=\dfrac{\sin sa}{s}e^{-sy}$.

Now, by Fourier Cosine Integral Theorem,

$$U(x,y)=\frac{2}{\pi}\int_0^\infty U_c(s)\cos sx\,ds$$

$$=\frac{2}{\pi}\int_0^\infty \frac{\sin sa}{s}e^{-sy}\cos sx\,ds$$

$$=\frac{1}{\pi}\int_0^\infty \frac{e^{-sy}}{s}(2\sin sa\cos sx)ds$$

$$=\frac{1}{\pi}\int_0^\infty \frac{e^{-sy}}{s}\{\sin(a+x)s+\sin(a-x)s\}ds$$

$$=\frac{1}{\pi}\left\{\int_0^\infty \frac{e^{-sy}\sin(a+x)s}{s}ds+\int_0^\infty \frac{e^{-sy}\sin(a-x)s}{s}ds\right\}$$

$$=\frac{1}{\pi}\left\{\tan^{-1}\frac{a+x}{y}+\tan^{-1}\frac{a-x}{y}\right\}$$

from the supplied result.

Ex. 3. *Solve the following boundary value problem :*

$$\frac{\partial^2 u}{\partial x^2}+\frac{\partial^2 u}{\partial y^2}=0,\quad -\infty<x<\infty,\quad y>0.$$

Boundary conditions are $u(x,0)=f(x)$, $-\infty<x<\infty$, $u(x,y)$ *is bounded as* $y\to\infty$ *and* $u(x,y)=\frac{\partial u}{\partial x}=0$ *at* $|x|=\infty$.

As $-\infty<x<\infty$, we take the Fourier Transform on both side of the partial differential equation

$$\frac{\partial^2 u}{\partial x^2}+\frac{\partial^2 u}{\partial y^2}=0 \text{ (considering function of } x)$$

$$\mathcal{F}\left(\frac{\partial^2 u}{\partial x^2}\right)+\mathcal{F}\left(\frac{\partial^2 u}{\partial y^2}\right)=0$$

Since $u(x,y)=0$ and $\frac{\partial u}{\partial x}=0$ at $|x|=\infty$ we get from above

$$-s^2U(s,y)+\int_{-\infty}^{\infty}\frac{\partial^2 u}{\partial y^2}e^{isx}dx=0$$

$$\text{or, } -s^2U(s,y)+\frac{\partial^2}{\partial y^2}\left\{\int_{-\infty}^{\infty}u(x,y)e^{isx}dx\right\}=0$$

$$\text{or, } \frac{\partial^2U(s,y)}{\partial y^2}-s^2U(s,y)=0 \quad \ldots (1)$$

Its auxiliary equation is $m^2-s^2=0$ $\therefore$ $m=\pm s$

$\therefore$ the general solution of (1) is $U(s,y)=Ae^{sy}+Be^{-sy}$

where A, B are independent of y but may depend on s.

Since $u(x,y)$ must be bounded as $y\to\infty$, $U(s,y)$ and its Fourier Transform also should be bounded as $y\to\infty$.

So $A=0$ if $s>0$ but $B=0$ if $s<0$. Thus for any s,

$$U(s,y)=\text{constant } e^{-|s|y} \quad \ldots (2)$$

Now taking Fourier Transform on the given condition $u(x,0)=f(x)$ we get $U(s,0)=\mathcal{F}(f(x))=F(s)$ where $F(s)$ is Fourier Transform of $f(x)$.

From (2) we get $U(s,0) = \text{constant} \cdot e^{0} = \text{constant}$

$\therefore\ F(s) = \text{constant}$.

Hence, from (2), $U(s,y) = F(s)\, e^{-|s|y} = \int_{-\infty}^{\infty} f(x) e^{-|s|y} e^{isx} dx$.

So, by Fourier Integral theorem,

$$u(x,y) = \frac{1}{2\pi}\int_{-\infty}^{\infty}\left\{\int_{-\infty}^{\infty} f(t) e^{-|s|y} e^{ist} dt\right\} e^{-isx} ds$$

$$= \frac{1}{2\pi}\int_{-\infty}^{\infty} f(t)dt \int_{-\infty}^{\infty} e^{s\{i(t-x)\}-|s|y} ds \qquad \ldots \quad (3)$$

But $\frac{1}{2\pi}\int_{-\infty}^{\infty} e^{s\{i(t-x)\}-|s|y} ds$

$$= \frac{1}{2\pi}\int_{-\infty}^{0} e^{s\{y+i(t-x)\}} ds + \frac{1}{2\pi}\int_{0}^{\infty} e^{-s\{y-i(t-x)\}} ds$$

$$= \frac{1}{2\pi}\left[\frac{e^{s\{y+i(t-x)\}}}{y+i(t-x)}\right]_{s=-\infty}^{0} - \frac{1}{2\pi}\left[\frac{e^{-s\{y-i(t-x)\}}}{y-i(t-x)}\right]_{s=0}^{\infty}$$

$$= \frac{1}{2\pi}\left\{\frac{1}{y+i(t-x)} + \frac{1}{y-i(t-x)}\right\} = \frac{1}{\pi}\frac{y}{(t-x)^2+y^2} \qquad \ldots \quad (4)$$

Substituting (4) into (3), the required solution is found to be

$$u(x,y) = \frac{y}{\pi}\int_{-\infty}^{\infty} \frac{f(t)dt}{(t-x)^2+y^2}.$$

Ex. 4. *Solve the partial differential equation*

$$\frac{\partial^2 u}{\partial x^2} + \frac{\partial^2 u}{\partial y^2} = 0, \quad -\infty < x < \infty, \quad y > 0$$

$$\left(\frac{\partial u}{\partial y}\right)_{(x,0)} = f(x), \quad -\infty < x < \infty$$

$u(x,y)$ *is bounded as* $y \to \infty$.

$u(x,y) = \frac{\partial u}{\partial x} = 0 \ at \ |x| = \infty$.

Let $\phi(x,y) = \frac{\partial u}{\partial y}$.

Then $\frac{\partial^2 \phi}{\partial x^2} + \frac{\partial^2 \phi}{\partial y^2} = \frac{\partial^2}{\partial x^2}\left(\frac{\partial u}{\partial y}\right) + \frac{\partial^2}{\partial y^2}\left(\frac{\partial u}{\partial y}\right) = \frac{\partial}{\partial y}\left(\frac{\partial^2 u}{\partial x^2} + \frac{\partial^2 u}{\partial y^2}\right) = 0$

$\therefore \phi(x,y)$ satisfies the Laplace diff equation $\frac{\partial^2 \phi}{\partial x^2} + \frac{\partial^2 \phi}{\partial y^2} = 0$ with boundary conditions $\phi(x,0) = \left(\frac{\partial u}{\partial y}\right)_{(x,0)} = f(x)$.

Thus the function $\phi(x,y)$ is a solution of the previous problem.

So, $\phi(x,y) = \frac{y}{\pi} \int_{-\infty}^{\infty} \frac{f(t)dt}{(t-x)^2 + y^2}$, $y > 0$.

However, $u(x,y) = \int \phi(x,y)dy$ [From the defn. of ϕ]

$= \frac{1}{\pi} \int_{-\infty}^{\infty} f(t) \int \frac{y\,dy}{(x-t)^2 + y^2} dt$.

Therefore, $u(x,y) = \frac{1}{2\pi} \int_{-\infty}^{\infty} f(t) \log\{(x-t)^2 + y^2\} dt +$ constant is the required solution.

EXERCISES

1. Solve the partial differential equation

$$k\frac{\partial^2 u}{\partial x^2} = \frac{\partial u}{\partial t}, \ 0 < x < \infty, \ t > 0$$

when $u(0,t) = u_0$, $t \geq 0$

$u(x,t) = 0$ at $x = \infty$

$\frac{\partial u}{\partial x} = 0$ at $x = \infty$

and $u(x,0) = 0$, $0 < x < \infty$.

[Hint : Here $f(x) = 0$, since $0 < x < \infty$ take Fourier sine transform on both side].

2. Solve the boundary value problem

$$\frac{\partial^2 u}{\partial x^2} = \frac{1}{k}\frac{\partial u}{\partial t}$$

with the conditions

$$u = 0 \text{ when } x \to \infty$$

$$u = 0 \text{ when } t = 0$$

using Laplace transform.

3. Solve the heat wave equation $\frac{\partial \theta}{\partial t} = c^2 \frac{\partial^2 \theta}{\partial x^2}$ $(t > 0)$ if the initial temperature of an infinite bar is given by

$$\theta(x) = \theta_0, \quad |x| < a$$

$$= 0, \quad |x| > a.$$

4. Use Fourier transform to solve the one dimensional wave equation $2\frac{\partial u}{\partial t} = \frac{\partial^2 u}{\partial x^2} (x > 0)$ if $u_x(0,t) = 0$, $u(x,0) = e^{-x}$ and $u(x,t)$ is bounded.

5. Solve the heat wave equation $k\frac{\partial^2 u}{\partial x^2} = \frac{\partial u}{\partial t}$, $-\infty < x < \infty$, $t > 0$

when $u(x,t) = 0$ at $x = \pm\infty$, $\frac{\partial u}{\partial x} = 0$ at $x = \pm\infty$

and $u(x,0) = 1$ when $x > 0$

$= 0$ when $x < 0$.

6. Find the temperature in the infinite bar if the initial temperature is

$$f(x) = k, \text{ when } |x| < 1$$

$$= 0, \text{ when } |x| > 1.$$

[Hint : Take the Heat wave equation, $\frac{\partial u}{\partial t} = c^2 \frac{\partial^2 u}{\partial x^2}$ with initial condition $u(x,0) = f(x)$].

7. Using Laplace transform solve $\frac{\partial u}{\partial t} = \frac{\partial^2 u}{\partial x^2}$, $x > 0, t > 0$ with $u(0,t) = 1$. $u(x,0) = 0$

8. A semi-infinite solid $x > 0$ is initially at temperature zero. At time $t = 0$, constant temperature $u_0 > 0$ is applied and maintained at the face $x = 0$. Show that the temperature at any point of the solid at $t = \frac{1}{4}$ is given by

$$u_0 crfc\left(\frac{x}{\sqrt{k}}\right)$$

9. Solve $\frac{\partial U}{\partial t} = \frac{\partial^2 U}{\partial x^2}$, $x > 0$, $t > 0$ subject to the conditions

(i) $U = 0$ at $x = 0$, $t > 0$

(ii) $U = 1$, $0 < x < 1$

$= 0$, $x \geq 1$ when $t = 0$

(iii) $U(x,t)$ is bounded [*W.B.U Tech 2002, 2004*]

[Hint : Take Fourier Sine Transform].

10. The temperature T in the semi-infinite bar $0 \leq x < \infty$ is determined by the differential equation $\frac{\partial T}{\partial t} = k\frac{\partial^2 T}{\partial x^2}$

subject to the condition

(i) $T = 0$ when $t = 0$, $x \geq 0$

(ii) $\frac{\partial T}{\partial x} = -\mu$ when $x = 0$, $t > 0$.

Making use of Cosine Transform, show that

$$U(x,t) = \frac{2\mu}{\pi}\int_0^\infty \frac{\cos sx}{s^2}\left(1 - e^{-ks^2 t}\right)ds.$$

11. Solve $\frac{\partial U}{\partial t} = 2\frac{\partial^2 U}{\partial x^2}$ if $U(0,t) = 0$, $U(x,0) = e^{-x}$, $x > 0$

$U(x,t)$ is bounded where $x > 0$, $t > 0$.

[Hint : Take Fourier sine transform] [*W.B.U.Tech,2003*]

12. Solve $\dfrac{\partial^2 U}{\partial x^2} = \dfrac{\partial U}{\partial t}$

if $\left(\dfrac{\partial U}{\partial x}\right)_{(0,t)} = 0$, $U(x,0) = x$, $0 \le x \le 1$

$= 0$, $x > 1$

and $U(x,t)$ is bounded where $x > 0$, $t > 0$.

[Hint : Take Fourier Cosine Transform].

13. Compute the displacement $u(x,t)$ of an infinite string using the method of Fourier Transform given that the string is initially at rest and that the initial displacement is $f(x)$, $-\infty < x < \infty$.

[Hint : This problem is represented by

$$\frac{\partial^2 u}{\partial t^2} = c^2 \frac{\partial^2 u}{\partial x^2}, \ -\infty < x < \infty, \ u(x,0) = f(x), \ -\infty < x < \infty; \ u_t(x,0) = 0$$

solve the problem independently].

14. An infinite string is initially at rest along the x axis. Its one end which is at $x = 0$, is given a periodic transverse displacement $a \sin wt$, $t > 0$. Show that the displacement of any point of the string at any time is given by

$$u(x,t) = a \sin w\left(t - \frac{x}{c}\right), \quad t > \frac{x}{c}$$

$$= 0, \quad t < \frac{x}{c}$$

where c is the wave velocity.

15. Solve the equation

$$\frac{\partial^2 u}{\partial t^2} = c^2 \frac{\partial^2 u}{\partial x^2}, t > 0, x > 0$$

with the conditions

(i) $u = p_0 \sin at$ at when $x = 0, t > 0$

(ii) u is finite when $x \to \infty$

(iii) $u = u_t = 0$ when $t = 0, x > 0$

16. Solve the following equation:

$$\frac{\partial^2 u}{\partial t^2} = a^2 \frac{\partial^2 u}{\partial x^2}, x > 0, t > 0$$

$u_n(0, t) = \sin ut,\ u(x, 0) = 0, |u(x_1\ t)| = 0$ and $(u(x, t)) < M$

17. Solve the wave equation

$$\frac{\partial^2 u}{\partial t^2} = a^2 \frac{\partial^2 u}{\partial x^2},\ (t > 0, x > 0)$$

where $u(x, 0) = 0, x \geq 0, u_t(x, 0) = 0, x > 0$

$$u(0, t) = t, u(\infty, t) = 0, t \geq 0$$

using Laplace transform

18. If $\frac{\partial^2 u}{\partial x^2} + \frac{\partial^2 u}{\partial y^2} = 0$ for $x \geq 0$ and $u = f(y)$ on $x = 0$, show by using the method of Fourier Transform that $u(x,y) = \frac{1}{\sqrt{2\pi}} \int_{-\infty}^{\infty} F(s) e^{-|s|x - isy} ds$

where $F(s)$ is the Fourier transform of $f(y)$. Further deduce that $u(x,y) = \frac{x}{\pi} \int_{-\infty}^{\infty} \frac{f(t)dt}{x^2 + (y-t)^2}$.

[Hint : For 2nd part : From 1st part we have

$$\mathcal{F}^{-1}\left(e^{-|s|x}\right) = \sqrt{\frac{2}{\pi}}\left(\frac{2}{x^2 + y^2}\right).$$

Applying Convolution theorem

$$u(x,y) = f(x) * \sqrt{\frac{2}{\pi}}\left(\frac{x}{x^2 + y^2}\right) = \frac{1}{\sqrt{2\pi}} \int_{-\infty}^{\infty} f(t) \sqrt{\frac{2}{\pi}} \frac{x}{x^2 + (y-t)^2} dt \text{].}$$

19. Using Fourier transform solve the partial differential equation

$$\frac{\partial^2 u}{\partial x^2} + \frac{\partial^2 u}{\partial y^2} = 0,\ -\infty < x < \infty,\ y > 0$$

where $u(x,0) = 0;\ x < 0$

$$= 1;\ x > 0.$$

20. Use Fourier transform to show that

$$V = \frac{1}{2\sqrt{xt}} \int_{-\infty}^{\infty} F(s) e^{\frac{x-s}{st}} ds$$

is the solution of the boundary value problem $\frac{\partial v}{\partial t}=\frac{\partial^2 v}{\partial x^2}$ $-\infty < x < \infty$, $t > 0$; $v = f(x)$ when $t = 0$.

[$\because$ range of x is $(-\infty, \infty)$ so we have to take Fourier Transform on both side].

ANSWERS

1. $u(x,t)=u_0\left(1-\frac{2}{\sqrt{\pi}}\int_0^{n/\sqrt{2kt}} e^{-u^2}\,du\right)$

2. $u(x,t)=\frac{cx}{\sqrt{k}}\frac{1}{2\sqrt{\pi t^3}}e^{-x^2/4kt}$

3. $\theta(x,t)=\frac{\theta_0}{\pi}\int_0^{\infty} e^{-v^2}\left\{\sin\frac{(a+x)v}{c\sqrt{t}}+\sin\frac{(a-x)v}{c\sqrt{t}}\right\}\frac{dv}{v}$ where $v^2=c^2s^2t$.

5. $u(x,t)=\frac{1}{\sqrt{\pi}}\int_{-x/2\sqrt{kt}}^{\infty} e^{-z^2 dz}$

6. $u(x,t)=\frac{k}{\sqrt{\pi}}\int_{-(1+x)/2c\sqrt{t}}^{(1-x)/2c\sqrt{t}} e^{-z^2 dz}$

7. $u(x,t)=crfc\left(\frac{x}{2\sqrt{t}}\right)$

8. $u_0 crfc\left(\frac{x}{\sqrt{k}}\right)$

9. $U(x,t)=\frac{2}{\pi}\int_0^{\infty}\frac{1-\cos s}{s}e^{-s^2 t}\sin sx\,ds$

11. $U(x,t)=\frac{2}{\pi}\int_0^{\infty}\frac{se^{-2s^2t}\sin sx}{1+s^2}ds$

12. $U(x,t)=\frac{2}{\pi}\int_0^{\infty}\left(\frac{\sin s}{s}+\frac{\cos s-1}{s^2}\right)e^{-s^2t}\cos sx\,ds$

13. $u(x,t)=\frac{1}{2}\{f(x+ct)+f(x-ct)\}$.

15. $u(n,t)=p_0\sin\left\{s\left(t-\frac{x}{c}\right)\right\}\times H\left(t-\frac{x}{c}\right)$

16. $u(x,t)=-\frac{ba}{w}\left\{1-\cos w\left(t-\frac{x}{a}\right)\right\}\times H\left(t-\frac{x}{a}\right)$

17. $\left(t-\frac{x}{a}\right)H\left(t-\frac{x}{a}\right)$

19. $u(x,y)=\frac{1}{\pi}\left\{\frac{\pi}{2}+\tan^{-1}\frac{x}{y}\right\}$

ORDINARY DIFFERENTIAL EQUATION (ODE)

4.3 SERIES SOLUTION OF ORDINARY DIFFERENTIAL EQUATION

4.3.1. Introduction.

Sometimes we face some differential equations which cannot be solved by the methods discussed so far. Such type of equations may be solved by assuming that an infinite convergent series arranged according to the ascending power of the independent variables satisfies the equation approximately.

4.3.2. Validity of Series Solution.

Consider a second order linear differential equation of the form

$$P_0(x)\frac{d^2y}{dx^2}+P_1(x)\frac{dy}{dx}+P_2(x)y=0 \quad \ldots \quad (1)$$

where $P_0(x), P_1(x), P_2(x)$ are polynomials in x. As every diffferntial equation of the form (1) may not have series solution, we now find the conditions under which the above equation has series solution.

If $P_0(a)\neq 0$, then the point $x=a$ is called **an ordinary point**; otherwise $x=a$ is called a **singular point** of (1).

Equation (1) can be written as

$$\frac{d^2y}{dx^2}+P(x)\frac{dy}{dx}+Q(x)y=0 \quad \ldots \quad (2)$$

where $P(x)=\dfrac{P_1(x)}{P_0(x)}$ and $Q(x)=\dfrac{P_2(x)}{P_0(x)}$.

If $(x-a)P(x)$ and $(x-a)^2Q(x)$ possess derivatives of all order in the neighbourhood of $x=a$, then $x=a$ is called a *regular singular point*.

Illustration.

(i) Consider the equation

$$(x-1)\frac{d^2y}{dx^2}+x\frac{dy}{dx}+\frac{1}{x}y=0.$$

which can also be written as $x(x-1)\dfrac{d^2y}{dx^2}+x^2\dfrac{dy}{dx}+y=0$.
Here $P_0(x)=x(x-1)$,

$\therefore P_0(0)=P_0(1)=0$ and so $x=0,1$ are singular points of the equation.

Here also $P(x)=\dfrac{P_1(x)}{P_0(x)}=\dfrac{x}{x-1}$, $Q(x)=\dfrac{P_2(x)}{P_0(x)}=\dfrac{1}{x(x_{-1})}$

$\therefore xP(x)=\dfrac{x^2}{x-1}$, $x^2Q(x)=\dfrac{x}{x-1}$.

since $xP(x)$ and $x^2Q(x)$ possess derivatives of all orders in the *nbd* of $x=0$, so $x=0$ is a regular singular point. Similarly $x=1$ is also a regular singular point.

(ii) Consider the equation $\left(1+x^2\right)\dfrac{d^2y}{dx^2}+x\dfrac{dy}{dx}-3y=0$.

Here $P_0(x)=1+x^2$.

$\therefore P_0(0)=1\neq 0$ and so $x=0$ is an ordinary point of the given equation.

Theorem. 1. When $x=a$ is an ordinary point of (1), its solution can be expressed as a series of the form

$$y=a_0+a_1(x-a)+a_2(x-a)^2+\cdots\cdots$$

$$=\sum_{n=0}^{\infty}a_n(x-a)^n$$

i.e. differential equation has a series solution.

Theorem. 2. When $x=a$ is a regular singular point of (1), its solution can be expressed as a series of the form

$$y=x^m\left(a_0+a_1(x-a)+a_2(x-a)^2+\cdots\cdots\right)$$

$$=\sum_{n=0}^{\infty}a_n(x-a)^{n+m}$$

where m may be positive or negative integer or a fraction i.e. the differential equation also has a series solution.

4.3.3. General Method to Solve an ordinary differential equation.

Here we discuss two methods, depending on the nature of the point $x=x_0$, about which the series solution is convergent.

Method. 1. *Series solution or power series solution when* $x=x_0$ *is an ordinary point.*

Let $y=\sum_{n=0}^{\infty}a_n(x-x_0)^n$ ... (1)

be the series solution of the p.d.e

$$P_0(x)\frac{d^2y}{dx^2}+P_1(x)\frac{dy}{dx}+P_2(x)y=0 \qquad \dots \qquad (2)$$

Then $\frac{dy}{dx}=\sum_{n=1}^{\infty} n a_n (x-x_0)^{n-1}$; $\frac{d^2y}{dx^2}=\sum_{n=2}^{\infty} n(n-1)a_n(x-x_0)^{n-2}$

Substituting these values in (2), we get an identity. Then equating the co-efficient of to zero we get a relation between a_n and a_{n+2} which is called the *recurrence relation*. From this relation we obtain the co-efficients of (1) in terms of a_0 and a_1. These are substituted in (1) to get the required solution.

Illustrative Examples.

Ex. 1. Solve in series the equation $(1+x^2)\frac{d^2y}{dx^2}+x\frac{dy}{dx}-y=0$.

Here $x=0$ is an ordinary point of the given equation

$$(1+x^2)\frac{d^2y}{dx^2}+x\frac{dy}{dx}-y=0 \qquad \dots \qquad (1)$$

So let $y=\sum_{n=0}^{\infty} a_n x^n \qquad \dots \qquad (2)$

be the series solution of (1). Then

$$\frac{dy}{dx}=\sum_{n=1}^{\infty} n a_n x^{n-1},\ \frac{d^2y}{dx^2}=\sum_{n=2}^{\infty} n(n-1)a_n x^{n-2}.$$

Substituting these values in (1), we get

$$(1+x^2)\sum_{n=2}^{\infty} n(n-1)a_n x^{n-2}+x\sum_{n=1}^{\infty} n a_n x^{n-1}-\sum_{n=0}^{\infty} a_n x^n=0$$

or, $$\sum_{n=2}^{\infty} n(n-1)a_n x^{n-2}+\sum_{n=2}^{\infty} n(n-1)a_n x^n+\sum_{n=1}^{\infty} n a_n x^n-\sum_{n=0}^{\infty} a_n x^n=0$$

or, $$\sum_{n=0}^{\infty}(n+2)(n+1)a_{n+2}x^n+\sum_{n=0}^{\infty} n(n-1)a_n x^n+\sum_{n=0}^{\infty} n a_n x^n-\sum_{n=0}^{\infty} a_n x^n=0$$

or, $$\sum_{n=0}^{\infty}(n+2)(n+1)a_{n+2}x^n+\sum_{n=0}^{\infty} a_n[n(n-1)+n-1]x^n=0$$

or, $$\sum_{n=0}^{\infty}(n+2)(n+1)a_{n+2}x^n+\sum_{n=0}^{\infty} a_n(n^2-1)x^n=0 \qquad \dots \qquad (3)$$

Equating the co-efficient of x^n to zero we have

$(n+2)(n+1)a_{n+2}+(n^2-1)a_n=0$

$$\therefore a_{n+2}=-\frac{n-1}{n+2}a_n, \qquad \cdots \quad (4)$$

Again equating to zero, the co-efficient of x^0 and x^1 we have from (3)

$2a_2=a_0,\ 3\cdot 2\cdot a_3=0$

$$\therefore a_2=\frac{1}{2}a_0,\ a_3=0.$$

Then putting $n=2,3,4,5,\cdots$ in (4) successively, we get

$$a_4=-\frac{1}{4}a_2=-\frac{1}{8}a_0,$$

$$a_5=-\frac{2}{5}a_3=0,$$

$$a_6=-\frac{1}{2}a_4=\frac{1}{16}a_0,\ a_7=-\frac{4}{7}a_5=0 \text{ and so on.}$$

Substituting these values of a_i 's in (2), we get

$$y=a_0\left(1+\frac{x^2}{2}-\frac{x^4}{8}+\frac{x^6}{16}-\cdots\right)+a_1x$$

which is the required general series solution; a_0, a_1 being arbitrary constants.

Ex. 2. *Find a power series solution of the equation*

$$\left(1-x^2\right)\frac{d^2y}{dx^2}+2y=0,\ \textit{given that}\ y(0)=4,\ y'(0)=5.$$

Here $x=0$ is an ordinary point of the given equation

$$\left(1-x^2\right)\frac{d^2y}{dx^2}+2y=0 \qquad \cdots \quad (1)$$

So let $y=\sum_{n=0}^{\infty}a_nx^n$ be the series solution of (1). $\qquad \cdots \quad (2)$

Then $\dfrac{dy}{dx}=\sum_{n=1}^{\infty}na_nx^{n-1}$,

$$\frac{d^2y}{dx^2}=\sum_{n=2}^{\infty}n(n-1)a_nx^{n-2}.$$

Substituting these values in (1) we get

$$\left(1-x^2\right)\sum_{n=2}^{\infty} n(n-1)a_n x^{n-2}+2\sum_{n=0}^{\infty} a_n x^n=0$$

or, $\sum_{n=2}^{\infty} n(n-1)a_n x^{n-2}-\sum_{n=0}^{\infty} n(n-1)a_n x^n+2\sum_{n=0}^{\infty} a_n x^n=0$

or, $\sum_{n=0}^{\infty}(n+2)(n+1)a_{n+2}x^n-\sum_{n=0}^{\infty} n(n-1)a_n x^n+2\sum_{n=0}^{\infty} a_n x^n=0$

or, $\sum_{n=0}^{\infty}(n+2)(n+1)a_{n+2}x^n-\sum_{n=0}^{\infty} a_n\left(n^2-n-2\right)x^n=0$ ... (3)

Equating the co-efficient of x^n to zero, we have

$$(n+2)(n+1)a_{n+2}-\left(n^2-n-2\right)a_n=0$$

or, $a_{n+2}=\dfrac{n^2-n-2}{(n+2)(n+1)}a_n$

$\therefore a_{n+2}=\dfrac{n-2}{n+2}a_n$ ... (4)

Again equating the co-efficint of x^0 and x^1 to zero, we have from (3),

$2a_2+2a_0=0$ i.e. $a_2=-a_0$

and $3\cdot 2\cdot a_3+2a_1=0$ i.e. $a_3=-\dfrac{1}{3}a_1$.

Then putting $n=2,3,4,5,\cdots$ in (4) successively, we get

$a_4=0,\ a_5=\dfrac{1}{5}a_3=-\dfrac{1}{15}a_1,$

$a_6=\dfrac{2}{6}a_4=0, a_7=\dfrac{3}{7}a_5=-\dfrac{1}{35}a_1$ and so on.

Substituting these values of a_i's in (2), we get

$y=a_0\left(1-x^2\right)+a_1\left(x-\dfrac{x^3}{3}-\dfrac{x^5}{15}-\dfrac{x^7}{35}-\cdots\right)$ ... (5)

$\therefore y'=-2a_0x+a_1\left(1-\dfrac{x^2}{2}-\dfrac{x^4}{3}-\dfrac{x^6}{5}\cdots\right)$ ... (6)

Putting $x=0$ in (5) and (6) and using $y(0)=4, y'(0)=5$

we get $a_0=4, a_1=5$.

$$\therefore y=4\left(1-x^2\right)+5\left(x-\frac{x^3}{3}-\frac{x^5}{15}-\frac{x^7}{35}\cdots\right)$$

which is the required series solution.

Ex. 3. *Solve in series the equation* $\frac{d^2y}{dx^2}+(x-1)\frac{dy}{dx}+y=0$ *in power of* $x-2$.

Obviously $x=2$ is an ordinary point of the equation

$$\frac{d^2y}{dx^2}+(x-1)\frac{dy}{dx}+y=0 \;\left[\because P_0(2)\neq 0\right]. \qquad \ldots (1)$$

So let its series solution be $y=\sum_{n=0}^{\infty} a_n(x-2)^n$ ... (2)

Take $t=x-2$ ... (3)

Then $\frac{dy}{dx}=\frac{dy}{dt}, \frac{d^2y}{dx^2}=\frac{d^2y}{dt^2}$

$\therefore$ (1) and (2) becomes

$$\frac{d^2y}{dt^2}+(t+1)\frac{dy}{dt}+y=0 \qquad \ldots (4)$$

and $y=\sum_{n=0}^{\infty} a_n t^n$ ... (5)

$$\therefore \frac{dy}{dt}=\sum_{n=1}^{\infty} n a_n t^{n-1}, \frac{d^2y}{dt^2}=\sum_{n=2}^{\infty} n(n-1)a_n t^{n-2}$$

Substituting these values in (4), we get

$$\sum_{n=2}^{\infty} n(n-1)a_n t^{n-2}+(t+1)\sum_{n=1}^{\infty} n a_n t^{n-1}+\sum_{n=0}^{\infty} a_n t^n=0$$

or, $\sum_{n=0}^{\infty}(n+2)(n+1)a_{n+2}t^n+\sum_{n=0}^{\infty} n a_n t^n+\sum_{n=0}^{\infty}(n+1)a_{n+1}t^n+\sum_{n=0}^{\infty} a_n t^n=0$.

Equating to zero, the co-efficient of t^n, we have

$$(n+2)(n+1)a_{n+2}+na_n+(n+1)a_{n+1}+a_n=0$$

or, $(n+2)(n+1)a_{n+2}+(n+1)a_n+(n+1)a_{n+1}=0$

$$\therefore a_{n+2} = -\frac{1}{n+2}(a_{n+1} + a_n) \quad \cdots \quad (6)$$

Putting $n = 0,1,2,\cdots$ in (6) successively, we get

$$a_2 = -\frac{1}{2}(a_1 + a_0)$$

$$a_3 = -\frac{1}{3}(a_2 + a_1) = -\frac{1}{3}\left\{-\frac{1}{2}(a_1 + a_0) + a_1\right\} = \frac{1}{6}(a_0 - a_1)$$

$$a_4 = -\frac{1}{4}(a_3 + a_2) = -\frac{1}{4}\left\{\frac{1}{6}(a_0 - a_1) - \frac{1}{2}(a_1 + a_0)\right\} = \frac{1}{12}(a_0 + 2a_1) \text{ and so on.}$$

Putting these values of a_i 's in (5), we get

$$y = a_0\left(1 - \frac{1}{2}t^2 + \frac{1}{6}t^3 - \cdots\right) + a_1\left(t - \frac{1}{2}t^2 - \frac{1}{6}t^3 - \cdots\right).$$

Then using (3), we get

$$y = a_0\left[1 - \frac{1}{2}(x-2)^2 + \frac{1}{6}(x-2)^3 - \cdots\right] + a_1\left[(x-2) - \frac{1}{2}(x-2)^2 - \frac{1}{6}(x-2)^3 - \cdots\right]$$

which is the required series solution, a_0, a_1 being arbitrary constants.

Method. 2. (Method of Frobenius) *Series Solution when* $x = x_0$ *is a Regular Singular Point* :

$$\text{Let } y = \sum_{n=0}^{\infty} a_n (x - x_0)^{m+n}, \quad a_0 \neq 0 \quad \cdots \quad (3)$$

be the series solution of the p.d.e

$$P_0(x)\frac{d^2y}{dx^2} + P_1(x)\frac{dy}{dx} + P_2(x)y = 0 \quad \cdots \quad (1)$$

$$\text{Then } \frac{dy}{dx} = \sum_{n=0}^{\infty} a_n (m+n)(x - x_0)^{m+n-1}$$

$$\frac{d^2y}{dx^2} = \sum_{n=0}^{\infty} a_n (m+n)(m+n-1)(x - x_0)^{m+n-2}$$

Substituting these values in (1) we get an identity. Then the co-efficient of the lowest power of x from both side we get a quadratic equation in m, known as **indicial equation**. Next equating the co-efficient of general power i.e. x^{m+n} or x^{m+n-1} whichever may be the lowest we get a relation between a_n, a_{n-2} or a_n, a_{n-1} which is called the *recurrence relation*. When the recurrence relation is between a_n

and a_{n-2}, a_1 will be determined by equating the co-efficient of the next higher power to 0 than already used for getting the indicial equation. This step is not necessary when the recurrence relation is between a_n and a_{n-1}. Then we obtain the co-efficients of (3) in terms of a_0 and get a series solution of (1) having a_0 as arbitrary constant. Obviously this solution is not the complete solution of (1), as the complete solution must have two independent arbitrary constant. The method to obtain the complete solution depends on the nature of roots of the indicial equation and so we consider the following cases :

Case (i). If the roots m_1, m_2 of the indicial equation are distinct and not differing by an integer, then the complete solution is given by

$$y = c_1(y)_{m_1} + c_2(y)_{m_2}.$$

Case (ii). If the roots m_1, m_2 of the indicial equation are equal, then the complete solution is given by

$$y = c_1(y)_{m_1} + c_2\left(\frac{\partial y}{\partial m}\right)_{m_1}.$$

Case (iii). If the roots m_1, m_2 $(m_1 < m_2)$ of the indicial equation are distinct and differ by an integer, then we consider the following two sub-cases :

(*a*) If some of the co-efficient of y series become infinite when $m = m_1$, then replacing a_0 by $b_0(m - m_1)$ in y we get the complete solution as

$$y = c_1(y)_{m_1} + c_2\left(\frac{\partial y}{\partial m}\right)_{m_1}$$

Here the solution corresponding to $m = m_2$ is a constant multiple of the above solution.

(*b*) If one of the co-efficient of y becomes indeterminate when $m = m_2$, say , then the complete solution is given by putting $m = m_2$ in y which contains two arbitrary constants.

The solution obtained by putting $m = m_1$ is a constant multiple of the above solution.

Illustrative Examples.

Ex.1.(*Against Case i) Solve, in series, the equation* $2x^2\frac{d^2y}{dx^2}+(2x^2-x)\frac{dy}{dx}+y=0.$

Here $x = 0$ is a regular singular point of the equation

$$2x^2\frac{d^2y}{dx^2}+(2x^2-x)\frac{dy}{dx}+y=0 \qquad \dots (1)$$

So let its series solution be

$$y=\sum_{n=0}^{\infty}a_n x^{m+n},\ a_0\neq 0 \qquad \dots \qquad (2)$$

$$\therefore \frac{dy}{dx}=\sum_{n=0}^{\infty}a_n(m+n)x^{m+n-1}$$

$$\frac{d^2u}{dx^2}=\sum_{n=0}^{\infty}a_n(m+n)(m+n-1)x^{m+n-2}$$

Substituting these values in (1) we get

$$2x^2\sum_{n=0}^{\infty}a_n(m+n)(m+n-1)x^{m+n-2}+\left(2x^2-x\right)\sum_{n=0}^{\infty}a_n(m+n)x^{m+n-1}+\sum_{n=0}^{\infty}a_n x^{m+n}=0$$

$$\text{or, } \sum_{n=0}^{\infty}\left[2(m+n)(m+n-1)-(m+n)+1\right]a_n x^{m+n}+\sum_{n=0}^{\infty}2(m+n)a_n x^{m+n+1}=0 \qquad (3)$$

Equating the co-efficient of the lower power of x i.e x^m to zero, we get the indicial equation as

$$2m(m-1)-m+1=0 \quad [\therefore a_0\neq 0]$$

$$\text{or, } (2m-1)(m-1)=0 \quad \therefore m=\tfrac{1}{2},\ 1.$$

So the roots of the indicial equation are distinct and not differing by an integer.

Next equating to zero, the co-efficient of x^{m+n}, we get the recurrence relation as

$$\left[2(m+n)(m+n-1)-(m+n)+1\right]a_n+2(m+n-1)a_{n-1}=0$$

$$\text{or, } (2m+2n-1)(m+n-1)a_n=-2(m+n-1)a_{n-1}$$

$$\therefore a_n=-\frac{2a_{n-1}}{2m+2n-1} \qquad \dots \qquad (4)$$

Putting $n=1,2,3,\cdots$ in (4) successively, we get

$$a_1=-\frac{2}{2m+1}a_0,\ a_2=-\frac{2}{2m+3}a_1=\frac{2^2}{(2m+1)(2m+3)}a_0$$

$$a_3=-\frac{2}{2m+5}=-\frac{2^3}{(2m+1)(2m+3)(2m+5)}a_0 \text{ and so on.}$$

Substituting these values of a_i 's in (2), we have

$$y = a_0 x^m \left(1 - \frac{2x}{2m+1} + \frac{2^2 x^2}{(2m+1)(2m+3)} - \frac{2^3 x^3}{(2m+1)(2m+3)(2m+5)} + \cdots\right)$$

$$\therefore (y)_{m=\frac{1}{2}} = a_0 x^{\frac{1}{2}}\left(1 - x + \frac{1}{2}x^2 - \frac{1}{6}x^3 + \cdots\right) \text{ and}$$

$$(y)_{m=1} = a_0 x\left(1 - \frac{2}{3}x + \frac{4}{15}x^2 - \frac{8}{105}x^3 + \cdots\right).$$

Hence the complete solution is

$$y = c_1 (y)_{m=\frac{1}{2}} + c_2 (y)_{m=1}$$

$$= a_1 x^{\frac{1}{2}}\left(1 - x + \frac{1}{2}x^2 - \frac{1}{6}x^3 + \cdots\right) + a_2 x\left(1 - \frac{2}{3}x + \frac{4}{15}x^2 - \frac{8}{105}x^3 + \cdots\right)$$

where $a_1 = a_0 c_1, a_2 = a_0 c_2$ are arbitrary constants.

Ex. 2. *(Against Case* (ii).) *Solve, in series, the equation* $x\frac{d^2y}{dx^2} + \frac{dy}{dx} + xy = 0$.

Here $x = 0$ is a regular singular point of the given equation

$$x\frac{d^2y}{dx^2} + \frac{dy}{dx} + xy = 0 \qquad \ldots \quad (1)$$

So let its series solution be

$$y = \sum_{n=0}^{\infty} a_n x^{m+n}, a_0 \neq 0 \qquad \ldots \quad (2)$$

$$\therefore \frac{dy}{dx} = \sum_{n=0}^{\infty} a_n (m+n) x^{m+n-1}, \frac{d^2y}{dx^2} = \sum_{n=0}^{\infty} a_n (m+n)(m+n-1) x^{m+n-2}$$

Substituting these values in (1), we get

$$x\sum_{n=0}^{\infty} a_n (m+n)(m+n-1) x^{m+n-2} + \sum_{n=0}^{\infty} a_n (m+n) x^{m+n-1} + x\sum_{n=0}^{\infty} a_n x^{m+n} = 0$$

$$\text{or, } \sum_{n=0}^{\infty} a_n \{(m+n)(m+n-1) + (m+n)\} x^{m+n-1} + \sum_{n=0}^{\infty} a_n x^{m+n+1} = 0$$

$$\text{or, } \sum_{n=0}^{\infty} a_n (m+n)^2 x^{m+n-1} + \sum_{n=0}^{\infty} a_n x^{m+n+1} = 0 \qquad \ldots \quad (3)$$

Equating the co-efficient of the lowest power of x i.e. x^{m-1} to zero, we get the indicial equation as

$$a_0 m^2 = 0 \therefore m = 0,0 \left[\because a_0 \neq 0\right].$$

Next equating the coefficient of x^{m+n-1} to zero, we get the recurrence relation as

$$a_n(m+n)^2 + a_{n-2} = 0$$

$$\therefore a_n = -\frac{a_{n-2}}{(m+n)^2} \qquad \ldots \qquad (4)$$

For getting a_1, equating the co-efficient of x^m to zero gives $a_1(m+1)^2 = 0$.

$\therefore\ a_1 = 0\ [\because m = 0]$.

$\therefore$ Putting $n = 3,5,7,\cdots$ successively in (4), we get

$$a_1 = a_3 = a_5 = \cdots = 0.$$

Again putting $n = 2,4,6,\cdots$ successively in (4) we get

$$a_2 = -\frac{a_0}{(m+2)^2}, a_4 = -\frac{a_2}{(m+4)^2} = \frac{a_0}{(m+2)^2(m+4)^2}$$

$$a_6 = -\frac{a_0}{(m+2)^2(m+4)^2(m+6)^2} \text{ and so on.}$$

Putting these a_i's in (2), we get

$$y = a_0 x^m \left\{1 - \frac{x^2}{(m+2)^2} + \frac{x^4}{(m+2)^2(m+4)^2} - \cdots\right\} \qquad \ldots \qquad (6)$$

$$\therefore \frac{\partial y}{\partial m} = a_0 x^m \log x \left\{1 - \frac{x^2}{(m+2)^2} + \frac{x^2}{(m+2)^2(m+4)^2} - \cdots\right\}$$

$$+ a_0 x^m \left\{\frac{2x^2}{(m+2)^3} + \frac{x^4}{(m+2)^2(m+4)^2}\left(\frac{-2}{m+2} - \frac{2}{m+4}\right) + \cdots\right\}$$

$$\therefore (y)_{m=0} = a_0\left(1 - \frac{x^2}{2^2} + \frac{x^4}{2^2 \cdot 4^2} - \cdots\right) = a_0 u \text{, say and}$$

$$\left(\frac{\partial y}{\partial m}\right)_{m=0} = a_0 \log x\left(1 - \frac{x^2}{2^2} + \frac{x^4}{2^2 \cdot 4^2} - \cdots\right) + a_0\left\{\frac{x^2}{2^2} - \frac{x^4}{2^2 \cdot 4^2}\left(1 + \frac{1}{2}\right) \cdots\right\}$$

$$= a_0\left[u\log x + \frac{x^2}{2^2} - \frac{x^4}{2^2 \cdot 4^2}\left(1+\frac{1}{2}\right) + \frac{x^6}{2^2 \cdot 4^2 \cdot 6^2}\left(1+\frac{1}{2}+\frac{1}{3}\right) + \cdots\right].$$

$= a_0 v$, say.

So the complete solution is

$$y = c_1(y)_{m=0} + c_2\left(\frac{\partial y}{\partial m}\right)_{m=0}$$

$$= a_1 u + a_2\left[u\log x + \frac{x^2}{2^2} - \frac{x^4}{2^2 \cdot 4^2}\left(1+\frac{1}{2}\right) + \cdots\right]$$

where $u = 1 - \frac{x^2}{2^2} - \frac{x^4}{2^2 \cdot 4^2} - \cdots$ and

$a_1 = a_0 c_1, a_2 = a_0 c_2$ are arbitrary constants.

Ex. 3. *(Against Case* (iii)) *Solve, in series, the equation*

$$x^2\frac{d^2y}{dx^2} + x\frac{dy}{dx} + (x^2-1)y = 0.$$

Here $x = 0$ is a regular singular point of the given equation

$$x^2\frac{d^2y}{dx^2} + x\frac{dy}{dx} + (x^2-1)y = 0 \qquad \cdots (1)$$

So let $y = \sum_{n=0}^{\infty} a_n x^{m+n}$, $a_0 \neq 0$ ... (2)

be the series solution of (1)

$$\because \frac{dy}{dx} = \sum_{n=0}^{\infty} a_n(m+n)x^{m+n-1}, \frac{d^2y}{dx^2} = \sum_{n=0}^{\infty} a_n(m+n)(m+n-1)x^{m+n-2}$$

Substituting these values in (1), we get

$$x^2\sum_{n=0}^{\infty} a_n(m+n)(m+n-1)x^{m+n-2} + x\sum_{n=0}^{\infty} a_n(m+n)x^{m+n-1} + (x^2-1)\sum_{n=0}^{\infty} a_n x^{m+n} = 0$$

or, $$\sum_{n=0}^{\infty} a_n(m+n)(m+n-1)x^{m+n} + \sum_{n=0}^{\infty} a_n(m+n)x^{m+n} + \sum_{n=0}^{\infty} a_n x^{m+n+2} - \sum_{n=0}^{\infty} a_n x^{m+n} = 0$$

or, $$\sum_{n=0}^{\infty} a_n\{(m+n)(m+n-1)+(m+n)-1\}x^{m+n} + \sum_{n=0}^{\infty} a_n x^{m+n+2} = 0$$

or, $$\sum_{n=0}^{\infty} a_n(m+n-1)(m+n+1)x^{m+n} + \sum_{n=0}^{\infty} a_n x^{m+n+2} = 0 \qquad \cdots (3)$$

Equating the co-efficient of the lower power of x i.e. x^m to zero we get the indicial equation as

$$a_0(m+1)(m-1)=0$$

$$\therefore m=1,-1(\because a_0 \neq 0).$$

Thus the roots of the indicial equation are unequal and differing by an integer. To find the recurrence relation, we equate the co-efficient of x^{m+n} to zero and get

$$a_n(m+n-1)(m+n-1)+a_{n-2}=0$$

$$\therefore a_n=-\frac{1}{(m+n-1)(m+n+1)}a_{n-2} \qquad \ldots \quad (4)$$

To determine a_1, we equate the co-efficient of x^{m+1} to zero and obtain

$$a_1(m+2)m=0$$

$$\therefore a_1=0 \text{ for both } m=1,-1.$$

Now putting $n=3,5,7,\cdots$ successively in (4), we get

$$a_1=a_3=a_5=\cdots=0$$

and then putting $n=2,4,6,\cdots$ successively in (4),

we get $a_2=-\frac{1}{(m+1)(m+3)}a_0$,

$$a_4=-\frac{1}{(m+3)(m+5)}a_2=\frac{1}{(m+1)(m+3)^2(m+5)}a_0$$

and so on.

Putting these values of a_i's in (2) we get

$$y=a_0x^m\left[1-\frac{x^2}{(m+1)(m+3)}+\frac{x^4}{(m+1)(m+3)^2(m+5)}-\cdots\right] \qquad \ldots \quad (5)$$

As some co-efficient become infinite for $m=-1$, we replace a_0 by $b_0(m+1)$ where $b_0 \neq 0$ and (5) becomes

$$y=b_0x^m\left[(m+1)-\frac{x^2}{m+3}+\frac{x^4}{(m+3)^2(m+5)}-\cdots\right]$$

$$\therefore \frac{\partial y}{\partial m} = b_0 x^m \log x \left[(m+1) - \frac{x^2}{m+3} + \frac{x^4}{(m+3)^2(m+5)} \right]$$

$$+ b_0 x^m \left[1 + \frac{x^2}{(m+3)^2} - \left\{ \frac{2}{(m+3)^3(m+5)} + \frac{1}{(m+3)^2(m+5)^2} \right\} x^4 - \cdots \right]$$

$$\therefore (y)_{m=-1} = \frac{b_0}{x} \left[-\frac{1}{2}x^2 + \frac{1}{2^2 \cdot 4}x^4 - \frac{1}{2^2 \cdot 4^2 \cdot 6}x^6 + \cdots \right]$$

$$= b_0\, u \text{, say}$$

and $$\left(\frac{\partial y}{\partial m}\right)_{m=-1} = \frac{b_0}{x} \log x \left[-\frac{1}{2}x^2 + \frac{1}{2^2 \cdot 4}x^4 - \cdots \right]$$

$$+ \frac{b_0}{x} \left[1 + \frac{1}{2^2}x^2 - \frac{1}{2^2 \cdot 4}\left(\frac{2}{2} + \frac{1}{4} \right) x^4 \cdots \right]$$

$$= b_0 u \log x + \frac{b_0}{x} \left[1 + \frac{1}{2^2}x^2 - \frac{1}{2^2 \cdot 4}\left(1 + \frac{1}{4} \right) x^4 - \cdots \right].$$

Hence the required general solution is given by

$$y = c_1 (y)_{m=-1} + c_2 \left(\frac{\partial y}{\partial x} \right)_{m=-1} = a_1 u + a_2 \left[u \log x + \frac{1}{x} + \frac{1}{2^2}x - \frac{1}{2^2 \cdot 4}\left(1 + \frac{1}{4} \right) x^3 \cdots \right]$$

where $u = -\frac{1}{2}x + \frac{1}{2^2 \cdot 4}x^3 - \frac{1}{2^2 \cdot 4^2 \cdot 6}x^5 + \cdots$ and $a_1 = b_0 c_1, a_2 = b_0 c_2$ are arbitrary contants.

Ex. 4. *(Against Case* (iii) (b)) *Solve, in series, the equation*

$$(2 + x^2)\frac{d^2 y}{dx^2} + x\frac{dy}{dx} + (1+x)y = 0 .$$

Here $x = 0$ is a regular singular point of the given equation.

$$(2 + x^2)\frac{d^2 y}{dx^2} + x\frac{dy}{dx} + (1+x)y = 0 \qquad \ldots \quad (1)$$

So let $y = \sum_{n=0}^{\infty} a_n x^{m+n}, \ a_0 \neq 0 \qquad \ldots \quad (2)$

be its series solution. Then

$$\frac{dy}{dx}=\sum_{n=0}^{\infty}a_n(m+n)x^{m+n-1},\frac{d^2y}{dx^2}=\sum_{n=0}^{\infty}a_n(m+n)(m+n-1)x^{m+n-2}$$

Substituting these values in (1), we get

$$(2+x^2)\sum_{n=0}^{\infty}(m+n)(m+n-1)a_nx^{m+n-2}+x\sum_{n=0}^{\infty}(m+n)a_nx^{m+n-1}$$
$$+(1+x)\sum_{n=0}^{\infty}a_nx^{m+n}=0$$

or, $$\sum_{n=0}^{\infty}\left[(m+n)(m+n-1)+(m+n)+1\right]a_nx^{m+n}$$

$$+\sum_{n=0}^{\infty}a_nx^{m+n+1}+\sum_{n=0}^{\infty}2(m+n)(m+n-1)a_nx^{m+n-2}=0 \quad \ldots \quad (3)$$

Equating the co-efficient of the lower power of x, i.e. x^{m-2} to zero we get the indicial equation as

$$a_0\cdot 2(m-1)m=0$$

$$\therefore\ m=0,1\left[\because a_0\neq 0\right].$$

So the roots are unequal and differ by an integer. To find the recurrence relation, we equat the co-efficient of x^{m+n} to zero and get

$$\{(m+n)(m+n-1)+(m+n)+1\}a_n+a_{n-1}+2(m+n+2)(m+n+1)a_{n+2}=0$$

or, $$2(m+n+2)(m+n+1)a_{n+2}=-a_{n-2}-\left\{(m+n)^2+1\right\}a_n$$

$$\therefore a_{n+2}=-\frac{a_{n-1}}{2(m+n+2)(m+n+1)}-\frac{(m+n)^2+1}{2(m+n+2)(m+n+1)}a_n \quad \ldots \quad (4)$$

Next equating the co-efficient of x^{m-1}, to zero we have

$$2a_1(m+1)m=0$$

which shows that a_1 is indeterminate for $m=0$.

So for $m=0$, (4) gives

$$a_{n+2}=-\frac{a_{n-1}}{2(n+2)(n+1)}-\frac{n^2+1}{2(n+2)(n+1)}a_n \quad \ldots \quad (5)$$

Finally equating the co-efficient of x^m to zero we have

$$a_0(m^2+1)+2a_2(m+2)(m+1)=0\,.$$

$$\therefore a_2=-\frac{m^2+1}{2(m+1)(m+2)}a_0$$

$$\therefore \text{ For } m=0,\ a_2=-\frac{a_0}{4}\,.$$

Now assuming that a_1 is finite and putting $n=1,2,3,\cdots$ successively in (5) we get

$$a_3=-\frac{a_0}{2\cdot3\cdot2}-\frac{2}{2\cdot3\cdot2}a_1=\frac{-a_0}{12}-\frac{a_1}{6}$$

$$a_4=-\frac{a_1}{2\cdot4\cdot3}-\frac{5}{2\cdot4\cdot3}a_2=-\frac{a_1}{24}-\frac{5}{24}\left(-\frac{a_0}{4}\right)$$

$$=-\frac{a_1}{24}+\frac{5}{96}a_0$$

and so on.

Putting these values of a_i in (2), we get

$$y=\sum_{n=0}^{\infty}a_nx^n\quad[\because m=0]$$

$$=a_0\left(1-\frac{1}{4}x^2-\frac{1}{12}x^3-\cdots\right)+a_1\left(x-\frac{1}{6}x^3-\frac{1}{24}x^4-\cdots\right)$$

which is the required general solution, a_0, a_1 being arbitrary constants.

EXERCISES

1. Determine whether the point $x=0$ is an ordinary or singular point of the equation

$$\left(1-x^2\right)\frac{d^2y}{dx^2}-x\frac{dy}{dx}+4y=0\,.$$

Also find all singular points of the equation.

2. Determine all the singular points of the equation

$$2x^2y''-xy'+(x+2)y=0\,.$$

3. Show that $x=0$ is a regular singular point but $x=2$ is not a regular singular point of the equation $x(x-2)^3y''+3(x-2)^3y'+4y=0$.

4. Determine all the singular points of the equation

$$x^2(x+1)^2\frac{d^2y}{dx^2}+(x^2-1)\frac{dy}{dx}+2y=0$$

5. Find the series solution of the following equation :

(i) $(2-x^2)\frac{d^2y}{dx^2}+2x\frac{dy}{dx}-2y=0$

(ii) $y''+xy'+y=0$.

(iii) $(x^2+1)y''+xy'-xy=0$.

(iv) $(x^2-1)y''+4xy'+2y=0$.

6. Find the power series solution in power of $x-1$ of the equation $xy''+y'+2y=0$ given that $y(1)=2, y'(1)=4$.

7. Solve $y''-xy'+2y=0$ near $x=1$.

8. Solve the following equations :

(i) $(x^2-1)\frac{d^2y}{dx^2}+3x\frac{dy}{dx}+xy=0$, given $y(2)=4, y'(2)=6$.

(ii) $x\frac{d^2y}{dx^2}+\frac{dy}{dx}+2y=0$, given $y(1)=1, y'(1)=2$.

9. Solve the following equations :

(i) $(2x+x^3)\frac{d^2y}{dx^2}-\frac{dy}{dx}-6xy=0$.

(ii) $9x(1-x)y''-12y'+4y=0$.

(iii) $(x-x^2)y''+(1-x)y'-y=0$.

(iv) $(x-x^2)y''+(1-5x)y'-4y=0$.

(v) $x^2y''+xy'+(x^2-4)y=0$.

(vi) $(1-x^2)y''+2xy'+y=0$.

(vii) $y''-x^2y'-y=0$.

(viii) $(x-x^2)y''+4y'+2y=0$.

(ix) $2x^2y''+xy'-(x+1)y=0$.

(x) $2x^2y''-xy'+(1-x^2)y=0$.

ANSWERS

1. $x=0$ is an ordinary point ; $x=\pm 1$ are the singular points.

2. $x=0$ is a regular singular point.

4. $x=0,-1$ are singular points.

5. (i) $y=a_0\left(1+\dfrac{x^4}{2!}+\dfrac{x^8}{4!}+\cdots\right)+a_1\left(x^2-\dfrac{x^6}{3!}+\dfrac{x^{10}}{5!}-\cdots\right)$.

(ii) $y=a_0\left(1-\dfrac{x^2}{2}+\dfrac{x^4}{2\cdot 4}-\dfrac{x^6}{2\cdot 4\cdot 6}+\cdots\right)+a_1\left(x-\dfrac{x^3}{3}+\dfrac{x^5}{3\cdot 5}-\dfrac{x^7}{3\cdot 5\cdot 7}+\cdots\right)$.

(iii) $y=a_0\left(1+\dfrac{1}{6}x^3-\dfrac{3}{40}x^5+\cdots\right)+a_1\left(x-\dfrac{1}{6}x^3+\dfrac{1}{12}x^4-\dfrac{3}{40}x^5+\cdots\right)$.

(iv) $y=a_0\left(1+x^2+x^4+\cdots\right)+a_1\left(x+x^3+x^5+\cdots\right)$.

6. $y=2\left\{1-(x-1)^2+\dfrac{2}{3}(x-1)^3\cdots\right\}+4\left\{(x-1)-\dfrac{1}{2}(x-1)^2+\dfrac{1}{12}(x-1)^4\cdots\right\}$.

7. $y=a_0\left\{1-(x-1)^2-\dfrac{1}{3}(x-1)^3-\cdots\right\}+a_1\left\{(x-1)+\dfrac{1}{2}(x-1)^2\cdots\right\}$.

8. (i) $y=4+6(x-2)-\dfrac{22}{3}(x-2)^2+\dfrac{169}{27}(x-2)^3-\cdots$.

(ii) $y=1+2(x-1)-2(x-1)^2+\dfrac{2}{3}(x-1)^3-\dfrac{1}{6}(x-1)^4-\cdots$.

9. (i) $y=a_1\left(1+3x^2+\dfrac{3}{5}x^4\cdots\right)+a_2x^{3/2}\left(1+\dfrac{3}{8}x^2-\dfrac{1\cdot 3}{8\cdot 16}x^4+\cdots\right)$.

(ii) $y=a_1\left(1+\dfrac{1}{3}x+\dfrac{1\cdot 4}{3\cdot 6}x^2+\cdots\right)+a_2x^{7/3}\left(1+\dfrac{8}{10}x+\dfrac{8\cdot 11}{10\cdot 13}x^2+\cdots\right)$.

(iii) $y=a_1u+a_2\left\{u\log x-2x-x^2-\dfrac{14}{27}x^3\cdots\right\}$

where $u=\left(1+x+\dfrac{1}{2}x^2+\dfrac{5}{18}x^3+\cdots\right)$

(iv) $y = a_1 u + a_2\left\{u\log x - 2\left(1\cdot 2x + 2\cdot 3x^2 + 3\cdot 4x^3 + \cdots\right)\right\}$ where

$u = \left(1 + 2^2 x + 3^2 x^2 + 4^2 x^3 + \cdots\right)$.

(v) $y = a_1 u + a_2\left\{u\log x + \frac{1}{x^2}\left(1 + \frac{x^2}{2^2} + \frac{x^4}{2^2\cdot 4^2} + \cdots\right)\right\}$ where

$u = -\frac{x^2}{2^2\cdot 4} + \frac{x^4}{2^3\cdot 4\cdot 6} - \frac{x^6}{2^3\cdot 4^2\cdot 6\cdot 8} + \cdots$.

(vi) $y = a_1\left(1 - \frac{1}{2}x^2 + \frac{1}{8}x^4 - \cdots\right) + a_2\left(x - \frac{1}{2}x^3 + \frac{1}{40}x^5 \cdots\right)$.

(vii) $y = a_1\left(1 + \frac{1}{2}x^2 + \frac{1}{24}x^4 + \frac{1}{20}x^5 + \cdots\right) + a_2\left(x + \frac{1}{6}x^3 + \frac{1}{12}x^4 + \frac{1}{12}x^5 + \cdots\right)$

(viii) $y = a_1\left(1 - \frac{x}{2} + \frac{x^2}{10} - \cdots\right) + \frac{a_2}{x^3}\left(1 - 5x + 10x^2 - 10x^3 + \cdots\right)$.

(ix) $y = c_1 x\left(1 + \frac{1}{5}x + \frac{1}{70}x^2 + \cdots\right) + a_2 x^{-1/2}\left(1 - x - \frac{1}{2}x^2 + \cdots\right)$.

(x) $y = a_1 x\left(1 + \frac{x^2}{2\cdot 5} + \frac{x^4}{2\cdot 4\cdot 5\cdot 9} + \cdots\right) + a_2 x^{1/2}\left(1 + \frac{x^2}{2\cdot 3} + \frac{x^4}{2\cdot 3\cdot 4\cdot 7} + \cdots\right)$.

4.4 BESSEL'S EQUATION AND BESSEL FUNCTIONS

4.4.1. Bessel's Equation.

The differential equation of the form

$$x^2\frac{d^2y}{dx^2}+x\frac{dy}{dx}+(x^2-n^2)y=0$$

or, $$\frac{d^2y}{dx^2}+\frac{1}{x}\frac{dy}{dx}+\left(1-\frac{n^2}{x^2}\right)y=0. \quad \ldots \quad (1)$$

is called Bessel's equation of order n, n being a non-negative constant.

Solution of Bessel's differential equation.

We now solve (1) in series by using the well known method of Frobenius.

Let the series solution of (1) be

$$y=\sum_{m=0}^{\infty} a_m x^{m+k} \quad [a_0 \neq 0.] \quad (2)$$

$$\therefore \quad \frac{dy}{dx}=\sum_{m=0}^{\infty} a_m (m+k)\, x^{m+k-1}.$$

Substituting these values in (1) we have

$$x^2\sum_{m=0}^{\infty} a_m (m+k)(m+k-1)\, x^{m+k-2}+x\sum_{m=0}^{\infty} a_m (m+k)\, x^{m+k-1}$$
$$+(x^2-n^2)\sum_{m=0}^{\infty} a_m x^{m+k}=0$$

or, $$\sum_{m=0}^{\infty} a_m\left\{(m+k)(m+k-1)+(m+k)-n^2\right\}x^{m+k}+\sum_{m=0}^{\infty} a_m x^{m+k+2}=0$$

or, $$\sum_{n=0}^{\infty} a_m (k+m+n)(k+m-n)x^{k+m}+\sum_{m=0}^{\infty} a_m x^{k+m+2}=0 \ldots \quad (3)$$

Equating to zero the smallest power of x, namely x^k, (3) gives the indicial equation $a_0(k+n)(k-n)=0$.

$\therefore \quad k=\pm n, \quad (\because a_0 \neq 0)$.

So the roots of the indicial equation are $n, -n$.

Next equating to zero the coefficient of x^{k+1} gives

$$a_1(k+1+n)(k+1-n)=0$$

$\therefore \quad a_1=0$ for $k=n,-n.$

Lastly equating to zero the coefficient of x^{k+m} gives

$$a_m(k+m+n)(k+m-n)+a_{m-2}=0.$$

$$\therefore \quad a_m=-\frac{a_{m-2}}{(k+m+n)(k+m-n)}. \quad \ldots \quad (4)$$

Putting $m = 3, 5, 7, \ldots$ in (4) and using $a_1 = 0$, we find

$$a_1=a_3=a_5=\ldots=0.$$

Now two cases arise :-

Case I. When $k=n.$

From (4), we have by putting succesively $m = 2, 4, 6, \ldots$

$$a_2=-\frac{1}{(k+2+n)(k+2-n)}a_0$$

$$a_4=-\frac{1}{(k+4+n)(k+4-n)}a_2$$

$$=\frac{1}{(k+4+n)(k+4-n)(k+2+n)(k+2-n)}a_0 \text{ and so on.}$$

Putting these values in (2), we get

$$y=a_0x^k\left[1-\frac{x^2}{(k+2+n)(k+2-n)}+\frac{x^4}{(k+4+n)(k+4-n)(k+2+n)(k+2-n)}\ldots\right]$$

$\therefore$ For $k = n$, we have

$$y=a_0x^n\left\{1-\frac{x^2}{4(1+n)}+\frac{x^4}{4.8(1+n)(2+n)}-\ldots\right\}.$$

If $a_0=\dfrac{1}{2^n\Gamma(n+1)}$, this solution is called Bessel's *function of the first kind of order n* and is denoted by $J_n(x)$, where $\Gamma(n)$ is Gamma Function.

$$\therefore \quad J_n(x) = \frac{x^n}{2^n \Gamma(n+1)} \left[1 + (-1) \frac{x^2}{2^2 \, \underline{|1} (n+1)} + (-1)^2 \frac{x^4}{2^4 \; 2!(n+1)(n+2)} + \dots \right]$$

$$= \frac{x^n}{2^n \Gamma(n+1)} \sum_{r=0}^{\infty} (-1)^r \frac{x^{2r}}{2^r \;\; r!(n+1)(n+2)\dots(n+r)}$$

$$= \sum_{r=0}^{\infty} (-1)^r \left(\frac{x}{2}\right)^{n+2r} \frac{1}{r!\,\Gamma(n+r+1)}.$$

Case II. When $k = -n$.

The series solution is obtained by replacing n by $-n$ in the value of J_n.

$$\therefore \quad J_{-n}(x) = \sum_{r=0}^{\infty} (-1)^r \left(\frac{x}{2}\right)^{-n+2r} \frac{1}{r!\,\Gamma(-n+r+1)}$$

When n is not integer $J_{-n}(x)$ is distinct from $J_n(x)$.

Note. The solution of Bessel's equation are called Bessel's functions.

General solution of Bessel's equation.

The general solution of Bessel's equation is

$$y = AJ_{-n}(x) + BJ_n(x)$$

where A, B are two arbitrary constants.

4.4.2. Series solution of Bessel's equation for $n = 0$.

The differential equation

$$\frac{d^2y}{dx^2} + \frac{1}{x}\frac{dy}{dx} + y = 0 \qquad \dots (1)$$

is known as Bessel's equation for $n = 0$.

Let the series solution be

$$y = \sum_{m=0}^{\infty} a_m x^{k+m}. \quad [a_0 \neq 0.] \qquad \dots (2)$$

$$\therefore \quad \frac{dy}{dx} = \sum a_m (k+m)\, x^{k+m-1}$$

$$\frac{d^2y}{dx^2} = \sum a_m (k+m)(k+m-1)\, x^{k+m-2}.$$

Substituting these values in (1), we get

$$\sum_{m=0}^{\infty} a_m (k+m)^2 x^{k+m-2} + \sum_{m=0}^{\infty} a_m x^{k+m} = 0 \qquad \ldots \quad (3)$$

which is an identity.

Equating to zero the coefficient of lowest power of x i.e., x^{k-2}. we have

$$a_0 k^2 = 0.$$

$$\therefore \quad k = 0, 0 \quad (\because\ a_0 \neq 0).$$

Now equating to zero the coefficient of next power of x i.e., x^{k-1}. we have

$$a_1 (k+1)^2 = 0.$$

$$\therefore \quad a_1 = 0, \text{ for } k = 0.$$

Again equating to zero the coefficient of the general term i.e., x^{k+m}, we have

$$a_{m+2} (k+m+2)^2 + a_m = 0.$$

$$\therefore \quad a_{m+2} = -\frac{a_m}{(k+m+2)^2}. \qquad \ldots \quad (4)$$

Putting m = 1, 3, 5, . . . and using $a_1 = 0$, we get

$$a_1 = a_3 = a_5 = \ldots = 0.$$

Next putting m = 0, 2, 4, . . . in (4) successively we get

$$a_2 = -\frac{a_0}{(k+2)^2}$$

$$a_4 = -\frac{a_2}{(k+4)^2} = \frac{a_0}{(k+2)^2 (k+4)^2} \text{ and so on.}$$

Hence from (2),

$$y = a_0 x^k \left(1 - \frac{x^2}{(k+2)^2} + \frac{x^4}{(k+2)^2 (k+4)^2} - \cdots \right)$$

When $k = 0$,

$$y = a_0 \left(1 - \frac{x^2}{2^2} + \frac{x^4}{2^2.4^2} \cdots \right).$$

If $a_0 = 1$, this solution is denoted by $J_0(x)$.

$$\therefore \quad J_0(x) = 1 - \frac{x^2}{2^2} + \frac{x^2}{2^2.4^2} - \frac{x^6}{2^2.4^2.6^2} + \cdots$$

where $J_0(x)$ is called Bessel's function of zeroth order.

4.4.3. Recurrence formulae.

(I) $\dfrac{d}{dx}(x^n J_n) = x^n J_{n-1}$ or $xJ_n'(x) = -nJ_n(x) + xJ_{n-1}(x)$.

Proof. We know that

$$J_n(x) = \sum_{r=0}^{\infty} \frac{(-1)^r \left(\frac{x}{2}\right)^{n+2r}}{\Gamma(r+1)\Gamma(n+r+1)}$$

$$\therefore \quad x^n J_n(x) = \sum_{r=0}^{\infty} \frac{(-1)^r x^{2(n+r)}}{2^{n+2r}\Gamma(r+1)\Gamma(n+r+1)}$$

$$\therefore \quad \frac{d}{dx}\{x^n J_n(x)\} = \sum_{r=0}^{\infty} \frac{(-1)^r 2(n+r)\, x^{2(n+r)-1}}{2^{n+2r}\Gamma(r+1)\Gamma(n+r+1)}$$

$$= x^n \sum_{r=0}^{\infty} \frac{(-1)^r \left(\frac{x}{2}\right)^{n-1+2r}}{\Gamma(r+1)\,\Gamma\{(n-1)+r+1\}}.$$

$$\therefore \quad \frac{d}{dx}\{x^n J_n(x)\} = x^n J_{n-1}(x).$$

$$\therefore \quad x^n J_n'(x) + nx^{n-1} J_n(x) = x^n J_{n-1}(x).$$

$$\therefore \quad xJ_n'(x) = -nJ_n(x) + xJ_{n-1}(x)$$

(II) $\dfrac{d}{dx}\left(x^{-n}J_n(x)\right)=-x^{-n}J_{n+1}(x)$ or $xJ_n(x)=nJ_n(x)-xJ_{n+1}(x)$.

Proof. We know that

$$J_n(x)=\sum_{r=0}^{\infty}\frac{(-1)^r\left(\dfrac{x}{2}\right)^{n+2r}}{\Gamma(r+1)\,\Gamma(n+r+1)}.$$

$$\therefore\quad x^{-n}J_n(x)=\sum_{r=0}^{\infty}\frac{(-1)^r x^{2r}}{2^{n+2r}\Gamma(r+1)\,\Gamma(n+r+1)}.$$

$$\therefore\quad \frac{d}{dx}\{x^{-n}J_n(x)\}=\sum_{r=0}^{\infty}\frac{(-1)^r 2r\, x^{2r-1}}{2^{n+2r}\Gamma(r+1)\,\Gamma(n+r+1)}.$$

$$=-x^{-n}\sum_{r=1}^{\infty}\frac{(-1)^{r-1}\left(\dfrac{x}{2}\right)^{(n+1)+2(r-1)}}{\Gamma\{(r-1)+1\}\,\Gamma\{(n+1)+(r-1)+1\}}.$$

[since $\Gamma(r)=\infty$ when $r=0$, so the term corresponding to $r=0$ vanishes]

$$=-x^{-n}\sum_{m=0}^{\infty}\frac{(-1)^m\left(\dfrac{x}{2}\right)^{n+1+2m}}{\Gamma(m+1)\,\Gamma\{(n+1)+m+1\}},\text{ by putting } m=r-1$$

$$=-x^{-n}J_{n+1}(x).$$

$$\therefore\quad \frac{d}{dx}\{x^{-n}J_n(x)\}=-x^{-n}J_{n+1}(x).$$

$$\therefore\quad x^{-n}J_n'(x)-nx^{-n-1}J_n(x)=-x^{-n}J_{n+1}(x).$$

$$\therefore\quad x\,J_n'(x)=n\,J_n(x)-x\,J_{n+1}(x).$$

(III). $2\,J_n'(x)=J_{n-1}(x)-J_{n+1}(x)$.

Proof. From recurrence relation (I) and (II), we have

$$xJ_n'(x)=-nJ_n(x)+x\,J_{n-1}(x) \qquad \ldots \quad (1)$$

$$xJ_n'(x)=nJ_n(x)-x\,J_{n+1}(x) \qquad \ldots \quad (2)$$

Adding (1) and (2), we get

$$2xJ'_n(x) = xJ_{n-1}(x) - xJ_{n+1}(x)$$

$$\therefore \quad 2J'_n(x) = J_{n-1}(x) - J_{n+1}(x).$$

(IV) $\quad 2nJ_n(x) = x\left[J_{n-1}(x) + J_{n+1}(x)\right].$

Proof. From recurrence relations (I) and (II), we have

$$xJ'_n(x) = -nJ_n(x) + xJ_{n-1}(x) \quad \ldots \quad (1)$$

$$xJ'_n(x) = nJ_n(x) - xJ_{n+1}(x) \quad \ldots \quad (2)$$

Subtracting (2) from (1), we get

$$O = 2nJ_n(x) - xJ_{n+1}(x) - xJ_{n-1}(x).$$

$$\therefore \quad 2nJ_n(x) = x\left[J_{n-1}(x) + J_{n+1}(x)\right]$$

Definition. Bessel's function of 2nd kind of order n or Neumann function is (given by)

$$Y_n(x) = J_n(x) \int \frac{dx}{x\left[J_n(x)\right]^2}.$$

Illustrative Examples.

Ex. 1. *Show that when n is a positive integer*

(i) $\quad J_{-n}(x) = (-1)^n J_n(x)$, *and* [*W.B.U.Tech'2003*]

(ii) $\quad J_n(-x) = (-1)^n J_n(x)$ *for +ve or –ve integers.*

(i) We have

$$J_{-n}(x) = \sum_{r=0}^{\infty} (-1)^r \left(\frac{x}{2}\right)^{-n+2r} \frac{1}{r!\,\Gamma(-n+r+1)}$$

since, if p is an integer, then $\Gamma(-p)$ is infinity for $p > 0$. So we get terms in J_{-n} equal to zero till $-n + r + 1 < 1$, i.e., $r < n$.

Hence we can write

$$J_{-n}(x) = \sum_{r=0}^{\infty} \frac{(-1)^r}{r!\,\Gamma(-n+r+1)} \left(\frac{x}{2}\right)^{-n+2r}$$

$$= \sum_{s=0}^{\infty} \frac{(-1)^{n+s}}{n+s!\,\Gamma(s+1)} \left(\frac{x}{2}\right)^{n+2s} \quad \text{by putting } r = n+s$$

$$= (-1)^n \sum_{s=0}^{\infty} \frac{(-1)^s}{\Gamma(n+s+1)\ \ s!} \left(\frac{x}{2}\right)^{n+2s}$$

$$= (-1)^n J_n(x).$$

(ii) We know that $J_n(x) = \sum_{r=0}^{\infty} (-1)^r \frac{1}{r!\,\Gamma(n+r+1)} \left(\frac{x}{2}\right)^{n+2r}$ ... (1)

Case (a). Let n be a +ve integer

Replacing x by $-x$ in (1), we have

$$J_n(-x) = \sum_{r=0}^{\infty} (-1)^r \frac{1}{r!\,\Gamma(n+r+1)} \left(-\frac{x}{2}\right)^{n+2r}$$

$$= (-1)^n \sum_{r=0}^{\infty} (-1)^r \frac{1}{r!\,\Gamma(n+r+1)} \left(\frac{x}{2}\right)^{n+2r} \quad \left[\because (-1)^{2r} = 1\right]$$

$$= (-1)^n J_n(x).$$

Case (b). Let n be a –ve integer.

So let $n = -m$ where m is a +ve integer.

$\therefore \quad J_n(x) = J_{-m}(x) = (-1)^m J_m(x)$ by (i)

Replacing x by $-x$, we have

$$J_n(-x) = (-1)^m J_m(-x)$$

$$= (-1)^m (-1)^m J_m(x) \text{ by case (a)}$$

$$= (-1)^m J_{-m}(x) = (-1)^{2m} (-1)^{-m} J_{-m}(x)$$

$$= (-1)^{-m} J_{-m}(x) \quad \left(\because (-1)^{2m} = 1\right) = (-1)^n J_n(x).$$

Hence $J_n(-x) = (-1)^n J_n(x)$ for +ve or –ve integers.

Ex. 2. *Prove that* $J_0' = -J_1$.

Putting $n = 0$ in the recurrence formula

$$xJ_n' = nJ_n - xJ_{n+1}$$

we get $xJ_0' = -xJ_1$

$\therefore \quad J_0' = -J_1.$

Ex. 3. *Show that*

(i) $J_{-1/2}(x) = \sqrt{\dfrac{2}{\pi x}}\cos x.$ [*W.B.U.Tech '2004*]

(ii) $J_{1/2}(x) = \sqrt{\dfrac{2}{\pi x}}\sin x$

(iii) $\left(J_{1/2}(x)\right)^2 + \left(J_{-1/2}(x)\right)^2 = \dfrac{2}{\pi x}.$

We know that

$$J_n(x) = \frac{x^n}{2^n\,\Gamma(n+1)}\left[1 - \frac{x^2}{2(2n+2)} + \frac{x^4}{2.4(2n+2)(2n+4)} + \cdots\right]$$

(i) Putting $n = -1/2$ in (1), we get

$$J_{-1/2}(x) = \frac{x^{-1/2}}{2^{-1/2}\,\Gamma(1/2)}\left[1 - \frac{x^2}{2.1} + \frac{x^4}{2.4.3.1} + \cdots\right]$$

$$= \sqrt{\frac{2}{\pi x}}\left[1 - \frac{x^2}{2!} + \frac{x^4}{4!} + \cdots\right]$$

$$= \sqrt{\frac{2}{\pi x}}\cos x.$$

(ii) Putting $x = 1/2$, in (i) we have

$$J_{1/2}(x) = \frac{x^{1/2}}{2^{1/2}\,\Gamma(3/2)}\left[1 - \frac{x^2}{2.3} + \frac{x^4}{2.4.3.5} + \cdots\right]$$

$$= \sqrt{\frac{x}{2}}.\frac{1}{\frac{1}{2}.\sqrt{\pi}}\frac{1}{x}\left(x - \frac{x^3}{3!} + \frac{x^5}{5!} + \cdots\right) = \sqrt{\frac{2}{\pi x}}\sin x.$$

(iii) Squaring and adding (i) and (ii) we get the result

Ex. 4. *Prove that*

(i) $J_{-3/2}(x)=\sqrt{\frac{2}{\pi x}}\left(-\frac{\cos x}{x}-\sin x\right)$

(ii) $J_{3/2}(x)=\sqrt{\frac{2}{\pi x}}\left(\frac{\sin x}{x}-\cos x\right)$.

We know that

$$J_{-1/2}(x)=\sqrt{\frac{2}{\pi x}}\cos x \qquad \dots (1)$$

$$J_{1/2}(x)=\sqrt{\frac{2}{\pi x}}\sin x \qquad \dots (2)$$

Also we have the recurrence relation

$$J_{n-1}(x)+J_{n+1}(x)=\frac{2n}{x}J_n(x). \qquad \dots (3)$$

(i) Putting $n=-\frac{1}{2}$ in (3), we get

$$J_{-3/2}(x)+J_{1/2}(x)=-\frac{1}{x}J_{-1/2}(x).$$

$$\therefore \quad J_{-3/2}(x)=\sqrt{\frac{2}{\pi x}}\left(-\frac{\cos x}{x}-\sin x\right), \text{ by (1), (2)}$$

(ii) Putting $n=\frac{1}{2}$ in (3) we get

$$J_{-1/2}(x)+J_{3/2}(x)=\frac{1}{n}J_{1/2}(x)$$

$$\therefore \quad J_{3/2}(x)=\sqrt{\frac{2}{\pi x}}\left(\frac{\sin x}{x}-\cos x\right).$$

Ex. 5. *Express $J_4(x)$ in terms of J_0 and J_1.*

From the recurrence relation

$$J_{n+1}(x)=\frac{2\pi}{x}J_n(x)-J_{n-1}(x). \qquad \dots (1)$$

we get by putting $n = 3, 2$.

$$J_4(x) = \frac{6}{n} J_3(x) - J_2(x)$$

$$J_3(x) = \frac{4}{n} J_2(x) - J_1(x).$$

$$\therefore \quad J_4(x) = \frac{6}{n}\left(\frac{4}{n} J_2 - J_1\right) - J_2 = \left(\frac{24}{n^2} - 1\right) J_2 - \frac{6}{n} J_1.$$

Again from (1),

$$J_2(x) = \frac{2}{n} J_1(x) - J_0(x) \qquad \ldots \quad (1)$$

$$\therefore \quad J_2(x) = \left(\frac{48}{n^3} - \frac{8}{n}\right) J_1(x) - \left(\frac{24}{n^2} - 1\right) J_0(x).$$

Ex. 6. *Show that :*

(i) $\cos x = J_0 - 2J_2 + 2J_4 - \ldots$, *and*

(ii) $\sin x = 2J_1 - 2J_3 + 2J_5 - \ldots$

We know that

$$e^{x\left(z - \frac{1}{z}\right)/2} = J_0 + \left(z - \frac{1}{z}\right) J_1 + \left(z^2 + \frac{1}{z^2}\right) J_2 + \left(z^3 - \frac{1}{z^3}\right) J_3 + \cdots$$

Putting $z = e^{i\theta}$, we have

$$e^{x\frac{e^{i\theta} - e^{-i\theta}}{2}} = J_0 + (e^{i\theta} - e^{-i\theta}) J_1 + (e^{2i\theta} + e^{-2i\theta}) J_2 + (e^{3i\theta} - e^{-3i\theta}) J_3 + \cdots$$

or, $e^{xi\sin\theta} = J_0 + (2i\sin\theta) J_1 + (2\cos 2\theta) J_2 + (2i\sin 3\theta) J_3 + \cdots$

or, $\cos(x\sin\theta) + i\sin(x\sin\theta) = (J_0 + 2\cos 2\theta\, J_2 + 2\cos 4\theta\, J_4 + \cdots)$
$\qquad + i\,(2\sin\theta\, J_1 + 2\sin 3\theta\, J_3 + \cdots)$

Equating real and imaginary parts we have

$$\cos(x\sin\theta) = J_0 + 2\cos 2\theta\, J_2 + 2\cos 4\theta\, J_4 + \cdots$$

and $\sin(x\sin\theta) = 2\sin\theta\, J_1 + 2\sin 3\theta\, J_3 + \cdots$

Putting $\theta = \pi/2$, we have

$$\cos x = J_0 - 2J_2 + 2J_4 - \cdots$$

and $\sin x = 2J_1 - 2J_3 + 2J_5 - \cdots$

Ex. 7. *Show that when n is integral,*

$$(a)\quad \pi J_n = \int_0^\pi \cos(n\theta - x\sin\theta)\, d\theta.$$

$$(b)\quad \pi J_0 = \int_0^\pi \cos(x\sin\phi)\, d\theta.$$

$$= \int_0^\pi \cos(x\sin\phi)\, d\phi.$$ *and hence deduce that*

$$J_0(x) = 1 - \frac{x^2}{2^2} + \frac{x^4}{2^2.4^2} - \frac{x^6}{2^2.4^2.6^2} + \cdots$$

$$= \sum_{r=0}^{\infty} \frac{(-1)^r x^{2r}}{(2^r r!)^2}.$$

(a) We shall use the following result :

$$\int_0^\pi \cos m\phi \cos n\phi\, d\phi = \int_0^\pi \sin m\phi \sin n\phi\, d\phi = \begin{cases} \pi/2 \text{ when } m = n \\ 0 \text{ when } m \neq n \end{cases} \quad ...(1)$$

$$\cos(x\sin\theta) = J_0 + 2J_2\cos 2\theta + \cdots + 2J_{2m}\cos 2m\theta + \cdots \quad ...\ (2)$$

$$\sin(x\sin\theta) = 2\sin\theta\, J_1 + 2\sin 3\theta\, J_3 + \cdots + 2J_{2m+1}\sin(2m+1)\theta + \cdots \quad ...\ (3)$$

Now multiplying both sides of (2) by $\cos 2m\theta$ and then integrating between the limits 0 to π, we get

$$\int_0^\pi \cos(x\sin\theta)\cos 2m\theta\, d\theta = J_0\int_0^\pi \cos 2m\theta d\theta + 2J_2\int_0^\pi \cos 2\theta\cos 2m\theta\, d\theta$$
$$+ \cdots + 2J_{2m}\int_0^\pi \cos^2 2m\theta\, d\theta + \cdots$$
$$= 2.J_{2m}.\pi/2, \text{ by (1)}$$
$$= \pi J_{2m}. \quad ...\ (4)$$

Similarly multiplying both sides of (2) by $\cos(2m+1)\theta$ and then integrating between the limits 0 to π and using (1), we have

$$\int_0^\pi \cos(x\sin\theta)\cos(2m+1)\theta\, d\theta = 0. \quad ...\ (5)$$

Again multiplying both sides of (3) by $\sin(2m+1)\theta$ and then integrating between the limits 0 to π, we get

Similarly we get

$$\int_0^\pi \sin(x\sin\theta)\sin 2m\theta\, d\theta = 0. \quad \ldots \quad (7)$$

Therefore $\int_0^\pi \cos(2m\theta - x\sin\theta)\, d\theta$

$$= \int_0^\pi \cos 2m\theta \cos(x\sin\theta)\, d\theta + \int_0^\pi \sin 2m\theta \sin(x\sin\theta)\, d\theta$$

$$= \pi J_{2m} + 0, \text{ by (4) \& (7)}$$

$$= \pi J_{2m}. \quad \ldots \quad (8)$$

Also $\int_0^\pi \cos[(2m+1)\theta - x\sin\theta]\, d\theta$

$$= \int_0^\pi \cos(2m+1)\theta\cos(x\sin\theta)\, d\theta + \int_0^\pi \sin(2m+1)\theta\sin(x\sin\theta)\, d\theta$$

$$= \pi J_{2m+1}, \text{ by (5) and (6)} \quad \ldots \quad (9)$$

So in virtue of (8) and (9), for all positive integral values of n, we have

$$\int_0^\pi \cos(n\theta - x\sin\theta)\, d\theta = \pi J_n. \quad \ldots \quad (10)$$

If n be a negative integer, say $n = -m$, where m is positive, then

$$\int_0^\pi \cos(n\theta - x\sin\theta)\, d\theta$$

$$= \int_0^\pi \cos(-m\theta - x\sin\theta)\, d\theta$$

$$= -\int_0^\pi \cos\{-m(\pi-\phi) - x\sin(\pi-\phi)\}\, d\phi, \text{ putting } \theta = \pi - \phi.$$

$$= \int_0^\pi \cos\{m\pi - (m\phi - x\sin\phi)\}\, d\phi$$

$$= (-1)^m \int_0^\pi \cos(m\phi - x\sin\phi)\, d\phi$$

$$= (-1)^m \pi J_m, \text{ by (10)}$$

$$= \pi J_{-m} \quad (\because J_{-m}(x) = (-1)^m J_m(x))$$

$$= \pi J_n. \quad \ldots \quad (11)$$

Hence, in virtue of (10) and (11), for all integral values of n,

$$\int_0^{\pi} \cos(n\theta - x\sin\theta)\, d\theta = \pi J_n.$$

(b) From (2), we have

$$\int_0^{\pi} \cos(x\sin\phi)\, d\phi = J_0 \int_0^{\pi} d\phi + 2J_2 \int_0^{\pi} \cos 2\phi\, d\phi + \cdots$$

$$= \pi J_0.$$

Now putting $\theta = \pi/2 + \phi$ in (2), we get

$$\cos(x\cos\phi) = J_0 - 2J_2 \cos 2\phi + 2J_4 \cos 4\phi - \cdots$$

$$\therefore \quad \int_0^{\pi} \cos(x\cos\phi)\, d\phi = J_0 \int_0^{\pi} d\phi - 2J_2 \int_0^{\pi} \cos 2\phi\, d\phi + \cdots$$

$$= \pi J_0.$$

Hence $J_0(x) = \frac{1}{\pi}\int_0^{\pi} \cos(x\cos\phi)\, d\phi$

$$= \frac{1}{\pi}\int_0^{\pi} \left(1 - \frac{x^2\cos^2\phi}{\lfloor 2} + \frac{x^4\cos^4\phi}{\lfloor 4} \cdots\right) d\phi$$

$$= \frac{1}{\pi}\left[\pi - \frac{x^2}{\lfloor 2}.\frac{1}{2}\pi + \frac{x^4}{\lfloor 4}\frac{1.3}{2.4}\pi - \frac{x^6}{\lfloor 6}\frac{1.2.5}{2.4.6}\pi + \cdots\right]$$

$$\left[\because \int_0^{\pi} \cos^{2r}\phi\, d\phi = \frac{1.3.5\ldots(2r-1)}{2.4.6\ldots 2r}\pi\right]$$

$$= 1 - \frac{x^2}{2^2} + \frac{x^4}{2^2.4^2} - \cdots$$

$$= 1 - \frac{x^2}{2^2} + \frac{x^4}{2^2(2!)^2} - \frac{x^6}{2^6(3!)^2} + \cdots$$

$$= \sum_{r=0}^{\infty} \frac{(-1)^n x^{2r}}{(2^r r!)^2}.$$

Ex. 8. *Show that $J_n(x)=0$ has no repeated roots except at $x=0$.*

If possible, suppose $J_n(x)=0$ has repeated roots. Then at least two roots must be equal, say α i.e., α is a double root of $J_n(n)$. Then from the theory of equations, we have

$$J_n(\alpha)=0 \text{ and } J_n'(\alpha)=0. \quad \ldots (1)$$

We have the recurrence relations,

$$J_{n-1}(x)=\frac{n}{x}J_n(x)+J_n'(x). \quad \ldots (2)$$

$$J_{n+1}(x)=\frac{n}{x}J_n(x)-J_n'(x). \quad \ldots (3)$$

Putting $x=\alpha$ in (2) and (3) and using (1), we get

$$J_{n+1}(\alpha)=0 \text{ and } J_{n-1}(\alpha)=0, \text{ except when } x=0.$$

Since two different power series have distinct sum functions, so $J_{n+1}(\alpha)=0=J_{n-1}(\alpha)$ must be absurd. Hence $J_n(x)=0$ has no repeated roots except at $x=0$.

EXERCISES

1. Show that $\sqrt{\frac{\pi x}{2}}J_{-3/2}(x)=-\left(\sin x+\frac{1}{x}\cos x\right)$.

2. Show that $J_{5/2}(x)=\sqrt{\frac{2}{\pi x}}\left(\frac{3-x^2}{x^2}\sin x-\frac{3}{x}\cos x\right)$.

3. Prove that $y=\frac{1}{\pi}\int_0^{\pi}\cos(x\cos\phi)\,d\phi$ satisfies the Bessel's differential equation of zeroth order.

4. Prove that :

(i) $J_nJ_{-n}'-J_n'J_{-n}=-\frac{2\sin n\pi}{\pi x}$.

(ii) $\frac{d}{dx}\left(\frac{J_{-n}}{J_n}\right)=-\frac{2\sin n\pi}{\pi x\,J_n^2}$.

5. Prove $J_n' = \frac{2}{x}\left[\frac{n}{2} J_n - (n+2)\, J_{n+2} + (n+4)\, J_{n+4} - \cdots\right]$.

6. Prove that :

(i) $J_{n+3} + J_{n+5} = \frac{2}{x}(n+4)\, J_{n+4}$.

(ii) $4J_n'' = J_{n-2} - 2J_n + J_{n+2}$.

7. Prove $\frac{d}{dx}\left(J_n^2 + J_{n+1}^2\right) = 2\left(\frac{n}{x} J_n^2 - \frac{n+1}{x} J_{n+1}^2\right)$.

8. Prove that :

(i) $J_0' = -J_1$.

(ii) $J_2 - J_0 = 2J_0''$.

(iii) $J_2 = J_0'' - \left(\frac{1}{x}\right) J_0$.

9. Evaluate $J_3(x)$ in terms of $J_0(x)$ and $J_1(x)$.

10. Establish the relation $2J_n'(x) = J_{n-1}(x) + J_{n+1}(x)$ for the Bessel's function $J_n(x)$. Hence deduce that

$$J_n''(x) = \frac{1}{4}\left[J_{n-1}(x) - 2J_n(x) + J_{n+1}(x)\right].$$

11. Prove that :

(i) $\frac{d}{dx}\{x\, J_1(x)\} = x\, J_0(x)$.

(ii) $\int_a^b x\, J_0(ax)dx = \frac{b}{a} J_1(ab)$.

ANSWERS

9. $J_3(x) = \left(\frac{8-x^2}{x^2}\right) J_1(x) - \frac{4}{x} J_0(x)$.

4.5

Legendre's Equation And Legendre Functions

4.5.1. Legendre's Equation.

The differential equation of the form

$$(1-x^2)\frac{d^2y}{dx^2}-2x\frac{dy}{dx}+n(n+1)y=0 \qquad \ldots \quad (1)$$

is called Legendre's equation where n is a positive integer.

This equation can also be written as

$$\frac{d}{dx}\left\{(1-x^2)\frac{dy}{dx}\right\}+n(n+1)y=0. \qquad \ldots \quad (2)$$

Solution of Legendre's equation.

Let the series solution of (1) be

$$y=\sum_{r=0}^{\infty}a_r x^{k-r},\ a_0\neq 0. \qquad \ldots \quad (3)$$

$$\therefore \quad \frac{dy}{dx}=\sum_{r=0}^{\infty}a_r(k-r)\,x^{k-r-1},$$

$$\frac{d^2y}{dx^2}=\sum_{r=0}^{\infty}a_r(k-r)(k-r-1)\,x^{k-r-2}.$$

Substituting the values of $y, \frac{dy}{dx}, \frac{d^2y}{dx^2}$ into (1), we have

$$(1-x^2)\sum_{r=0}^{\infty}a_r(k-r)(k-r-1)\,x^{k-r-2}-2x\sum_{r=0}^{\infty}a_r(k-r)\,x^{k-r-1}$$

$$+n(n+1)\sum_{r=0}^{\infty}a_r x^{k-r}=0$$

or, $$\sum_{r=0}^{\infty}a_r(k-r)(k-r-1)x^{k-r-2}+\sum_{r=0}^{\infty}a_r(n-k+r)(n+k-r+1)x^{k-r}=0.$$

Now (4) is an identity. To get the indicial equation we equate to zero the coefficient of the highest power of x, namely x^k in (4) and obtain

$$a_0(n-k)(n+k+1)=0 \qquad \ldots \quad (4)$$

$$\therefore \quad k=n,-(n+1) \quad (\because\ a_0\neq 0).$$

So the roots of the indicial equation are unequal and differ by an integer.

Now equating to zero the coefficient of x^{k-r} in (4), we have

$$a_{r-2}(k-r+2)(k-r+1)+a_r(n-k+r)(n+k-r+1)=0$$

$$\therefore \quad a_r=-\frac{(k-r+2)(k-r+1)}{(k-n+r)(k+n-r+1)}a_{r-2}. \quad \ldots \quad (5)$$

Lastly equating to zero the coefficient of x^{k-1}, we have

$$a_1(n-k+1)(n+k)=0.$$

$$\therefore \quad a_1=0 \text{ for both } k=-(n+1), n.$$

Putting $r=3, 5, 7, \ldots$ in (5) and noting that $a_1=0$, we have

$$a_1=a_3=a_5=\cdots=0. \quad \ldots \quad (6)$$

which hold good for both $k=n, -(n+1)$.

To obtain $a_2, a_4, a_6, \ldots$ etc., we consider two cases.

Case I. When $k=n$, then from (5), we have

$$a_r=-\frac{(n-r+2)(n-r+1)}{r(2n-r+1)}a_{r-2}. \quad \ldots \quad (7)$$

Putting $r=2, 4, 6, \ldots$ in (7) successively, we get

$$a_2=-\frac{n(n-1)}{2(2n-1)}a_0$$

$$a_4=-\frac{(n-2)(n-3)}{4(2n-3)}a_2$$

$$=\frac{n(n-1)(n-2)(n-3)}{2.4(2n-1)(2n-3)}a_0 \text{ and so on.}$$

Rewriting (3), we have for $k=n$,

$$y=a_0x^n+a_1x^{n-1}+a_2x^{n-2}+\cdots$$

Using the value of a_i s we have

$$y=a_0\left[x^n-\frac{n(n+1)}{2(2n-1)}x^{n-2}+\frac{n(n-1)(n-2)(n-3)}{2.4(2n-1)(2n-3)}x^{n-4}\ldots\right] \quad \ldots \quad (8)$$

which is one solution of Legendre's equation.

Case II. When $k=-(n+1)$, then replacing n by $-(n+1)$ in (8), we get

$$y = a_0\left[x^{-n-1} + \frac{(n+1)(n+2)}{2(2n+3)}x^{-n-3} + \frac{(n+1)(n+2)(n+3)(n+4)}{2.4(2n+3)(2n+5)} \times x^{-n-5} + \cdots\right] \quad \ldots (9)$$

which is other solution of Legendre's equation.

4.5.2. Definition of $P_n(x)$ and $Q_n(x)$.

The solution of Legendre's equation is called Legendre's function or Legendre's polynomials.

When n is a positive integer and $a_0 = \frac{1.3.5\ldots(2n-1)}{n!}$, then solution (8) of (1) is denoted by $P_n(x)$ and is called *Legendre's function of the first kind or Legendre's polynomial of degree n.*

$$\therefore\ P_n(x) = \frac{1.3.5\ldots(2n-1)}{n!}\left[x^n - \frac{n(n-1)}{2(2n-1)}x^{n-2} + \frac{n(n-1)(n-2)(n-3)}{2.4(2n-1)(2n-3)} \times x^{n-4} - \ldots\right]$$

$$= \sum_{r=0}^{(n/2)} (-1)^r \frac{\lfloor 2n-r}{2^n \lfloor r \lfloor n-2r \lfloor n-r} x^{n-2r}$$

where $\left(\frac{n}{2}\right) = \begin{cases} \frac{n}{2} & \text{if } n \text{ is even} \\ \frac{n-1}{2} & \text{if } n \text{ is odd.} \end{cases}$

Here $P_n(x)$ is a terminating series, as n is a positive integer.

Again when n is a positive integer and $a_0 = \frac{n!}{1.3.5\ldots(2n+1)}$, the solution (9) of (1) is denoted by $Q_n(x)$ and is called Legendre's function of the second kind.

$$\therefore\quad Q_n(x) = \frac{n}{1.3.5\ldots(2n+1)}\left[x^{-n-1} + \frac{(n+1)(n+2)}{2.(2n+3)}x^{-n-3} + \frac{(n+1)(n+2)(n+3)(n+4)}{2.4(2n+3)(2n+5)}x^{-n-5} + \cdots\right]$$

Here $Q_n(n)$ is an infinite or non-terminating series, as n is not positive integer.

General solution of Legendre's equation.

The general solution of Legendre's equation is

$y = AP_n(x) + BQ_n(x)$ where A and B are arbitrary constants.

4.5.3. Generating function for Legendre polynomials.

Theorem. $P_n(x)$ is the coefficient of h^n in the expansion of $(1-2xh+h^2)^{-1/2}$ in ascending powers of h, where $|x| \le 1, |h| < 1$.

Proof. Since $|x| \le 1$, $|h| < 1$, we have

$$(1-2xh+h^2)^{-1/2} = [1-h(2x-h)]^{-1/2}$$

$$= 1 + \frac{1}{2}h(2x-h) + \frac{1.3}{2.4}h^2(2x-h)^2 + \cdots + \frac{1.3\ldots(2n-3)}{2.4\ldots(2n-2)} \times$$

$$h^{n-1}(2x-h)^{n-1} + \frac{1.3\ldots(2n-1)}{2.4\ldots 2n}h^n(2x-h)^n + \cdots$$

Therefore coefficient of h^n

$$= \frac{1.3\ldots(2n-1)}{2.4\ldots 2n}(2x)^n + \frac{1.3\ldots(2n-3)}{2.4\ldots(2n-2)}\,{}^{n-1}c_1(-1).(2x)^{n-2}$$

$$+ \frac{1.3\ldots(2n-5)}{2.4\ldots(2n-4)}\,{}^{n-2}c_2(-1)^2.(2x)^{n-4} + \cdots \quad \ldots \quad (1)$$

$$= \frac{1.3\ldots(2n-1)}{n!}\left[x^n - \frac{2n}{2n-1}(n-1).\frac{x^{n-2}}{2^2} + \frac{2n(2n-1)}{(2n-1)(2n-3)} \times \right.$$

$$\left. \frac{(n-2)(n-3)}{2!}.\frac{x^{n-4}}{2^4} - \ldots\right]$$

$$= \frac{1.3\ldots(2n-1)}{n!}\left[x^n - \frac{n(n-1)}{2(2x-1)}x^{n-2} + \frac{n(n-1)(n-2)(n-3)}{2.4(2n-1)(2n-2)}x^{n-4} - \ldots\right]$$

$$= P_n(x)$$

Thus we can say that

$$\sum_{n=0}^{\infty} h^n P_n(x) = (1-2xh+h^2)^{-1/2}.$$

Note. $(1-2xh+h^2)^{-1/2}$ is called the generating function of the Legendre polynomials.

4.5.4. Orthogonal Properties of Legendre's Polynomials.

(i) $\int_{-1}^{+1} P_m(x)P_n(x)dx = 0$ if $m \neq n$. [W.B.U.Tech '2003]

(ii) $\int_{-1}^{+1} [P_n(x)]^2 dx = \dfrac{2}{2n+1}$ if $m = n$.

Proof. (i) Legendre's equation may be written as

$$\frac{d}{dx}\left\{(1-x^2)\frac{dy}{dx}\right\} + n(n+1)y = 0.$$

$$\therefore \quad \frac{d}{dx}\left\{(1-x^2)\frac{dP_n}{dx}\right\} + n(n+1)y = 0. \quad \ldots \quad \text{(i)}$$

and $$\frac{d}{dx}\left\{(1-x^2)\frac{dP_m}{dx}\right\} + m(m+1)P_m = 0. \quad \ldots \quad \text{(ii)}$$

Multiplying (i) by P_m and (ii) by P_n and then subtracting, we have

$$P_m \frac{d}{dx}\left\{(1-x^2)\frac{dP_n}{dx}\right\} - P_n \frac{d}{dx}\left\{(1-x^2)\frac{dP_m}{dx}\right\} + \{n(n+1) - m(m+1)\}P_n P_m = 0.$$

or, $$\frac{d}{dx}\{(1-x^2)(P_n' P_m - P_m' P_n)\} + (m-n)(m+n-1) \times P_m P_n = 0.$$

Integrating between the limits -1 to $+1$, we have

$$(m-n)(m+n-1)\int_{-1}^{+1} P_m P_n dx = \left[(1-x^2)(P_n' P_m - P_n' P_n)\right]_{-1}^{1} = 0.$$

$$\therefore \quad \int_{-1}^{+1} P_m(x)P_n(x)dx = 0 \text{ if } m \neq n.$$

(ii) We have

$$(1-2xh+h^2)^{-1/2} = \sum_{n=0}^{\infty} h^n P_n(x).$$

Squaring both sides, we have

$$(1-2xh+h^2)^{-1} = \sum_{n=0}^{\infty} h^{2n}\{P_n(x)\}^2 + 2\sum_{\substack{m,n=0\\ m\neq n}}^{\infty} h^{m+n} P_m(x)P_n(x).$$

Integrating between limits –1 to +1, we have

$$\sum_{n=0}^{\infty}\int_{-1}^{+1} h^{2n}\{P_n(x)\}^2\,dx + 2\sum_{\substack{m,n=0\\ m\neq n}}^{\infty}\int_{-1}^{+1} h^{m+n}P_m(x)P_n(x)\,dx$$

$$=\int_{-1}^{+1}\frac{dx}{1-2xh+h^2}$$

or, $\displaystyle\sum_{n=0}^{\infty}\int_{-1}^{+1} h^{2n}\{P_n(x)\}^2dx = \int_{-1}^{+1}\frac{dx}{1-2xh+h^2}$, since other integral on L.H.S. is zero by (1), as $m \neq n$.

$$= -\frac{1}{2h}\left\{\log(1-2xh+h^2)\right\}_{-1}^{+1} = \frac{1}{2h}\log\frac{(1+h)^2}{(1-h)^2}$$

$$= \frac{1}{h}\log\frac{1+h}{1-h} = \frac{2}{h}\left\{h+\frac{h^3}{3}+\frac{h^5}{5}+\cdots\right\}.$$

$$= 2\left\{1+\frac{h^2}{3}+\frac{h^4}{5}+\cdots+\frac{h^{2n}}{2n+1}+\cdots\right\} = \sum_{n=0}^{\infty}\frac{2h^{2n}}{2n+1}.$$

Equating the coefficient of h^{2n}, we have

$$\int_{-1}^{+1}\left[P_n(x)\right]^2 dx = \frac{2}{2n+1}.$$

4.5.5. Recurrence formulae.

(1) $(2n+1)\,xP_n = (n+1)\,P_{n+1} + n\,P_{n-1}$. [*W.B.U.Tech '2002, '2004*]

Proof. We have

$$(1-2xh+h^2)^{-1/2} = \sum_{n=0}^{\infty} h^n P_n(x).$$

Differentiating both sides w.r.t. 'h' we have

$$-\frac{1}{2}(1-2xh+h^2)^{-3/2}\times(-2x+2h) = \sum_{n=0}^{\infty} nh^{n-1}P_n(x)$$

or, $(x-h)(1-2xh+h^2)^{-1/2} = (1-2xh+h^2)\sum_{n=0}^{\infty} nh^{n-1}P_n(x)$

or, $(x-h)\sum_{n=0}^{\infty} h^n P_n(x) = (1-2xh+h^2)\sum_{n=0}^{\infty} nh^{n-1}P_n(x)$.

Equating the coefficients of h^n from both sides, we have

$$xP_n(x) - P_{n-1}(x) = (n+1)P_{n+1}(x) - 2xnP_n(x) + (n-1)P_{n-1}(x)$$

$\therefore \quad (2n+1)xP_n = (n+1)P_{n+1} + nP_{n-1}$.

(2) $nP_n = xP_n' - P_{n-1}'$ [*W.B.U.Tech '2002*]

Proof. We have $(1-2xh+h^2)^{-1/2} = \sum_{n=0}^{\infty} h^n P_n(x)$. . . . (i)

Differentiating (i) w.r.t. h, we have

$$(x-h)(1-2xh+h^2)^{-3/2} = \sum_{n=0}^{\infty} nh^{n-1}P_n(x) \quad \ldots \quad \text{(ii)}$$

Again differentiating (i) w.r.t. x, we have

$$h(1-2xh+h^2)^{-3/2} = \sum_{n=0}^{\infty} h^n P_n'(x).$$

$\therefore \quad h(x-h)(1-2xh+h^2)^{-3/2} = (x-h)\sum_{n=0}^{\infty} h^n P_n'(x) \quad \ldots \quad$ (iii)

From (ii) and (iii), we have

$$h\sum_{n=0}^{\infty} nh^{n-1}P_n(x) = (x-h)\sum_{n=0}^{\infty} h^n P_n'(x).$$

Equating the coefficient of h^n from both sides, we have

$$nP_n(x) = xP_n'(x) - P_{n-1}'(x)$$

$\therefore \quad nP_n = xP_n' - P_{n-1}'$

(3) $(2n+1)P_n = P_{n+1}' - P_{n-1}'$.

Proof. From recurrence formula (1), we have

$$(2n+1)xP_n = (n+1)P_{n+1} + nP_{n-1}.$$

Differentiating w.r.t. x, we have

$$(2n+1)xP_n' + (2n+1)P_n = (n+1)P_{n+1}' + nP_{n-1}' \quad \ldots \quad \text{(i)}$$

Again from recurrence formula, we have

$$xP_n' = nP_n + P_{n-1}'. \qquad \ldots \qquad \text{(ii)}$$

From (i) and (ii), we have

$$(2n+1)(nP_n + P_{n-1}') + (2n+1)P_n = (n+1)P_{n+1}' + nP_{n-1}'$$

or, $(2n+1)(n+1)P_n = (n+1)P_{n+1}' + nP_{n-1}' - (2n+1)P_{n-1}'$

$$= (n+1)P_{n+1}' - (n+1)P_{n-1}'$$

$\therefore \quad (2n+1)P_n = P_{n+1}' - P_{n-1}'.$

(4) $(n+1)P_n = P_{n+1}' - xP_n'.$

Proof. From recurrence formula (2) and (3), we have

$$nP_n = xP_n' - P_{n-1}' \qquad \ldots \qquad \text{(i)}$$

$$(2n+1)P_n = P_{n+1}' - P_{n-1}'. \qquad \ldots \qquad \text{(ii)}$$

Subtracting (i) from (ii), we have

$\therefore \quad (n+1)P_n = P_{n+1}' - xP_n'.$

4.5.6. Rodrigue's formula.

$$P_n(x) = \frac{1}{n!\ 2^n}\frac{d^n}{dx^n}(x^2-1)^n. \qquad \textit{[W.B.U.Tech M 303]}$$

Proof. Let $y = (x^2-1)^n$

$\therefore \quad \frac{dy}{dx} = n(x^2-1)^{n-1}.2x$

$\therefore \quad (x^2-1)\frac{dy}{dx} = 2nxy.$

Differentiating (n + 1) times by Leibnitz's theorem, we have

$$(x^2-1)\frac{d^{n+2}y}{dx^{n+2}} + (n+1).\frac{d^{n+1}y}{dx^{n+1}}.2x + \frac{(n+1)n}{2}.\frac{d^n y}{dx^n}.2 = 2n\left[x\frac{d^{n+1}y}{dx^{n+1}} + (n+1).\frac{d^n y}{dx^n}.1\right]$$

$$\therefore \quad (1-x^2)\frac{d^{n+2}y}{dx^{n+2}} - 2x\frac{d^{n+1}y}{dx^{n+1}} + n(n+1)\frac{d^n y}{dx^n} = 0.$$

Putting $z = \frac{d^n y}{dx^n}$, we have

$$\therefore \quad (1-x^2)\frac{d^2 z}{dx^2} - 2x\frac{dz}{dx} + n(n+1)z = 0.$$

which is Legendre's equation. Hence its solution is

$z = cP_n(x)$ where c is a constant.

$$\therefore \quad \frac{d^n y}{dx^n} = c\,P_n(x) \qquad \cdots \quad (1)$$

Putting $x = 1$, we have

$$c = \left(\frac{d^n y}{dx^n}\right)_{x=1}, \text{ since } P_n(1) = 1.$$

Now $y = (x^2-1)^n = (x+1)^n (x-1)^n$.

Differentiating n times by Leibnitz's theorem, we have

$$\frac{d^n y}{dx^n} = (x-1)^n\, n! + {}^nC_1\, \frac{n!}{1!}(x+1)n(x-1)^{n-1} + \cdots$$

$$+ n\, n\,(x+1)^{n-1} . \frac{n!}{1!}(x-1) + (x+1)^n\, n$$

$$\therefore \quad \left(\frac{d^n y}{dx^n}\right)_{x=1} = (1+1)^n\, n! = 2^n\, n!$$

$$\therefore \quad c = 2^n\, n!.$$

Therefore from (1), we have $P_n(x) = \frac{1}{c}\frac{d^n y}{dx^n}$

$$\therefore \quad P_n(x) = \frac{1}{2^n\, n!}\, \frac{d^n}{dx^n}(x^2-1)^n.$$

This is Rodrigue's formula.

Illustrative Examples.

Ex. 1. *Show that*

$$P_0(x) = 1, P_1(x) = x, P_2(x) = \frac{3x^2-1}{2},$$

$$P_3(x) = \frac{5x^3-3x}{2}, P_4(x) = \frac{35x^4-30x^2+3}{8}.$$

We know that

$$\sum_{n=0}^{\infty} h^n P_n(x) = (1-2xh+h^2)^{-1/2}$$

$$= \{1-h(2x-h)\}^{-1/2}$$

$$= 1 + \frac{h}{2}(2x-h) + \frac{1.3}{2.4}h^2(2x-h)^2 + \frac{1.3.5}{2.4.6}h^3(2x-h)^3$$

$$+ \frac{1.3.5.7}{2.4.6.8}h^4(2x-h)^4 + \cdots$$

or, $P_0(x) + hP_1(x) + h^2P_2(x) + h^3P_3(x) + h^4P_n(n) + \cdots$

$$= 1 + xh + \frac{1}{2}(3x^2-1)h^2 + \frac{1}{2}(5x^3-3x)h^3 + \frac{1}{8}(35x^4-30x^2+3) + \cdots$$

Equating the coefficient of like powers of h, we have

$$\therefore \quad P_0(x) = 1, P_1(x) = x, P_2(x) = \frac{1}{2}(3x^2-1),$$

$$P_3(x) = \frac{1}{2}(5x^3-3x), P_4(x) = \frac{1}{8}(35x^4-30x^2+3).$$

Note. The above result can also be obtained from Rodrigue's formula

$$P_n(x) = \frac{1}{2^n \, n!}\frac{d^n}{dx^n}(x^2-1)^n.$$

Ex. 2. *Express* $P(x) = x^4 + 2x^3 + 2x^2 - x - 3$ *in terms of Legendre's polynomials.*

We have $P_0(x) = 1, P_1(x) = x, P_2(x) = \frac{1}{2}(3x^2-1),$

$$P_3(x) = \frac{1}{2}(5x^3-3x), P_4(x) = \frac{1}{8}(35x^4-30x^2+3).$$

From $P_4(x)$, $x^4 = \frac{8}{35}P_4(x) + \frac{6}{7}x^2 - \frac{3}{35}.$

From $P_3(x)$, $x^3 = \frac{2}{5}P_3(x) + \frac{3}{5}x.$

From $P_2(x)$, $x^2 = \frac{2}{3}P_2(x) + \frac{1}{3}$ and $x = P_1(x), 1 = P_0(x).$

Substituting the values of x^4, x^3, x^2, we have

$$P(x) = \frac{8}{35}P_4(x) + \frac{6}{7}x^2 - \frac{3}{35} + 2x^3 + 2x^2 - x - 3$$

$$= \frac{8}{35}P_4(x) + 2\left(\frac{2}{5}P_3(x) + \frac{3}{5}x\right) + \frac{20}{7}x^2 - x - \frac{108}{35}$$

$$= \frac{8}{35}P_4(x) + \frac{4}{5}P_3(x) + \frac{20}{7}\left(\frac{2}{3}P_2(x) + \frac{1}{3}\right) + \frac{1}{5}x - \frac{108}{35}$$

$$= \frac{8}{35}P_4(x) + \frac{4}{5}P_3(x) + \frac{40}{21}P_2(x) + \frac{1}{5}P_1(x) - \frac{224}{105}P_0(x).$$

Ex. 3. *Show that*

(i) $P_n(1) = 1$. *(ii)* $P_n(-x) = (-1)^n P_n(x)$.

Hence deduce that $P_n(-1) = (-1)^n$.

(i) We know that

$$\sum_{n=0}^{\infty} h^n P_n(x) = (1 - 2xh + h^2)^{-1/2}.$$

Putting $x = 1$,

$$\sum_{n=0}^{\infty} h^n P_n(1) = (1 - 2h + h^2)^{-1/2} = (1-h)^{-1}$$

$$1 + h + h^2 + \cdots = \sum_{n=0}^{\infty} h^n.$$

Equating the coefficient of h^n, we have $P_n(1) = 1$.

(ii) We have

$$(1 - 2xh + h^2)^{-1/2} = \sum_{n=0}^{\infty} h^n P_n(x). \quad \ldots \quad (1)$$

Replacing h by $(-h)$, we have

$$(1 + 2xh + h^2)^{-1/2} = \sum_{n=0}^{\infty} (-h)^n P_n(x)$$

$$= \sum_{n=0}^{\infty} (-1)^n h^n P_n(x). \quad \ldots \quad (2)$$

Replacing x by $-x$ in (1), we have

$$(1+2xh+h^2)^{-1/2} = \sum_{n=0}^{\infty} h^n P_n(-x). \quad \ldots \quad (3)$$

From (2) and (3), we have

$$\sum_{n=0}^{\infty} h^n P_n(-x) = \sum_{n=0}^{\infty} (-1)^n h^n P_n(x).$$

Equating the coefficient of h^n, we have

$$P_n(-x) = (-1)^n P_n(x).$$

Next putting $x = 1$, we have

$$P_n(-1) = (-1)^n P_n(1)$$

$$\therefore \quad P_n(-1) = (-1)^n \quad [\because P_n(1) = 1.]$$

Note. If n is even, $P_n(x)$ is even function of x.

and if n is odd, $P_n(x)$ is odd function of x.

Ex. 4. *Show that all the roots of $P_n(x) = 0$ are real and lie between -1 and $+1$.*

We have $P_n(x) = \dfrac{1}{2^n \; n} \dfrac{d^n}{dx^n}(x^2-1)^n$.

Now $(x^2-1)^n = (x-1)^n (x+1)^n$.

Hence $(x^2-1)^n$ vanishes n times at $x = +1$ and n times at $x = -1$.

Therefore by the theory of equations $\dfrac{d^n}{dx^n}(x^2-1)^n = 0$ will have n roots all real and lying between -1 and $+1$.

Hence $P_n(x) = \dfrac{1}{2^n \; n!} \dfrac{d^n}{dx^n}(x^2-1)^n = 0$ has n roots all real and lying between -1 and $+1$.

Ex. 5. *Prove that all the root of $P_n(x) = 0$ are distinct.*

If the roots of $P_n(x) = 0$ are not all different, then at least two of them must be equal.

Let α be their common value.

$\therefore \quad P_n(\alpha) = 0$ . . . (1)

and $P_n'(\alpha) = 0.$ . . . (2)

But $P_n(x)$ is the solution of Legendre's equation.

$$\therefore \quad (1-x^2)\frac{d^2}{dx^2}P_n(x) - 2x\frac{d}{dx}P_n(x) + n(n+1)P_n(x) = 0.$$

Differentiating r times by Leibnitz's theorem, we have

$$\left[(1-x^2)\frac{d^{r+2}}{dx^{r+2}}P_n(x) - 2x\, {}^rC_1\frac{d^{r+1}}{dx^{r+1}}P_n(x) - 2.{}^rC_2\frac{d^r}{dx^r}P_n(x)\right]$$

$$-2\left[x\frac{d^{r+1}}{dx^{r+1}}P_n(x) + 1.{}^rC_1\frac{d^r}{dx^r}P_n(x)\right] + n(n+1)\frac{d^r}{dx^r}P_n(x) = 0$$

or, $$(1-x^2)\frac{d^{r+2}}{dx^{r+2}}P_n(x) - 2x(r+1)\frac{d^{r+1}}{dx^{r+1}}P_n(x)$$

$$-\left\{r^2 + r - n(n+1)\right\}\frac{d^r}{dx^r}P_n(x) = 0. \quad \ldots (3)$$

Putting $r = 0$ and $x = \alpha$,

$$(1-\alpha^2)P_n''(\alpha) - 2\alpha P_n'(\alpha) + n(n+1)P_n(\alpha) = 0$$

$\therefore \quad P_n''(\alpha) = 0$, by (1) and (2).

Similarly writing r = 1, 2, 3, . . . in (3) and simplifying stepwise, we have $\quad P_n'''(\alpha) = P_n^{iv}(\alpha) = \ldots = P_n^n(\alpha) = 0.$

But, since $P_n(x) = \dfrac{1.3.\ldots(2n-1)}{n!}\left[x^n - \dfrac{n(n-1)}{2(2n-1)}x^{n-2} - \ldots\right]$

$$\therefore \quad P_n^n(\alpha) = \frac{1.3.5.\ldots(2n-1)}{n!}\,n!$$

$$= 1.3.5.\ldots(2n-1).$$

Hence $P_n^n(\alpha) \neq 0.$

Therefore our assumption that $P_n(x) = 0$ has a repeated root is not correct.

Hence all the roots of $P_n(x) = 0$ are distinct.

EXERCISES

1. Prove that (*i*) $P_n(1) = 1.$ (*ii*) $P_n(-1) = (-1)^n.$

(*iii*) $P_n'(1) = \frac{1}{2}n(n+1)$. (*iv*) $P_n'(-1) = (-1)^{n-1}\frac{n(n+1)}{2}$

2. Prove that $P_n(-x) = (-1)^n P_n(x)$. Deduce $P_n(-1) = (-1)^n$

3. Prove that $|P_n(\cos\theta)| \le 1$ when θ is real.

4. Express $x^4 - 3x^2 + x$ as a series in Legendre's polynomials.

5. Express $f(x) = 4x^3 + 6x^2 + 7x + 2$, in terms of Legendre's Polynomials. [*W.B.U.Tech '2004*]

6. Prove that for all x, $x^4 = \frac{8}{35}P_4(x) + \frac{4}{7}P_2(x) + \frac{1}{5}P_0(x)$.

7. Prove that $\int_{-1}^{1} P_n(x)\,dx = 2$ if $n = 0$

$= 0$ if $n \ge 1$.

8. Using Rodrigues' formula, prove that, $P'_{n+1} - P'_{n-1} = (2n+1)\,P_n$.

9. Prove that $P'_n - P'_{n-1} = (2n-1)P_{n-1}$.

10. Show that $xP'_9 = P'_8 - 2P_9$.

11. Prove that $P_1(x) = \frac{1}{\pi}\int_0^{\pi}\left\{x + \sqrt{(x^2-1)}\cos\theta\right\}d\theta$.

12. Prove that $(1 - 2xz + z^2)^{-1/2}$ is a solution of the equation

$$z\frac{\partial^2}{\partial z^2}(zu) + \frac{\partial}{\partial x}\left\{(1-x^2).\frac{\partial u}{\partial x}\right\} = 0.$$

13. Prove that

$$P'_{n+1} + P'_n = P_0 + 3P_1 + 5P_2 + \cdots + (2n+1)\,P_n = \sum_{r=1}^{n}(2r+1)P_r(x).$$

ANSWERS

4. $\frac{8}{35}P_4 - \frac{10}{7}P_2 + P_1 - \frac{4}{5}P_0$ **5.** $\frac{8}{5}P_3 + 4P_2 + \frac{47}{5}P_1 + 4P_0$

INDEX

APPENDIX

Ex. *Find the function whose Fourier transform is*

$$F(s)=\frac{1}{s^2+is+2}$$

Solution. Let $f(x)$ be the required function. Then

$$F(s)=\frac{1}{s^2+is+2} \text{ corresponds to } f(x)$$

By Fourier integral theorem,

$$\frac{1}{2\pi}\int_{-\infty}^{\infty} F(s)e^{-isx}ds=f(x)$$

or, $$f(x)=\frac{1}{2\pi}\int_{-\infty}^{\infty}\frac{1}{s^2+is+2}e^{-isx}ds$$

$$=\frac{1}{2\pi}\int_{-\infty}^{\infty}\frac{1}{s^2+2\cdot\frac{i}{2}\cdot s+\left(\frac{i}{2}\right)^2+2+\frac{1}{4}}e^{-isx}ds$$

$$=\frac{1}{2\pi}\int_{-\infty}^{\infty}\frac{1}{\left(s+\frac{i}{2}\right)^2+\frac{9}{4}}e^{-isx}ds \qquad \left[\text{Put } s+\frac{i}{2}=w \text{ i.e. } ds=dw\right]$$

$$=\frac{1}{2\pi}\int_{-\infty}^{\infty}\frac{1}{w^2+\left(\frac{3}{2}\right)^2}e^{-i\left(w-\frac{i}{2}\right)x}dw$$

$$=e^{\frac{x}{2}}\frac{1}{2\pi}\int_{-\infty}^{\infty}\frac{1}{w^2+\left(\frac{3}{2}\right)^2}e^{-iwx}dw$$

$$=e^{\frac{x}{2}}\frac{1}{2\pi}\cdot\frac{1}{3}\int_{-\infty}^{\infty}\frac{2\cdot\frac{3}{2}}{w^2+\left(\frac{3}{2}\right)^2}e^{-iwx}dw$$

$$=\frac{1}{3}e^{\frac{x}{2}}\mathcal{F}^{-1}\left(\frac{2\cdot\frac{3}{2}}{w^2+\left(\frac{3}{2}\right)^2}\right)=\frac{1}{3}e^{\frac{x}{2}}.e^{-\frac{3}{2}|x|}$$

$\therefore$ the required function,

$$f(x) = \frac{1}{3}e^{-x}, \quad x \geq 0$$

$$= \frac{1}{3}e^{2x}, \quad x \leq 0$$

Ex. 2. *Find the function whose Fourier transform is*

$$\frac{1}{s^2 + 4s + 13}$$

$$\left[\text{Ans. } \frac{e^{2it}.e^{-3|t|}}{6}\right]$$

Ex. 1. *Find the poles (mentioning order) of the function*

$$f(z)=\frac{1}{\left(1-z^2\right)^3 z}.$$

Solution. We know $f(z)$ has pole where $\frac{1}{f(z)}$ i.e. $\left(1-z^2\right)^3 z$ has zero.

Now, $\left(1-z^2\right)^3 z=0 \Rightarrow (1-z)^3(1+z)^3 z=0$

$$\Rightarrow z=1,1,1,-1,-1,-1,0$$

$\therefore$ $f(z)$ has pole of order 3 at $z=1$ and at $z=-1$. Also $f(z)$ has pole of order 1 at $z=0$.

Ex. 2. *Find the poles of the function* $f(z)=\frac{z^4}{(z-5)(z-3)}$. [Ans. $z = 5, 3$]

Ex. 3. *Find the pole (mentioning its order) of* $f(z)=\frac{e^{3z}}{(z-3)^6}$

[Ans. z=3 of order 6]

Ex. 4. *Find the pole (mentioning its order) of* $f(z)=\cot z$.

Soln. $f(z)$ has pole where $\phi(z)=\frac{1}{f(z)}=\tan z=0$

i.e. at $z=n\pi$, where n is integer

Now, $\phi'(z)=\frac{d}{dz}(\tan z)=\sec^2 z$ and we see $\phi'(n\pi)=\sec^2 n\pi \neq 0$

$\therefore$ $z=n\pi$ is a zero of order 1 of $\phi(z)$

$\therefore$ $z=n\pi$ is a pole of order 1 of $f(z)$.

Ex. 5. *Find the value of k so that the function* $k\log\left(x^2+y^2\right)$ *is harmonic.*

Solution. Let $u(x,y)=k\log\left(x^2+y^2\right)$

$$\therefore \quad \frac{\partial u}{\partial x}=\frac{k.2x}{x^2+y^2}, \quad \frac{\partial u}{\partial y}=\frac{k.2y}{x^2+y^2}$$

$$\frac{\partial^2 u}{\partial x^2}=\frac{\partial}{\partial x}\left(\frac{k.2x}{x^2+y^2}\right)=2k\frac{\partial}{\partial x}\left(\frac{x}{x^2+y^2}\right)$$

$$=2k.\frac{1.(x^2+y^2)-2x.x}{(x^2+y^2)^2}=2k\frac{y^2-x^2}{(x^2+y^2)^2}$$

and $$\frac{\partial^2 u}{\partial y^2}=\frac{\partial}{\partial y}\left(\frac{k.2y}{x^2+y^2}\right)=2k\frac{\partial}{\partial y}\left(\frac{y}{x^2+y^2}\right)$$

$$=2k\frac{1.(x^2+y^2)-y.2y}{(x^2+y^2)^2}=2k.\frac{x^2-y^2}{(x^2+y^2)^2}$$

Now $u(x, y)$ will be Harmonic if $\frac{\partial^2 u}{\partial x^2}+\frac{\partial^2 u}{\partial y^2}=0$

or, $$2k\frac{y^2-x^2}{(x^2+y^2)^2}+2k.\frac{x^2-y^2}{(x^2+y^2)^2}=0$$

or, $$2k.\left(\frac{0}{(x^2+y^2)^2}\right)=0$$

which is satisfied for all values of k.

$\therefore$ the given function is Harmonic for all values of k.

Ex. 6. Find the value of k for which $u(x,y)=x^3-kxy^2$ is harmonic.

[$k=3$]

Ex. 7. Find the value of k for which $u=e^{-x}(x\sin y+ky\cos y)$ is harmonic.

[$k=-1$]

Ex. 8. *Expand* $f(z)=\frac{z+1}{(z-3)(z-4)}$ *in a Taylor's series about the point* $z=2$. *Find also the region of convergence.*

Let $z-2=u$

$\therefore \quad z=u+2$

$$\therefore \quad \frac{z+1}{(z-3)(z-4)} = \frac{u+2+1}{(u+2-3)(u+2-4)}$$

$$= \frac{u+3}{(u-1)(u-2)} = \frac{A}{u-1} + \frac{B}{u-2}, \text{ say}$$

$$\therefore \quad u+3 = A(u-2) + B(u-1)$$

Putting $u = 1, 2$, we get $A = -4$, $B = 5$

$$\therefore \quad \frac{z+1}{(z-3)(z-4)} = \frac{-4}{u-1} + \frac{5}{u-2} = 4(1-u)^{-1} - \tfrac{5}{2}\left(1 - \frac{u}{2}\right)^{-1}$$

$$= 4\left(1 + u + u^2 + \cdots\right) - \tfrac{5}{2}\left(1 + \frac{u}{2} + \frac{u^2}{2^2} + \cdots\right)$$

$$= \left(4 - \tfrac{5}{2}\right) + \left(4 - \frac{5}{2^2}\right)u + \left(4 - \frac{5}{2^3}\right)u^2 + \cdots$$

$$= \left(4 - \tfrac{5}{2}\right) + \left(4 - \frac{5}{2^2}\right)(z-2) + \left(4 - \frac{5}{2^3}\right)(z-2)^2 + \cdots$$

Here the region of convergence is

$$|z-2| < 1$$

Ex. 9. *Using Residue theorem, evaluate the integral*

$$\int_C \frac{3z^2 + z - 1}{\left(z^2 - 1\right)(z-3)}\, dz \text{ about the circle } |z| = 2.$$

Let $f(z) = \dfrac{3z^2 + z - 1}{\left(z^2 - 1\right)(z-3)}$

The poles of $f(z)$ are given by $\left(z^2 - 1\right)(z-3) = 0$

$$\therefore \quad z = \pm 1, 3$$

These are all simple pole. But only $z = \pm 1$ lie within the circle $|z| = 2$. Hence by Residue theorem,

$$\int_C f(z)\,dz = 2\pi i\{\text{Res}(-1) + \text{Res}(1)\}$$

Now $\text{Res}(1) = \lim_{z\to 1} (z-1)f(z) = \lim_{z\to 1} \dfrac{3z^2+z-1}{(z+1)(z-3)} = \dfrac{3.1^2+1-1}{(1+1)(1-3)} = -\frac{3}{4}$

$\text{Res}\ (-1) = \lim_{z\to -1} (z+1)f(z) = \lim_{z\to -1} \dfrac{3z^2+z-1}{(z-1)(z-3)} = \dfrac{3(-1)^2+(-1)-1}{(-1-1)(-1-3)} = \frac{1}{8}$

$\therefore \quad \int_C f(z)dz = 2\pi i\left(-\frac{3}{4}+\frac{1}{8}\right) = -\dfrac{5\pi i}{4}$

Ex. 10. *Evaluate* $\displaystyle\int_0^{2\pi} \frac{d\theta}{1+\frac{1}{2}\sin\theta}$ *using contour integration method.*

Let $I = \displaystyle\int_0^{2\pi} \frac{d\theta}{1+\frac{1}{2}\sin\theta}$

$= \displaystyle\int_C \frac{1}{1+\frac{1}{4i}\left(z-\frac{1}{z}\right)} \frac{dz}{iz}$, putting $z = e^{i\theta}$ so that $d\theta = \dfrac{dz}{iz}$ and

$\sin\theta = \dfrac{1}{2i}\left(z-\frac{1}{z}\right)$ where C is the circle $|z| = 1$

$= \displaystyle\int_C \frac{4dz}{4iz+z^2-1} = 4\int \frac{dz}{(z+2i)^2+\left(\sqrt{3}\right)^2}$

$= 4\displaystyle\int \frac{dz}{\left\{z+i\left(2+\sqrt{3}\right)\right\}\left\{z+i\left(2-\sqrt{3}\right)\right\}}$

Let $f(z) = \dfrac{1}{(z+ia)\left(z+\frac{i}{a}\right)}$ where $a = 2+\sqrt{3}$

Then poles of $f(z)$ are $-ia$, $-\dfrac{i}{a}$ which are simple poles. But only $-\dfrac{i}{a}$ lies within the circle C. Hence by residue theorem

$\int_C f(z)dz = 2\pi i\ \text{Res}\left(-\dfrac{i}{a}\right)$

Now $\text{Res}\left(-\frac{i}{a}\right) = \lim_{z\to -\frac{i}{a}}\left(z+\frac{i}{a}\right)f(z)$

$$= \lim_{z\to -\frac{i}{a}} \frac{1}{z+ia} = \frac{1}{-\frac{i}{a}+ia}$$

$$= \frac{1}{i}\frac{1}{\left(a-\frac{1}{a}\right)} = \frac{1}{i}\frac{1}{2+\sqrt{3}-2+\sqrt{3}} = \frac{1}{2i\sqrt{3}}$$

$\therefore \quad \int_C f(z)dz = 2\pi i \times \frac{1}{2i\sqrt{3}} = \frac{\pi}{\sqrt{3}}$

$\therefore \quad \int_0^{2\pi} \frac{d\theta}{1+\frac{1}{2}\sin\theta} = 4.\frac{\pi}{\sqrt{3}} = \frac{4\pi}{\sqrt{3}}$

Ex. 11. *Find the bilinear transformation which maps the points* $z=1,i,-1$. *into the points* $w=i,o,-i$.

Let the bilinear transformation be

$$\frac{(w-w_1)(w_2-w_3)}{(w-w_3)(w_2-w_1)} = \frac{(z-z_1)(z_2-z_3)}{(z-z_3)(z_2-z_1)}$$

Here $z_1=1$, $z_2=i$, $z_3=-1$ and

$w_1=i$, $w_2=0$, $w_3=-i$

$\therefore \quad \frac{(w-i)(0+i)}{(w+i)(0-i)} = \frac{(z-1)(i+1)}{(z+1)(i-1)}$

or, $\quad \frac{(w-i)i}{(w+i)(-i)} = \frac{(z-1)(i+1)^2}{(z+1)(i^2-1)}$

or, $\quad -\frac{w-i}{w+i} = \frac{(z-1)\times 2i}{(z+1)(-2)}$

or, $\quad \frac{w-i}{w+i} = \frac{(z-1)i}{z-1}$

or, $wz - w - iz + i = wzi - wi - z + 1$

or, $w(z - 1 - zi + i) = -z + 1 + iz + i$

or, $w\{(i+z)(1-i)\} = (i-z)(1-i)$

$\therefore \quad w = \dfrac{i-z}{i+z}.$

Ex. 12. *Locate and name the singularities of* $f(z) = \dfrac{z}{(z^2+4)^2}$

The singularities of $\dfrac{z}{(z^2+4)^2}$ is given by $(z^2+4)^2 = 0$

$\therefore \quad z = \pm 2i, \pm 2i$

Then $2i$, $-2i$ are pole of order 2.

Ex. 13. *Evaluate* $\int_{(0,3)}^{(2,4)} (2y+x^2)dx + (3x-y)dy$ *along the paths.*

(i) *the parabola* $x = 2t,\ y = t^2 + 3$

(ii) *the straight line joining (0, 3) and (2, 4)*

(i) $\int_{(0,3)}^{(2,4)} \{(2y+x^2)dx + (3x-y)dy\}$

$$= \int_{t=0}^{1} [\{2(t^2+3) + 4t^2\} 2dt + (3.2t - t^2 - 3) 2tdt]$$

$$= \int_{0}^{1} 12(t^2+1)dt + 2t(6t - t^2 - 3)dt$$

$$= 12\left[\frac{t^3}{3} + t\right]_0^1 + 2\left[6\frac{t^3}{3} - \frac{t^4}{4} - 3\frac{t^2}{2}\right]_0^1$$

$$= 12\left(\frac{1}{3} + 1\right) + 2\left(2 - \frac{1}{4} - \frac{3}{2}\right) = \frac{33}{2}.$$

(ii) The equation of the straight line is

$$\frac{y-3}{x-0}=\frac{4-3}{2-0} \text{ or, } \frac{y-3}{x}=\frac{1}{2} \therefore x=2y-6$$

$$\therefore \quad \int_{(0,3)}^{(2,4)} \{(2y+x^2)dx+(3x-y)dy\}$$

$$=\int_{y=3}^{4} \{2y+(2y-6)^2\}2dy+\{3(2y-6)-y\}dy$$

$$=\int_{3}^{4} 2(4y^2-22y+36)dy+(5y-18)dy$$

$$=4\left[2\frac{y^3}{3}-11.\frac{y^2}{2}+18y\right]_3^4+\left[\frac{5y^2}{2}-18y\right]_3^4$$

$$=\frac{47}{3}$$

Ex. 14. *Using Rasidue theorem, evaluate*

$$\int_C \frac{e^z dz}{(z^2+\pi^2)^2} \text{ where } C \text{ is the circle } |z|=4$$

Let $f(z)=\dfrac{e^z}{(z^2+\pi^2)^2}$

So the poles of $f(z)$ are given by

$$(z^2+\pi^2)^2=0$$

$\therefore \quad z=\pm\pi i, \pm\pi i$

So $z=\pi i, -\pi i$ are poles of order 2 and both lie within the circle $|z|=4$.

$\therefore$ By residue theorem, we have

$$\int_C f(z)dz=2\pi i \ \{\text{Res}(\pi i)+\text{Res}(-\pi i)\}$$

Now $\text{Res}(\pi i)=\lim\limits_{z\to\pi i}\frac{d}{dz}\left\{(z-\pi i)^2 f(z)\right\}=\lim\limits_{z\to\pi i}\frac{d}{dz}\ \frac{e^z}{(z+\pi i)^2}$

$$=\lim_{z\to\pi i}\left\{\frac{e^z}{(z+\pi i)^2}-\frac{2e^z}{(z+\pi i)^3}\right\}$$

$$=\frac{c^{\pi i}}{-4\pi^2}+\frac{2e^{\pi i}}{8\pi^3 i}=-\frac{e^{\pi i}}{4\pi^2}-\frac{ie^{\pi i}}{4\pi^3}$$

Similarly $\text{Res}(-\pi i)=-\frac{e^{-\pi i}}{4\pi^2}+\frac{ie^{-\pi i}}{4\pi^3}$

$$\therefore \quad \int_C f(z)dz=2\pi i\left\{-\frac{1}{4\pi^2}\left(e^{\pi i}+e^{-\pi i}\right)-\frac{i}{4\pi^3}\left(e^{\pi i}-e^{-\pi i}\right)\right\}$$

$$=2\pi i\left\{-\frac{1}{4\pi^2}2\cos\pi-\frac{2i^2}{4\pi^3}\sin\pi\right\}$$

$$=2\pi i\left(\frac{1}{2\pi^2}+0\right)=\frac{i}{\pi}$$

Ex. 15. *Find the analytic function* $f(z)=u+iv$ *where* $v=e^x(x\sin y+y\cos y)$.

Here $v=e^x(x\sin y+y\cos y)$

$$\therefore \quad \frac{\partial v}{\partial x}=e^x(x\sin y+y\cos y)+e^x\sin y$$

$$\frac{\partial^2 v}{\partial x^2}=e^x(x\sin y+y\cos y)+2e^x\sin y$$

$$\frac{\partial v}{\partial y}=e^x(x\cos y+\cos y-y\sin y)$$

$$\therefore \quad \frac{\partial^2 v}{\partial y^2}=e^x(-x\sin y-2\sin y-y\cos y)$$

$$\therefore \quad \frac{\partial^2 v}{\partial x^2}+\frac{\partial^2 v}{\partial y^2}=0$$

$\therefore$ v is a harmonic function.

Let u be the conjugate harmonic of v.

$$\therefore \quad \frac{\partial u}{\partial x}=\frac{\partial v}{\partial y}, \quad \frac{\partial u}{\partial y}=-\frac{\partial v}{\partial x}$$

Now $du=\frac{\partial u}{\partial x}dx+\frac{\partial u}{\partial y}dy=\frac{\partial v}{\partial y}du-\frac{\partial v}{\partial x}dy$

$$=e^x(x\cos y+\cos y-y\sin y)dx-e^x(x\sin y+y\cos y+\sin y)dy$$

$$=d\{e^x(x\cos y-y\sin y)\}$$

$$\therefore \quad u=e^x(x\cos y-y\sin y)+C$$

Hence the corresponding analytic function $f(z)$ is given by

$$f(z)=u+iv$$

$$=e^x(x\cos y-y\sin y)+ie^x(x\sin y+y\cos y)+ic$$

$$=ze^z+ic$$

Ex. 16. *Find the bilinear transformation which maps the points $z=1,0,-1$ onto the points $w=i,0,-i$ and find the fixed points of the transformation.*

Let the bilinear transformation be

$$\frac{(w-w_1)(w_2-w_3)}{(w-w_3)(w_2-w_1)}=\frac{(z-z_1)(z_2-z_3)}{(z-z_3)(z_2-z_1)}$$

Here $z_1=1, z_2=0, z_3=-1$ and $w_1=i$, $w_2=0$, $w_3=-i$.

$$\therefore \quad \frac{(w-i)(0+i)}{(w+i)(0-i)}=\frac{(z-1)(0+1)}{(z+1)(0-1)} \quad \text{or,} \quad \frac{w-i}{w+i}=\frac{z-1}{z+1}$$

$$\text{or,} \quad \frac{w-i+w+i}{w-i-w-i}=\frac{z-1+z+1}{z-1-z-1}$$

$$\text{or,} \quad \frac{2w}{-2i}=\frac{2z}{-2}$$

$$\therefore \quad w=zi$$

For a fixed point we have $w=z$

$$\therefore \quad z=zi$$

$$\text{or,} \quad z(1-i)=0$$

$$\therefore \quad z=0$$

So $z=0$ is the only fixed point.

Ex. 17. *Prove that* $u(x,y)=\frac{1}{2}\log\left(x^2+y^2\right)$ *is harmonic and find its conjugate harmonic function* $v(x,y)$ *so that* $f=u+iv$ *is analytic.*

Here $u=\frac{1}{2}\log\left(x^2+y^2\right)$

$$\therefore \quad \frac{\partial u}{\partial x}=\frac{x}{x^2+y^2},\ \frac{\partial^2 u}{\partial x^2}=\frac{1.\left(x^2+y^2\right)-x.2x}{\left(x^2+y^2\right)^2}=\frac{y^2-x^2}{\left(x^2+y^2\right)^2}$$

$$\frac{\partial u}{\partial y}=\frac{y}{x^2+y^2},\ \frac{\partial^2 u}{\partial y^2}=\frac{x^2-y^2}{\left(x^2+y^2\right)^2}$$

$$\therefore \quad \frac{\partial^2 u}{\partial x^2}+\frac{\partial^2 u}{\partial y^2}=0$$

$\therefore$ **u is a harmonic function.**

Let v be the corresponding conjugate of u. Then

$$\frac{\partial u}{\partial x}=\frac{\partial v}{\partial y},\ \frac{\partial u}{\partial y}=-\frac{\partial v}{\partial x}$$

Now $dv=\frac{\partial v}{\partial x}dx+\frac{\partial v}{\partial y}dy=-\frac{\partial u}{\partial y}dx+\frac{\partial u}{\partial x}dy$

$$=-\frac{y}{x^2+y^2}dx+\frac{x}{x^2+y^2}dy$$

$$=\frac{x\,dy-y\,dx}{x^2+y^2}$$

$$=\frac{d\left(\frac{y}{x}\right)}{1+\frac{y^2}{x^2}}$$

On integration we get

$$v=\tan^{-1}\left(\frac{y}{x}\right)+c.$$

Ex. 1. *A and B are two events such that* $P(A \cup B) = 7/8$, $P(A \cap B) = 1/4$ *and* $P(A^C) = \frac{5}{8}$. *Find the conditional probability of A on the hypothesis that B does not occur.* [WBUT 2003]

Solution. $P(A^C) = \frac{5}{8}$ or, $1 - P(A) = \frac{5}{8}$

$\therefore \quad P(A) = \frac{3}{8}$;

$P(A \cup B) = \frac{7}{8}$

or, $P(A) + P(B) - P(A \cap B) = \frac{7}{8}$

or, $\frac{3}{8} + P(B) - \frac{1}{4} = \frac{7}{8}$

$\therefore \quad P(B) = \frac{3}{4}$

Required conditional probability

$$= P(A/B^C) = \frac{P(A \cap B^C)}{P(B^C)} = \frac{P(A \cap (S - B))}{1 - P(B)} = \frac{P(A \cap S - A \cap B)}{1 - P(B)}$$

$$= \frac{P(A - A \cap B)}{1 - P(B)} = \frac{P(A) - P(A \cap B)}{1 - P(B)}$$

$$= \frac{\frac{3}{8} - \frac{1}{4}}{1 - \frac{3}{4}} = \frac{\frac{1}{8}}{\frac{1}{4}} = \frac{1}{2}$$

Ex. 2. *In a certain class 20% of the students failed in English, 30% failed in Physics, 8% failed in both English and Physics. A student is selected at random.*

(*i*) *If he failed in Physics, find the probability that he failed in English.*

(*ii*) *If he failed in English, find the probability that he failed in Physics.*

(*iii*) *Find the probability that he failed both in English and Physics.*

Solution. Let A = A Student passed in English,

B = A Student passed in Physics

$$\therefore \quad P(\overline{A}) = \frac{20}{100} = \tfrac{1}{5} \text{ or, } 1 - P(A) = \tfrac{1}{5} \quad \therefore \quad P(A) = \tfrac{4}{5}$$

$$P(\overline{B}) = \frac{30}{100} = \frac{3}{10} \text{ or, } 1 - P(B) = \frac{3}{10} \quad \therefore \quad P(B) = \frac{7}{10}$$

and $P(\overline{A} \cap \overline{B}) = \dfrac{8}{100} = \dfrac{2}{25}$

(i) Required probability

$$= P(\overline{A}/\overline{B}) = \frac{P(\overline{A} \cap \overline{B})}{P(\overline{B})} = \frac{\frac{2}{25}}{\frac{3}{10}} = \frac{4}{15}$$

(ii) Required probability $= P(\overline{B}/\overline{A}) = \dfrac{P(\overline{B} \cap \overline{A})}{P(\overline{A})}$

$$= \frac{\frac{2}{25}}{\frac{1}{5}} = \frac{2}{5}$$

(iii) Required probability $= P(A \cap B) = \dfrac{1}{10}$.

Ex. 3. *Two newspapers X and Y are published in a certain city. It is estimated from a survey that 16% read X, 14% read Y and 5% read both the newspapers. Find the probabilities that a randomly selected person*

(*i*) *does not read any newspaper* (*ii*) *reads only Y.* [WBUT 2004]

Solution. Let A = a randomly selected person reads newspaper X

B = a randomly selected person reads newspaper Y

$$\therefore \quad P(A) = \frac{16}{100}, \quad P(B) = \frac{14}{100}, \quad P(A \cap B) = \frac{5}{100}$$

(i) Required probablity

$$= P(\overline{A \cup B}) = 1 - P(A \cup B) = 1 - \{P(A) + P(B) - P(A \cap B)\}$$

$$= 1 - \left\{\frac{16}{100} + \frac{14}{100} - \frac{5}{100}\right\} = 1 - \frac{25}{100} = 1 - \frac{1}{4} = \frac{3}{4}$$

(ii) Required probability

$= P(B - A\cap B) = P(B) - P(A\cap B) \quad [\because\ A\cap B \subset B]$

$= \frac{14}{100} - \frac{5}{100} = \frac{9}{100}$

Ex. 4. *If* $P(A\cap B\cap C) = 0$ *prove that*

$$P(A\cup B/C) = P(A/C) + P(B/C)$$

Solution. Let $P(A\cup B/C) = \frac{P((A\cup B)\cap C)}{P(C)} = \frac{P\{(A\cap C)\cup(B\cap C\}}{P(C)}$

Now, $P\{(A\cap C)\cap(B\cap C)\} = P(A\cap B\cap C) = 0$

$\therefore$ $A\cap C$ and $B\cap C$ are mutuallly exclusive

$\therefore$ $P\{(A\cap C)\cup(B\cap C)\} = P(A\cap C) + P(B\cap C)$

$\therefore$ from above, $P(A\cup B/C) = \frac{P(A\cap C) + P(B\cap C)}{P(C)}$

$$= \frac{P(A\cap C)}{P(C)} + \frac{P(B\cap C)}{P(C)}$$

$$= P(A/C) + P(B/C)$$

Ex. 5. *If A and B are two events such that* $P\left(A^C \cup B^C\right) = \frac{5}{6}$, $P(A) = \frac{1}{2}$ *and* $P\left(B^C\right) = \frac{2}{3}$, *show that A and B are independent.*

Solution. *From D'Morgan's Rule we know,*

$A^C \cup B^C = (A\cap B)^C$

$\therefore$ $P\left(A^C \cup B^C\right) = P(A\cap B)^C = 1 - P(A\cap B)$

or, $\frac{5}{6} = 1 - P(A\cap B)$ $\therefore$ $P(A\cap B) = \frac{1}{6}$

Again we see $P(A)P(B) = \frac{1}{2}\left(1 - \frac{2}{3}\right) = \frac{1}{2} \times \frac{1}{3} = \frac{1}{6}$

So, $P(A\cap B) = P(A)P(B)$

$\therefore$ A and B are independent.

Ex. 6. *The probabilities of X, Y and Z becoming a principal of a college are respectively 0.3, 0.5 and 0.2. The probabilities that 'Student Aid–Fund' will be introduced in the college if X, Y and Z becomes principal, are 0.4, 0.6 and 0.1 respectively. Given that 'Student Aid Fund' has been introduced, find the probability that Y has been appointed as the principal.*

[WBUT 2004]

Solution. $P(X)=0.3$, $P(Y)=0.5$, $P(Z)=0.2$

Let A = student aid fund is introduced. By problem $P(A/X)=0.4$, $P(A/Y)=0.6$, $P(A/Z)=0.1$, $P(Y/A)=?$

By Baye's theorem,

$$P(A)=P(X)P(A/X)+P(Y)P(A/Y)+P(Z)P(A/Z)$$

$$=0.3\times0.4+0.5\times0.6+0.2\times0.1=0.44$$

Again by Baye's Theorem,

$$P(Y/A)=\frac{P(Y)P(A/Y)}{P(A)}=\frac{0.5\times0.6}{0.44}$$

$$=\frac{\frac{1}{2}\times\frac{6}{10}}{\frac{44}{100}}=\frac{\frac{3}{10}}{\frac{44}{100}}=\frac{30}{44}=\frac{15}{22}$$

Ex. 7. *Three boxes contain respectively 7 Red, 7 white, 9 black; 8 Red, 2 white, 1 black and 8 red, 8 white and 5 black balls. One ball is chosen at ramdom from each box. Find the probability that the balls are* (a) *of same colour* (b) *one red and two white balls* (c) *not of same colour.*

Solution. Here $n(s)={}^{23}C_1\times{}^{11}C_1\times{}^{21}C_1=23\times11\times21$

(a) Let A_1 = 'all the balls are Red'

A_2 = 'all the balls are White'

A_3 = 'all the balls are black'

Now, $n(A_1)={}^{7}C_1\times{}^{8}C_1\times{}^{8}C_1=448$

$$n(A_2)={}^{7}C_1\times{}^{2}C_1\times{}^{8}C_1=112$$

$$n(A_3)={}^{9}C_1\times{}^{1}C_1\times{}^{5}C_1=45$$

So if A = 'the balls are of same colour' then

$n(A)=n(A_1)+n(A_2)+n(A_3)$ as A_1, A_2, A_3 are pairwise mutually disjoint.

$\therefore \quad n(A)=448+112+45=605$

$$\therefore \quad P(A)=\frac{605}{23\times 11\times 21}=\frac{605}{5313}=\frac{55}{483}$$

(b) Let B_1 = ball drawn from 1st box is red, ball drawn from 2nd is white, ball drawn from 3rd box is white.

B_2 = 'Ball drawn from 1st box is white, ball drawn from 2nd box is red and ball drawn from 3rd box is white.

B_3 = 'Ball drawn from 1st box is white, ball drawn from 2nd box is white and the ball drawn from 3rd box is red'.

$\therefore$ If B = 'one red, two white' then

$B=B_1\cup B_2\cup B_3$ where B_1, B_2 and B_3 are pairwise muchually exclusive

$$\therefore \quad n(B)=n(B_1)+n(B_2)+n(B_3)$$

$$={}^7C_1\times{}^2C_1\times{}^8C_1+{}^7C_1\times{}^8C_1\times{}^8C_1+{}^7C_1\times{}^2C_1\times{}^8C_1$$

$$=7\times 2\times 8+7\times 8\times 8+7\times 2\times 8$$

$$=112+448+112=672$$

$$\therefore \text{ Required probability }=\frac{n(B)}{n(S)}=\frac{672}{23\times 11\times 21}=\frac{672}{5313}=\frac{32}{253}$$

(c) We found $P(A)=\frac{55}{483}$ where A = 'all the balls are of same colour.'.

$\therefore$ its complement A^c = 'the balls are not of same colour'.

$$\therefore \text{ required probability }=P(A^c)=1-P(A)=1-\frac{55}{483}=\frac{428}{483}$$

Ex. 8. *It A and B are two independent events, show that* $P(A\cup B)=1-P(\overline{A})P(\overline{B})$. [WBUT 2002]

Solution.

We know, by D'Morgan's law,

$$\overline{A\cup B}=\overline{A}\cap\overline{B}$$

$$\therefore \quad P(\overline{A\cup B})=P(\overline{A}\cap\overline{B})$$

$$=P(\overline{A})P(\overline{B})$$

[as independence of *A*, *B* imply independence of *A* and *B*] or, $1-P(A\cup B)=P(\overline{A})P(\overline{B})$

or, $P(A\cup B)=1-P(\overline{A})P(\overline{B})$.

Ex. 9. *Four dice are thrown. Find the probability that the sum of the numbers will be 16.*

Hint: Since four dice are thrown so $n(S)=6\times6\times6\times6=1296$

Let A = "sum of the numbers is 16"

$\therefore$ $n(A)=$ No. of solution of $x_1+x_2+x_3+x_4=16$

[where each x_i is integer and $1\le x_i\le 6$]

= coefficient of x^{16} in the product,

$$(x+x^2+x^3+x^4+x^5+x^6)(x+x^2+x^3+x^4+x^5+x^6)$$

$$\times(x+x^2+x^3+x^4+x^5+x^6)\times(x+x^2+x^3+x^4+x^5+x^6)$$

= coefficient of x^{16} in $(x+x^2+x^3+x^4+x^5+x^6)^4$

= coefficient of x^{16} in $\left\{\dfrac{x(x^6-1)}{x-1}\right\}^4$

= coefficient of x^{16} in $x^4(1-x^6)^4(1-x)^{-4}$

= coefficient of x^{16} in $x^4(1-{}^4C_1x^6+{}^4C_2x^{12}-{}^4C_3x^{18}+x^{24})$

$$\times\left\{1+\sum_{r=1}^{\infty}\frac{4.5....(3+r)}{r!}x^r\right\}$$

= coefficient of x^{16} in $(x^4-{}^4C_1x^{10}+{}^4C_2x^{16}-{}^4C_3x^{22}+x^{28})$

$$\times\left\{1+\sum_{r=1}^{\infty}\frac{4.5....(3+r)}{r!}x^r\right\}$$

Ex. 10. *A box contains 4 red balls and 10 black balls. 7 balls are drawn at random in succession without replacement. What is the probability that the 7th ball is the 4th red ball.*

Solution. There are 14 balls in the box. 7 balls can be drawn in succession on without replacement in $14\times13\times12\times11\times10\times9\times8$ ways.

If 7th ball is 4th red then from 1st draw to 6th draw 3 red balls will come and 3 blacks.

$$\therefore \quad n(A) = {}^4C_3 \times {}^{10}C_3 \times 6! = 4 \times 120 \times 720$$

$$\therefore \quad \text{Required probability } P(A) = \frac{4 \times 120 \times 720}{14 \times 13 \times 12 \times 11 \times 10 \times 9 \times 8}.$$

Ex. 11. *There were three candidates A, B and C for the position of a manager whose chances of getting the appointment are in the proportion 4:2:3 respectively. The probability that A, if selected, would launch a new product in the market is 0.3. The probabilities of B and C doing the same are 0.5 and 0.8 respectively. What is the probability that the new product was launched in the market by C?* [WBUT 2006]

Solution. Wet A = Candidate A gets appointment

B = Candidate B gets appointment

C = Candidate C gets appointment

X = The new product is launched

By problem, $P(A) = \frac{4}{9}, \quad P(B) = \frac{2}{9}, \quad P(C) = \frac{3}{9}$

$$P(X/A) = 0.3, \quad P(X/B) = 0.5, \quad P(X/C) = 0.8$$

$$P(X) = P(A)P(X/A) + P(B)P(X/B) + P(C)P(X/C)$$

$$= \frac{4}{9} \times 0.3 + \frac{2}{9} \times 0.5 + \frac{3}{9} \times 0.8 = \frac{23}{45}$$

Ex. 12. *When two dice are thrown, find the probability that difference of the points on the dice is 2 or 3.* [WBUT 2006]

Solution. $n(S) = 6 \times 6 = 36$

A = "Difference of two faces is 3"

$$= \{(1,4),(2,5),(3,6),(4,1),(5,2),(6,3)\}$$

$$\therefore \quad n(A) = 6$$

$$\therefore \quad P(A) = \frac{n(A)}{n} = \frac{6}{36} = \frac{1}{6}$$

Let B = "Difference is 2"

$$=\{(1,3),(2,4),(3,5),(4,2),(5,3),(6,4),(3,1),(4,6)\}$$

$$\therefore \quad n(B)=8 \therefore \quad P(B)=\frac{8}{36}=\frac{2}{9}$$

We see $A \cap B = \phi$.

$$\therefore \quad P(A \cup B) = P(A) + P(B) = \frac{1}{6} + \frac{2}{9} = \frac{7}{18}.$$

Ex. 13. *A bag contains 5 white and 4 black balls. If 3 balls are drawn at random, what are the probabilities of the following:*

(i) 2 of them are white

(ii) atmost one of them is white

(iii) at least two are white. [WBUT 2007]

Solution. The bag contains 5 + 4 = 9 balls.

$$\therefore \quad n(S) = {}^9C_3 = 84$$

(i) A = "2 of the three drawn balls are white"

$$\therefore \quad n(A) = {}^5C_2 \times {}^4C_1 = 40$$

$$\therefore \quad P(A) = \frac{40}{84} = \frac{10}{21}$$

(ii) B = "atmost one of the three is white"

$$= B_1 \cup B_2$$

where B_1 = none of the three is white

B_2 = 1 one of the three is white

$$\therefore \quad n(B_1) = {}^4C_3 = 4; \quad n(B_2) = {}^5C_1 \times {}^4C_2 = 30$$

$$\therefore \quad P(B_1) = \frac{4}{84}; \; P(B_2) = \frac{30}{84}$$

$$\therefore \quad P(B) = \frac{4}{84} + \frac{30}{84} = \frac{34}{84} = \frac{17}{42}$$

(iii) $C =$ 'at least two are white' $= C_1 \cup C_2$

where $C_1 =$ Two of the three are white

$C_2 =$ Three of the three are white

$\therefore \quad n(C_1) = {}^5C_2 \times {}^4C_1 = 40$; $n(C_2) = {}^5C_3 = 10$

$$\therefore \quad P(C) = \frac{40}{84} + \frac{10}{84} = \frac{25}{42}.$$

Ex. 14. *A box contains 5 defective and 10 non defective lamps. 8 lamps are drawn at ramdom in succession without replacement. What is the probability that the 8th lamp is the 5th defective?* [WBUT 2008, 2007]

Solution. There are $5+10=15$ lamps in the box. 8 lamps can be drawn in $15\times14\times13\times12\times11\times10\times9\times8$ ways.

If 8th drawn lamp is 5th defective then from '1st drawn lamp' to '7th drawn lamp' 4 defective and 3 non-defective lamp will come.

$\therefore \quad n(A) = {}^5C_4 \times {}^{10}C_3 \times 7!$

$$\therefore \quad P(A) = \frac{{}^5C_4 \times {}^{10}C_3 \times 7!}{15\times14\times13\times12\times11\times10\times9\times8\times} = \frac{5}{429}.$$

Ex. 15. *It is observed that a cricket player becomes out within 10 runs in 3 out of 10 innings. If he played 4 innings, what is the probability that he will become* (i) *out twice* (ii) *out at least once within 10 runs*!

[WBUT 2007]

Solution. Let success = The player becomes out within 10 runs.

$\therefore \quad p =$ Probability of success in a trial $= \frac{3}{10}$

Let $X =$ No. of success in four trials.

$\therefore \quad X$ has Binomial distribution with $n=4$, $p = \frac{3}{10}$

$$\text{(i)} \quad P(X=2) = {}^4C_2\, p^2 (1-p)^{4-2} = 6\times\left(\frac{3}{10}\right)^2\left(1-\frac{3}{10}\right)^2$$

$$= 6\times\frac{9}{100}\times\frac{49}{100} = \frac{2646}{10000} = \frac{1323}{5000} = 0.2646.$$

(ii) $P(X \geq 1) = 1 - P(X = 0) = 1 - {}^4C_0 \, p^0 (1-p)^{4-0}$

$$= 1 - \left(1 - \frac{3}{10}\right)^4 = 1 - \frac{7^4}{10000} = 1 - \frac{2401}{10000} = 0.7599 .$$

Ex. 16. *A pair of dice is thrown. Find the probability of getting a sum of 7, when it is known that the digit in the 1st die is greater than that of the second.* [W.BUT 2009]

Solution. Here $n(S) = 6 \times 6 = 36$

Let A = 'sum of the two faces is 7'

$$= \{(1,6),(2,5),(3,4),(4,3),(5,2),(6,1)\}$$

B = 'Digit in 1st die is greater than that of 2nd

$$= \{(6,5),(6,4),(6,3),(6,2),(6,1),(5,4),(5,3),(5,2),(5,1),$$

$$(4,3),(4,2),(4,1),(3,2),(3,1),(2,1)\}$$

$\therefore \; n(B) = 15 \;\; \therefore \; P(B) = \frac{15}{36} = \frac{5}{12}$

Now, $A \cap B = \{(6,1),(5,2),(4,3)\}$

$\therefore \; n(A \cap B) = 3 \; \therefore \; P(A \cap B) = \frac{3}{36} = \frac{1}{12}$

Now, required probability $= P(A/B)$

$$= \frac{P(A \cap B)}{P(B)} = \frac{\frac{1}{12}}{\frac{5}{12}} = \frac{1}{5}$$

Ex. 17. *The manufacturing process of an article consists of two parts x and y. The probabilities of defect in parts x and y are 10% and 15% respectively. What is the probabllity that the assembled product will not have any defect?* [WBUT 2009]

Solution. Let A = Defect in part x; B = Defect in part y.

$\therefore \; P(A) = \frac{10}{100} = \frac{1}{10}, \;\; P(B) = \frac{15}{100} = \frac{3}{20}$

Now, $A \cup B =$ The assembled product will have at least one defect

$$\therefore \quad P(A\cup B)=P(A)+P(B)-P(A\cap B)$$

$$=P(A)+P(B)-P(A)P(B) \quad [\because \ A, B \text{ are independent}]$$

$$=\frac{1}{10}+\frac{3}{20}-\frac{1}{10}\cdot\frac{3}{20}=\frac{47}{200}$$

$\therefore$ Required probability $=P(\overline{A\cup B})=1-\dfrac{47}{200}=\dfrac{153}{200}$.

Ex. 18. *A die is tossed five times. A success is getting 1 or 6 on a toss. Find the mean and variance of the number of success.* [WBUT 2005]

Solution. s = success = "getting 1 or 6"

$\therefore$ p = prob. of success $=\frac{2}{6}=\frac{1}{3}$

$\therefore$ $q=1-\frac{1}{3}=\frac{2}{3}$

Let X = Number of success in n = 5 trials.

$\therefore$ X follows Binomial distribution with parameter, $n=5$ and $p=\frac{1}{3}$.

$\therefore$ Mean of $X=np=\frac{5}{3}$

Variance of $X=npq=5\times\dfrac{1}{3}\times\dfrac{2}{3}=\dfrac{10}{9}$.

Ex. 19. *A and B play a game in which their chances of winning are in the ratio.* 3 : 2. *Find A's chances of winning at least three games out of the five games played.* [WBUT 2004]

Solution. If A wins in a game then we suppose this is success. So probability of success, $p=\frac{3}{5}$.

Let X = Number of success in 5 trials

$\therefore$ Probability that A wins at least three $=P(X\geq 3)$

X has binomial distribution with $p=\frac{3}{5}$, $n=5$

$$\therefore \quad f_i=P(X=i)={}^5C_i\left(\tfrac{3}{5}\right)^i\left(1-\tfrac{3}{5}\right)^{5-i}={}^5C_i\left(\tfrac{3}{5}\right)^i\left(\tfrac{2}{5}\right)^{5-i}={}^5C_i\cdot\frac{1}{5^5}\cdot 3^i\cdot 2^{5-i}$$

∴ Required probability

$$= P(X \geq 3) = 1 - \{P(X=0) + P(X=1) + P(X=2)\}$$

$$= 1 - \{f_0 + f_1 + f_2\}$$

$$= 1 - \left\{{}^5C_0 \cdot \frac{1}{5^5} \cdot 2^5 + {}^5C_1 \cdot \frac{1}{5^5} \cdot 3.2^4 + {}^5C_2 \cdot \frac{1}{5^5} \cdot 3^2.2^3\right\}$$

$$= 1 - \frac{1}{5^5}\{32 + 5\times 3\times 16 + 10\times 9\times 8\} = 1 - \frac{992}{5^5}$$

$$= 1 - \frac{992}{3125} = \frac{2133}{3125}.$$

Ex. 20. *Five balls are drawn from an urn containing 4 white and 6 black balls. Find the probability distribution of the number of white balls drawn without replacement.* [WBUT 2004]

Solution. Let X = Number of white balls drawn.

So, X : 0 1 2 3 4

Now, $P(X=0)$ = Probability of no white balls

$P(X=1)$ = Probability of drawing 1 white ball

$$= \frac{{}^4C_1 \times {}^6C_4}{{}^{10}C_5} = \frac{5}{21}$$

$P(X=2)$ = Probability of drawing 2 white balls

$$= \frac{{}^4C_2 \times {}^6C_3}{{}^{10}C_5} = \frac{6\times 20}{252} = \frac{10}{21}$$

$P(X=3)$ = Probability of drawing 3 white balls

$$= \frac{{}^4C_3 \times {}^6C_2}{{}^{10}C_5} = \frac{5}{21}$$

$$P(X=4) = \frac{{}^4C_4 \times {}^6C_1}{{}^{10}C_5} = \frac{1}{42}$$

∴ the required distribution is

X :	0	1	2	3	4
f_i :	$\frac{1}{42}$	$\frac{5}{21}$	$\frac{10}{21}$	$\frac{5}{21}$	$\frac{1}{42}$

Ex. 21. *For what value of k, the function*

$$f(x)=ke^{-b(x-a)},\ a\le x<\infty(a,b,x>0)$$

$$=0,\ \text{elsewhere}$$

will be a probability density function? [WBUT 2003]

Solution. From fundamental property of pdf we have

$$\int_{-\infty}^{\infty} f(x)dx=1$$

or, $$\int_{-\infty}^{a} 0\,dx+\int_{a}^{\infty} k\,e^{-b(x-a)}dx=1$$

or, $$k\int_{a}^{\infty} e^{-b(x-a)}dx=1$$

or, $$k\int_{0}^{\infty} e^{-bz}dz=1 \qquad \text{putting } x-a=z$$

or, $$k\lim_{X\to\infty}\int_{0}^{X} e^{-bz}dz=1$$

or, $$k\lim_{X\to\infty}\left[\frac{e^{-bz}}{-b}\right]_0^X=1$$ or, $$-\frac{k}{b}\lim_{X\to\infty}\left[\frac{1}{e^{bz}}\right]_0^X=1$$

or, $$-\frac{k}{b}\lim_{X\to\infty}\left(\frac{1}{e^{bX}}-1\right)=1$$ or, $$-\frac{k}{b}(0-1)=1$$

or, $$\frac{k}{b}=1$$

or, $k=b$

Ex. 22. *Show that* $f(x)=x,\ 0\le x<1$

$$=k-x,\ 1\le x\le 2$$

$$=0,\ \textit{elsewhere}$$

is a pdf of a random variable X for a suitable value of k which you are to determine. Then find the distribution function of the ramdom variable X.

[WBUT 2005

Solution. By fundamental property of pdf we know

$$\int_{-\infty}^{\infty} f(x)dx = 1$$

or, $$\int_{-\infty}^{0} 0\,dx + \int_{0}^{1} x\,dx + \int_{1}^{2} (k-x)dx + \int_{2}^{\infty} 0\,dx = 1$$

or, $$\left[\frac{x^2}{2}\right]_0^1 + \left[kx - \frac{x^2}{2}\right]_1^2 = 1$$

or, $$\frac{1}{2} + \left\{\left(2k - \frac{4}{2}\right) - \left(k - \frac{1}{2}\right)\right\} = 1$$

or, $\frac{1}{2} + 2k - 2 - k + \frac{1}{2} = 1$ or, $k - 2 = 0$

$\therefore$ $k = 2$.

Now the distribution function, $F(x) = \int_{-\infty}^{x} f(x)dx$

If $x < 0$, $F(x) = \int_{-\infty}^{x} 0\,dx = 0$

If $0 \le x < 1$, $F(x) = \int_{-\infty}^{0} 0\,dx + \int_{0}^{x} x\,dx = \frac{x^2}{2}$

If $1 \le x \le 2$, $F(x) = \int_{-\infty}^{0} 0\,dx + \int_{0}^{1} x\,dx + \int_{1}^{x} (k-x)dx$

$$= \left[\frac{x^2}{2}\right]_0^1 + \left[kx - \frac{x^2}{2}\right]_1^x$$

$$= \frac{1}{2} + \left\{\left(2x - \frac{x^2}{2}\right) - \left(2 - \frac{1}{2}\right)\right\}$$

$$= \frac{1}{2} + 2x - \frac{x^2}{2} - \frac{3}{2} = 2x - \frac{x^2}{2} - 1$$

If $2<x$, $F(x)=\int_{-\infty}^{0} 0\,dx+\int_{0}^{1} x\,dx+\int_{1}^{2}(k-x)dx+\int_{2}^{\infty} 0\,dx$

$=1$ (putting $k=2$)

Thus the distribution function is

$$F(x)=0,\ -\infty<x<0$$

$$=\frac{x^2}{2},\ 0\le x<1$$

$$=2x-\frac{x^2}{2}-1,\ 1\le x\le 2$$

$$=1,\ x>2.$$

Ex. 23. *The weight of students in a college is normally distributed with* m = 40 *kg,* $\sigma=5$ *kg. Find the percentage of the students that have weight*

(i) *greater than 40 kg.*
(ii) *greater than 50 kg.*
(iii) between 38 kg. and 52 kg.

[Given : $\frac{1}{\sqrt{2\pi}}\int_{-\infty}^{2} e^{-\frac{t^2}{2}}dt=0.9772$, $\frac{1}{\sqrt{2\pi}}\int_{-\infty}^{0.4} e^{-\frac{t^2}{2}}dt=0.6554$

and $\frac{1}{\sqrt{2\pi}}\int_{-\infty}^{2.4} e^{-\frac{t^2}{2}}dt=0.9918$.] [WBUT 2003]

Solution. Let X = Weight of students. X has normal distribution with $m=40$, $\sigma=5$.

$\therefore\ Z=\frac{X-40}{5}$ has standard normal distribution.

(i) When $X=40$, $Z=\frac{40-40}{5}=0$

$$\therefore\ P(X>40)=P(Z>0)=\int_{0}^{\infty}\frac{1}{\sqrt{2\pi}}e^{-\frac{z^2}{2}}dz=0.5$$

$\therefore$ Required % of students $=0.5\times 100=50$

(ii) When $X=50,\ Z=\frac{50-40}{5}=\frac{10}{5}=2$

$$\therefore\quad P(X>50)=P(Z>2)=\int_{2}^{\infty}\frac{1}{\sqrt{2\pi}}e^{-\frac{z^2}{2}}dz$$

$$=1-\int_{-\infty}^{2}\frac{1}{\sqrt{2\pi}}e^{-\frac{z^2}{2}}dz=1-0.9772=0.0228$$

∴ Required % of students = 0228 × 100 = 2.28.

(iii) When $X=38,\ Z=\frac{38-40}{5}=-0.4$

and when X = 52, $Z=\frac{52-40}{5}=2.4$

$$\therefore\quad P(38<X<52)=P(-0.4<Z<2.4)$$

$$=\int_{-0.4}^{2.4}\frac{1}{\sqrt{2\pi}}e^{-\frac{z^2}{2}}dz$$

$$=\int_{-\infty}^{2.4}\frac{1}{\sqrt{2\pi}}e^{-\frac{z^2}{2}}dz-\int_{-\infty}^{-0.4}\frac{1}{\sqrt{2\pi}}e^{-\frac{z^2}{2}}dz$$

$$=0.9918-\int_{0.4}^{\infty}\frac{1}{\sqrt{2\pi}}e^{-\frac{z^2}{2}}dz$$

$$=0.9918-\left\{1-\int_{-\infty}^{0.4}\frac{1}{\sqrt{2\pi}}e^{-\frac{z^2}{2}}dz\right\}$$

$$=0.9918-1+0.6554=0.6472$$

∴ Required % of students = 0.6472 × 100 = 64.72.

Ex. 24. *If the random variable X takes the values* 1, 2, 3 *and* 4 *such that* $2P(X=1)=3P(X=2)=P(X=3)=5P(X=4)$, *find the probability distribution of X.* [WBUT 2006]

Solution. Let $P(X=i)=f_i$

$$\therefore\quad 2f_1=3f_2=f_3=5f_4=k(\text{say})$$

$$\therefore\quad f_1=\frac{k}{2},\ f_2=\frac{k}{3},\ f_3=k,\ f_4=\frac{k}{5}.$$

Now, $f_1+f_2+f_3+f_4=1$

or, $\frac{k}{2}+\frac{k}{3}+k+\frac{k}{5}=1$

or, $\frac{61k}{30}=1$

$\therefore\ k=\frac{30}{61}$

$\therefore\ f_1=\frac{1}{2}\times\frac{30}{61}=\frac{15}{61}\ ;\ f_2=\frac{1}{3}\times\frac{30}{61}=\frac{10}{61}$

$f_3=\frac{30}{61}\ ;\ f_4=\frac{1}{5}\times\frac{30}{61}=\frac{6}{61}$

$\therefore$ the distribution of X is

X :	1	2	3	4
f_i :	$\frac{15}{61}$	$\frac{10}{61}$	$\frac{30}{61}$	$\frac{6}{61}$

Ex. 25. *There are 500 misprints in a book of 500 pages. What is the probability that a given page will contain at most 3 misprints?*

[WBUT 2007]

Solution. Average number of misprint in a page $=\frac{500}{500}=1$

Let X = No. of misprint in the page

$\therefore$ X has poisson distribution with parameter $\mu=1$.

$\therefore\ P(X=i)=e^{-\mu}\frac{\mu^i}{i!}$

Required Probability $=P(X\leq 3)$

$=P(X=0)+P(X=1)+P(X=2)+P(X=3)$

$=e^{-\mu}.\frac{\mu^0}{0!}+e^{-\mu}.\frac{\mu^1}{1!}+e^{-\mu}\frac{\mu^2}{2!}+e^{-\mu}.\frac{\mu^3}{3!}$

$=e^{-\mu}\left(1+1+\frac{1}{2}+\frac{1}{6}\right)=e^{-1}\times\frac{8}{3}=\frac{8}{3e}$

Ex. 26. *If the chance of being killed by flood during a year is* $\frac{1}{3000}$, *use poisson distribution to calculate probability that out of 3000 persons living in a village, at least one will die in flood in a year.* [WBUT 2009]

Solution. Let X = No. of persons dyeing in flood. X has Bonomial distribution with parameter $n = 3000, \quad p = \frac{1}{3000}$.

Since n is large, X is approximately poisson variate with parameter, $\mu = np = 1$

$\therefore$ The probability mass, $f_i = P(X = i) = \frac{e^{-\mu}\mu^i}{i!}$

Required probability $= P(X \geq 1)$

$$= 1 - P(X = 0) = 1 - f_0 = 1 - \frac{e^{-\mu}.\mu^0}{0!}$$

$$= 1 - e^{-\mu} = 1 - e^{-1}.$$

Ex. 27. *If* $A_1, A_2, \ldots A_n$ *be n number of events connected to a random experiment then show that*

$$P(A_1 \cap A_2 \cap \ldots \cap A_n) \geq \sum_{i=1}^{n} P(A_i) - (n-1).$$ [WBUT 2003]

Solution. We have shown in Art 3.1.5

$$P(A_1 \cap A_2 \cap \ldots \cap A_n) \geq 1 - \sum_{i=1}^{n} P(\overline{A}_i) \quad \ldots \quad (1)$$

Now, $\sum_{i=1}^{n} P(\overline{A}_i) = \sum_{i=1}^{n} \{1 - P(A_i)\}$

$$= \sum_{i=1}^{n} 1 - \sum_{i=1}^{n} P(A_i) = n - \sum_{i=1}^{n} P(A_i)$$

$\therefore$ From (1) we get

$$P(A_1 \cap A_2 \cap \ldots \cap A_n) \geq 1 - \left\{n - \sum_{i=1}^{n} P(A_i)\right\}$$

$$= \sum_{i=1}^{n} P(A_i) - (n-1).$$

BASIC CONCEPT OF PARTIAL DIFFERENTIAL EQUATION

.1. Origin of PDE & its Order, Degree.

Partial differential equations (p.d.e) arise in study when the number of independent variables is two or more. Most of the problems of applied mathematics, physics, engineering sciences are involved with partial differential equations.

A differential equation which involves the independent variables, dependent variable and the partial derivative w.r.t the independent variables is called the **partial differential equation (p.d.e).**

For example, $x\frac{\partial z}{\partial x}+y\frac{\partial z}{\partial y}=z$ ... (1)

$$\frac{\partial^2 z}{\partial x^2}+xy\frac{\partial^2 z}{\partial x\partial y}+y^2\frac{\partial^2 z}{\partial y^2}=x^2 \quad ... \quad (2)$$

$$\frac{\partial^2 p}{\partial x\partial y}+\left(\frac{\partial p}{\partial x}\right)^2=0 \quad ... \quad (3)$$

$$\left(\frac{\partial z}{\partial x}\right)^2+\left(\frac{\partial z}{\partial y}\right)^2=xy \quad ... \quad (4)$$

$$\frac{\partial^2 u}{\partial x^2}+\frac{\partial^2 u}{\partial y^2}+\frac{\partial^2 u}{\partial z^2}=0 \quad ... \quad (5)$$

$$\sqrt{\frac{\partial z}{\partial x}}+\sqrt{\frac{\partial z}{\partial y}}=x-y \quad ... \quad (6)$$

are partial differential equations.

As in ordinary differential equation, the order of the highest partial derivative in the equation is called the order of the p.d.e and the degree of this derivative is called the degree of the p.d.e.

For example, the order and degree of (1) is 1and 1, that of (2) is 2 and 1; that of (4) is 1 and 2.

If the dependent variable and its derivatives occur linearly i.e. power or product of the dependent and its derivatives presents with degree one then p.d.e is called *linear*, otherwise *non-linear*.

For example, equations (1), (2) and (5) are linear whereas equations (3), (4) and (6) are non-linear.

In the present and subsequent chapters we shall use the following notation for the partial derivatives of z w.r.t the independent variables x and y :

$$\frac{\partial z}{\partial x}=p, \frac{\partial z}{\partial y}=q, \frac{\partial^2 z}{\partial x^2}=r, \frac{\partial^2 z}{\partial x \partial y}=s, \frac{\partial^2 z}{\partial y^2}=t .$$

Partial differential equations can be derived either by the elimination of arbitrary constants or by the elimination arbitrary functions. If the number of arbitrary constants is more than the number of independent variables then the p.d.e thus obtained will be of higher order than the first. Otherwise the p.d.e will be of first order. In general the order of p.d.e obtained by elimination of arbitrary functions is equal to the number of arbitrary functions eliminated.

The following examples illustrate the method :

Ex. 1. *Eliminating the arbitrary constants from the following equations form p.d. equation :*

(i) $az+b=a^2x+y$

(ii) $z=ax+a^2y^2+b$ *(iii)* $z=ax^2+bxy+cy^2$.

(i) Differentiating w.r.t x and y we get

$$a\frac{\partial z}{\partial x}+0=a^2+0 \quad \therefore ap=a^2 \quad \therefore p=a .$$

and $a\frac{\partial z}{\partial y}+0=1 \quad \therefore aq=1 \quad \therefore q=\frac{1}{a}$

$\therefore pq=a\cdot\frac{1}{a}=1 \quad \therefore pq=1$ which is the required p.d.e.

(ii) Differentiating partially w.r.t x and y we get

$p=a$ and $q=a^2 2y$.

$\because$ eliminating a between these results we get

$q=2p^2y$ which is the required p.d.e.

(iii) Differentiating z partially w.r.t x and y we get

$$\frac{\partial z}{\partial x}=p=2ax+by,\quad \frac{\partial z}{\partial x}=q=bx+2cy,$$

$$\therefore\ \frac{\partial^2 z}{\partial x^2}=r=2a, \frac{\partial^2 z}{\partial x\partial y}=s=b, \frac{\partial^2 z}{\partial y^2}=t=2c.$$

Thus $a=\frac{r}{2}, b=s, c=\frac{t}{2}$

Substituting these in the given equation we get

$$z=\frac{r}{2}x^2+sxy+\frac{t}{2}y^2$$

$$\therefore\ 2z=rx^2+2sxy+ty^2$$

which is the required p.d.e.

Ex. 2. *Form the p.d. equation by eliminating the arbitrary functions from the following equations :*

(i) $z=F\left(x^2+y^2\right)$

(ii) $z=f_1(x+iy)+f_2(x-iy)$

(iii) $z=xf(x+y)+g(x+y)$.

(i) Differentiating z partially w.r.t x and y, we have

$$\frac{\partial z}{\partial x}=f_1'(x+iy)+f_2'(x-iy)$$

$$\frac{\partial^2 z}{\partial x^2}=f_1''(x+iy)+f_2''(x-iy) \quad \ldots (1)$$

$$\frac{\partial z}{\partial y}=i f_1'(x+iy)-i f_2'(x-iy)$$

$$\frac{\partial^2 z}{\partial y^2}=i^2 f_1''(x+iy)+i^2 f_2''(x-iy)$$

$$=-f_1''(x+iy)-f_2''(x-iy). \quad \ldots (2)$$

Adding (1) and (2) we get

$$\therefore\ \frac{\partial^2 z}{\partial x^2}+\frac{\partial^2 z}{\partial y^2}=0$$ which is the required p.d.e equation.

(ii) Differentiating z partially w.r.t x and y we have

$$p=\frac{\partial z}{\partial x}=F'\left(x^2+y^2\right)\cdot 2x \quad \ldots (1)$$

$$q=\frac{\partial z}{\partial y}=F'\left(x^2+y^2\right)\cdot 2y \quad \ldots (2)$$

From (1) and (2), we get $\dfrac{p}{2x}=\dfrac{q}{2y}$

$\therefore\ yp=qx$ i.e. $yp-qx=0$ which is the required p.d.e.

(iii) We have $z=x\,f(x+y)+g(x+y)$

$$\therefore \frac{\partial z}{\partial x}=f(x+y)+x\,f'(x+y)+g'(x+y)$$

$$\therefore \frac{\partial^2 z}{\partial x^2}=2f'(x+y)+x\,f''(x+y)+g''(x+y) \quad \ldots (1)$$

$$\frac{\partial z}{\partial y}=xf'(x+y)+g'(x+y)$$

$$\therefore \frac{\partial^2 z}{\partial y^2}=x\,f''(x+y)+g''(x+y) \quad \ldots (2)$$

Also $\dfrac{\partial^2 z}{\partial x\partial y}=f'(x+y)+x\,f''(x+y)+g''(x+y) \quad \ldots (3)$

Adding (1) and (2), we get

$$\frac{\partial^2 z}{\partial x^2}+\frac{\partial^2 z}{\partial y^2}=2\left\{x\,f''(x+y)+f'(x+y)+g''(x+y)\right\}=2\frac{\partial^2 z}{\partial x\partial y}$$

$$\therefore \frac{\partial^2 z}{\partial x^2}+\frac{\partial^2 z}{\partial y^2}-2\frac{\partial^2 z}{\partial x\partial y}=0$$

which is the required p.d. equation.

1.2. Solution of Partial Differential Equations.

The solution of a p.d.e. in a region R is a function z of independent variables x, y which satisfies the equation identically at every point of R.

Complete Solution or Complete Integral.

A relation of the form $f(x,y,z,a,b)=0$ which contains as many arbitrary constants as there are independent variables in the p.d.e. of the form

$F(x,y,z,p,q)=0$ is known as complete solution.

General Solution or General Integral.

Any relation of the form $\phi(u,v)=0$ involving an arbitrary function ϕ connecting the known function u and v of x,y,z is said to be a general solution of a first order p.d.e.

Particular Solution or Particular Integral.

When the arbitrary constants or function of a complete or general solution are specified by given conditions, the solution is called a particular solution.

1.3. Ordinary Simultaneous Differential Equations.

The general form of an ordinary simultaneous differential equation is

$$\frac{dx}{P}=\frac{dy}{Q}=\frac{dz}{R}$$

where P,Q,R are functions of x,y,z.

A solution of this equation is a two parametric family of curves, namely $u(x,y,z)=c_1, v(x,y,z)=c_2$, c_1 and c_2 being arbitrary constants.

Note. $u(x,y,z)=c_1, v(x,y,z)=c_2$ are also said to be integral surfaces, satisfying the simultaneous equations.

Methods of Solving the Simultaneous Ordinary Differential Equations.

Method. 1. First write the equation as

$$\frac{dx}{P}=\frac{dy}{Q}=\frac{dz}{R}=\frac{\ell\,dx+m\,dy+n\,dz}{\ell P+mQ+nR}$$ where ℓ,m,n are functions of x,y,z or constants.

Choose ℓ,m,n such that either $\ell P+mQ+nR=0$ and $\ell dx+mdy+ndz$ is an exact differential of some function $u(x,y,z)$, say i.e.

$$\ell\,dx+mdy+ndz=d(u(x,y,z))$$

or, if $\ell P+mQ+nR=\phi(x,y,z)$, then

$$\ell dx+mdy+ndz=d(\phi(x,y,z)).$$

Then procceding as usual we obtain the integral surface.

Illustration.

(i) Consider the simultaneous equations

$$\frac{dx}{y-z}=\frac{dy}{z-x}=\frac{dz}{x-y}.$$

Then we can write the equation as

$$\frac{dx}{y-z}=\frac{dy}{z-x}=\frac{dz}{x-y}=\frac{dx+dy+dz}{(y-z)+(z-x)+(x-y)}=\frac{d(x+y+z)}{0}$$

$\therefore d(x+y+z)=0$

which on integration gives $x+y+z=c_1$, c_1 being an arbitrary constant.

Again $$\frac{dx}{y-z}=\frac{dy}{z-x}=\frac{dz}{x-y}=\frac{x\,dx+y\,dy+z\,dz}{x(y-z)+y(z-x)+z(x-y)}$$

$$=\frac{\frac{1}{2}d\left(x^2+y^2+z^2\right)}{0}$$

$\therefore d\left(x^2+y^2+z^2\right)=0$

which on integration gives $x^2+y^2+z^2=c_2$, c_2 being an arbitrary constant.

Thus the general solution of the given equation is

$x+y+z=c_1, x^2+y^2+z^2=c_2$, c_1,c_2 being arbitrary constants.

Note. The solution can also be written as $\phi\left(x+y+z, x^2+y^2+z^2\right)=0$ where ϕ is an arbitrary function.

(ii) Next consider the simultaneous equations

$$\frac{dx}{y+z}=\frac{dy}{z+x}=\frac{dz}{x+y} \quad \ldots \text{(a)}$$

we can write (a) as

$$\frac{dx}{y+z}=\frac{dy}{z+x}=\frac{dz}{x+y}=\frac{dx+dy+dz}{y+z+z+x+x+y}=\frac{d(x+y+z)}{2(x+y+z)} \quad \ldots \text{(b)}$$

Also from (a)

$$\frac{dx}{y+z}=\frac{dy}{z+x}=\frac{dx-dy}{(y+z)-(z+x)}=\frac{d(x-y)}{-(x-y)} \quad \dots \quad \text{(c)}$$

Again from (a),

$$\frac{dy}{z+x}=\frac{dz}{x+y}=\frac{dy-dz}{(z+x)-(x+y)}=\frac{d(y-z)}{-(y-z)} \quad \dots \quad \text{(d)}$$

Thus, in virtue of (b), (c) and (d),

$$\frac{d(x-y)}{-(x-y)}=\frac{d(y-z)}{-(y-z)}=\frac{d(x+y+z)}{2(x+y+z)} \quad \dots \quad \text{(e).}$$

From (e), $\dfrac{d(x-y)}{x-y}=\dfrac{d(y-z)}{y-z}$

which on integration gives

$\log(x-y)=\log(y-z)+\log c_1$

i.e. $x-y=c_1(y-z)$, c_1 being an arbitrary constant.

Again from (e),

$$\frac{d(x+y+z)}{2(x+y+z)}+\frac{d(x-y)}{x-y}=0$$

which on integration gives

$(x+y+z)(x-y)^2=c_2$, c_2 being an arbitrary constant.

Thus the solution of the given equation is

$x-y=c_1(y-z),(x+y+z)(x-y)^2=c_2$, c_1,c_2 being arbitrary constants.

Method. 2. If possible, first find an integral surface $u(x,y,z)=c_1$, say of the equation

$$\frac{dx}{P}=\frac{dx}{Q}=\frac{dz}{R}$$

as usual procedure. Then using this, we obtain an ordinary differential equation of the form

$\frac{dx}{P'}=\frac{dy}{Q'}$ or, $\frac{dy}{Q'}=\frac{dz}{R'}$ or, $\frac{dx}{P'}=\frac{dz}{R'}$ such that $\frac{dx}{P'},\frac{dy}{Q'},\frac{dz}{R'}$ are integrabl
and procceding as usual we obtain second integral surface.

Illustration. (iii). Consider the simultaneous equations

$$\frac{dx}{xy}=\frac{dy}{y^2}=\frac{dz}{x-yz} \quad \text{...} \quad \text{(a)}$$

From (a), $\frac{dx}{xy}=\frac{dy}{y^2}$ i.e. $\frac{dx}{x}=\frac{dy}{y}$

which on integration gives

$$x=c_1y, \quad \text{...} \quad \text{(b)}$$

c_1 being an arbitrary constant.

Again from (a)

$$\frac{dx}{xy}=\frac{dz}{x-yz}$$

or, $\frac{dx}{xy}=\frac{dz}{c_1y-yz}$, using (b)

or, $\frac{dx}{x}=\frac{dz}{c_1-z}$

which on integration gives

$$x(z-c_1)=c_2, \quad \text{...} \quad \text{(c)}$$

c_2 being an arbitrary constant.

Now from (b) and (c)

$$x\left(z-\frac{x}{y}\right)=c_2 \;\therefore\; x(yz-x)=c_2y.$$

Thus the solution of the equation is

$x=c_1y, x(yz-x)=c_2y, c_1, c_2$ being arbitrary constant.

.4. Linear Partial Differential Equations of First Order.

A linear partial differential equation of first order is of the form

$$Pp+Qq=R \quad \dots \quad (1)$$

where P, Q, R are functions of x, y, z. This equation is also known as *agrange's Equation*.

The general solution of this equation is

$\phi(u,v)=0$ or $u=\phi(v)$

where ϕ is an arbitrary function and $u(x,y,z)=c_1$ and $v(x,y,z)=c_2$ are ndependent solutions of the simultaneous ordinary differential equations

$$\frac{dx}{P}=\frac{dy}{Q}=\frac{dz}{R} \quad \dots \quad (2)$$

Note. Equations (2) are called *Lagrange's auxiliary equation or subsidiary equations*.

llustration.

Consider the p.d. equation

$(y-z)p+(z-x)q=x-y$

whose subsidiary equations are

$$\frac{dx}{y-z}=\frac{dy}{z-x}=\frac{dz}{x-y}.$$

Then from Illustration (i) of previous Article we have $x+y+z=c_1$ and $x^2+y^2+z^2=c_2$ are two independent solutions of this subsidiary equation. Hence the general solution of the p.d. equation is

$$\phi(x+y+z, x^2+y^2+z^2)=0$$

where ϕ is an arbitrary differentiable function. This solution can also be written as $x^2+y^2+z^2=\phi(x+y+z)$.

Particular Solution Passing Through a Given Curve.

Let $\phi(u,v)=0$ be the general solution of a p.d.e where $u(x,y,z)=c_1$ and $v(x,y,z)=c_2$ are two independent solutions i.e. the integral surfaces of the corresponding subsidiary equation. If we are given that the integral surface passes through a given curve $C:\{x(t),y(t),z(t)\}$ with t as a parameter, then we must have

$$u\{x(t),y(t),z(t)\}=c_1 \text{ and } v\{x(t),y(t),z(t)\}=c_2 .$$

Now eliminating t between these two equations we get a relation between c_1 and c_2. Then replacing c_1 with $u(x,y,z)$ and c_2 with $v(x,y,z)$ in this relation we get the particular solution of the given p.d.e.

Illustration.

We now find the particular solution of the p.d.e $(y-z)p+(z-x)q=x-y$ which passes through the curve $xy=4, z=0$.

The two integral surfaces of this p.d.e is

$$x+y+z=c_1 \quad \dots \quad (1)$$

$$x^2+y^2+z^2=c_2 \quad \dots \quad (2)$$

The parametric form of the given curve is

$x=t, y=\frac{4}{t}, z=0$, t being a parameter.

As this curve lies on (1) and (2), we get

$$t+\frac{4}{t}=c_1 \quad \dots \quad (3)$$

$$t^2+\frac{16}{t^2}=c_2$$

or, $\left(t+\frac{4}{t}\right)^2-8=c_2$

or, $c_1^{\,2}=8+c_2$, by (3)

$\therefore (x+y+z)^2=8+x^2+y^2+z^2$, using (1) and (2)

i.e. $xy+yz+zx-4=0$.

which is the required particular solution.

Illustrative Examples.

Ex. 1. *Find the general solution of the following p.d. equations :*

(i) $(y+zx)p-(x+yz)q=x^2-y^2$

(ii) $(z^2-2yz-y^2)p+(xy+zx)q=xy-zx$

(iii) $z(p-q)=z^2+(x+y)^2$

(iv) $(xy^3-2x^4)p+(2y^4-x^3y)q=9z(x^3-y^3)$

(v) $p\cos(x+y)+q\sin(x+y)=z$.

(i) The subsidiary equations of the given p.d.e are

$$\frac{dx}{y+zx}=\frac{dy}{-x-yz}=\frac{dz}{x^2-y^2}$$

$\therefore$ Each fraction $=\dfrac{x\,dx+y\,dy-z\,dz}{x(y+zx)+y(-x-yz)-z(x^2-y^2)}$

$$=\frac{\frac{1}{2}d(x^2+y^2+z^2)}{0}$$

$\therefore d(x^2+y^2+z^2)=0$

which on integration gives $x^2+y^2+z^2=c_1$, c_1 being on arbitrary constant.

Again each fraction $=\dfrac{y\,dx+x\,dy}{y^2-x^2}$

$$\therefore \frac{dz}{x^2-y^2}=\frac{y\,dx+x\,dy}{y^2-x^2}$$

i.e. $y\,dx+x\,dy+dz=0$

i.e. $d(xy)+dz=0$

which on integration gives $xy+z=c_2$, c_2 being an arbitrary constant.

Thus the general solution of equation is

$\phi(x^2+y^2+z^2, xy+z)=0$ where ϕ is an arbitrary differentiable function.

(ii) The subsidiary equations of the given p.d.e are

$$\frac{dx}{z^2-2yz-y^2}=\frac{dy}{xy+zx}=\frac{dz}{xy-zx}$$

$\therefore$ Each fraction $=\dfrac{x\,dx+y\,dy+z\,dz}{0}=\dfrac{\frac{1}{2}d\left(x^2+y^2+z^2\right)}{0}$

$\therefore d\left(x^2+y^2+z^2\right)=0$

which on integration gives $x^2+y^2+z^2=c_1$, c_1 being an arbitrary constant

Again from last two fractions, we have

$$\frac{dy}{y+z}=\frac{dz}{y-z}$$

i.e. $y\,dy-\left(z\,dy+y\,dz\right)-z\,dz=0$.

i.e. $\frac{1}{2}d\left(y^2\right)-d\left(yz\right)-\frac{1}{2}d\left(z^2\right)=0$

which on integration gives

$$\frac{1}{2}y^2-yz-\frac{1}{2}z^2=\frac{c_2}{2}$$

i.e. $y^2-2yz-z^2=c_2$, c_2 being an arbitrary constant.

So the general solution of the given equation is

$$\phi\left(x^2+y^2+z^2,y^2-2yz-z^2\right)=0$$

where ϕ is an arbitrary differentiable function.

(iii) The subsidiary equations are

$$\frac{dx}{z}=\frac{dy}{-z}=\frac{dz}{z^2+\left(x+y\right)^2}.$$

From the first two ratios, we have

$$dx=-dy$$

which on integration gives $x+y=c_1$.

Again from first and third ratios, we have

$$\frac{dx}{z}=\frac{dz}{z^2+(x+y)^2}$$

i.e. $dx=\dfrac{z\,dz}{z^2+{c_1}^2}$

which on integration gives

$$\log\left(z^2+c_1^2\right)=2x+\log c_2$$

i.e. $\left\{z^2+(x+y)^2\right\}e^{-2x}=c_2$.

So the general solution of the equation is

$\phi\left(x+y,\left(z^2+(x+y)^2e^{-2x}\right)\right)=0$ where ϕ is an arbitrary differentiable function.

(iv) The subsidiary equations of the given p.d.e are

$$\frac{dx}{xy^3-2x^4}=\frac{dy}{2y^4-x^3y}=\frac{dz}{9z\left(x^3-y^3\right)}$$

From first two fractions, we have

$$\left(2y^4-x^3y\right)dx=\left(xy^3-2x^4\right)dy$$

or, $\dfrac{1}{x^2}dy-\dfrac{2y}{x^3}dx+\dfrac{1}{y^2}dx-\dfrac{2x}{y^3}dy=0$

or, $d\left(\dfrac{y}{x^2}\right)+d\left(\dfrac{x}{y^2}\right)=0$

which on integration gives $\dfrac{y}{x^2}+\dfrac{x}{y^2}=c_1$.

Again each fraction $=\dfrac{\dfrac{1}{x}dx+\dfrac{1}{y}dy+\dfrac{1}{3z}dz}{0}$

$$\therefore\ \frac{1}{x}dx+\frac{1}{y}dy+\frac{1}{3z}dz=0$$

which on integration gives

$$\log x+\log y+\frac{1}{3}\log z=\log c_2$$

$$\therefore\ xyz^{1/3}=c_2.$$

So the general solution of the equation is

$$f\left(\frac{y}{x^2}+\frac{x}{y^2}, xyz^{1/3}\right)=0$$

where f is an arbitrary differentiable function.

(v) The subsidiary equations are

$$\frac{dx}{\cos(x+y)}=\frac{dy}{\sin(x+y)}=\frac{dz}{z}$$

$$\therefore \frac{dx+dy}{\cos(x+y)+\sin(x+y)}=\frac{dz}{z}$$

or, $\dfrac{du}{\sin u+\cos u}=\dfrac{dz}{z}$, putting $x+y=u$

or, $$\frac{\frac{1}{\sqrt{2}}du}{\sin\left(\pi/4+u\right)}=\frac{dz}{z}$$

or, $\operatorname{cosec}\left(\pi/4+u\right)du=\sqrt{2}\dfrac{dz}{z}$

which on integration gives

$$\sqrt{2}\log z=\log\tan\frac{1}{2}\left(u+\pi/4\right)+\log c_1$$

$$\therefore \quad z^{\sqrt{2}}=c_1\tan\left(\frac{u}{2}+\frac{\pi}{8}\right)$$

$$\therefore \quad z^{\sqrt{2}}\cot\left(\frac{x}{2}+\frac{y}{2}+\frac{\pi}{8}\right)=c_1.$$

Also from first two fractions we have

$$\frac{dx+dy}{\cos(x+y)+\sin(x+y)}=\frac{dx-dy}{\cos(x+y)-\sin(x+y)}$$

or, $$\frac{\cos(x+y)-\sin(x+y)}{\cos(x+y)+\sin(x+y)}d(x+y)=d(x-y)$$

which on integration gives

$$\log\{\cos(x+y)+\sin(x+y)\} = x - y + \log c_2$$

$$\therefore \cos(x+y)+\sin(x+y) = c_2\, e^{x-y}.$$

Thus the general solution of the equation is

$$\phi\left\{z^{\sqrt{2}}\cot\left(\frac{x}{2}+\frac{y}{2}+\frac{\pi}{8}\right), \left(\cos(x+y)+\sin(x+y)\right)e^{y-x}\right\} = 0.$$

where ϕ is an arbitrary differentiable function.

Ex. 2. *Find the integral surface satisfying the p.d.e* $(x-y)p+(y-x-z)q = z$ *and passing through the circle* $x^2+y^2 = 1, z = 1$.

The subsidiary equation of the given p.d.e are

$$\frac{dx}{x-y} = \frac{dy}{y-x-z} = \frac{dz}{z} \quad \ldots \quad (1)$$

$$\therefore \text{Each fraction} = \frac{dx+dy+dz}{0} = \frac{d(x+y+z)}{0}$$

$$\therefore d(x+y+z) = 0$$

which on integration gives

$$x+y+z = c_1, \quad \ldots \quad (2)$$

c_1 being on arbitrary constant.

$$\text{Again each fraction} = \frac{dx-dy+dz}{2(x-y+z)} = \frac{d(x-y+z)}{2(x-y+z)}$$

$$\therefore \frac{dz}{z} = \frac{d(x-y+z)}{2(x-y+z)}$$

$$\text{or, } 2\frac{dz}{z} = \frac{d(x-y+z)}{x-y+z}$$

which on integration gives

$$\log(x-y+z) = 2\log z + \log c_2$$

$$\therefore x-y+z = c_2 z^2, \quad \ldots \quad (3)$$

c_2 being an arbitrary constant.

The parametric form of the given equation is

$$x=\cos\theta, y=\sin\theta, z=1 \quad \ldots (4)$$

From (2) and (4), we get

$$\cos\theta+\sin\theta+1=c_1$$

$$\therefore \cos\theta+\sin\theta=c_1-1 \quad \ldots (5)$$

Again from (3) and (4), we get

$$\cos\theta-\sin\theta+1=c_2.$$

$$\therefore \cos\theta-\sin\theta=c_2-1 \quad \ldots (6)$$

Squaring and adding (5) and (6), we get

$$2=(c_1-1)^2+(c_2-1)^2$$

$$\therefore c_1^2+c_2^2-2c_1-2c_2=0.$$

Hence from (1) and (3), we get

$$(x+y+z)^2+\left(\frac{x-y+z}{z^2}\right)^2-2(x+y+z)-2\left(\frac{x-y+z}{z^2}\right)=0$$

$$\therefore z^4(x+y+z)^2+(x-y+z)^2-2z^4(x+y+z)-2z^2(x-y+z)=0$$

which is the required integral surface.

1.5. Non-linear Partial Differential Equations of First Order.

A p.d.e. containing first order partial derivatives p and q with degree greater than one and products of p, q is called a non-linear p.d.e of first order. The general form of this equation is $f(x,y,z,p,q)=0$. The general method for solving such equation is ***Charpit's Method***. Before detail discussion of this method, we discuss some special methods for solving some standard types of this p.d.e.

Solutions of Some Standard Types of Equations.

(a) Equations of the form $f(p,q)=0$ i.e. equations involving only p and q but no x, y, z.

The complete solution of this type of equation is

$z = ax + by + c$ where $f(a,b) = 0$.

From $f(a,b) = 0$, we get $b = \phi(a)$, say.

Thus the complete solution is $z = ax + \phi(a)y + c$ where a and c are arbitrary constants.

llustration.

Consider the equation $p + q = pq$.

Let the complete solution be $z = ax + by + c$ where

$$a + b = ab \therefore b = \frac{a}{a-1}.$$

So the complete solution is

$$z = ax + \frac{a}{a-1}y + c$$ where a, c are arbitrary constants.

(b) *Equations of the form* $f(z,p,q) = 0$.

Let $z = \phi(u)$ where $u = x + ay$ be a trial solution of this type of equation. Then

$$p = \frac{\partial z}{\partial x} = \phi'(u) \cdot \frac{\partial u}{\partial x} = \frac{dz}{du}$$

$$q = \frac{\partial z}{\partial y} = \phi'(u) \cdot a = a\frac{dz}{du}.$$

So the given equation becomes

$$f\left(z, \frac{dz}{du}, a\frac{dz}{du}\right) = 0$$

which is an first order ordinary differential equation and so the solution of this equation is of the form $F(z,u) = 0$, i.e. $F(z, x + ay) = 0$

which is the required complete solution.

Illustration.

Consider the p.d. equation

$z^2(p^2 + q^2 + 1) = c^2$.

Let $z = \phi(u)$ where $u = x + ay$ be a trial solution. Then

$$p = \frac{dz}{du}, q = a\frac{dz}{du}.$$

So the equation becomes

$$z^2\left\{\left(\frac{dz}{du}\right)^2+\left(a\frac{dz}{du}\right)^2+1\right\}=c^2$$

or, $z^2\left(1+a^2\right)\left(\frac{dz}{du}\right)^2=c^2-z^2$

or, $\frac{dz}{du}=\pm\frac{\sqrt{c^2-z^2}}{z\sqrt{1+a^2}}$

or, $\pm\sqrt{1+a^2}\,\frac{z}{\sqrt{c^2-z^2}}dz=du$

which on integration gives

$\pm\sqrt{1+a^2}\sqrt{c^2-z^2}=u+c_1$, c_1 being an arbitrary constant

$$\therefore \left(1+a^2\right)\left(c^2-z^2\right)=\left(u+c_1\right)^2$$

i.e. $\left(1+a^2\right)\left(c^2-z^2\right)=\left(x+ay+c_1\right)^2$

which is the required complete solution.

(c) Equations of the form $f_1(x,p)=f_2(y,q)$.

Let $f_1(x,p)=f_2(y,q)=a$, a constant. Then

$f_1(x,p)=a$, $f_2(y,q)=a$. Solving these equations we get $p=\phi_1(x), q=\phi_2(y)$, say. Then from $dz=\frac{\partial z}{\partial x}dx+\frac{\partial z}{\partial y}dy=pdx+q\,dy$ we get $dz=\phi_1(x)dx+\phi_2(y)dy$.

which on integration gives the required solution.

Illustration.

Consider the equation $x(1+y)p=y(1+x)q$ which can be written as

$\frac{xp}{1+x}=\frac{yq}{1+y}=a$, say.

$$\therefore \frac{xp}{1+x}=a, \frac{yq}{1+y}=a$$

which gives $p=a\left(\frac{1+x}{x}\right), q=a\left(\frac{1+y}{y}\right)$.

Putting these values of p and q in $dz = pdx + qdy$ we get

$$dz = a\left(\frac{1+x}{x}\right)dx + a\left(\frac{1+y}{y}\right)dy$$

$$= a\left(1+\frac{1}{x}\right)dx + a\left(1+\frac{1}{y}\right)dy$$

which on integration gives

$z = a(x+\log x) + a(y+\log y) + c_1$, a, c_1 being arbitrary constant which is the required complete solution.

(d) Equations of the form $z = px + qy + f(p,q)$.

This type of equation is known as Clairaut's Equaion and the complete solution is $z = ax + by + f(a,b)$ which is obtained just replacing p by a and q by b.

Illustration.

Consider the equation $z = px + qy - 2\sqrt{pq}$.

The complete solution of this equation is

$z = ax + by - 2\sqrt{ab}$.

General Method For Solving a Non-linear Equation : Charpit's Method.

Let $f(x,y,z,p,q) = 0$ ... (1)

be a non-linear p.d.e.

Let us assume that there exist a relation $F(x,y,z,p,q) = 0$ ... (2)

involving x,y,z,p,q such that the equation (1) and (2) determine p and q as explicit functions of x,y,z. Then substituting these values of p and q in

$$dz = p\,dx + q\,dy$$

and integrating we get the required solution of the equation (1).

The relation (2) can be found out from the equations

$$\frac{dx}{-\frac{\partial f}{\partial p}} = \frac{dy}{-\frac{\partial f}{\partial q}} = \frac{dz}{-p\frac{\partial f}{\partial p} - q\frac{\partial f}{\partial q}} = \frac{dp}{\frac{\partial f}{\partial x} + p\frac{\partial f}{\partial z}} = \frac{dq}{\frac{\partial f}{\partial y} + q\frac{\partial f}{\partial z}} = \frac{dF}{0}.$$

These equations are known as *Charpit's equations* or *Charpit's auxiliary equations*.

Note. This method is to be applied only when the other methods fail to solve the equation.

Illustration.

Consider the p.d. equation $z^2\left(p^2z^2+q^2\right)=1$

i.e. $z^2\left(p^2z^2+q^2\right)-1=0$ ... (1)

Let $f(x,y,z,p,q)=z^2\left(p^2z^2+q^2\right)-1$

$$\therefore \frac{\partial f}{\partial x}=0, \frac{\partial f}{\partial y}=0, \frac{\partial f}{\partial z}=4z^3p^2+2zq^2, \frac{\partial f}{\partial p}=2pz^4, \frac{\partial f}{\partial q}=2qz^2$$

So the Charpit's equations of the given p.d. equation are

$$\frac{dx}{-2pz^4}=\frac{dy}{-2qz^2}=\frac{dz}{-2p^2z^4-2q^2z^2}=\frac{dp}{0+p\left(4z^3p^2+2zq^2\right)}=\frac{dq}{0+q\left(4z^3p^2+2zq^2\right)}$$

$$\therefore \frac{dp}{p}=\frac{dq}{q}$$

which on integration gives $\log p=\log q+\log c$

$\therefore p=qc$, c being an arbitrary constant

$$\therefore \text{From (1), } q^2=\frac{1}{z^2\left(c^2z^2+1\right)}$$

$$\therefore q=\frac{1}{z\sqrt{1+c^2z^2}} \text{ and } p=\frac{c}{z\sqrt{1+c^2z^2}}.$$

Then from $dz=p\,dx+q\,dy$, we get

$$dz=\frac{c}{z\sqrt{1+c^2z^2}}dx+\frac{1}{z\sqrt{1+c^2z^2}}dy$$

or, $z\sqrt{1+c^2z^2}\,dz=c\,dx+dy$

which on integration gives

$$\frac{1}{3c^2}\left(1+c^2z^2\right)^{3/2} = cx + y + d$$

$$\therefore \left(1+c^2z^2\right)^3 = 9c^4\left(cx+y+d\right)^2$$

which is the required solution.

Illustrative Examples.

Ex. 1. *Solve (a)* $\sqrt{p}+\sqrt{q}=1$ *(b)* $x^2p^2+y^2q^2=z$.

(a) The equation is of the form $f(p,q)=0$. So let the complete solution be $z=ax+by+c$ where

$\sqrt{a}+\sqrt{b}=1$ i.e. $b=\left(1-\sqrt{a}\right)^2$.

So the required complete solution is

$z=ax+\left(1-\sqrt{a}\right)^2 y+c$ where a,c are arbitrary constants.

(b) The given equation can be written as

$$\frac{x^2}{z}\left(\frac{\partial z}{\partial x}\right)^2+\frac{y^2}{z}\left(\frac{\partial z}{\partial y}\right)^2=1$$

$$\text{i.e. } \left(\frac{x}{\sqrt{z}}\frac{\partial z}{\partial x}\right)^2+\left(\frac{y}{\sqrt{z}}\frac{\partial z}{\partial y}\right)^2=1 \qquad \dots (1)$$

Put $\frac{1}{x}\partial x=\partial X, \frac{1}{y}\partial y=\partial Y$ and $\frac{1}{\sqrt{z}}\partial z=\partial Z$,

so that $\log x = X, \log y = Y$ and $2\sqrt{z}=Z$.

Then $\frac{x}{\sqrt{z}}\frac{\partial z}{\partial x}=\frac{\partial Z}{\partial X}, \frac{y}{\sqrt{z}}\frac{\partial z}{\partial y}=\frac{\partial Z}{\partial Y}$.

Thus (1) becomes

$$\left(\frac{\partial Z}{\partial X}\right)^2+\left(\frac{\partial Z}{\partial Y}\right)^2=1 \text{ i.e. } P^2+Q^2=1 \qquad \dots (2)$$

where $P=\frac{\partial Z}{\partial X}, Q=\frac{\partial Z}{\partial Y}$.

Let the complete solution of (2) be $z = aX + bY + c$ where $a^2 + b^2 = 1$ i.e. $b = \sqrt{1-a^2}$.

So the required solution is

$$z = aX + \sqrt{1-a^2}\,Y + c$$

i.e. $2\sqrt{z} = a\log x + \sqrt{1-a^2}\log y + c$

$$= \log\left\{x^a y^{\sqrt{1-a^2}} / c_1\right\} \text{ where } c_1 = -\log c$$

$$\therefore x^a y^{\sqrt{1-a^2}} = c_1 e^{2\sqrt{z}}$$

where a, c_1 are arbitrary constants.

Ex. 2. *Solve (a)* $\frac{1}{p} + \frac{1}{q} = z$ *(b)* $z^2\left(p^2x^2 + q^2\right) = 1$.

(a) The given equation can be written as

$$2pq = p + q \quad \cdots \quad (1)$$

which is of the form $f(p,q,z) = 0$. So let $z = \phi(u)$ where $u = x + ay$ be a trial solution.

Then $p = \frac{dz}{du}, q = a\frac{dz}{du}$. So equation (1) becomes

$$z \cdot \frac{dz}{du}\, a\frac{dz}{du} = \frac{dz}{du} + a\frac{dz}{du}$$

or, $az\frac{dz}{du} = 1 + a$

or, $az\,dz = (1+a)du$

which on integration gives

$$\frac{az^2}{2} = (1+a)u + \frac{c}{2}$$

$$\therefore az^2 = 2(1+a)u + c = 2(1+a)(x+ay) + c$$

So the required complete solution is

$az^2 = 2(1+a)(x+ay) + c$, where a, c are arbitrary constants.

(b) The given equation can be written as

$$z^2\left\{\left(x\frac{\partial z}{\partial x}\right)^2+\left(\frac{\partial z}{\partial y}\right)^2\right\}=1 \quad \ldots \quad (1)$$

Put $\frac{1}{x}\partial x=\partial X$, so that $X=\log x$. Then

$x\frac{\partial z}{\partial x}=\frac{\partial z}{\partial X}$. Thus equation (1) becomes

$$z^2\left\{\left(\frac{\partial z}{\partial X}\right)^2+\left(\frac{\partial z}{\partial y}\right)^2\right\}=1.$$

i.e. $z^2\left(p^2+q^2\right)=1 \quad \ldots \quad (2)$

where $p=\frac{\partial z}{\partial X}$.

Let $z=\phi(u)$ where $u=X+ay$ be a trial solution.

Then $p=\frac{dz}{du}, q=a\frac{dz}{du}$. So equation (2) becomes

$$z^2\left\{\left(\frac{dz}{du}\right)^2+a^2\left(\frac{dz}{du}\right)^2\right\}=1.$$

or, $\left(1+a^2\right)z^2\left(\frac{dz}{du}\right)^2=1$

or, $\sqrt{1+a^2}\cdot z\,dz=\pm du$

which on integration gives

$$\sqrt{1+a^2}\,\frac{z^2}{2}=\pm u+c/2$$

i.e. $\sqrt{1+a^2}\,z^2=\pm 2(X+ay)+c$

$\therefore z^2\sqrt{1+a^2}=\pm 2(\log x+ay)+c$

which is the required solution.

Ex. 3. *Solve (a)* $yp - \log q = 2xy$ *(b)* $z(xp - yq) = y^2 - x^2$.

(a) The given equation can be written as

$$p - \frac{1}{y}\log q = 2x$$

i.e. $p - 2x = \dfrac{1}{y}\log q$

which is of the form $f_1(x,p) = f_2(y,q)$.

Let $p - 2x = \dfrac{1}{y}\log q = a$

$\therefore p - 2x = a$ i.e. $p = a + 2x$

and $\dfrac{1}{y}\log q = a$ i.e. $q = e^{ay}$.

Putting these values of p, q in $dz = pdx + q\,dy$ we get

$$dz = (a + 2x)dx + e^{ay}dy$$

which on integration gives

$$z = ax + x^2 + \frac{1}{a}e^{ay} + c$$

where a,c are arbitrary constant.

(b) The given equation can be written as

$$x\left(z\frac{\partial z}{\partial x}\right) - y\left(z\frac{\partial z}{\partial y}\right) = y^2 - x^2 \qquad \dots (1)$$

Put $z\partial z = \partial Z$ so that $\dfrac{1}{2}z^2 = Z$.

Then $z\dfrac{\partial z}{\partial x} = \dfrac{\partial Z}{\partial x}, z\dfrac{\partial z}{\partial y} = \dfrac{\partial Z}{\partial y}$. Thus equation (1) becomes

$$x\frac{\partial Z}{\partial x} - y\frac{\partial Z}{\partial y} = y^2 - x^2$$

i.e. $xP - yQ = y^2 - x^2$ where $P = \dfrac{\partial Z}{\partial x}, Q = \dfrac{\partial Z}{\partial y}$

i.e. $xP + x^2 = yQ + y^2$.

Let $xP+x^2=yQ+y^2=a$, a constant.

$\therefore\ xP+x^2=a$ i.e. $P=\dfrac{a-x^2}{x}$

and $yQ+y^2=a$ i.e. $Q=\dfrac{a-y^2}{y}$.

Putting these values of P, Q in $dZ=Pdx+Qdy$ we get

$$dZ=\frac{a-x^2}{x}dx+\frac{a-y^2}{y}dy=\left(\frac{a}{x}-x\right)dx+\left(\frac{a}{y}-y\right)dy$$

which on integration gives

$$Z=a\log x-\frac{x^2}{2}+a\log y-\frac{y^2}{2}+\frac{1}{2}c$$

i.e. $\frac{1}{2}z^2=a(\log x+\log y)-\frac{1}{2}\left(x^2+y^2-c\right)$

$\therefore z^2=2a\log xy-\left(x^2+y^2\right)+c$

which is the required solution.

Ex. 4. *Solve : (a)* $(p+q)(z-xp-yq)=1$

(b) $4xyz=pq+2px^2y+2qxy^2$.

(a) The given equation can be written as

$$z-px-qy=\frac{1}{p+q}$$

$$\therefore z=px+qy+\frac{1}{p+q}$$

which is the clairaut's equation. So the complete solution of this equation is $z=ax+by+\dfrac{1}{a+b}$ where a,b being arbitrary constants.

(b) The given equation can be written as

$$z=\frac{1}{2x}\frac{\partial z}{\partial x}\cdot\frac{1}{2y}\frac{\partial z}{\partial y}+x^2\cdot\left(\frac{1}{2x}\frac{\partial z}{\partial x}\right)+y^2\left(\frac{1}{2y}\frac{\partial z}{\partial y}\right). \quad \cdots \quad (1)$$

Put $2x\,\partial x = \partial X, y\partial y = \partial Y$, so that

$x^2 = X, y^2 = Y$. Then

$\frac{1}{2x}\frac{\partial z}{\partial x} = \frac{\partial z}{\partial X}, \frac{1}{2y}\frac{\partial z}{\partial y} = \frac{\partial z}{\partial Y}$. Thus equation (1) becomes

$$z = \frac{\partial z}{\partial X}\frac{\partial z}{\partial Y} + X\frac{\partial z}{\partial X} + Y\frac{\partial z}{\partial Y}$$

$\therefore z = XP + YQ + PQ$ where $P = \frac{\partial z}{\partial X}, Q = \frac{\partial z}{\partial Y}$

which is the Clairaut's equation. So the complete solution of this equation is

$z = aX + bY + ab$

i.e. $z = ax^2 + by^2 + ab, a, b$ being arbitrary constants which is the required solution.

Ex. 5. *Using Charpit's method find the complete integral of the following equations.*

(a) $p^2x + qy = z$ (b) $xp(1+q) = (y+z)q$

(c) $\frac{\partial z}{\partial y} = x\frac{\partial z}{\partial x} + \left(\frac{\partial z}{\partial x}\right)^2$.

(a) The given p.d.e is $p^2x + qy = z$

i.e. $p^2x + qy - z = 0$ ... (1)

Let $f(x, y, z, p, q) = p^2x + qy - z$

$$\therefore \frac{\partial f}{\partial x} = p^2, \frac{\partial f}{\partial y} = q, \frac{\partial f}{\partial z} = -1, \frac{\partial f}{\partial p} = 2px, \frac{\partial f}{\partial q} = y.$$

So the Charpit's equation of the given p.d.e are

$$\frac{dx}{-\frac{\partial f}{\partial p}} = \frac{dy}{-\frac{\partial f}{\partial q}} = \frac{dz}{-p\frac{\partial f}{\partial p} - q\frac{\partial f}{\partial q}} = \frac{dp}{\frac{\partial f}{\partial x} + p\frac{\partial f}{\partial z}} = \frac{dq}{\frac{\partial f}{\partial y} + q\frac{\partial f}{\partial z}}$$

i.e. $\frac{dx}{-2px} = \frac{dy}{-y} = \frac{dz}{-(2p^2x + yq)} = \frac{dp}{p^2 - p} = \frac{dq}{0}$

$\therefore \; dq = 0$

which on integration gives $q = a, a$ being an arbitrary constant.

$\therefore$ From (1), $p^2 x = z - ay$

$$\therefore p = \sqrt{\frac{z - ay}{x}}$$

Substituting these values of p and q in $dz = pdx + qdy$ we get.

$$dz = \sqrt{\frac{z - ay}{x}}dx + ady$$

$$\text{or, } dz - ady = \sqrt{\frac{z - ay}{x}}dx$$

$$\text{or, } \frac{d(z - ay)}{\sqrt{z - ay}} = \frac{dx}{\sqrt{x}}$$

which on integration gives

$\sqrt{z - ay} = \sqrt{x} + b$, b being an arbitrary constant.

Thus the complete solution of the equation is

$\sqrt{z - ay} = \sqrt{x} + b$ where a, b are arbitrary constants.

(b) Let $f(x,y,z,p,q) = xp(1+q) - (y+z)q = 0$

$$\therefore \frac{\partial f}{\partial x} = p(1+q), \frac{\partial f}{\partial y} = -q, \frac{\partial f}{\partial z} = -q, \frac{\partial f}{\partial p} = x(1+q), \frac{\partial f}{\partial q} = xp - y - z.$$

So the Charpit's equations are

$$\frac{dx}{-x(1+q)} = \frac{dy}{y - xp + z} = \frac{dz}{-px - 2pqx + qy + qz} = \frac{dp}{p(1+q) - pq} = \frac{dq}{-q - q^2}$$

$$\therefore \frac{dx}{-x(1+q)} = \frac{dq}{-q(1+q)}$$

$$\text{i.e. } \frac{dx}{x} = \frac{dq}{q}$$

which on integration gives
$q = cx$, c being an arbitrary constant.

$$\therefore \text{From (1), } xp(1+cx) - (y+z)cx = 0 \quad \therefore p = \frac{c(y+z)}{1+cx}.$$

Then $dz = pdx + q\,dy$ becomes

$$dz = \frac{c(y+z)}{1+cx}dx + cx\,dy$$

or, $(1+cx)dz = cy\,dx + cz\,dx + cx(1+cx)dy$

or, $(1+cx)dz = yd(1+cx) + zd(1+cx) + \left\{(1+cx)^2 - (1+cx)\right\}dy$

or, $\dfrac{(1+cx)dz - zd(1+cx)}{(1+cx)^2} = \dfrac{-[(1+cx)dy - yd(1+cx)]}{(1+cx)^2} + dy$

or, $d\left(\dfrac{z}{1+cx}\right) + d\left(\dfrac{y}{1+cx}\right) = dy$

which an integration gives

$\dfrac{z}{1+cx} + \dfrac{y}{1+cx} = y + c'$, c' being an arbitrary constant.

i.e. $y + z = (y + c')(1+cx)$

$\therefore z = cxy + c'(1+cx)$

which is the required complete integral.

(c) The given equation can be written as

$q = xp + p^2$

i.e. $xp + p^2 - q = 0$ ... (1)

Let $f(x,y,z,p,q) = xp + p^2 - q$.

Then $\dfrac{\partial f}{\partial x} = p, \dfrac{\partial f}{\partial y} = 0 = \dfrac{\partial f}{\partial z}, \dfrac{\partial f}{\partial p} = x + 2p, \dfrac{\partial f}{\partial q} = -1$.

So the Charpit's equations are

$$\frac{dx}{-x-2p} = \frac{dy}{1} = \frac{dz}{-xp-2p^2+q} = \frac{dp}{p} = \frac{dq}{0}$$

$\therefore \dfrac{dp}{p} = dy$

which on integration gives $\log{}^{p}\!/_{c} = y$

i.e. $p = ce^y$

$\therefore$ From (1), $q = cxe^y + c^2e^{2y}$.

$\therefore$ From $dz = pdx + qdy$ we get

$$dz = ce^y dx + \left(cxe^y + c^2e^{2y}\right)dy$$

or, $dz = cd\left(xe^y\right) + c^2 e^{2y} dy$

which on integration gives

$z = cxe^y + \frac{1}{2}c^2e^{2y} + c'$, c' being an arbitrary constant.

So the required complete integral is

$$z = cxe^y + \frac{1}{2}c^2e^{2y} + c'.$$

Exercise 1

1. Form the partial differetial equation by eliminating the arbitrary constants from the following relations :

(i) $2z = \frac{x^2}{a^2} + \frac{y^2}{b^2}$ (ii) $z = cx + by + ab$

(iii) $z = a(x+y) + b$ (iv) $z = axe^{\alpha} + \frac{1}{2}a^2e^{2\alpha} + b$.

2. Eliminating the arbitrary functions from the following relations form the partial differetial equations :

(i) $z = e^{my}\phi(x-y)$ (ii) $z = f_1(x+ay) + f_2(x-ay)$

(iii) $f\left(x+y+z, x^2+y^2+z^2\right) = 0$ (iv) $z = y^2 + 2g\left(\frac{1}{x} + \log y\right)$.

3. Solve the following linear partial differential equations :

(i) $yzp + zxq = xy$.

(ii) $(mz - ny)p + (nx - lz)q = ly - mx$.

(iii) $x^2(y-z)p + y^2(z-x)q = z^2(x-y)$

(iv) $\left(y^2 + z^2 - x^2\right)p - 2xyq + 2zx = 0$

(v) $\left(x^2-yz\right)p+\left(y^2-zx\right)q=z^2-xy$.

(vi) $(xy-zx)p+(yz-xy)q=xz-yz$

(vii) $p+3q=5z+\tan(y-3x)$

(viii) $x\left(y^2+z\right)p-y\left(x^2+z\right)q=z\left(x^2-y^2\right)$

(ix) $p\tan x+q\tan y=\tan z$

(x) $(x+2z)p+(4zx-y)q=x^2+y$

(xi) $x(z-y)p+y(x-z)q=(y-x)z$

(xii) $y^2p-xyq=x(z-2y)$

(xiii) $x\left(z^2-y^2\right)p+y\left(x^2-z^2\right)q=z\left(y^2-x^2\right)$

(xiv) $z(xp-yq)=y^2-x^2$

(xv) $x^2p+y^2q=(x+y)z$

4. Find the integral surface of the p.d.e $x\left(y^2+z\right)p-y\left(x^2+z\right)q=\left(x^2-y^2\right)z$ which contains the straight line $x+y=0, z=1$.

5. Find the particular solution of the p.d.e $(2xy-1)p+\left(z-2x^2\right)q=2(x-yz)$ through the line $x=1, y=0$.

6. Find the general integral of the p.d.e $2y(z-3)p+(2x-z)q=y(2x-3)$ and find the particlulur integral through the circle $x^2+y^2-2x=0, z=0$.

7. Find the complete integral of the following non-linear p.d equation :

(i) $p^2+q^2=1$

(ii) $p^2=zq$

(iii) $pz=1+q^2$

(iv) $p\left(1+q^2\right)=q(z-a)$

(v) $\sqrt{p}+\sqrt{q}=2x$

(vi) $p^2+q^2=x+y$

(vii) $z=px+qy-p^2q$

(viii) $z=px+qy+\log pq$

8. Using Charpit's method solve the following non-linear p.d.equation :

(i) $\left(p^2+q^2\right)y=qz$

(ii) $p=(qy+z)^2$

(iii) $p^2x+q^2y=z$

(iv) $zpq=p+q$

(v) $2z+p^2+qy+2y^2=0$.

Answers 1

1. (i) $2z=px+qy$ (ii) $z=px+qy+pq$ (iii) $p-q=0$

(iv) $q=px+p^2$.

2. (i) $p+q=mz$ (ii) $\dfrac{\partial^2 z}{\partial y^2}=a^2\dfrac{\partial^2 z}{\partial x^2}$ (iii) $(y-z)p+(z-x)q=x-y$

(iv) $x^2p+yq=2y^2$.

3. (i) $f\left(x^2-y^2,x^2-z^2\right)=0$ (ii) $x^2+y^2+z^2=f(lx+my+nz)$.

(iii) $f\left(\dfrac{1}{x}+\dfrac{1}{y}+\dfrac{1}{z},xyz\right)=0$. (iv) $x^2+y^2+z^2=zf\left(\dfrac{x}{z}\right)$.

(v) $f\left(\dfrac{x-y}{y-z},\dfrac{z-x}{y-z}\right)=0$ (vi) $f(x+y+z,xyz)=0$.

(vii) $f\left(y-3x,e^{-5x}\{5z+\tan(y-3x)\}\right)=0$ (viii) $f\left(x^2+y^2-2z,xyz\right)=0$.

(ix) $\dfrac{\sin z}{\sin y}=f\left(\dfrac{\sin x}{\sin y}\right)$ (x) $f\left(xy-z^2,x^2-y-z\right)=0$.

(xi) $xyz=f(x+y+z)$ (xii) $f\left(x^2+y^2,yz-y^2\right)=0$.

(xiii) $f\left(x^2+y^2+z^2, xyz\right)=0$. (xiv) $f\left(xy, x^2+y^2+z^2\right)=0$.

(xv) $f\left(\frac{xy}{z}, \frac{1}{y}-\frac{1}{x}\right)=0$.

4. $x^2+y^2+2xyz-2z+2=0$. **5.** $x^2+y^2-xz-y-z-1=0$.

6. $x^2+y^2-z^2-2x+4z=0$.

7. (i) $z=ax+\sqrt{1-a^2}\,y+c$ (ii) $z=be^{ax+a^2y}$

(iii) $z^2 \mp \left[z\sqrt{z^2-4a^2}-4a^2\log\left\{z+\sqrt{z^2-4a^2}\right\}\right]=4x+4ay+2c$.

(iv) $4a(z-a)=4+(x+ay+b)^2$ (v) $z=\frac{1}{6}(a+2x)^3+a^2y+b$.

(vi) $z=\frac{2}{3}(x+a)^{3/2}+\frac{2}{3}(y-a)^{3/2}+b$. (vii) $z=ax+by-a^2b$.

(viii) $z=ax+by+\log ab$.

8. (i) $z^2=a^2y^2+(ax+b)^2$. (ii) $yz=ax+2\sqrt{ay}+b$.

(iii) $\sqrt{\{(1+a)z\}}=\sqrt{ax}+\sqrt{y}+b$. (iv) $z^2=2(a+1)\left(x+y/a\right)+b$.

(v) $y^2\left\{(x-a)^2+y^2+2z\right\}=b$.

SOLUTION OF HOMOGENEOUS LINEAR PDE

2.1. Introduction.

The general form of a homogeneous linear p.d.e of order two with constant co-efficient is

$$a_0 \frac{\partial^2 z}{\partial x^2} + a_1 \frac{\partial^2 z}{\partial x \partial y} + a_2 \frac{\partial^2 z}{\partial y^2} = Q(x,y) \quad \cdots \quad (1)$$

where a_0, a_1, a_2 are constants. This equation can also be written as

$$\left(a_0 D^2 + a_1 DD' + a_2 D'^2\right) z = Q(x,y)$$

i.e. $F(D,D')z = Q(x,y)$, where

$$D = \frac{\partial}{\partial x}, D^2 = \frac{\partial^2}{\partial x^2}, D' = \frac{\partial}{\partial y}, D'^2 = \frac{\partial^2}{\partial y^2} \text{ and}$$

$$F(D,D') = a_0 D^2 + a_1 DD' + a^2 D'^2.$$

As in the case of ordinary linear differential equation, the complete solution of (1) is $z = CF + PI$

where CF i.e. complemetery function is the complete solution of the equation $F(D,D')z = 0$ and PI i.e. the particular integral is $\frac{Q(x,y)}{F(D,D')}$.

In the next article we find the CF and PI of the second order PDE.

2.2. Rules for Finding C F of a Second Order PDE.

Let $z = f(y + mx)$ be a trial solution of the equation

$$F(D,D')z = 0 \text{ i.e. } \left(a_0 D^2 + a_1 DD' + a_2 D'^2\right) z = 0 \quad \cdots \quad (2)$$

Then $Dz = \frac{\partial z}{\partial x} = mf'(y+mx), D^2 z = m^2 f''(y+mx)$

$$DD'z = \frac{\partial^2 z}{\partial x \partial y} = mf''(y+mx), D'^2 z = \frac{\partial^2 z}{\partial y^2} = f''(y+mx).$$

Substituting these values in (2) we get

$$\left(a_0m^2+a_1m+a_2\right)f''(y+mx)=0.$$

which is possible if $a_0m^2+a_1m+a_2=0$ ... (3)

This equation (3) is known as the *auxiliary equation* (AE) and is obtained by replacing D by m and D' by 1 in $F(D,D')=0$ i.e. in $a_0D^2+a_1DD'+a_2D'^2=0$.

Let m_1 and m_2 be two roots of (3). Then we consider the following two cases :

Case I. When m_1 and m_2 are distinct, the complete solution of (2) is

$$z=f_1(y+m_1x)+f_2(y+m_2x)$$

where f_1 and f_2 are two arbitrary functions.

Case II. When m_1 and m_2 are equal then the complete solution of (2) is

$$z=f_1(y+m_1x)+x\,f_2(y+m_1x)$$

Illustration.

(i) Consider the p.d.e $\dfrac{\partial^2 z}{\partial x^2}-5\dfrac{\partial^2 z}{\partial x\partial y}+6\dfrac{\partial^2 z}{\partial y^2}=0$

i.e. $\left(D^2-5DD'+6D'^2\right)z=0$. Its AE is

$m^2-5m+6=0$ i.e. $(m-3)(m-2)=0$

$\therefore m=2,3$

$\therefore$ So the complete solution is

$$z=f_1(y+2x)+f_2(y+3x).$$

(ii) Consider the p.d.e $\left(D^2+DD'+D'^2\right)z=0$.

whose AE is $m^2+m+1=0$

i.e. $m=\dfrac{-1\pm\sqrt{3}i}{2}=\omega,\omega^2$

So the complete solution is $z=f_1(y+\omega x)+f_2\left(y+\omega^2x\right)$.

(iii) Consider the p.d.e. $\left(25D^2 - 20DD' + 4D'^2\right)z = 0$

whose AE is $25m^2 - 20m + 4 = 0$

i.e. $(5m-2)^2 = 0 \therefore m = \frac{2}{3}, \frac{2}{5}$.

So the complete solution is $z = f_1\left(y + \frac{2}{5}x\right) + x f_2\left(y + \frac{2}{5}x\right)$ which may be written as $z = \phi_1(5y + 2x) + x\,\phi_2(5y + 2x)$.

2.3. Rules for finding PI.

A. General Method. (1) Let $F(D, D') = D - mD'$. Then

$$P \cdot I = \frac{Q(x, y)}{D - mD'} = \int Q(x, c - mx)\,dx$$

where constant c will be replaced by $y + mx$ after integration.

(2) Let $F(D, D') = (D - m_1 D')(D - m_2 D')$

Then $$PI = \frac{1}{D - m_2 D'} \frac{Q(x, y)}{D - m_1 D'}$$

$$= \frac{1}{D - m_2 D'} \int Q(x, c_1 - m_1 x)\,dx$$

where c_1 will be replaced by $y + m_1 x$.

Again applying the above method, we can find PI.

Illustration.

We now evaluate, $$PI = \frac{e^{2x+y}}{D^2 - 4DD' + 4D'^2}$$

$$\therefore PI = \frac{e^{2x+y}}{(D - 2D')^2} = \frac{1}{D - 2D'} \cdot \frac{1}{D - 2D'} e^{2x+y}$$

$$= \frac{1}{D - 2D'} \int e^{2x + (c - 2x)}\,dx \quad [\because m = 2]$$

$$= \frac{1}{D - 2D'} \int e^c\,dx = \frac{1}{D - 2D'} x e^c = \frac{1}{D - 2D'} x e^{y+2x} \, [\because c = y + 2x]$$

$$= \int x e^{(c-2x)+2x}\,dx = e^c \int x\,dx = \frac{1}{2} x^2 e^{y+2x}.$$

B. Some short Methods.

1. When $Q(x,y) = e^{ax+by}$, then

$$PI = \frac{e^{ax+by}}{F(D,D')} = \frac{e^{ax+by}}{F(a,b)}$$

(replacing D by a and D' by b), provided $F(a,b) \neq 0$.

Note. When $F(a,b) = 0$ then this method fails and we apply the general method.

Illustration.

We now evaluate, $PI = \dfrac{e^{2x-5y}}{D^2 - 2DD' + 3D'^2}$.

Here $PI = \dfrac{e^{2x-5y}}{2^2 - 2\cdot 2(-5) + 3(-5)^2} = \dfrac{e^{2x-5y}}{99}$.

2. When $Q(x,y) = \sin(ax+by)$ or $\cos(ax+by)$, then

$$PI = \frac{\sin(ax+by)}{F(D,D')} = \frac{\sin(ax+by)}{f(D^2, DD', D'^2)}$$

[here $F(D,D') = f(D^2, DD', D'^2)$]

$= \dfrac{\sin(ax+by)}{f(-a^2, -ab, -b^2)}$, (replacing D^2 by $-a^2$, DD' by $-ab$, D'^2 by $-b^2$)

Similarly for $\cos(ax+by)$.

Illustration.

(i) We now evaluate, $PI = \dfrac{\sin(x-3y)}{5D^2 - DD' + 6D'^2}$.

Here $PI = \dfrac{\sin(x-3y)}{5(-1^2) - (-1\cdot 3) + 6(-3^2)} = \dfrac{\sin(x-3y)}{-56}$.

(ii) Now $PI = \dfrac{\cos(2x+7y)}{2D^2 - 5DD' + 2D'^2}$

$$= \frac{\cos(2x+7y)}{2(-2^2) - 5(-2\cdot 7) + 2(-7^2)}$$

$$= \frac{\cos(2x+7y)}{-36}.$$

3. When $Q(x,y) = x^m y^n$, then

$$PI = \frac{x^m y^n}{F(D,D')} = \left[F(D,D')\right]^{-1} x^m y^n.$$

Then we expand $\{F(D,D')\}^{-1}$ in powers of $\dfrac{D}{D'}$ if $m < n$ and in powers of $\dfrac{D'}{D}$ if $m > n$.

In this regard we remember that $\dfrac{1}{D}Q(x,y) = \int Q(x,y)dx$, treating y as a constant and $\dfrac{1}{D'}Q(x,y) = \int Q(x,y)dy$, treating x as a constant.

Illustration.

Now $PI = \dfrac{x^3y^2}{2D^2 + 5D'^2}$

$$= \frac{1}{2D^2}\left(1 + \frac{5}{2}\frac{D'^2}{D^2}\right)^{-1} x^3y^2$$

$$= \frac{1}{2D^2}\left(1 - \frac{5}{2}\frac{D'^2}{D^2} + \frac{25}{4}\frac{D'^4}{D^4}\cdots\right)x^3y^2$$

$$= \frac{1}{2D^2}\left(x^3y^2 - \frac{5}{2}\frac{2x^3}{D^2}\right) = \frac{1}{2D^2}\left(x^3y^2 - \frac{x^5}{4}\right)$$

$$= \frac{1}{2}\left(\frac{x^5y^2}{20} - \frac{x^7}{168}\right) = \frac{x^5y^2}{40} - \frac{x^7}{336}.$$

Illustrative Examples.

Ex. 1. *Solve* $\frac{\partial^2 z}{\partial x^2}+\frac{\partial^2 z}{\partial x \partial y}-6\frac{\partial^2 z}{\partial y^2}=\sin(x+y)$.

The given p.d. equation can be written as

$(D^2+DD'-6D'^2)z=\sin(x+y)$.

Then auxiliary equation (A.E) is $m^2+m-6=0$ i.e. $(m+3)(m-2)=0$

$\therefore m=-3,2$.

$\therefore C\cdot F=f_1(y-3x)+f_2(y+2x)$.

Now $PI=\frac{\sin(x+y)}{D^2+DD'-6D'^2}=\frac{\sin(x+y)}{-1^2+(-1\cdot 1)-6(-1^2)}$

$=\frac{\sin(x+y)}{4}$.

So the complete solution is

$y=f_1(y-3x)+f_2(y+2x)+\frac{1}{4}\sin(x+y)$

where f_1, f_2 are two arbitrary functions.

Ex. 2. *Solve* $\frac{\partial^2 z}{\partial x^2}-\frac{\partial^2 z}{\partial t^2}=xt$.

The given p.d.e can be written as

$(D^2-D'^2)z=xt$

whose A.E. is $m^2-1=0 \therefore m=\pm 1$.

$\therefore C\cdot F=f_1(t+x)+f_2(t-x)$.

Now $P\cdot I=\frac{xt}{D^2-D'^2}=\frac{1}{D^2}\left(1-\frac{D'^2}{D^2}\right)^{-1}xt$

$=\frac{1}{D^2}\left(1-\frac{D'^2}{D^2}+\cdots\right)xt$

$=\frac{1}{D^2}\left(xt-\frac{0}{D^2}\right)=\frac{1}{D^2}xt=\frac{x^3 t}{6}$.

So the complete solution is

$$z = f_1(t+x) + f_2(t-x) + \frac{1}{6}x^3 t.$$

Ex. 3. *Solve* $(4D^2 - 4DD' + D'^2)z = \log(x+2y)$.

The AE. of the given p.d.e is $4m^2 - 4m + 1 = 0$

i.e. $(2m-1)^2 = 0 \therefore m = \frac{1}{2}, \frac{1}{2}$.

$$\therefore CF = f_1\left(y + \frac{1}{2}x\right) + x f_2\left(y + \frac{1}{2}x\right)$$

which may be written as

$$CF = g_1(x+2y) + x g_2(x+2y).$$

Now $PI = \dfrac{\log(x+2y)}{4D^2 - 4DD' + D'^2} = \dfrac{1}{4}\dfrac{1}{D - \frac{D'}{2}} \cdot \dfrac{1}{D - \frac{D'}{2}} \log(x+2y)$

$$= \frac{1}{4}\frac{1}{D - \frac{D'}{2}} \int \log\left\{x + 2\left(c - \frac{1}{2}x\right)\right\} dx \quad \left[\because m = \frac{1}{2}\right]$$

$$= \frac{1}{4}\frac{1}{D - \frac{D'}{2}} \int \log 2c \, dx$$

$$= \frac{1}{4}\frac{x\log 2c}{D - \frac{D'}{2}} = \frac{1}{4}\frac{x\log(2y+x)}{D - \frac{D'}{2}} \qquad \left[\because c = y + \frac{1}{2}x\right]$$

$$= \frac{1}{4}\int x \log\left\{2\left(c - \frac{1}{2}x\right) + x\right\} dx$$

$$= \frac{1}{4}\int x \log 2c \, dx = \frac{1}{4}\log 2c \int x \, dx = \frac{x^2}{8}\log 2c$$

$$= \frac{x^2}{8}\log 2\left(y + \frac{1}{2}x\right) = \frac{x^2}{8}\log(x+2y).$$

So the complete solution is

$$z = g_1(x+2y) + x\,g_2(x+2y) + \frac{x^2}{8}\log(x+2y).$$

Ex. 4. *Solve* $r + 2s + t = x\cos y$.

The given equation can be written as

$$\left(D^2 + 2DD' + D'^2\right)z = x\cos y.$$

The AE of this equation is

$m^2 + 2m + 1 = 0$ i.e. $(m+1)^2 = 0 \quad \therefore m = -1, -1$.

$\therefore C\cdot F = f_1(y-x) + x f_2(y-x)$.

Now $PI = \dfrac{x\cos y}{D^2 + 2DD' + D'^2} = \dfrac{1}{D+D'}\dfrac{x\cos y}{D+D'}$

$= \dfrac{1}{D+D'}\int x\cos(c+x)\,dx \qquad [\because m = -1]$

$= \dfrac{1}{D+D'}\{x\sin(c+x) + \cos(c+x)\}$

$= \dfrac{1}{D+D'}(x\sin y + \cos y) \qquad [\because c \text{ is replaced by } y-x]$.

$= \int\{x\sin(c+x) + \cos(c+x)\}\,dx$

$= -x\cos(c+x) + \sin(c+x) + \sin(c+x)$

$= 2\sin(c+x) - x\cos(c+x)$

$= 2\sin y - x\cos y$.

So the complete solution is

$$z = f_1(y-x) + xf_2(y-x) + 2\sin y - x\cos y.$$

Ex. 5. *Solve* $\left(D^2 - 2DD'\right)z = e^{2x} + x^2 y$.

The AE is $m^2 - 2m = 0 \;\therefore m = 0, 2$.

$\therefore\; CF = f_1(y + 0\cdot x) + f_2(y + 2\cdot x)$

$= f_1(y) + f_2(y+2x)$.

Now $PI = \frac{e^{2x} + x^2 y}{D^2 - 2DD'} = \frac{e^{2x}}{D^2 - 2DD'} + \frac{x^2 y}{D^2 - 2DD'}$

$$= \frac{e^{2x}}{2^2 - 2\cdot 2\cdot 0} + \frac{1}{D^2}\left(1 - \frac{2D'}{D}\right)^{-1} x^2 y$$

$$= \frac{e^{2x}}{4} + \frac{1}{D^2}\left(1 + \frac{2D'}{D} + \frac{4D'^2}{D^2} + \cdots\right) x^2 y$$

$$= \frac{e^{2x}}{4} + \frac{1}{D^2}\left(x^2 y + \frac{2x^2}{D}\right)$$

$$= \frac{e^{2x}}{4} + \frac{1}{D^2}\left(x^2 y + \frac{2x^3}{3}\right)$$

$$= \frac{e^{2x}}{4} + \left(\frac{x^4 y}{12} + \frac{x^5}{30}\right).$$

So the complete solution is

$$z = f_1(y) + f_2(y + 2x) + \frac{e^{2x}}{4} + \frac{x^4 y}{12} + \frac{x^5}{30}$$

Ex. 6. *Solve* $\left(D^2 - 3DD' + 2D'^2\right)z = e^{2x-y} + \cos(x + 2y)$.

The AE is $m^2 - 3m + 2 = 0$ i.e. $(m-2)(m-1) = 0 \quad \therefore m = 1, 2$.

$\therefore C\cdot F = f_1(y + x) + f_2(y + 2x)$.

Now $PI = \frac{e^{2x-y} + \cos(x + 2y)}{D^2 - 3DD' + 2D'^2}$

$$= \frac{e^{2x-y}}{D^2 - 3DD' + 2D'^2} + \frac{\cos(x + 2y)}{D^2 - 3DD' + 2D'^2}$$

$$= \frac{e^{2x-y}}{2^2 - 3\cdot 2\cdot(-1) + 2(-1)^2} + \frac{\cos(x + 2y)}{-1^2 - 3(-1\cdot 2) + 2\left(-2^2\right)}$$

$$= \frac{e^{2x-y}}{12} + \frac{\cos(x + 2y)}{-3}.$$

So the complete solution is

$$z = f_1(y+x) + f_2(y+2x) + \frac{e^{2x-y}}{12} - \frac{1}{3}\cos(x+2y)$$

where f_1, f_2 being arbitrary functions.

Ex. 7. *Solve* $r + (a+b)s + abt = xy$.

The given equation can be written as

$$\frac{\partial^2 z}{\partial x^2} + (a+b)\frac{\partial^2 z}{\partial x \partial y} + ab\frac{\partial^2 z}{\partial y^2} = xy$$

i.e. $\left(D^2 + (a+b)DD' + abD'^2\right)z = xy$.

The A.E is $m^2 + (a+b)m + ab = 0$

i.e. $(m+a)(m+b) = 0 \quad \therefore m = -a, -b$.

$\therefore CF = f_1(y-ax) + f_2(y-bx)$.

Now $PI = \dfrac{xy}{D^2 + (a+b)DD' + abD'^2}$

$$= \frac{1}{D^2}\left[1 + (a+b)\frac{D'}{D} + ab\frac{D'^2}{D^2}\right]^{-1} xy$$

$$= \frac{1}{D^2}\left[1 - (a+b)\frac{D'}{D} - ab\frac{D'^2}{D^2} - \cdots\right]xy$$

$$= \frac{1}{D^2}\left[xy - \frac{a+b}{D}x\right]$$

$$= \frac{1}{D^2}\left[xy - (a+b)\frac{x^2}{2}\right] = \frac{x^3 y}{6} - \frac{(a+b)x^4}{24}.$$

So the required complete solution is

$$z = f_1(y-ax) + f_2(y-bx) + \frac{1}{6}x^3 y - \frac{1}{24}(a+b)x^4$$

where f_1, f_2 are arbitrary constant.

8. *Solve* $r+5s+6t=\dfrac{1}{y-2x}$.

The given equation can be written as

$$\left(D^2+5DD'+6D'^2\right)z=\frac{1}{y-2x}.$$

The AE is $m^2+5m+6=0$ i.e. $(m+3)(m+2)=0$

$\therefore m=-2,-3$.

$\therefore CF=f_1(y-2x)+f_2(y-3x)$.

$$\text{Now } PI=\frac{(y-2x)^{-1}}{D^2+5DD'+6D'^2}=\frac{1}{D+2D'}\frac{(y-2x)^{-1}}{(D+3D')}$$

$$=\frac{1}{D+2D'}\int(c+3x-2x)^{-1}dx \quad [\because m=-3]$$

$$=\frac{1}{D+2D'}\int\frac{1}{c+x}dx=\frac{1}{D+2D'}\log(c+x)$$

$$=\frac{1}{D+2D'}\log(y-3x+x)=\frac{\log(y-2x)}{D+2D'}.$$

$$=\int\log(c+cx-2x)dx \qquad [\because m=-2]$$

$$=\log c\int dx=x\log c=x\log(y-2x).$$

So the complete solution is $z=f_1(y-2x)+f_2(y-3x)+x\log(y-2x)$

where f_1,f_2 are arbitrary functions.

4. Euler – Cauchy Equations.

The equation of the type

$$F(xD,yD')z=Q(x,y) \qquad \ldots \qquad (1)$$

where $F(xD, yD')$ is polynomial in xD, yD' are said to be Euler – Cauchy equation. The general form of a 2nd order Euler – Cauchy p.d.e is

$$a_0x^2\frac{\partial^2 z}{\partial x^2}+a_1xy\frac{\partial^2 z}{\partial x\partial y}+a_2y^2\frac{\partial^2 z}{\partial y^2}+a_3x\frac{\partial z}{\partial x}+a_4y\frac{\partial z}{\partial y}=Q(x,y)$$

i.e. $\left(a_0x^2D^2+a_1xyDD'+a_2y^2D'^2+a_3xD+a_4yD'\right)z=Q(x,y)$.

To reduce this equation into the form $F(D,D')z=Q(x,y)$, with constant co-efficient, let us substitute $u=\log x, v=\log y$. Then

$$(xD)z=x(Dz)=x\frac{\partial z}{\partial x}=x\left[\frac{\partial z}{\partial u}\cdot\frac{\partial u}{\partial x}+\frac{\partial z}{\partial v}\cdot\frac{\partial v}{\partial x}\right]$$

$$=x\frac{\partial z}{\partial u}\cdot\frac{1}{x}=\frac{\partial z}{\partial u}=\theta z\text{, where }\theta\equiv\frac{\partial}{\partial u}.$$

Similarly $(yD')z=\theta' z$, where $\theta'\equiv\frac{\partial}{\partial v}$.

$D^2z=\theta(\theta-1)z, D'^2z=\theta'(\theta'-1)z$. Hence (1) reduces to

$F(\theta,\theta')z=Q(e^u,e^v)$ with constant co-efficients.

Illustrative Examples.

Ex. 1. *Find a general solution of the p.d.e*

$$x^2\frac{\partial^2 z}{\partial x^2}-y^2\frac{\partial^2 z}{\partial y^2}+x\frac{\partial z}{\partial x}-y\frac{\partial z}{\partial y}=x^2y \qquad \dots (1)$$

Let us substitute $u=\log x, v=\log y$. Then the given equation reduces to

$\{\theta(\theta-1)-\theta'(\theta'-1)+\theta-\theta'\}z=(e^u)^2e^v$, where $\theta\equiv\frac{\partial}{\partial u},\theta'\equiv\frac{\partial}{\partial v}$

i.e. $\left(\theta^2-\theta'^2\right)z=e^{2u+v}$ ... (2)

Then the AE of (2) is $m^2-1=0$ i.e. $m=\pm 1$.

$\therefore CF=f_1(v+u)+f_2(v-u)$.

Now $PI = \dfrac{e^{2u+v}}{\theta^2 - \theta'^2} = \dfrac{e^{2u+v}}{2^2 - 1^2} = \dfrac{1}{3}e^{2u+v}$.

So the general solution of (2) is

$$z = f_1(v+u) + f_2(v-u) + \frac{1}{3}e^{2u+v}.$$

Hence the general solution of the given p.d.e is

$$z = f_1(\log y + \log x) + f_2(\log y - \log x) + \frac{1}{3}x^2 y$$

$$= f_1(\log xy) + f_2\left(\log \frac{y}{x}\right) + \frac{1}{3}x^2 y$$

where f_1, f_2 being arbitrary functions.

Ex. 2. *Solve* $x^2 r - 3xys + 2y^2 t + px + 2qy = x + 2y$.

The given equation can be written as

$$\left(x^2D^2 - 3xyDD' + 2y^2D'^2 + xD + 2yD'\right)z = x + 2y \quad \ldots \quad (1)$$

Let us substitute $u = \log x, v = \log y$ i.e. $x = e^u, y = e^v$ so that the given equation (1) reduces to

$$\left(\theta(\theta-1) - 3\theta\theta' + 2\theta'(\theta'-1) + \theta + 2\theta'\right)z = e^u + 2e^v \text{ where } \theta \equiv \frac{\partial}{\partial u}, \theta' \equiv \frac{\partial}{\partial v}$$

i.e. $\left(\theta^2 - 3\theta\theta' + 2\theta'^2\right)z = e^u + 2e^v \quad \ldots \quad (2)$

Then the A.E of (2) is

$m^2 - 3m + 2 = 0$ i.e. $(m-2)(m-1) = 0$. $\therefore m = 1, 2$

$\therefore CF = f_1(v+u) + f_2(v+2u)$.

Now $PI = \dfrac{e^u + 2e^v}{\theta^2 - 3\theta\theta' + 2\theta'^2}$

$$= \frac{e^u}{\theta^2 - 3\theta\theta' + 2\theta'^2} + \frac{2e^v}{\theta^2 - 3\theta\theta' + 2\theta'^2}$$

$$= \frac{e^u}{1^2 - 0 + 0} + \frac{2e^v}{0 - 0 + 2\cdot 1^2}$$

$$= e^u + e^v.$$

So the general solution of (2) is

$$z = f_1(v+u) + f_2(v+2u) + e^u + e^v.$$

Hence the general solution of the given equation is

$$z = f_1(\log y + \log x) + f_2(\log y + 2\log x) + x + y$$

$$= f_1(\log xy) + f_2\left(\log x^2 y\right) + x + y$$

where f_1, f_2 being arbitrary functions.

2.5. Second Order Partial Differential Equation with Variable Co-efficients.

The general form of 2nd order p.d.e with variable co-efficients is

$$Rr + Ss + Tt = V$$

where R, S, T and V are function of x, y, z, p, q. The general solution of these equations will contain two arbitrary functions. Here we study a method due to Monge's to solve such type of equations.

Monge's Method.

The given equation is $Rr + Ss + Tt = V \cdots$ ... (1)

we have $dp = \dfrac{\partial p}{\partial x}dx + \dfrac{\partial p}{\partial y}dy = \dfrac{\partial^2 z}{\partial x^2}dx + \dfrac{\partial^2 z}{\partial x \partial y}dy = rdx + sdy$ ... (2)

and $dq = \dfrac{\partial q}{\partial x}dx + \dfrac{\partial q}{\partial y}dx = sdx + tdy$ ... (3)

From (2) and (3), we have

$$r = \frac{dp - sdy}{dx}, t = \frac{dq - sdx}{dy}$$

Substituting these values of *r* and *t* in (1), we get

$$R\frac{dp - sdy}{dx} + Ss + T\frac{dq - sdx}{dy} = V$$

i.e. $(Rdp\,dy + Tdq\,dx - Vdx\,dy) - s\left(Rdy^2 - sdx\,dy + Tdx^2\right) = 0$ ... (4)

If there exist a relation among x, y, z, p and q such that

$$Rdp\,dy + Tdq\,dx - Vdx\,dy = 0 \quad \ldots \quad (5)$$

$$Rdy^2 - sdx\,dy + Tdx^2 = 0 \quad \ldots \quad (6)$$

then that relation must satisfy (4) as well as (1).

These equations (5) and (6) are called *Monge's subsidiary equations*. Let us suppose that the equation (6) be resolved into two factors of the form

$$dy - \lambda_1 dx = 0 \quad \ldots \quad (7)$$

and $dy - \lambda_2 dx = 0$ ... (8)

Now equations (5), (7) and if necessary with $dz = pdx + q\,dy$, we obtain two integrals of the form $u_1 = c_1$ and $v_1 = c_2$. Then the relation

$$v_1 = f_1(u_1) \quad \ldots \quad (9)$$

is an integral i.e. solution and is called the first integral where f_1 is an arbitrary function.

Similarly from (5) and (8) we obtain another integral of the type

$$v_2 = f_2(u_2) \quad \ldots \quad (10)$$

where f_2 is an arbitrary functions.

Solving (9) and (10) we get p, q in terms of x, y and then substituting in $dz = pdx + qdy$ and integrating we get the solution of the given equation.

Note. (i) When equation (6) is a perfect square, then $\lambda_1 = \lambda_2$ and we get only one integral of the form $Pp + Qq = R$ which can be solved with the help of Lagrange's method (see ex. 3).

(ii) When equation (6) neither gives two factors nor a perfect square we can obtain an integral as usual using $dz = pdx + qdy$. Then integrating this integral we get the required solution (see ex. 4).

Illustrative Examples.

Ex. 1. *Solve* $r = a^2 t$.

Comparing the given equation $r = a^2 t$ ... (1)

with the standard equation $Rr + Ss + Tt = V$.

we get $R=1, T=-a^2, S=0, V=0$. So the Monge's subsidiary equations are

$$dp\,dy - a^2 dq\,dx = 0 \qquad \dots (2)$$

$$dy^2 - a^2 dx^2 = 0 \qquad \dots (3)$$

From (3), we have $dy - a dx = 0$. ... (4)

and $dy + a dx = 0$ ... (5)

Now from (2) and (4), we obtain $dp(a dx) - a^2 dx\,dq = 0$

i.e. $dp - a\,dq = 0$

which on integration gives $p - aq = c_1$ ... (6)

Again integrating (4), we get $y - ax = c_2$ ... (7)

Thus the first integral is $p - aq = f_1(y-ax)$ ... (8)

Similarly from (2) and (5) we obtain another integral $p + aq = f_2(y+ax)$ (9)

Solving (8) and (9) for p and q we get

$$p = \frac{1}{2}\left[f_1(y-ax) + f_2(y+ax)\right]$$

$$q = \frac{1}{2}a\left[f_2(y+ax) - f_1(y-ax)\right].$$

Substituting these values of p and q in $dz = p dx + q\,dy$ we get

$$dz = \frac{1}{2}\left[f_1(y-ax) + f_2(y+ax)\right]dx + \frac{1}{2a}\left[f_2(y+ax) - f_1(y-ax)\right]$$

$$= \frac{1}{2a}(dy + a dx) f_2(y+ax) - \frac{1}{2a}(dy - a dx) f_1(y-ax)$$

$$= \frac{1}{2a} f_2(y+ax) d(y+ax) - \frac{1}{2a} f_1(y-ax) d(y-ax)$$

which on integration gives

$$z = \frac{1}{2a} f(y+ax) - \frac{1}{2a} F(y-ax)$$

i.e. $z = \phi(y+ax) + \psi(y-ax)$

which is the required solution.

Ex. 2. *Solve* $t - r\sec^4 y = 2q\tan y$.

Here $R = -\sec^4 y, S = 0, T = 1, V = 2q\tan y$.

So the Monge's subsidiary equations are

$$-\sec^4 y\,dp\,dy + dq\,dx - 2q\tan y\,dx\,dy = 0 \quad \ldots \quad (1)$$

and $-\sec^4 y dy^2 + dx^2 = 0$ ... (2)

From (2), we get $dx - \sec^2 y dy = 0$ ... (3)

and $dx + \sec^2 y dy = 0$ ... (4)

Now from (1) and (3), we get

$$-\sec^4 y dp dy + dq \cdot \sec^2 y dy - 2q\tan y \cdot \sec^2 y dy \cdot dy = 0$$

or, $dq - \sec^2 y dp - 2q\tan y dy = 0$

or, $dq\cos^2 y - dp - 2q\sin y\cos y\,dy = 0$

or, $dp - d\left(q\cos^2 y\right) = 0$

which on integration gives $p - q\cos^2 y = c_1$ ... (5)

Agan integrating (3), we get $x - \tan y = c_2$ ... (6)

So the first integral is $p - q\cos^2 y = f_1(x - \tan y)$ ... (7)

Similarly from (1) and (4), we get another integral as

$$p + q\cos^2 y = f_2(x + \tan y) \quad \ldots \quad (8)$$

Solving (7) and (8) for p and q we get

$$p = \frac{1}{2}\left[f_1(x - \tan y) + f_2(x + \tan y)\right]$$

$$q = \frac{1}{2}\sec^2 y\left[f_2(x + \tan y) - f_1(x - \tan y)\right]$$

Substituting these values of p and q in $dz = pdx + qdy$ we get

$$dz = \frac{1}{2}\left[f_1(x-\tan y) + f_2(x+\tan y)\right]dx + \frac{1}{2}\sec^2 y\left[f_2(x+\tan y) - f_1(x-\tan y)\right]dy$$

$$= \frac{1}{2}\left(dx + \sec^2 y\,dy\right)f_2(x+\tan y) + \frac{1}{2}\left(dx - \sec^2 y\,dy\right)f_1(x-\tan y)$$

$$= \frac{1}{2}f_2(x+\tan y)d(x+\tan y) + \frac{1}{2}f_1(x-\tan y)d(x-\tan y)$$

which on integration gives

$z = \phi(x+\tan y) + \psi(x-\tan y)$ is the required solution.

Ex. 3. *Solve* $y^2 r - 2ys + t = p + 6y$.

Here $R = y^2, S = -2y, T = 1, V = p + 6y$

So Monge's subsidiary equations are

$$y^2 dpdy + dqdx - (p+6y)dxdy = 0 \qquad \dots (1)$$

$$y^2 dy^2 + 2y\,dy\,dx + dx^2 = 0 \qquad \dots (2)$$

From (2), we have $(y\,dy + dx)^2 = 0$ i.e. $ydy + dx = 0$

i.e. $dx = -ydy \qquad \dots (3)$

Hence from (1), $y^2 dpdy + dq(-ydy) - (p+6y)(-ydy)dy = 0$

or, $ydp - dq + (p+6y)dy = 0$

or, $(ydp + pdy) - dq + 6y\,dy = 0$

i.e. $d(yp) - dq + 6y\,dy = 0$

which on integration gives

$$yp - q + 3y^2 = c_1 \qquad \dots (4)$$

Again integrating (3) we get

$$x = -\frac{1}{2}y^2 + \frac{c_2}{2}$$

i.e. $y^2 + 2x = c_2 \qquad \dots (5)$

Thus one integral of the given equation is

$$yp-q+3y^2=f\left(y^2+2x\right)$$

i.e. $yp-q=f\left(y^2+2x\right)-3y^2$ ... (6)

which is of the form $Pp+Qq=R$. So the Lagrange's subsidiary equations are

$$\frac{dx}{y}=\frac{dy}{-1}=\frac{dz}{f\left(y^2+2x\right)-3y^2} \quad \ldots \quad (7)$$

From first two ratio's of (7), we get

$dx+y\,dy=0$.

which on integration gives $x+\frac{y^2}{2}=\frac{c_3}{2}$

i.e. $2x+y^2=c_3$ ... (8)

Invirtue of (5), (8), $c_2=c_3$

$\therefore 2x+y^2=c_2$ ... (9)

Again from last two ratios,

$$dz+\left\{f\left(y^2+2x\right)-3y^2\right\}dy=0$$

i.e. $dz+\left\{f\left(c_2\right)-3y^2\right\}dy=0$

which on integration gives

$z+yf\left(c_2\right)-y^3=c_4$

i.e. $z+yf\left(2x+y^2\right)-y^3=c_4$ ... (10)

In virtue of (9), (10), the required general solution is

$$z+yf\left(y^2+2x\right)-y^3=F\left(y^2+2x\right)$$

when f, F are two arbitrary functions.

Ex. 4. *Solve* $pt-qs=q^3$.

Here $R=0, S=-q, T=p, V=q^3$.

So Monge's subsidiary equations are

$$p\,dq\,dx-q^3dx\,dy=0$$

and $q\,dxdy + p\,dx^2 = 0$

i.e. $p\,dq - q^3 dy = 0$ ... (1)

and $pdx + q\,dy = 0$ ... (2)

From (2), we have $dy = -\frac{pdx}{q}$.

Then from (1), $pdq - q^3\left(-\frac{pdx}{q}\right) = 0$

i.e. $\frac{dq}{q^2} + dx = 0$

which on integration gives $-\frac{1}{q} + x = c_1$ ... (3)

Using $dz = pdx + qdy$, and (2) we get

$dz = 0$

$\therefore z = c_2$ ... (4)

From (3) and (4), we get one integral of (1) as

$$-\frac{1}{q} + x = f(z)$$

i.e. $\frac{\partial y}{\partial z} = x - f(z)$ $\left[\because q = \frac{\partial z}{\partial y}\right]$.

Integrating w.r.t. z treating x as constant, we get

$y = xz - \int f(z)dz + \phi(x)$

$\therefore y = xz - F(z) + \phi(x)$, which is the required solution where F, ϕ are arbitrary functions and $F(z) = \int f(z)dz$.

Exercise 2

1. Solve the following equations :

(i) $2\frac{\partial^2 z}{\partial x^2} + 5\frac{\partial^2 z}{\partial x \partial y} + 2\frac{\partial^2 z}{\partial y^2} = 0$.

(ii) $25r - 40s + 16t = 0$.

(iii) $\left(2D^2 - 5DD' + 2D'^2\right)z = 24(y - x)$

(iv) $\left(D^2-6DD'+9D'^2\right)z=12x^2+36xy$.

(v) $\dfrac{\partial^2 z}{\partial x^2}+3\dfrac{\partial^2 z}{\partial x\partial y}+2\dfrac{\partial^2 z}{\partial y^2}=6(x+y)$.

(vi) $\left(D^2-2DD'+D'^2\right)=e^{x+2y}$.

(vii) $\left(2D^2-DD'-3D'^2\right)z=5e^{x-y}$.

(viii) $\dfrac{\partial^2 z}{\partial x^2}-5\dfrac{\partial^2 z}{\partial x\partial y}+4\dfrac{\partial^2 z}{\partial y^2}=\sin(4x+y)$.

(ix) $\left(D^2-DD'-2D'^2\right)z=(y-1)e^x$.

(x) $\left(D^2+DD'-6D'^2\right)z=y\sin x$.

(xi) $r+2s+t=2(y-x)+\sin(x-y)$.

(xii) $\left(D^2+2DD'+D'^2\right)z=2\cos y-x\sin y$.

2. Solve the following p.d. equation :

(i) $x^2\dfrac{\partial^2 z}{\partial x^2}+2xy\dfrac{\partial^2 z}{\partial x\partial y}+y^2\dfrac{\partial^2 z}{\partial y^2}=0$.

(ii) $x^2\dfrac{\partial^2 z}{\partial x}-y^2\dfrac{\partial^2 z}{\partial y^2}=x^2y$.

(iii) $x^2\dfrac{\partial^2 z}{\partial x^2}-4xy\dfrac{\partial^2 z}{\partial x\partial y}+4y^2\dfrac{\partial^2 z}{\partial y^2}+6y\dfrac{\partial z}{\partial y}=x^3y^4$.

(iv) $x^2r+2y^2t+px-3xys+2qy=x+2y$.

3. Solve the following equation by Monge's method :

(i) $r=t$. (ii) $r=2y^2$. (iii) $x^2r+2xys+y^2t=0$.

(iv) $r-t\cos^2 x+p\tan x=0$. (v) $r+(a+b)s+abt=xy$.

(vi) $y^2r-2yrs+t=p+6y$. (vii) $q\left(q^2+s\right)=pt$.

(viii) $y^2r+2xys+x^2t+px+qy=0$. (ix) $r-2s+t=\sin(2x+3y)$.

(x) $z(qs-pt)=pq^2$.

Answers 2

1. (i) $z = f_1(2y - x) + f_2(y - 2x)$. (ii) $z = f_1(5y + 4x) + x f_2(5y + 4x)$.

(iii) $z = f_1(2y + x) + f_2(y + 2x) + 6x^2 y + 3x^3$.

(iv) $z = f_1(y + 3x) + x f_2(y + 3x) + 10x^4 + 6x^3 y$.

(v) $z = f_1(y - x) + f_2(y - 2x) - \frac{1}{3}x^3 + \frac{1}{2}x^2 y$.

(vi) $y = f_1(y + x) + x f_2(y + x) + e^{x+2y}$.

(vii) $z = f_1(2y + 3x) + f_2(y - x) + x e^{x-y}$.

(viii) $z = f_1(y + x) + f_2(y + 4x) - \frac{1}{3} x \cos(4x + y)$.

(ix) $z = f_1(y + 2x) + f_2(y - x) + y e^x$.

(x) $z = f_1(y - 3x) + f_2(y + 2x) - (y \sin x + \cos x)$.

(xi) $z = f_1(y - x) + x f_2(y - x) + x^2 y - x^3 + \frac{1}{2}x^2 \sin(x - y)$.

(xii) $z = f_1(y - x) + x f_2(y - x) + x \sin y$.

2. (i) $y = f_1\left(y/x\right) + x f_2\left(y/x\right)$. (ii) $z = x f_1\left(y/x\right) + f_2(xy) + \frac{1}{2}x^2 y$.

(iii) $y = f_1(yx^2) + x f_2(yx^2) + \frac{1}{30}x^3 y^4$. (iv) $z = f_1(xy) + f_2(x^2 y) + x + y$.

3. (i) $z = f_1(x + y) + f_2(y - x)$. (ii) $z = x^2 y^2 + x f(y) + F(y)$.

(iii) $z = y f_1\left(y/x\right) + f_2\left(y/x\right)$. (iv) $z = \phi_1(y - \sin x) + \phi_2(y + \sin x)$.

(v) $z = \frac{1}{6}x^3 y - \frac{1}{24}(a + b)x^4 \psi_1(y - bx) + \psi_2(y - ax)$.

(vi) $z = y^3 - y f(y^2 + 2x) + g(y^2 + 2x)$. (vii) $y = xz + g(z) + h(z)$.

(viii) $z = f(x^2 - y^2)\log(y + x) + F(x^2 - y^2)$.

(ix) $z = f(x + y) + xF(x + y) - \sin(2x + 3y)$. (x) $z = a\log(y - b) + c$.

IMPORTANT WORKED OUT PROBLEMS

Ex. 1. Express $f(x)=x^4+3x^3-x^2+5x-2$ in terms of Legendre's olynomials. [WBUT 2010]

We have $P_0(x)=1,\ P_1(x)=x,\ P_2(x)=\frac{1}{2}(3x^2-1),$

$P_3(x)=\frac{1}{2}(5x^3-3x),\ P_4(x)=\frac{1}{8}(35x^4-30x^2+3)$

From $P_4(x)$, $x^4=\dfrac{8}{35}.P_4(x)+\dfrac{6}{7}x^2-\dfrac{3}{35}$

From $P_3(x)$, $x^3=\dfrac{2}{5}P_3(x)+\dfrac{3}{5}x$

From $P_2(x)$, $x^2=\dfrac{2}{3}P_2(x)+\dfrac{1}{3}$

and $x=P_1(x),\ 1=P_0(x)$

Substituting the values of x^4, x^3, x^2, we have

$$f(x)=\frac{8}{35}P_4(x)+\frac{6}{7}x^2-\frac{3}{35}+3\left\{\frac{2}{3}P_3(x)+\frac{3}{5}x\right\}-x^2+5x-2$$

$$=\frac{8}{35}P_4(x)+\frac{6}{5}P_3(x)-\frac{1}{7}x^2+\frac{34}{5}x-\frac{73}{35}$$

$$=\frac{8}{35}P_4(x)+\frac{6}{5}P_3(x)-\frac{1}{7}\left\{\frac{2}{3}P_2(x)+\frac{1}{3}\right\}+\frac{34}{5}P_1(x)-\frac{73}{35}$$

$$=\frac{8}{35}P_4(x)+\frac{6}{5}P_3(x)-\frac{2}{21}P_2(x)+\frac{34}{5}P_1(x)-\frac{32}{15}P_0(x)$$

Ex. 2. Use method of Frobenius to find the solution of the differential quation

$$x(1-x)y''-3xy'-y=0 \text{ near } x=0$$

Here $x=0$ is a regular singular point of the given equation

$$x(1-x)y''-3xy'-y=0 \quad .... \quad (1)$$

So let $y=\sum_{n=0}^{\infty} a_n x^{m+n},\ a_0 \neq 0 \quad .... \quad (2)$

be the series solution of (1).

$$\therefore \quad \frac{dy}{dx}=\sum_{n=0}^{\infty} a_n (m+n) x^{m+n-1},$$

$$\frac{d^2y}{dx^2}=\sum_{n=0}^{\infty} a_n (m+n)(m+n-1) x^{m+n-2},$$

Substituting these values in (1), we get

$$(x-x^2)\sum_{n=0}^{\infty} a_n (m+n)(m+n-1) x^{m+n-2}$$

$$-3x\sum_{n=0}^{\infty} a_n (m+n) x^{m+n-1} - \sum_{n=0}^{\infty} a_n x^{m+n}=0$$

or, $$\sum_{n=0}^{\infty} a_n (m+n)(m+n-1) x^{m+n-1} - \sum_{n=0}^{\infty} a_n (m+n)(m+n-1) x^{m+n}=0$$

$$-3\sum_{n=0}^{\infty} a_n (m+n) x^{m+n} - \sum_{n=0}^{\infty} a_n x^{m+n}=0$$

or, $$\sum_{n=0}^{\infty} a_n (m+n)(m+n-1) x^{m+n-1}$$

$$-\sum_{n=0}^{\infty} a_n [(m+n)(m+n-1)+3(m+n)+1] x^{m+n}=0$$

or, $$\sum_{n=0}^{\infty} a_n (m+n)(m+n-1) x^{m+n-1} - \sum_{n=0}^{\infty} a_n (m+n+1)^2 x^{m+n}=0$$

Equating to zero the coefficient of the smallest power of x, i.e., x^{m-1}, gives the indicial equation as

$$a_0 . m(m-1)=0$$

$$\therefore \quad m=0,1 \; [\because \; a_0 \neq 0] \cdot$$

Then the roots of the indicial equation are unequal and differ by an integer.

Again equating to zero the coefficient of x^{m+n-1} and we get

$$a_n(m+n)(m+n-1)-a_{n-1}(m+n)^2=0$$

$$\therefore \quad a_n=\frac{m+n}{m+n-1}a_{n-1}$$

Putting $n=1, 2, 3, \cdots$ we get

$$c_1=\frac{m+1}{m}c_0$$

$$c_2=\frac{m+2}{m+1}c_1=\frac{m+2}{m}c_0$$

$$c_3=\frac{m+3}{m+2}c_2=\frac{m+3}{m}c_0$$

and so on

Putting these values of a_i 's in (2) we get

$$y=a_0x^m\left[1+\frac{m+1}{m}x+\frac{m+2}{m}x^2+\cdots\right] \quad .... \quad (3)$$

As some coefficient become infinite for $m = 0$, we replace a_0 by b_0m where $b_0 \neq 0$ and (3) becomes

$$y=b_0x^m\left[m+(m+1)x+(m+2)x^2+\cdots\right]$$

$$\therefore \quad \frac{\partial y}{\partial m} = b_0 x^m \log x\left[m+(m+1)x+(m+2)x^2+\cdots\right]$$

$$+b_0 x^m\left[1+x+x^2+\cdots\right]$$

$$\therefore \quad (y)_{m=0} = b_0\left(x+2x^2+\cdots\right) = b.u \text{, say}$$

and $$\left(\frac{\partial y}{\partial m}\right)_{m=0} = b_0 \log x\left[x+2x^2+\cdots\right] + b_0\left(1+x+x^2+\cdots\right)$$

$$= b_0 u \log x + b_0\left(1+x+x^2+\cdots\right)$$

Hence the required general solution is given by

$$y = c_1 (y)_{m=0} + c_2\left(\frac{\partial y}{\partial m}\right)_{m=0}$$

$$= a_1 u + a_2\left[x\log x + 1 + x + x^2 + \cdots\right]$$

where $a_1 = c_1 b_0$, $a_2 = c_2 b_0$ are arbitrary constants and

$$u = x + 2x^2 + 3x^3 \cdots$$

Ex. 3. Solve $\dfrac{\partial u}{\partial t} = k\dfrac{\partial^2 u}{\partial x^2}$, $x>0$, $t>0$

if $u(0,t)=0$, $u(x,0)=e^{-x}$, $x>0$, $u(x,t)$ is bounded.

[WBUT 2009, 2010]

Taking Fourier sine transform on both side of the given equation, we get

$$\Im s\left(\frac{\partial u}{\partial t}\right) = k\,\Im s\left(\frac{\partial^2 u}{\partial x^2}\right)$$

or, $$\frac{\partial Us}{\partial t}=k\int_0^\infty \frac{\partial^2 u}{\partial x^2}\sin sx\,dx$$

$$=k\left\{\left[\sin sx\frac{\partial u}{\partial x}\right]_0^\infty - s\int_0^\infty \cos sx\frac{\partial u}{\partial x}dx\right\}$$

$$=k\left[0-s\left\{\left([\cos sx.u]_0^\infty-\int_0^\infty(-s\sin sx.u\,dx)\right)\right\}\right]$$

[by the given condition]

$$=-ks\left\{-0+s\int_0^\infty u\sin sx\,dx\right\}$$

$=-ks^2\,Us$, where Us is Fourier sine transform of $u(x,t)$

$$\therefore \quad \frac{\partial Us}{\partial t}+ks^2\,Us=0 \quad .... \quad (1)$$

Again taking Fourier sine transform on the condition

$u(x,0)=e^{-x}$ we get

$$Us=\int_0^\infty e^{-x}\sin sx\,dx$$

$$=\lim_{X\to\infty}\int_0^X e^{-x}\sin sx\,dx$$

$$=\lim_{X\to\infty}\left[\frac{e^{-x}}{1+s^2}(-\sin sx-s\cos sx)\right]_0^X$$

$$=\frac{s}{1+s^2}$$

$$\therefore \quad Us=\frac{s}{1+s^2} \text{ at } t=0 \quad .... \quad (2)$$

Solving equation (1) and (2) we get

$$Us = \frac{s}{1+s^2} e^{-ks^2 t}$$

Then by Fourier sine integral theorem we get

$$u(x,t) = \frac{2}{\pi} \int_0^\infty \frac{s}{1+s^2} e^{-ks^2 t} \sin sx \, ds$$

which gives the required solution.

Ex. 4. Solve the two dimensional Laplace's equation

$$\frac{\partial^2 u}{\partial x^2} + \frac{\partial^2 u}{\partial y^2} = 0, \; 0 \le x \le \pi, \; 0 \le y \le \pi,$$

subject to the conditions

$$u(x,0) = x^2, \; u(x,\pi) = 0$$

$$\frac{\partial u}{\partial x}(0,y) = \frac{\partial u}{\partial x}(\pi,y) = 0$$ [WBUT 2002, 2003]

Here only possible solution of the Laplace's equation is

$$u(x,y) = (a_1 x + a_2)(a_3 y + a_4) \quad .... \quad (1)$$

$$u(x,y) = (c_1 \cos px + c_2 \sin px)\left(c_3 e^{py} + c_4 e^{-py}\right) \quad .... \quad (2)$$

which can be written in the form

$$u(x,y) = (c_1 \cos px + c_2 \sin px)(c_3 \cos h\, py + c_4 \sin h\, py) \quad ..(3)$$

Using $u(x,\pi) = 0$, we get from (2)

$$(a_1 x + a_2)(a_3 \pi + a_4 = 0)$$

$$\therefore \quad a_4 = -a_3 \pi$$

Also $\dfrac{\partial u}{\partial x}=a_2\,(a_3y+a_4)$

$\therefore \quad \dfrac{\partial u}{\partial x}(0,y)=0$ gives

$a_2\,(a_3y+a_4)=0$

$\therefore \quad a_2=0$

In this case,

$$u(x,y)=a_1\,(a_3y-a_3\pi)$$

$$=A(y-\pi) \text{ where } A=a_1a_3 \quad .... \quad (4)$$

Using the condition $u(x,\pi)=0$ we get from (3)

$$(c_1\cos px+c_2\sin px)(c_3\cosh p\pi+c_4\sinh p\pi)=0$$

or, $c_3\cosh p\pi+c_4\sinh p\pi=0$

$\therefore \quad c_3=-c_4y\tanh p\pi \quad .... \quad (3)$

From (3), we get

$$\frac{\partial u}{\partial x}=(-c_1p\sin px+c_2p\cos px)(c_3\cosh py+c_4\sinh py) \quad ..(5)$$

Using the condition

$\dfrac{\partial u}{\partial x}(0,y)=\dfrac{\partial u}{\partial x}(\pi,y)=0$, we get from (5)

$c_2p(c_3\cosh py+c_4\text{sech } py)=0$

$\therefore \quad c_2=0$

and $(-c_1p\sin p\pi+c_2p\cos p\pi)(c_3\cosh py+c_4\sinh py)=0$

or, $-c_1p\sin p\pi(c_3\cosh py+c_4\sinh py)=0$

or, $\sin p\pi=0 \quad [\because c_1=0\Rightarrow u(x,y)=0 \ \forall x,y$ which is impossible]

$\therefore \quad p\pi=n\pi$

$\therefore \quad p=n,\ n=1,2,3,\cdots$

Then (1) takes the form

$$u_n(x,y)=a_n\cos nx\,\frac{\sin\text{h}\,n(\pi-y)}{\sinh n\pi};\ n=1,2,3,\cdots$$

where $a_n=c_1c_4$.

We now consider the most general solution as

$$u(x,y)=A(y-\pi)+\sum_{n=1}^{\infty} a_n \cos nx \sinh n(\pi-y)/\cosh n\pi \qquad \text{...(6)}$$

Then using the condition $u(x,0)=x^2$ we get

$$x^2=-A\pi+\sum_{n=1}^{\infty} a_n \cos nx$$

which is the Fourier cosine series

$$\therefore \quad a_n=\frac{2}{\pi}\int_0^{\pi} x^2 \cos nx\, dx$$

$$=\frac{2}{\pi}\left[x^2\frac{\sin nx}{n}+2x\frac{\cos nx}{n^2}+2\cdot\frac{\sin nx}{n^3}\right]_0^{\pi}$$

$$=\frac{2}{\pi}\left[0+2\pi\frac{(-1)^n}{n^2}+0\right]$$

$$=\frac{4}{n^2}(-1)^n$$

$$-A\pi=\frac{1}{\pi}\int_0^{\pi} x^2 dx$$

$$=\frac{1}{\pi}\left[\frac{x^3}{3}\right]_0^{\pi}$$

$$=\frac{\pi^2}{3}$$

$$\therefore \quad A=-\frac{\pi}{3}$$

Hence the complete solution of (1) is

$$u(x,y)=\frac{\pi}{3}(\pi-y)+4\sum_{n=1}^{\infty}(-1)^n\frac{\sinh n(\pi-y)\cos nx}{n^2\sinh n\pi}$$

Question Papers
of
WBUT Examinations
(*Arranged as per Modules*)

QUESTIONS & SOLUTION OF WBUT
MODULE-I
MULTIPLE CHOICE QUESTIONS
YEAR-2004

1. (c) The Fourier Transform of a function $f(x)$ defined on the interval $(-\infty, \infty)$ is ————————. [Ans.

(e) The period of $\cos 2\pi x$ is

(i) 2π (ii) 1 (iii) 2 (iv) none of these.

[*Ans.* 1]

(f) If $f(x) = x + x^2, -\pi < x < \pi$, be represented in Fourier series as $\frac{a_0}{2} + \sum_{n=1}^{\infty}(a_n \cos nx + b_n \sin nx)$, then $a_0 =$ ———————— [*Ans* $2\pi^2/3$]

(g) $F\left[e^{iax} f(x)\right] = F(s+a)$, where F stands for Fourier transform,

(i) True, (ii) False. [*Ans*]

YEAR-2005

1. (c) Show that every function is a sum of an even function and an odd function.

YEAR-2006

1. (ii) Obtain the period of sin 3x. [*Ans.* $\frac{2\pi}{3}$]

(iii) Define an odd function with an example.

YEAR-2007

1. (i) $F\left[e^{iax} f(x)\right] = F(s+a)$, where F stands for Fourier Transform. The statement is

(a) True (b) False [*Ans.* a]

(iii) The period of cos $2\pi x$ is

(a) 2π (b) 1 (c) 2 (d) none of these

[*Ans.*b]

(iv) If $f(x)=x\sin x, -\pi \le x \le \pi$, be presented in Fourier series as $\frac{a_0}{2}+\sum_{n=1}^{\infty}(a_n \cos nx + b_n \sin nx)$, then the value of a_0 will be

(a) 2 (b) 0 (c) 4 (d) 1

[*Ans.*a]

(xii) The period of the function $f(x)=|\sin x|$ is

(a) 2π (b) $\frac{\pi}{2}$ (c) 3π (d) π

[*Ans.*d]

YEAR-2008

1. (xii) A function $f(x)=x^2, -\pi \le x \le \pi$ is represented by a Fourier series $\frac{a_0}{2}+\sum_{n=1}^{\infty}(a_n \cos nx + b_n \sin nx)$. Then the value of b_n is

(a) $\frac{2\pi^2}{3}$ (b) $\frac{4(-1)^n}{3}$ (c) 0 (d) none of these

[*Ans.*c]

(xiii) The period of the function $f(x)=\sin 2\pi x$ is

(a) $\frac{1}{2}$ (b) 1 (c) 0 (d) $\frac{1}{3}$ [*Ans.*b]

YEAR-2009

1. (x) The period of the function $f(x)=\sin 2\pi x$ is

(a) $\frac{1}{2}$ (b) 1 (c) 0 (d) $\frac{1}{3}$

(xi) If $f(x)=x\sin x, -\pi \le x \le \pi$ be presented in Fourier series as $\frac{a_0}{2}+\sum_{n=1}^{\infty}(a_n \cos nx + b_n \sin nx)$, then the value of a_0 will be

(a) 2 (b) 0 (c) 4 (d) 1

YEAR-2010

1. (i) If $F|f(x)|=F(s)$ represents the Fourier tranform of the function $f(x)$, then $F[f(ax)]$('a' being a constant) equals

(a) $F(s/a)$ (b) $aF(s)$ (c) $(1/|a|)F(s/a)$ (d) $(1/a^2)F(as)$

(ii) A function $f(x), a<x<b$, can be expanded in a Fourier series

(a) only if it is continuous everywhere

(b) even if it is discontinuous at a finite number of points in (*a*, *b*)

(c) even if it is unboudede in (*a*, *b*)

(d) only if it is both continuous & bounded in (*a*, *b*)

(vii) If $f(x)$ is an odd function th n $F(f(x))$ is given by

(a) $F(s)=2F_s(s)$ (b) $F(s)=2iF_s(s)$

(c) $F(s)=0.5iF_s(s)$ (d) $2F(s)=iF_s(s)$

SHORT TYPE & LONG TYPE QUESTIONS

YEAR-2002

1. (a) Obtain the Fourier series to represent 6

$f(x)=x^2$ in $-\pi\le x\le\pi$. Hence show that

$\frac{1}{1^2}+\frac{1}{2^2}+\frac{1}{3^2}+\frac{1}{4^2}+\cdots\ \cdots\ \cdots=\frac{\pi^2}{6}$. [*Ans.* $\frac{2\sin s}{s},\ s\neq 0$]

(b) Find the Fourier transform of

$$f(x)=\begin{cases}1 & for\ |x|\le 1\\ 0 & for\ |x|>1\end{cases}$$

Hence show that $\int_0^\infty \frac{\sin x}{x}dx=\pi/2$

YEAR-2003

1. (a) Write the Fourier series of the function

$$f(x)=0,\quad -5<x<0$$
$$=3,\quad 0<x<5$$

and $f(x)$ is a periodic function of period 10.

How should $f(x)$ be defined at $x=-5, x=0$ and $x=5$ so that its Fourier series converges to $f(x)$ for a $-5\le x\le 5$?

[*Ans.* $\frac{6}{\pi}\left\{\sin\frac{\pi x}{5}+\frac{1}{3}\sin\frac{3\pi x}{5}+\frac{1}{5}\sin\frac{5\pi x}{5}+\cdots\right\}$

We should define $f(-5)=\frac{3}{2},\ f(5)=\frac{3}{2}$.]

(b) State Parseval's Identity on Fourier sine transform.

Find the Fourier sine transform of the functio $f(x)=e^{-x},\ x>0$.

Using Parseval's Identity, find the value of

$$\int_0^{\infty}\frac{x^2dx}{\left(x^2+1\right)^2}$$

YEAR-2004

6. (a) Find the Fourier cosine series of the function

$f(x)=\pi-x$ on $0<x<\pi$.

Hence prove that $\displaystyle\sum_{r=0}^{\infty}\frac{1}{(2r+1)^2}=\frac{\pi^2}{8}$.

$$\left[Ans.\ \pi-x=\frac{\pi}{2}+\frac{4}{\pi}\left(\frac{\cos x}{1^2}+\frac{\cos 3x}{3^2}+\cdots\right)\right.$$

(b) The function $f(x)$ is defined as $f(x)=x,\ 0<x<2$. Extenc the definition of $f(x)$ to that of the even function of period 4 and show that its graph becomes Saw-Toothed.

7. (a) Find the Fourier Transform of the function

$$f(x)=1,\quad |x|<a$$
$$=0,\quad |x|>a$$

Hence evaluate $\displaystyle\int_{-\infty}^{\infty}\frac{\sin\alpha a\cos ax}{\alpha}d\alpha$

$$\left[Ans.\ F(s)=\frac{\sin sa}{s},\quad s\neq 0\right.$$
$$=a,\quad s=0$$
$$\pi \text{ when } |x|<a$$
$$=\frac{\pi}{2} \text{ when } |x|=a$$
$$\left.=0 \text{ when } |x|>a\right]$$

YEAR-2005

5. (a) Find the Fourier series expansion of the periodic function of period 2π, $f(x)=x^2,\ -\pi\le x\le\pi$. Hence deduce

$$\frac{1}{1^2}-\frac{1}{2^2}+\frac{1}{3^2}-\frac{1}{4^2}+\cdots+=\frac{\pi^2}{12}.$$

$$[Ans.\ x^2=\frac{\pi^2}{3}+4\sum_{n=1}^{\infty}(-1)^n\frac{\cos nx}{n^2}]$$

(b) Let $f(x)=\frac{1}{4}-x,\ 0<x\le\frac{1}{2}$

$=x-\frac{3}{4},\ \frac{1}{2}<x<1$

Find half-range Fourier sine series of $f(x)$.

$$[Ans.\ \left(\frac{1}{\pi}-\frac{4}{\pi^2}\right)\sin\pi x+\left(\frac{1}{3\pi}+\frac{4}{3^2\pi^2}\right)\sin 3\pi x+\left(\frac{1}{5\pi}-\frac{4}{5^2\pi^2}\right)\sin 5\pi x+\cdots]$$

YEAR-2006

. Find the Fourier transform of

$f(x)=x^2, |x|<a$

$=0, |x|\ge a.$ $\quad$ $[Ans.\ \frac{2}{s^3}\left\{\left(a^2s^2-2\right)\sin as+2as\cos as\right\}]$

1. (b)Obtain Fourier series for $f(x)=x^2$ in $-\pi<x\le\pi$, by extending $f(x)=f(x+2\pi)$ and deduce therefrom

$$\frac{\pi^2}{6}=1+\frac{1}{2^2}+\frac{1}{3^2}+\cdots$$ 7+8

$$[Ans.\ \frac{\pi^2}{3}+4\sum_{n=1}^{\infty}\frac{\cos n\pi\cos nx}{n^2}]$$

YEAR-2007

. Find a Fourier series of the function

$f(x)=-x, -2<x\le 0$

$=x, 0\le x\le 2$

$$[Ans.\ 1+\sum_{n=1}^{\infty}\frac{4}{n^2\pi}\{\cos n\pi-1\}\sin\frac{n\pi x}{2}]$$

8. (a) Show that $\int_0^{2\pi} e^{\cos\theta}\cdot\cos(n\theta-\sin\theta)d\theta=\frac{2\pi}{n!}$.

(c) Find the Fourier sine transform of the function

$$F(x)=1,\ \ 0\le x\le\pi$$
$$=0,\ x>\pi$$

and hence evaluate the integral

$$\int_0^{\infty}\frac{1-\cos\pi s}{s}\cdot\sin sx\,ds$$ 5+3+7

9. (b) Show that the Fourier series corresponding to $f(x)=x^2$ in $-\pi\le x\le\pi$ is $\frac{\pi^2}{3}+4\sum_{n=1}^{\infty}(-1)^n\frac{\cos nx}{n^2}$ and hence deduce that

$$1-\frac{1}{2^2}+\frac{1}{3^2}-\frac{1}{4^2}+\cdots\cdots=\frac{\pi^2}{12}$$

$$1+\frac{1}{3^2}+\frac{1}{5^2}+\frac{1}{7^2}+\cdots=\frac{\pi^2}{8}$$ 8+7

YEAR-2008

5. (b) Find the Fourier sine transform of $f(x)=\frac{1}{x}$. 3+2

[*Ans.* $\pi/2$]

7. Find the Fourier series to represent $x-x^2$ from $x=-\pi$ to $x=\pi$ and hence find the value of $\frac{1-1}{4}+\frac{1}{9}-\frac{1}{16}+\cdots\cdots$

[*Ans.* $-\frac{\pi^2}{3}+4\sum_{n-1}^{\infty}\frac{(-1)^{n+1}}{n^2}\cos nx+2\sum_{n=1}^{\infty}\frac{(-1)^{n+1}}{n}\sin nx$]

12. (c) Find the Fourier Transform of the function $f(x)=1,\ |x|\le a$

$$=0,\ |x|>a$$

Hence evaluate $\int_{-\infty}^{\infty} \frac{\sin sa \cos sx}{s} ds$.

$$[Ans.(c)\ F(s) = \frac{2\sin sa}{s}, \quad s \neq 0$$
$$= 2a, \quad s = 0$$
$$\int_{-\infty}^{\infty} \frac{\sin a \cos sx}{s} ds = \pi \text{ when } |x| < a$$
$$= \frac{\pi}{2} \text{ when } |x| = a$$
$$= 0 \text{ when } |x| > a\ ]$$

YEAR-2009

3. Find the Fourier sine transform of $\frac{e^{-ax}}{x}$

8. (a) Find the Fourier series expansion of the periodic function of period 2π.

$f(x) = x^2, -\pi \leq x \leq \pi$. Hence deduce

$$\frac{1}{1^2} - \frac{1}{2^2} + \frac{1}{3^2} - \frac{1}{4^2} + ... + ... = \frac{\pi^2}{12}.$$

12. (c) Find the Fourier sine transform of the function

$$f(x) = \begin{cases} 1 & for \quad 0 < x \leq x \\ 0 & for \quad x > \pi \end{cases}$$

and hence evaluate the integral

$$\int_0^{\infty} \frac{1 - \cos p\pi}{p} \sin pxdp.$$ 5+5+5

YEAR-2010

4. Expand the following function in a Fourier series in $[-\pi, \pi]$

$$f(x) = \begin{cases} -\frac{1}{2}(\pi + x) & when -\pi \leq x < 0 \\ \frac{1}{2}(\pi - x) & when\ 0 \leq x < \pi \end{cases}$$

7. (b) Using Parseval's identity of Fourier transform prove that

$$\int_0^\infty (1-\cos x/x)^2\,dx = \pi/2$$

11. (a) If $f(x)=\begin{cases}0 & -\pi\le x\le 0\\ \sin x & 0\le x\le\pi\end{cases}$, prove that

$$f(x)=\frac{1}{\pi}+\frac{1}{2}\sin x-\frac{2}{\pi}\sum_{n-1}^{\infty}\frac{\cos 2nx}{4n^2-1}$$

Hence show that

$$\frac{1}{1.3}+\frac{1}{3.5}+\frac{1}{5.7}+\ldots\ldots\ldots=\frac{1}{2}.$$

MODULE-II

MULTIPLE CHOICE QUESTIONS

YEAR-2004

1. (a) The poles of the function $f(z)=z^2/[(z-1)^2(z+2)]$ are

______________.

(b) The residue of the function $f(z)$ at $z=1$ is 5/9.

(n) If C is circle with center 'a' and radius 'r' and $n\in z$, then $\int_c \frac{dz}{(z-a)^n}=2\pi i$, if $n=$ ______________. [*Ans.* $n=1$]

YEAR-2005

1. (a) State Cauchy's residue theorem.

(b) Evaluate $\int_c \frac{3z^2-2}{z-1}\,dz$ where C is the circle given by $|z|=\frac{1}{2}$.

(f) Determine the poles of the function cot z.

YEAR-2006

1. (ix) Obtain singularities of $f(z)=\frac{1}{\sin z-\cos z}$

[*Ans.* $z=n\pi+\frac{\pi}{4};n=0,\pm1,+2,\cdots$]

1. (i) The value of *m* for which $2x - x^2 + my$ may be harmonic is

(a) 0 (b) 1 (c) 2 (d) 3.

(ii) The value of the integral $\int_C \frac{dz}{z^2 - 2z}$, where the circle $C:|z-2|=1$ is traversed in the counter clockwise sense, is

(a) $-\pi i$ (b) $2\pi i$ (c) πi (d) none of these

(ix) The residue of a function can be evaluated, only if the pole is an isolated singularity.

(a) True (b) False

(xiii) The residue of $\frac{z^2}{z^2 + a^2}$ at $z = ia$ is

(a) $-\frac{1}{2}(ia)$ (b) $\frac{1}{2}(ia)$ (c) a (d) ia

YEAR-2007

1. (ix) If $f(z) = \frac{1}{z^4 - 2z^3}$, then $z = 0$ is a pole of order

(a) 3 (b) 2 (c) 1 (d) 4

[*Ans.*a]

(x) The value of the integral $\int_C \frac{e^z}{z-2} dz$ where $C: |z-2| = 4$ is

(a) $2\pi i$ (b) $2\pi i e^2$ (c) πi (d) $\pi i e$

[*Ans.*b]

(xi) Residue of $f(z) = \frac{2 + 3\sin \pi z}{z(z-1)^2}$ at $z = 0$, is

(a) 1 (b) 3 (c) 2 (d) i

[*Ans.*c]

1. (i) The value of *m* for which $2x - 2x^2 + my^2$ may be harmonic is

(a) 0 (b) 1 (c) 2 (d) 3.

(ii) The residue of a function can be evaluated, only if the pole is an isolated singularity.

(a) False (b) True.

(iv) The residue of $\frac{z^2}{z^2+3^2}$ at $z=i3$ is

(a) $-i.3/2$ (b) $i.3/2$ (c) 3 (d) none of these.

(ix) The function $f(z)=\bar{z}$ is

(a) continuous at $z=0$ (b) differentiable at $z=0$

(c) analytic (d) none of these.

(x) The function $f(z)=\frac{e^{z^2}}{z^4}$ has

(a) an essential singularity at $z=0$

(b) a pole of order 4 at $z=0$

(c) a simple pole at $z=0$

(d) no singularity at $z=0$.

YEAR-2008

1. (i) If $f(z)=\bar{z}$, then $f'(0)$ is

(a) 1 (b) –1 (c) 0 (d) does not exist

[*Ans.*a]

(ii) If $f(z)=u(x,y)+w(x,y)$, then $f'(z)$ is

(a) $\frac{\partial v}{\partial y}-i\frac{\partial u}{\partial y}$ (b) $\frac{\partial u}{\partial x}-i\frac{\partial v}{\partial x}$

(c) $\frac{\partial u}{\partial y}+i\frac{\partial v}{\partial x}$ (d) $\frac{\partial u}{\partial x}+i\frac{\partial v}{\partial y}$ [*Ans.*a]

(viii) Evaluate $\int_C \frac{2}{z-\alpha}dz$, where C is the circle whose equation is $|z-\alpha|=\rho$

(a) $4\pi i$ (b) $2\pi i$ (c) $\frac{\pi i}{2}$ (d) πi [*Ans.*a]

(xiv) The order of the pole $z=0$ of function $\frac{\sin z}{z^3}$ is

(a) 1 (b) 2 (c) 3 (d) 4

YEAR-2009

1. (v) The value of m such that $3y-5x^2+my^2$ is a harmonic function is

(a) 5 (b) –5 (c) 0 (c) 3

(vii) The value of $\int_C \frac{dz}{z+3}$ where C is a circle $|z|=1$ is

(a) 0 (b) 1 (c) 2 (d) –1

(viii) If $f(z)=\frac{1}{z^4-2z^3}$, then $z=0$ is a pole of order

(a) 3 (b) 2 (c) 1 (d) 4

(ix) The order of the pole $z=0$ of the function $\frac{\sin z}{z^3}$ is

(a) 1 (b) 2 (c) 3 (d) 4

1. (ii) The value of m for which $2x-x^2+my$ may be harmonic is

(a) 0 (b) 1 (c) 2 (d) 3

(iii) The value of the integral $\int_C \frac{1}{z^2-2z}dz$, where the circle $C:|z-2|=1$ is tranversed in the counter clockwise sense, is

(a) $-\pi i$ (b) $2\pi i$ (c) πi (d) none of these

(iv) The function $f(z)=\bar{z}$ is

(a) continuous at $z=0$ (b) differntiable at $z=0$

(c) analytic (d) none of these

(viii) The function $f(z)=\frac{e^{z^2}}{z^4}$ has

(a) and essential singularity at $z=0$

(b) a pole of order 4 at $z=0$

(c) a simple pole at $z=0$

(d) no singularity at $z=0$.

(xii) The residue of $\frac{z^2}{z^2+a^2}$ at $z = ia$ is

(a) $-\frac{1}{2}(ia)$ (b) $\frac{1}{2}(ia)$ (c) a (d) ia.

YEAR-2010

1. (viii) The order of pole $z = 0$ of the function $\frac{\cos z}{z^3}$ is

(a) 2 (b) 1 (c) 3 (d) 4

1. (i) If $f(z) = u(x,y) - iv(x,y)$ is analytic, then $f'(z)$ equals to

(a) $\frac{\partial u}{\partial x} - i\frac{\partial u}{\partial y}$ (b) $\frac{\partial u}{\partial x} + i\frac{\partial u}{\partial y}$

(c) $\frac{\partial v}{\partial y} - i\frac{\partial v}{\partial x}$ (d) none of these

(viii) The value of the integral $\oint C \frac{2z}{z-4}$ where $C: |z-1| = 2$ is

(a) $2\pi i$ (b) πi (c) $4\pi i$ (d) 0

(ix) The residue of $\frac{z^2}{z^2+3^2}$ at $z = 3i$ is

(a) $-i\frac{3}{2}$ (b) $i\frac{3}{2}$ (c) 3 (d) none of these

(x) The function $f(z) = \bar{z}$ is

(a) continuous at $z = 0$ (b) differentiable at $z = 0$

(c) analytic at $z = 0$ (d) continuous at $z \neq 0$

(xi) The value of $\oint_C \frac{z}{z-2z}; C: |z-2| = 1$ is tranformed in the clock-wise sense, is

(a) πi (b) $2\pi i$ (c) $-\pi i$ (d) $2\pi i$

(xii) The function $f(z)=\frac{e^{z^2}}{z^4}$ at $z=0$ has

(a) an essential singularity (b) a pole of order 4

(c) a simple pole (d) non-singularity

SHORT & LONG TYPE QUESTIONS

YEAR-2002

3. (a) Prove that the function defined by

$$f(z)=\begin{cases}\frac{x^3(1+i)-y^3(1-i)}{x^2+y^2}, & z\neq 0\\ 0, & z=0\end{cases}$$

is continuous and Cauchy-Riemann equations are satisfied at the origin, yet $f'(0)$ does not exist.

(b) Determine the analytic function $f(z)=u(x,y)+iv(x,y)$; $(z=x+iy)$ whose imaginary part is $v=x^4-6x^2y^2+y^4$.

[*Ans.* iz^4+c]

(c) Expand $f(z)=\frac{z+1}{(z-3)(z-4)}$ in a Taylor's series about the point $z=2$. Find also the region of convergence.

4. (a) State the residue theorem and use it to evaluate the integral

$$\int_c \frac{3z^2+z-1}{(z^2-1)(z-3)}dz$$

around the circle $c:|z|=2$. [*Ans.* $-\pi i/4$]

(b) Evaluate $\int_0^{2\pi}\frac{d\theta}{1+1/2\sin\theta}$ using contour integration method.

(c) Find the bilinear transformation which maps the points $z=1$, $z=i$ and $z=-1$ into the points $w=i, w=0$ and $w=-1$.

[*Ans.* $W=\frac{i-z}{i+z}$]

1. (a) Prove that $\lim\limits_{z \to 0} \frac{\bar{z}}{z}$ does not exist. 4

(b) Define analytic function in a region R of the complex plane. State a set of necessary conditions for a function $f(z) = u(x, y) + iv(x, y)$ to be analytic in R. 1+2

(c) Prove that $u(x, y) = e^{-x}(x\sin y - y\cos x)$ is harmonic. Find $v(x, y)$ such that $f(z) = u + iv$ is analytic. 5

(d) Locate and name the singularities in the finite z -plane for the function $f(z) = \frac{z}{(z^2+4)^2}$.

2. (a) Evaluate $\int_{(0,3)}^{(2,4)} (2y + x^2)dx + (3x - y)dy$ along the paths 2+2

(i) the parabola $x = 2t,\ y = t^2 + 3$.

(ii) the straight line joining (0, 3) and (2,4).

(b) State Cauchy's Integral Formula. Use this to evaluate 1+5

$$\int_c \frac{e^{2z}}{(z+1)^4} dz \text{ where } c \text{ is the circle } |z| = 3.$$

(c) State the Residue theorem. Use this to evaluate 1+3

$$\int_C \frac{e^z dz}{(z^2 + \pi^2)^2} \text{ where } C \text{ is the circle } |z| = 4.$$

YEAR-2003

3. (a) Find the analytic function $f(z) = u + iv$ where

$$v = e^x(x\sin y + y\cos y)$$

(b) Find the bilinear transformation which maps the points $z = 1, 0, -1$, onto the points $w = i, 0, -i$ and find the fixed points of the transformation.

(c) Prove that $u(x, y) = \frac{1}{2}\log(x^2 + y^2)$ is harmonic and find its conjugate harmonic function $v(x, y)$ so that $f = u + iv$ is analytic.

4. (a) State Cauchy's Integral formula. Use it to evaluate $\int_c \frac{z}{z^2-1}dz$ where c is the the circle $|z|=2$. [*Ans.*0]

(b) State Cauchy's Residue Theorem. Use it to evalute

$$\int_0^{2\pi}\frac{d\theta}{5+4\cos\theta}\cdot$$ [*Ans.* $2\pi/3$]

1. (a) Find the general value of $\log(-i)$. 3

(b) Find the singularities of the function 4

$$f(z)=\sec\frac{1}{z}$$

in the finite z plane and give the nature of singularities.

(c) Prove that the function 4

$$u=x^3-3xy^2+3x^2-3y^2+1$$

satisfies Laplace's equation and determine the corresponding analytic function $u+iv$.

(d) Show that the function $f(z)=\sqrt{|xy|}$ 3

is not analytic at the origin, although the Cauchy-Riemann equations are satisfied at that point.

2. (a) Using contour integration show that 4+2

$$\int_0^{\infty}\frac{dx}{x^6+1}=\pi/3.$$

(b) Expand $f(z)=\sin z$ in Taylor series about $z=\pi/4$. 4

(c) Evaluate : $\oint_C \frac{\sin\pi z^2+\cos\pi z^2}{(z-1)(z-2)}dz$ 4

YEAR-2004

2. (a) Find the analytic function $f(z)=u(x,y)+iv(x,y)$, $[z=x+iy]$ where $u-v=(x-y)(x^2+4xy+z^2)$. [*Ans.* $-iz^3+c$]

(b) Expand the function $f(z)=\frac{1}{(z-1)(z-2)}$ between the annular region $z=1$ and $z=2$. [*Ans.* $f(z)=-\sum_{n=0}^{\infty}\frac{z^n}{2^{n+1}}-\sum_{n=0}^{\infty}\frac{1}{z^{n+1}}$]

3. (a) Find the poles and residues at the poles for the function $\frac{z}{\cos z}$.

[*Ans.*Res $\left\{\frac{(2n+1)\pi}{2}\right\} = \frac{(-1)^n(2n+1)\pi}{2}$ $\quad n = 0, \pm 1, \pm 2, \cdots$]

(b) Evaluate $\int_0^{2\pi} \frac{d\theta}{2+\cos\theta}$. [*Ans.* $\frac{2\pi}{\sqrt{3}}$]

1. (a) Show that $W = \bar{Z}$ is no where analytic.

(b) Define harmonic function in a region R of the complex plane. If $u - v = (x-y)(x^2+4xy+y^2)$ and $f(z) = u + iv$ is an analytic function of $z = x + iy$, find $f(z)$ in terms of z.

(c) Evaluate $\int_C (z^2+3z+2)dz$, where C is the arc of the cycloid $x = a(\theta + \sin\theta)$, $y = a(1-\cos\theta)$ between the points $(0,0)$ and $(\pi a, 2a)$.

2. (a) State residue theorem and use it to evaluate the integral

$$\int_C \frac{4.37}{z(z-1)(z-2)}dz, \text{ where } C \text{ is the circle } |z| = \frac{3}{2}.$$

(b) Find the poles and the residue at each pole of the function

$$f(z) = \frac{\cot \pi z}{(z-a)^2}.$$

(c) Use the complex variable technique to find value of the integral

$$\int_0^{2\pi} \frac{d\theta}{2+\cos\theta}.$$

YEAR-2005

2. (a) Evaluate $\int_C \frac{e^z dz}{(z^2+\pi^2)}$ where C is the circle $|z| = 4$.

(b) Prove the $u = 4xy - 3x + 2$ is harmonic. Find the conjugate harmonic function v and express $u + iv$ as an analytic function of z.

4. (a) Find a Laurent Series for $f(z) = \frac{1}{(z+1)(z+3)}$, valid in $1 < |z| < 3$.

(b) Use residue theorem to evaluate $\int_{|z|=\frac{1}{2}} \frac{(z+1)dz}{z^3-2z^2}$.

. (a) Define an analytic function.

(b) Evaluate $\int_0^{2+i} z^2 dz$ along $y=2x$.

(c) Test whether the following function

$$f(z)=\frac{x^2y^5(x+iy)}{x^4+y^{10}}, \quad z\neq 0, \ f(0)=0$$

Satisfies Cauchy-Riemann equation at the origin or not.

(d) Determine the singular point of the following equation :

$$x^2(x-2)^2 y''+2(x-2)y'+(x+3)y=0.$$

(f) Define pole of a function $f(z)$ at a point a.

2. (a) Show that the function $u=\frac{1}{2}\log(x^2+y^2)$ is harmonic and find its harmonic conjugate.

(b) Determine the poles of the function $f(z)=\frac{z^2}{(z-1)^2(z+2)}$ and the residue at each pole.

3. (a) Find the Taylor series expansion of a function of the complex variable $f(z)=\frac{1}{(z-1)(z-3)}$ about the point $z=4$. Find its region of convergence.

(b) Evaluate $\oint_C \frac{z+1}{z^2-2z}dz$, where C is the circle $|z|=5$.

4. (a) Use residue calculus to evaluate the following integral :

$$\int_0^{2\pi} \frac{1}{5-4\sin\theta}d\theta.$$

(b) State Cauchy's Integral Formula and use it to evaluate $\int_C \frac{\cos\pi z}{z^2-1}dz$ around a rectangle with vertices $2\pm i, \ -2\pm i$.

YEAR-2006

2. If $f(z)$ is an analytic function of z, show that

$$\left(\frac{\partial}{\partial x}|f|\right)^2+\left(\frac{\partial}{\partial y}|f|^2\right)=|f'|^2.$$ [*Ans.* Wrong Question

3. Evaluate $\int_C |z|^2 dz$ around the square with vertices at (0, 0), (1, 0), (1, 1) and (0, 1).

8. Evaluate : $\int_{|z|=1} \frac{e^{3z}dz}{(4z-\pi i)^3}$. [*Ans.* $\frac{9\pi(1+i)}{64}$]

9. (a) If $u(x, y)=4xy-x^3+3xy^2$, verify that u is a Harmonic function and obtain its conjugate $v(x, y)$ so that $f=u+iu$ is an analytic function.

[*Ans.* $f=-z^3-2iz^2+c,\ v=y^3-3x^2y-2x^2+2y^2+c,$]

(b) State Cauchy's Integral Theorem and if $f(z)=5z^2-4z+3$, show that $\int_r \frac{f(z)dz}{z+3i}=-\pi(6+84i)$, where r is the ellipse $16x^2+9y^2=144$. 8+7

[*Ans.* Wrong.]

11. (a) Evaluate $\int_0^{2\pi}\frac{d\theta}{1+a^2-2a\cos\theta}; 0<a<1$ by contour integration.

[*Ans.* $\frac{2\pi}{1-a^2}$]

2. Find the analytic function $f(z)=u+iv$ of which the real part is $u=-\sin x \sin hy$.

6. Find the residue of the function $\frac{e^{-z}}{z^2-4}$ at any one pole.

11. (a) If $u-v=(x-y)(x^2+4xy+y^2)$ and $f(z)=u+iv$ is an analytical function of $z=x+iy$, find $f(z)$ in terms of z. 8

(b) Evaluate $\int_C \frac{dz}{z^2 - 1}$, where C is the circle $x^2 + y^2 = 4$.

YEAR-2007

4. Find the harmonic conjugate $v(x, y)$ of the function

$$u(x, y) = y^3 - 3x^2 y$$

[*Ans.* $x^3 - 3xy^2 + c$]

8. (b) Prove that $\frac{1}{2\pi i} \int_C \frac{e^{zt}}{z^2 + 1} dz = \sin t$, where $t > 0$ and C is the circle $|z| = \pi$.

[*Ans.* $\frac{1 - \cos \pi s}{S}$; $\frac{\pi}{2}$, $0 < x < \pi$

0, $x > \pi$

10. (a) Prove that the function $f(z)$ defined by

$$f(z) = \frac{x^3(1+i) - y^3(1-i)}{x^2 + y^2} (z \neq 0);\ f(0) = 0$$

is continuous & the C-R equations are satisfied at the origin, yet $f'(0)$ does not exist.

(b) Evaluate $\int_C \frac{\sin \pi z^2 + \cos \pi z^2}{(z-1)^2 (z-2)} dz$, where C is the circle $|z| = 3$.

8+7

2. Test whether the following function

$$f(z) = \frac{x^2 y^5 (x + iy)}{x^4 + y^8}, z \neq 0$$

$$f(0) = 0$$

satisfies Cauchy-Riemann equation at the origin or not.

3. Evaluate $\int_0^{1+i} z^2 dz$ along $y = x$.

8. (a) Find the Taylor series expansion of a function of complex variable $f(z)=\frac{1}{(z-1)(z-2)}$ about the point $z=3$. Find the region of convergence.

(b) State Cauchy's Integral formula and use it to evaluate $\int \frac{\cos(\pi z)}{z^2-1}dz$, around a rectangle with vertices $2\pm i, -2\pm i$ 8+7

12. (a) Find $\int_c \frac{z^2+2z+1}{(z-1).z^2.(z-2)}dz, C:|z|=6$.

(b) Show that $U=4xy-x^3+3xy^2$ is harmonic and find its conjugated harmonic function. 8+7

13. (a) Expand $f(z)=\sin z$ in a Taylor's series about $z=\pi/2$.

(b) Find $\oint_C \frac{e^{2z}dz}{(z-1)(z-2)}; C:|z|=3$. 7+8

YEAR-2008

2. Expand $f(z)=\sin z$ in a Taylor series about $z=\frac{\pi}{4}$.

[*Ans.* $\frac{\sqrt{2}}{2}\left\{1+\left(z-\frac{\pi}{4}\right)-\frac{\left(z-\frac{\pi}{4}\right)^2}{2!}-\frac{\left(z-\frac{\pi}{4}\right)^3}{3!}+\ldots\ldots\right\}$]

4. (b) Determine the poles and residue at each pole of the function $f(z)=\cot z$. 3+2

11. (a) Evaluate $\int_0^{2\pi}\frac{\cos 2\theta}{5+4\cos\theta}$ using Cauchy's Residue theorem.

(c) Expand $f(z)=\frac{1}{(z+1)(z+3)}$ in a Laurent's series valid for $0<|z+1|<2$. 7+3+5

12. (b) Show that $u(x, y) = x^3 - 3xy^2$ is harmonic in C and find a function $v(x, y)$ such that $f(z) = u + iv$ is analytic.

[*Ans.* $v = 3x^2y - y^3 + c$]

YEAR-2009

4. Evaluate $\int_c \frac{3z^2 - 2}{z-1} dz$, where c is the circle $|z| = \frac{1}{2}$.

11. (a) Use residue theorem to evaluate $\int_C \frac{3z^2 + z - 1}{(z^2-1)(z-3)} dz$ around the circle $|z| = 2$.

(b) Expand the function $f(z) = \frac{1}{(z^2+1)(z+2)}$ in the region $|z| < 1$.

(c) Show that the function $f(z) = \begin{cases} \frac{3xy^2}{x^2+y^2} & \text{for } z \neq 0 \\ 0 & \text{for } z = 0 \end{cases}$ is continuous at $z = 0$.

5+7+3

4. Find the analytic function $f(z) = u(x, y) + v(x, y)$ of which the real part is $u(x, y) = -\sin x \sinh x$.

6. State Cauchy's integral formula. Use this to evaluate

$\int_C \frac{e^{2z}}{(z+1)^4} dz$, where C is the circle $|z| = 3$.

8. (c) If $f(x)$ is analytic, prove that $\left(\frac{\partial^2}{\partial x^2} + \frac{\partial^2}{\partial y^2}\right)|f(z)|^2 = 4\,|f'(x)|^2$

10. (a) Use residue theorem to evaluate $\int_0^{2\pi} \frac{1}{5 - 4\sin\theta} d\theta$.

(b) If $f(z) = u(x, y) + iv(x, y)$ is and analytic functions such that $u - v = \frac{\cos x + \sin x - e^{-y}}{2\cos x - e^y - e^{-y}}$ and $f\left(\frac{\pi}{2}\right) = 0$ find the function $f(z)$.

7+8

YEAR-2010

2. If $f(z)=\dfrac{xy^2(x+iy)}{x^2+y^2}$, $z\neq 0$ & $f(0)=0$, then prove that

$\dfrac{f(z)-f(0)}{z}\to 0$ as $z\to 0$ along any radius vector but not as $z\to 0$ in any manner.

3. If f is analytic function then show that $\nabla^2|f(z)|^2=4\dfrac{\partial(u,v)}{\partial(x,y)}$ where $f(z)=u+iv$ and $z=x+iy$.

11. (b) Evaluate $\int_C \dfrac{4-3z}{(z-1)z(z-3)}dz$, where C is the circle $|z|=\dfrac{5}{2}$.

(c) Show that $u(x,y)=x^3-3xy^2$ is harmonic in C and find a function $v(x,y)$ such that $f(z)=u+iv$ is analytic.

2. If $f(z)$ is a analytic function of z, prove that $\left(\dfrac{\partial^2}{\partial x^2}+\dfrac{\partial^2}{\partial y^2}\right)|f(z)|^2=4|f'(z)|^2$

3. Evaluate $\int_C \dfrac{e^{2x}}{(z+1)^4}dz$, $C:|z|=3$

4. Find the Taylor's series expansion of $\dfrac{1}{(z-1)(z-3)}$ when $0<|z-1|<2$.

9. (a) If $u-v=(x-y)(x^2+4xy+y^2)$ and $f(z)=u+iv$ is an analytic function of $z=x+iy$, find $f(z)$ in terms of z.

MODULE-III

MULTIPLE CHOICE QUESTIONS

YEAR-2004

1. (a) There is 80% chance that a problem will be solved by a statistics student and 60% chance that the same problem will be solved by a mathematics student. The probability that at least the problem will be solved is

(i) 0.48 (ii) 0.92 (iii) 0.10 (iv) 0.75 [*Ans.* 0.92]

(b) If A and B be two events such that $P(A \cap B) = \frac{1}{2}$, $P(A^c \cap B^c) = \frac{1}{3}$ and $P(A) = P(B) = p$, then p is ———.

[*Ans.* 7/12]

(c) If $p(x) = \frac{x}{12}$ for $x = 1(1)6$

$= 0$, otherwise,

then $P(x = 2 \text{ or } 3)$ is ———. [*Ans.* 5/12]

(d) A random variable x has the density function

$f(x) = \frac{1}{4}$, $-2 < x < 2$

$= 0$, otherwise.

Then $P(2X + 3 > 5)$ is ———. [*Ans.* 1/4]

(f) If the random variable x has the p.d.f.

$f(x) = 3x$, $0 < x < 1$

$= 0$, otherwise.

then the p.d.f. of $y = 4x + 3$ is ———. [*Ans.* $3y - 9/16$]

(g) If for a random varaible X, $\text{var}(X) = 1$, then

$\text{var}(2X + 3)$ is ———. [*Ans.* 4]

(h) The mean of the Poisson distribution is μ. Then its standard deviation is ———. [*Ans.* $\sqrt{\mu}$]

(k) The median and mode of the normal distribution

$N(m, \sigma)$ are equal and equal to ———. [*Ans.* m]

(n) If a Poisson variate X is such that $P(X = 1) = P(X = 2)$,

then $P(X = 4)$ is ———. [*Ans.* $\frac{2}{3}e^{-2}$]

(o) For any two events A and B,

$P(AB) \le P(A) \le P(A + B) \le P(A) + P(B)$.

(i) True (ii) False. [*Ans.* True]

1. (h) The A.M. of the first n natural numbers is ———.

[*Ans.* $\frac{n+1}{2}$]

(i) If A and B be two events with $P(A)=0.4$, $P(B)=0.3$ an $P(A \cap B)=0.2$, then $P(A^c \cap B)$ is ———.

YEAR-2005

1. (d) A die is tossed five times. A success is getting 1 or 6 on toss. Find the mean and variance of the number of successes.

YEAR-2006

1. (i) The condition for independence of two events A and B is

(a) $P(A \cap B) = P(A)P(B)$ (b) $P(A+B) = P(A)P(B)$

(c) $P(A-B) = P(A)P(B)$ (d) $P(A \cap B) = P(A)P(B/A)$

[Ans.a

(ii) The probability that a leap year selected at random will conta 53 Wednesdays is

(a) $\frac{3}{4}$ (b) $\frac{2}{7}$ (c) $\frac{1}{3}$ (d) $\frac{3}{8}$ [Ans.b

(iii) The variance of a random variable X is

(a) $[E(X)]^2$ (b) $E(X^2)$

(c) $E(X^2)-[E(X)]^2$ (d) $E(X^2)-[E(X)]$

[Ans.c

(iv) The probability $P(a \le X \le b)$ [where $F(x)$ is the distributio function of the random variable X] is defined by

(a) $F(b)-F(a)$ (b) $F(b)+F(a)$

(c) $F(a)-F(b)$ (d) $F(a)F(b)$ [Ans.a

(v) The mean and standard deviation of a Binomial distribution ar respectively 4 and $\sqrt{\frac{8}{3}}$ The values of n and p are [where n an p are the parameters of the probability distribution]

(a) $11, \frac{3}{4}$ (b) $12, \frac{2}{7}$ (c) $12, \frac{1}{3}$ (d) $11, \frac{4}{3}$ [Ans.c

(x) The mean of a uniform distribution with parameters a and b is

(a) $b-a$ (b) $b+a$ (c) $\frac{a+b}{2}$ (d) $\frac{b-a}{2}$ [Ans.c

(xi) The distribution for which mean and variance are equal is

(a) Poisson (b) Normal

(c) Binomial (d) Exponential [*Ans.*a]

(v) If X and Y are independent random variables, then

(a) $E(XY) = E(X) + E(Y)$ (b) $E(XY) = E(X) - E(Y)$

(c) $E(XY) = E(X)E(Y)$ (d) $E(XY) = E(X)/E(Y)$

[*Ans.*c]

(x) If A and B are two events with $P(A) = 0.4, P(B) = 0.3$ and $P(A \cap B) = 0.2$, find $P(A^C \cap B)$. [*Ans.*0.1]

(xii) The mean and median of a normal distribution are equal

(State True/ False) [*Ans.*True]

(xiii) Find the value of k so that $f(x) = kx(1-x), 0 < x < 1$

$= 0$, otherwise

shall be a probability density function. [*Ans.* $k = 6$]

(xiv) If A and B are mutually disjoint events, then $P(A \cup B)$ is equal to

(a) $P(A)P(B)$ (b) $P(A) + P(B)$

(c) $P(A) - P(B)$ (c) none of these. [*Ans.*b]

YEAR-2007

1. (i) The chance that a leap year selected at random will contain 53 Wednesday is

(a) $\frac{2}{7}$ (b) 0 (c) 1 (d) $\frac{6}{7}$ [*Ans.*a]

(ii) If $P(A) = \frac{1}{2}$, $P(B) = \frac{1}{3}$, $P(AB) = \frac{1}{4}$, then the value of $P(A \cup B)$ is

(a) $\frac{6}{7}$ (b) $\frac{3}{7}$ (c) 1 (d) $\frac{7}{12}$ [*Ans.*d]

(iii) The mean of Binomial distribution Bin (n, p) (where n and p ar the Number of trials and probability of success) is

(a) $\frac{n}{p}$ (b) 0 (c) np (d) 1 [Ans.c

(iv) If X is normally distributed with zero mean and unit variance, then the expectation of X^2 is

(a) 1 (b) 2 (c) 8 (d) 0 [Ans.a

(vii) The mean of the binomial distribution Bin $\left(10, \frac{2}{5}\right)$ is

(a) 4 (b) 6 (c) 5 (d) 0 [Ans.a

(viii) A coin is tossed. Event {H}, {T} are

(a) mutually exclusive (b) Independent events

(c) dependent events (d) both (a) and (c). [Ans.d

(ix) A random variable X has the following *pdf*:

$$f(x) = e^{-x}, \quad x \geq 0$$

$$= 0, \quad \text{otherwise.}$$

Then the value of $E(X)$ is

(a) 0 (b) 1 (c) 2 (d) −1 [Ans.b

(x) A random variable X has the following *pdf*:

$$f(x) = k, \quad -2 < x < 2$$

$$= 0, \quad \text{otherwise.}$$

Then the value of the constant k is

(a) $\frac{1}{8}$ (b) $\frac{1}{2}$ (c) $\frac{1}{4}$ (d) $\frac{1}{12}$ [Ans.c

(xi) A random variable X has Possion distribution such that $P(1) = P(2)$. Then the standard deviation of X is

(a) 0 (b) 2 (c) $\sqrt{2}$ (d) − 2. [Ans.c

1. (viii) The mean of Poisson variate is

(a) greater than its variance (b) less than its variance

(c) equal to its variance (d) twice its variance

[*Ans.*c]

YEAR-2008

1. (i) For any two events A and B,

$$P(AB) \le P(A) \le P(A+B) \le P(A) + P(B)$$

(a) True (b) False [*Ans.*a]

(v) The expected value of the sample vairance of size n drawn from a population with mean m and standard deviation σ is

(a) σ^2 (b) $n\sigma^2$ (c) $\frac{n-1}{n}\sigma^2$ (d) $\frac{\sigma^2}{n}$ [*Ans.*c]

(vi) If $F(x)$ is the distribution function of a random variable, then

(a) $F(x)$ is continuous at all points

(b) $F(x)$ is monotonic decreasing function

(c) $F(-\infty) = 1$

(d) $F(+\infty) = 1$ [*Ans.*d]

(viii) Three unbiased coins are tossed simultaneously. This is repeated four time. The probability of getting at least one head each time is

(a) $\left(\frac{1}{8}\right)^4$ (b) $\left(\frac{2}{8}\right)^4$

(c) $\left(\frac{3}{8}\right)^4$ (d) $\left(\frac{7}{8}\right)^4$. [*Ans.*d]

(ix) Let A and B two events, then $P(A \cap B) \ge$

(a) $P(A) + P(B)$ (b) $P(A)$

(c) $P(B)$ (d) $P(A) + P(B) - 1$ [*Ans.*d]

(x) If X is a discrete random variable, then

(a) $E(|X|) \le |E(X)|$ (b) $E(|X|) \ge |E(X)|$

(c) $E(|X|) = |E(X)|$ (d) none of these [*Ans.*b]

(xi) If $A \subset B$ then $P(B-A)$ is equal to

(a) $P(A)-P(B)$ (b) $P(B)-P(A)$

(c) $P(A)-P(AB)$ (d) none of these [*Ans.*b]

(xii) The value of k for which $f(x)=kx(1-x), 0<x<1$

$= 0,$ otherwise

will be the p.d.f. of a random variable X is

(a) 6 (b) 2 (c) 1 (d) 3 [*Ans.*a]

(xiii) The mean of the Poisson distribution is μ, then its standard deviation is

(a) $\frac{1}{\sqrt{\mu}}$ (b) $\sqrt{\mu}$ (c) μ (d) $\frac{1}{\mu}$ [*Ans.*b]

1. (iii) The probability $P(a \le X \le b)$

[where $F(x)$ is distribution function of a continuous random variable X] is defined by

(a) $F(a)-F(b)$ (b) $F(b)+F(a)$

(c) $F(b)-F(a)$ (d) $F(a)F(b)$. [*Ans.*c]

(vi) Given $P(A)=\frac{1}{2}, P(B)=\frac{1}{3}, P(AB)=\frac{1}{4}$. The value of $P(\overline{A}\overline{B})$ is

(a) $\frac{5}{12}$ (b) $\frac{1}{12}$ (c) $\frac{3}{4}$ (d) $\frac{7}{12}$ [*Ans.*a]

(ix) A purse contains 4 copper coins. 3 silver coins and an another purse contains 6 copper coins and 2 silver coins. A purse is chosen at random and a coin is taken out of it. The probability that it is a copper coin is

(a) 4/7 (b) 3/4 (c) $\frac{3}{7}$ (d) $\frac{37}{56}$ [*Ans.*d]

YEAR-2009

1. (i) If A and B are any two events such that $P(A \cap B) = \frac{1}{2}$.

$P(A^C \cap B^C) = \frac{1}{3}$ and $P(A) = P(B) = p$, then the value of p is

(a) $\frac{7}{12}$ (b) $\frac{5}{6}$ (c) $\frac{1}{3}$ (d) $\frac{1}{2}$

(iii) The variance of the binomial distribution $B(n, p)$ is

(a) np (b) npq (c) $(n-1)p$ (d) $n(p-1)$.

(iv) The mean of a Poisson distribution with parameter μ is

(a) μ (b) μ^2 (c) $-\mu$ (d) $-\mu^2$

(v) If A and B are any two events, then

(a) $P(A+B) = P(A) + P(B)$

(b) $P(A+B) = P(A) + P(B) - P(AB)$

(c) $P(A+B) = P(A) + 1$

(d) $P(A+B) = 1 - P(B)$

(vi) If A and B are any two events such that A is the subset of B, then

(a) $P(A) = P(B)$ (b) $P(A) \le P(B)$

(c) $P(A) \ge P(B)$ (d) $P(A) = P(B) = 1$

(xiii) The probability $P(a < x \le b)$ (where $F(x)$ is the distribution function of the random variable x) is given by

(a) $F(b) - F(a)$ (b) $F(b) + F(a)$

(c) $F(a) - F(b)$ (d) $F(a)F(b)$

1. (i) The probability that a leap-year selected at random will contain 53 sundays is

(a) $\frac{3}{7}$ (b) $\frac{2}{7}$ (c) $\frac{5}{7}$ (d) $\frac{4}{9}$

(ii) If a coin is tossed 6 times in succession, the probability of getting at least one head is

(a) $\frac{63}{64}$ (b) $\frac{3}{64}$ (c) $\frac{7}{63}$ (d) None of these

(iii) The propability that the 4 children of a family have different birthdays

(a) 0.9836 (b) 0.4735 (c) 0.9 (d) 0.757

(vi) Let X and Y be two random variables such that $Y = a + bX$ where a and b are constants. Then, $Var(y)$ is

(a) $b^2 Var(X)$ (b) $Var(X)$

(c) $a^2 Var(X)$ (d) $(b/a)Var(X)$

YEAR-2010

1. (i) If $P(A) = \frac{1}{3}, P(B) = \frac{1}{4}, P(A \cup B) = \frac{1}{2}$ then $P(B/A)$ is

(a) $\frac{3}{4}$ (b) $\frac{4}{3}$ (c) $\frac{1}{4}$ (d) $\frac{1}{3}$

(ii) The random variable X has distribution function F. If

$F(0) = \frac{1}{6}, F(1) = \frac{1}{2}, F(3) = \frac{3}{4}$ then $P(0 < X \leq 3)$ is

(a) $\frac{3}{4}$ (b) $\frac{1}{3}$ (c) $\frac{7}{12}$ (d) none of these

(iii) A box contains 6 white and 4 black balls. One ball is drawn. What is the probability that it is white ?

(a) $\frac{2}{5}$ (b) $\frac{3}{5}$ (c) $\frac{1}{5}$ (c) none of these

(iv) If $\overline{A}$ is the complement of the event A, then

(a) $P(\overline{A}) = 1 - P(A)$ (b) $P(\overline{A}) = P(A)$

(c) $P(\overline{A}) = 1 + P(A)$ (d) none of these

(v) If the random variable X has Binomial distribution with parameters n and p, then the mean and variance are respectively

(a) np & $np(1-p)$ (b) $n/0$ & $p2n(1-p)$

(c) $np(1-p)$ & np (d) none of these

(viii) For two random variables X and Y $E(XY) = E(X).E(Y)$ hold if X and Y are

(a) independent (b) Uncorrelated

(c) continuous variate (d) none of these

(ix) The distribution function $F(X)$ of a random variable X is given by where $(-\infty < X \le \infty)$

(a) $P(-\infty < X < \infty)$ (b) $P(-\infty < X \le x)$

(c) $P(-\infty \le X < \infty)$ (d) none of these

. (iii) Three unbiased coins are tossed simultaneously. This is repeated four times. Then the probabilit of getting at least one head each time is

(a) $(1/8)^4$ (b) $(2/8)^4$ (c) $(7/8)^4$ (d) $(3/8)^4$

(iv) For a Poisson distribution $P(X)$ is $P(1) = P(2)$, then $P(0)$ is

(a) $1/e$ (b) $1/e^2$ (c) $1/e^3$ (d) none of these

(ix) If X is normally distributed with zero mean and unit variance, then the expectation of X^2, is

(a) 1 (b) 0 (c) 8 (d) 2

SHORT TYPE AND LONG TYPE QUESTIONS

YEAR-2002

1. (a) What is the chance that a leap year selected at random, will contain 53 Wednesday ? [*Ans.* $\frac{2}{7}$]

(b) In a certain class 25% of the students failed in Mathematics, 15% failed in Chemistry, 10% failed in both Mathematics and Chemistry. A student is selected at random.

(i) If he failed in Chemistry, what is the probability that he failed in Mathematics ? [*Ans.* $\frac{2}{3}$]

(ii) If he failed in Mathematics, what is the probability that he failed in Chemistry ? [*Ans.* $\frac{2}{5}$]

(iii) What is the probability that he failed both in Mathematics and Chemistry ? [*Ans.* $\frac{1}{10}$]

(c) Two persons A and B throw alternatively with a pair of dice A wins if he throws 6 before B throws 7 and B wins if he throws 7 before A throws 6. If A begins, find his probability of winning.

[*Ans.* $\frac{30}{61}$

2. (a) If events A and B are independent, prove that A and B are also independent

(b) State and prove Baye's Theorem.

(c) If $P(A \cap B \cap C) = 0$,show that $P(A \cap B / C) = P(A/C) + P(B/C)$

4. (a) Obtain Poisson distribution as the limiting case of Binomial distribution. Hence obtain mean and variance of Poisson distribution.

(b) Define normal distribution.

(c) The marks obtained by 1000 students in a final examination are found to be approximately normally distributed with mean 70 and standard deviation 5. Extimate the number of students whose marks will be between 60 and 75 both inclusive. Given that area under the normal curve $\phi(z) = \frac{1}{\sqrt{2\pi}} \exp\left(-\frac{z^2}{2}\right)$ between $z = 0$ and $z = 2$ is 0.4772 and between $z = 0$ and $z = 1$ is 0.3413.

5. (a) A continuous variate X has the p.d.f

$$f(x) = x/2, \; 0 \le x \le 1$$
$$= 1/2, 1 < x \le 2$$
$$= (3 - x)/2, \; 2 < x \le 3$$

Find the mean of the distribution. [*Ans.* $\frac{3}{2}$]

5. (b) If A and B are two independent events, show that $P(A \cup B) = 1 - P(\overline{A}) P(\overline{B})$.

(c) In a bolt factory, the machines M_1, M_2 and M_3 manufactur respectively 25, 35 and 40 percent of the total product, of their outpu 5, 4, and 2 percent respectively are defective bolts. One bolt is drawr at random from the product and is found to be defective. What is the probability that it was manufactured by Machine M_3 ? [*Ans.* $\frac{16}{69}$

6. (a) The distribution function $F(x)$ of a random variable x is defined as follows :

$$F(x) = A, \ -\alpha < x < 1$$
$$= B, \ -1 \le x < 0$$
$$= C, \ 0 \le x < 2$$
$$= D, \ 2 \le x < \infty$$

where A, B, C, D are constants. Determine values of A, B, C and D, it being given that

$$P(x=0) = \frac{1}{6} \text{ and } P(x>1) = \frac{2}{3}.$$

[*Ans.* $A = 0,\ B = \frac{1}{6}, C = \frac{1}{3}, D = 1$]

(c) If 20% of the articles produced by a machine are defective, determine the probability that out of the 4 articles chosen at random less than 2 articles will be defective.

7. (a) A discrete random variable X has the mean 6 and variance 2. Assuming the distribution is binomial, find the probability that $5 \le X \le 7$.

(b) If X is normally distributed with zero mean and unit variance, find the expectation of X^2.

(c) If X is Poisson variate such that $P(X=1) = 0.2$ and $P(X=2) = 0.2$ Find $P(X=0)$.

YEAR-2003

1. (a) If A, B, C are mutually independent events show that

$$P(A)P(B+C) = P\{A(B+C)\}$$

(b) For what value of k, the function

$$f(x) = ke^{-b(x-a)}, a \le x < \infty, \ a,b,x > 0$$
$$= 0, \qquad \text{elsewhere}$$

will be a prabability density function ? [*Ans.* $k = b$]

(c) State law of large numbers.

(d) Find the mean of Binomial distribution with parameters n and p.

[*Ans.* np]

2. (a) If $F(x)=0, \quad -\infty<x<0$

$$=\frac{1}{5}, \quad 0\le x<1$$

$$=\frac{3}{5}, \quad 1\le x<3$$

$$=1, \quad 3\le x<\infty$$

Show that $F(x)$ is a possible distribution function and find the corresponding p.m.f. [*Ans.* $f_0=\frac{1}{5}, f_1=\frac{2}{5}, f_3=\frac{2}{5}$]

(b) Show that a function *f* given by $f(x)=0 \quad 0<x<1$

$$=(k-x) \quad 1\le x<2$$

$$=0 \quad \text{elsewhere}$$

is a p.d.f. for a suitable value of *k*. Find $P\left(\frac{1}{2}<x<\frac{3}{2}\right)$.

[*Ans.* $k=2, \ \frac{3}{4}$]

3. (a) State and prove the Bayes' theorem for conditional probability.

(b) In a bolt factory, machines *A*, *B* and *C* manufacture respectively 25%, 35% and 40% of the total. Of their output 5%, 4% and 2% are defective bolts. A bolt is drawn at random from the product and is found to be defective. What are the probalities that it was maufactured by machines *A*, *B*, and *C* ? [*Ans.* $\frac{25}{69}, \frac{28}{69}, \frac{16}{69}$]

4. (a) Four dice are thrown. Find the probability that the sum of numbers will be 18. [*Ans.* $\frac{80}{64}$]

(b) If $X_1, X_2, \cdots \cdots \cdots X_n$ be any *n* events connected to a random experiment *E*, then show that

(i) $P(X_1, X_2, \cdots \cdots \cdots, X_n)\ge 1-\sum_{i=1}^{n} P(\overline{X}_i)$

(ii) $P(X_1, X_2, \cdots \cdots \cdots, X_n)\ge \sum_{i=1}^{n} P(X_i)-(n-1)$

5. (a) Obtain the Poisson approximation to the Binomial law.

(b) A care hire firm has two cars, which it hires out by the day. The number of demands for a car on each day is Poisson distributed with parameter 1.5. Calculate the proportion of days on which neither of the cars is used and the proportion of days on which some demand can not be met for lack of cars.

[Given $e^{-1.5} = 0.223$] [*Ans.*0.2231; 0.1923]

6. (b) The weight of students in a college is normally distributed with $m = 40$ kg., $\sigma = 5$ kg. Find the percentage of the students that have weight

(i) greater that 40 kg. [*Ans.*50]

(ii) greater that 50 kg. [*Ans.*2]

(iii) between 38 kg and 52 kg [*Ans.*65]

[Given : $\frac{1}{\sqrt{2\pi}}\int_{-\infty}^{2} e^{-t^2/2}dt = 0.9772$, $\frac{1}{\sqrt{2\pi}}\int_{-\infty}^{2} e^{-t^2/2}dt = 0.6554$

$\frac{1}{\sqrt{2\pi}}\int_{-\infty}^{2.4} e^{-t^2/2}dt = 0.9918$

7. (a) In a Normal Distribution, 31% of the items are under 45 and 8% are above 64. Find the mean and standard deviation.

[Given $P(0 < z < 1.405) = 0.42$, $P(-0.496 < z < 0) = 0.19$]

[*Ans.* 49.46, 9.99]

(b) If a random variable follows Poisson Distribution such that $P(1) = P(2)$ find

(i) mean of the distribution [*Ans.*2]

(ii) $P(4)$. [*Ans.* $\frac{2}{3}e^{-2} = 0.09$]

8. (a) A normal population has a mean 0.1 and standard deviation 2.1.

Find the probability that the mean of a sample of size 900 will be negative. Given that $P(|z| < 1.43) = 0.847$ [*Ans.*0.076]

5. (a) Define conditional probability of an event. A and B are two events such that $P(A \cup B) = 7/8$, $P(A \cap B) = 1/4$ and $P(A^c) = 5/8$.

Find the conditional probability of A on the hypothesis that B does not occur.

(b) If A and B are independent find whether
(i) A, B^c are independent.
(ii) A^c, B^c are independent.

(c) For the pdf

$$f(x)=\frac{2(a+x)}{b(a+b)}, \quad -b\le x\le 0$$
$$=\frac{2(a-x)}{a+b}, \quad 0\le x\le a$$

find the mean and variance. [*Ans.* $\frac{a-b}{3}, \frac{1}{18}\left(a^2+ab+b\right)$]

YEAR-2004

2. (a) Show that the probability that exactly one of the events A and B occur is $P(A)+P(B)-2P(AB)$.

(b) Two newspapers X and Y are published in a certain city. It is estimated from a survey that 16% read X, 14% read Y and 5% read both the newspapers. Find the probabilities that a randomly selected person

(i) does not read any newspaper [*Ans.*3/4]

(ii) reads only Y. [*Ans.*9/100]

3. (a) If A and B are two events such that

$$P\left(A^c\cup B^c\right)=\frac{5}{6},\ P(A)=\frac{1}{2}$$

and $P\left(B^c\right)=\frac{2}{3}$, show that A and B are independent.

[*Ans.* $\frac{1}{6}$]

(b) The probabilities of X, Y and Z becoming the principal of a college are respectively 0.3, 0.5 and 0.2. The probabilities that "Student Aid-Fund" will be introduced in the college if X, Y and Z become Principal, are 0.4, 0.6 and 0.1 respectively. Given that "Student Aid-Fund" has been introduced, find the probability that Y has been appointed as the principal. [*Ans.* $\frac{15}{22}$]

4. (a) A discrete random variable X has the following probability function:

Values of X

(x)	:	0	1	2	3	4	5	6	7
$p(x)$	:	0	K	$2K$	$2K$	$3K$	K^2	$2K^2$	$7K^2+K$

(i) Find K; (ii) Evaluate $P(0<X<5)$ and $P(X\geq 6)$.

[*Ans.*(i) $\frac{1}{10}$, (ii) $\frac{4}{5}$, $\frac{19}{100}$]

(b) The distribution function $F(x)$ of a random variable X is defined as follows :

$$F(x)=A,\ -\infty<x<-1$$
$$=B,\ -1\leq x<0$$
$$=C,\ 0\leq x<2$$
$$=D,\ 2\leq x<\infty$$

where A, B, C, D are constants. Determine the values of A, B, C, D, it being given that $P(X=0)=\frac{1}{6}$ and $P(X>1)=\frac{2}{3}$.

[*Ans.* 0, $\frac{1}{6}$, $\frac{1}{3}$, 1]

5. (a) A and B play a game in which their chances of winning are in the ratio 3 : 2. Find A's chances of winning at least three games out of the five games played. [*Ans.*2133/3125]

(b) In a certain factory turning razor blades, there is a small chance, 1/500 for any blade to be defective. The blades are in packets of 10. Use Poisson distribution to calculate the approximate number of packet containing

(i) no defective (ii) one defective and

(iii) two defective blades respectively in one consignment of 10,000 packets. [*Ans.*(i) 9802, (ii) 196,(iii) $2\approx 1.7$]

6. (a) The probability density of a continuous distribution is given by

$f(x)=\frac{3}{4}x(2-x),\ 0<x<2$. Compute mean and variance.

[*Ans.* 1, $\frac{1}{5}$]

(b) If the weekly wage of 10,000 workers in a factory follow normal distribution with mean and s.d. Rs. 70 and Rs. 5 respectively, find the expected number of workers whose weekly wages are

(i) between Rs. 66 and Rs. 72, (ii) less than Rs. 66 and

(iii) more than Rs. 72.

[Given that $\frac{1}{\sqrt{2\pi}}\int_0^z e^{-t^2/2}dt = 0.1554$ and 0.2881, according as $Z = 0.4$ and $Z = 0.8$] [*Ans.*(i) 4435 (ii) 2119 (iii) 3446]

4. (b) The probability that a certain electronic component fails when used is 0.10. If it does not fail immediately, the probability that it lasts for one year is 0.99. What is the probability that a new component will last one year ?

(c) Five balls are drawn from an urn containing 4 white and 6 black balls. Find the probability distribution of the number of white balls drawn without replacement.

[*Ans.*

x	:	0	1	2	3	4
Prob	:	$\frac{1}{42}$	$\frac{5}{21}$	$\frac{10}{21}$	$\frac{5}{21}$	$\frac{1}{42}$

]

5. (a) The length of bolts produced by a machine is normally distributed with mean 4 and s.d 0.5. A bolt is defective if its length does not lie in the interval (3.8, 4.3). Find the percentage of defective bolts produced by the machine.

[$\frac{1}{\sqrt{2\pi}}\int_{-\infty}^{0.6} e^{-t^2/2}\,dt = 0.7257$, $\frac{1}{\sqrt{2\pi}}\int_{-\infty}^{0.4} e^{-t^2/2}\,dt = 0.6554$] [*Ans.*61.89%]

11. (a) A random vairable X has the following probability function VALUES OF X:

x :	0	1	2	3	4	5	6	7
$p(x)$:	0	K	$2K$	$2K$	$3K$	K^2	$2K^2$	$7K^2+K$

(i) Find K, (ii) evealuate $P(X<6)$, $P(X \geq 6)$ and $P(0 < X < 5)$.

(iii) If $P(X \leq 9) > \frac{1}{2}$, find the minimum value of a, and
(iv) determine the distribution function of X.

[*Ans.*(i) $k = \frac{1}{10}$ (ii) $P(X < 6) = \frac{81}{100}$

$P(X \geq 6) = \frac{19}{100}$, $P(0 < X < 5) = \frac{4}{5}$

(iv)
$$\begin{aligned} F(x) &= 0, & x < 1 \\ &= \frac{1}{10}, & 1 \leq x < 2 \\ &= \frac{3}{10}, & 2 \leq x < 3 \\ &= \frac{1}{2}, & 3 \leq x < 4 \\ &= \frac{4}{5}, & 4 \leq x < 5 \\ &= \frac{81}{100}, & 5 \leq x < 6 \\ &= \frac{83}{100}, & 6 \leq x < 7 \\ &= 1, & 7 \leq x. \end{aligned}$$
]

(b) The random variable X has the probability function

$$\begin{aligned} f(x) &= K, \text{ if } x = 0 \\ &= 2K, \quad \text{if } x = 1 \\ &= 3K, \text{if } x = 2 \\ &= 0, \text{ else where} \end{aligned}$$

determine

(i) value of K, (ii) $P(X<2)$, (iii) $P(X\le 2)$, (iv) The smallest value of K for which $P(X\le K)>\frac{1}{2}$.

[*Ans.*(i) $k=\frac{1}{81}$ (ii) $P(X<2)=\frac{1}{9}$ (iii) $P(X\le 2)=\frac{6}{81}$ (iv) 5]

YEAR-2005

1. (a) Give the axiomatic definition of probability.

(b) If two events A and B are independent, then prove that the events A^c and B^c are also independent.

(c) Find the mean of a Poisson distribution from its moment generating function. [Ans. μ or *m*]

2. (a) A box contains 5 defective and 10 non-defective lamps. 8 lamps are drawn at random in succession without replacement. What is the probability that the 8-th lamp is the 5-th defective ?

[*Ans.* $5/429$]

(b) A box contains 10 pairs of shoes. If 8 shoes are randomly selected, what is the probability that there is

(i) no complete pair, (ii) exactly one complete pair ?

[*Ans.*(i) $\frac{{}^{10}C_8\times 2^8}{{}^{20}C_8}$ (ii) $\frac{10\times{}^{9}C_6\times 2^6}{{}^{20}C_8}$]

3. (a) Define probability density function of a random variable.

(b) Show that

$$\begin{aligned} f(x) &= x, && 0\le x<1 \\ &= k-x, && 1\le x\le 2 \\ &= 0, && \text{elsewhere} \end{aligned}$$

is a p.d.f of a random varaible X for a suitable value of k which you are to determine. Then find the distribution function of the random variable X.

[*Ans.* $k=2$; $F(x)=0$, $-\infty<x<0$

$$\begin{aligned} &= \frac{x^2}{2}, && 0\le x<1 \\ &= 2x-\frac{x^2}{2}-1, && 1\le x\le 2 \\ &= 1, && x>2 \end{aligned}$$]

5. (a) 100 unbiased coins are tossed. Using normal approximation to binomial distribution calculate the probability to getting (i) exactly 40 heads, (ii) 55 heads or more. Given

$$\Phi(2.1) = 0.9821, \Phi(1.9) = 0.9713, \Phi(0.9) = 0.8159 .$$

[*Ans.*(i) 0.0108 (ii) 0.1841]

(b) If X and Y are independent Poisson variates with means λ_1 and λ_2 respectively, then find the probability that (i) $X+Y=k$ where k is a constant, (ii) $X=Y$.

[*Ans.*(i) $\dfrac{e^{-(\lambda_1+\lambda_2)}(\lambda_1+\lambda_2)^k}{k!}$ (ii) $e^{-(\lambda_1+\lambda_2)}$]

11. (b) The length of the life of a type manufactured by a company follows a continuous distribution given by the density function

$$f(x) = \frac{k}{x^3}, \; 1000 \le x \le 1500$$

$$= 0, \quad \text{elsewhere}$$

Find k and find the probability that a randomly selected tyre would function for at least 1200 hours.

[*Ans.* $k = 36 \times 10^5$, $P(X \ge 1200) = \dfrac{9}{20}$.]

3. (b) A random variable X has the following probability function :

$X = x$:	-2	-1	0	1	2	3
$p(x)$:	0.1	k	0.2	$2k$	0.3	$3k$

(i) Find k.

(ii) Evaluate $P(X<2), P(X \le 2), P(-2<X<2)$.

(iii) Determine the distribution function of $F(X)$ of X.

8. (a)Three groups of children contain respectively 3 girls and 1 boy, 2 girls and 2 boys and 1 girl and 3 boys. One child is selected at random from each group. Show that the chance that the three selected children consisting of 1 girl and 2 boys is $\dfrac{13}{32}$.

(b) Find the Mean and Standard deviation of a Binomial distribution.

YEAR-2006

2. Show that the probability of occurrence of only one of the events A and B is $P(A)+P(B)-2P(AB)$

3. Find the standard deviation of the Poison distribution with parameter μ. [*Ans.* $\sqrt{\mu}$]

8. (a) A can hit a target 4 times in 5 shots; B, 3 times in 4 shots ; C twice in 3 shots. They fire a target. What is the probability that at least two shots are hit ? [*Ans.* $\frac{5}{6}$]

(b) There were three candidates A, B and C for the position of a manager whose chances of getting the appoinment are in the proportion 4 : 2 :3 respectively. The probability that A, if selected, would launch a new product in the market is 0.3. The probabilities of B and C doing the same are 0.5 and 0.8 respectively. What is the probability that the new product was launched in the market by C ? [*Ans.* $\frac{12}{23}$]

(c) A and B throw alternatively with a pair of dice. A wins if he throws 8 before B throws 5 and B wins if he throws 5 before A throws 8. Find the probability of A wins. [*Ans.* $\frac{45}{76}$]

9. (a) If the random variable X takes the values 1, 2, 3 and 4 such that $2P(X=1)=3P(X=2)=P(X=3)=5P(X=4)$, find the probability distribution of X.

[*Ans.*

X	:	1	2	3	4
P	:	$\frac{15}{61}$	$\frac{10}{61}$	$\frac{30}{61}$	$\frac{6}{61}$]

(b) A random variable X has the following probability distribution :

x	:	-2	-1	0	1	2	3
$P(x)$	:	0.1	k	0.2	2k	0.3	3k

Find $P(-2 < X < 2)$ and the mean of X. [*Ans.* $k=\frac{1}{15}$; $\frac{2}{5}$; $\frac{16}{15}$]

(c) Show that a function :

$$f(x)=|x| \qquad -1<x<1$$
$$=0 \qquad \text{elsewhere}$$

is a possible probability density function and hence find the corresponding distribution function.

[*Ans.* $F(x)=0,\ x\le -1$

$=\frac{1}{2}\left(1-x^2\right),\ -1<x\le 0$

$=\frac{1}{2}\left(1+x^2\right),\ 0<x\le 1$

$=1,\qquad 1<x$]

10.(a) A car hire firm has two cars which it hires out day by day. The number of demands for a car on each day is distributed as a Poisson distribution with average number of demand per day 1.5. Calculate the proportion of days on which neither car is used and the proportion of days on which some demand is refused. [*Ans.*0.2231 ; 0.19126]

(b) If the weekly wages of 10,000 workers in a factory follow a Normal Distribution with mean and standard deviation Rs. 70 and Rs. 5 respectively, find the expected number of workers whose wages are (i) between Rs. 66 and Rs. 72, (ii) less than Rs. 66. (iii) more than Rs. 72.

[*Ans.*(i) 4435 (ii) 2119 (iii) 3446]

11. (a) If a random variable X has Binomial distribution, find the mean and variance of with parameters (n, p). [*Ans.*3; 4]

5. A random variable X has the following probability function :

x	0	1	2	3	4	5	6	7
$p(x)$:	0	K	$2K$	$2K$	$3K$	K^2	$2K^2$	$7K^2+K$

Obtain the value of K and estimate $P(X<6)$ and $P(0<X<5)$.

[*Ans.* $k=\frac{1}{10}$, $P(X<6)=\frac{81}{100}$, $P(0<x<5)=\frac{4}{5}$]

6. Show that $f(x)$ given by $f(x)=x:0<x<1$

$=k-x;1<x<2$

$=0$ elsewhere.

is a probability density function for a suitable value of k.

Calculate the probability that random variable lies between $\frac{1}{2}$ and $\frac{3}{2}$.

[*Ans.* $k=2,\ P\left(\frac{1}{2}<X<\frac{3}{2}\right)=\frac{3}{4}$]

12. (a) When two dice are thown, find the probability that difference of the points on the dice is 2 or 3. [*Ans.* $\frac{7}{18}$]

(b) Two urns contain 3 white, 7 red, 15 black and 10 white, 6 reds, 9 black balls respectively. One ball is drawn from each urn. Find the probability that both the balls are of the some colour. 7+8

[*Ans.* $\frac{207}{625}$]

14. (a). If X has a binomial distribution with parameters p and n, show that $E(X)=np$ and $Var(X)=npq(q=1-p)$.

YEAR-2007

2. If A and B are not mutually exclusive events, then show that

$$P(A\cup B)=P(A)+P(B)-P(A\cap B).$$

3. The p.d.f of a random variable X is $f(x)=k(x-1)(2-x)$, for $1\le x\le 2$.

Determine (i) the value of k; (ii) the distribution function $F(x)$

(iii) $P\left(\frac{5}{4}\le X\le\frac{3}{2}\right)$

[*Ans.*(i) $k=6$

(ii) $F(x)=0, \quad x<1$

$$=6\left(\frac{3x^2}{2}-\frac{x^3}{3}-2x+\frac{5}{6}\right), 1\le x\le 2$$

$=1, \quad x>2$

(iii) $\frac{11}{32}$]

4. If X is normally distributed with mean 3 and s.d. 2, find c such that $P(X > c) = 2P(X \le c)$. Given that $\int_{-\infty}^{0.43} \phi(t)dt = 0.6666$. [*Ans.* $2 \cdot 14$]

8. (a) If the probability density function of a random variable X is given by $f(x) = c.e^{-(x-2)^2/2}$ $-\infty < x < \infty$, find the value of c, the expectation and variance of the distribution.

[*Ans.* $c = \frac{1}{\sqrt{2\pi}}$, Expectation =2, variance =1]

[Hint : $\int_{-\infty}^{\infty} ce^{-\frac{(x-2)^2}{2}} dx = 1$

Putting $Z = \frac{x-2}{\sqrt{2}}$ we get

$c\int_{-\infty}^{\infty} e^{-Z^2} \sqrt{2}dZ = 1$

or $\sqrt{2}c \times 2\int_{0}^{\infty} e^{-Z^2} dZ = 1$

Putting $Z^2 = u$ we get

$2\sqrt{2}c\int_{0}^{\infty} \frac{1}{2}e^{-u}u^{-\frac{1}{2}} du = 1$

or $\sqrt{2}c\int_{0}^{\infty} e^{-u}u^{-\frac{1}{2}} du = 1$

or $\sqrt{2}c\Gamma\left(\frac{1}{2}\right) = 1$ or $\sqrt{2}c \cdot \sqrt{\pi} \cdot 1$ $\therefore c = \frac{1}{\sqrt{2\pi}}$

(b) If the mean of a binomial distribution is 3 and the variance is $\frac{3}{2}$, find the probability of obtaining atmost 3 success. 10 + 5

[*Ans.* $\frac{21}{32}$]

9. (a) Let X be a Poisson distributed random variable with the parameter μ ; then show that $E(X)=\mu$ and $Var(X)=\mu$.

(b) A bag contains 5 white and 4 black balls. If 3 balls are drawn at random, what are the probabilities of the following :

(i) 2 of them are white

(ii) atmost one of them is white

(iii) at least two are white.

[*Ans.*(i) $\frac{10}{21}$ (ii) $\frac{17}{42}$ (iii) $\frac{25}{42}$]

11. (a) One ticket is selected at random from 100 tickets numbered 00, 01, 02, ... 98, 99. If X and Y denote the sum and product of the digit on the tickets respectively. Find $P(X=9/Y=0)$. [*Ans.* $\frac{2}{19}$]

[Hint : $(X=9)=\{09,18,27,36,45,54,63,72,81,90\}$

$(Y=0)=\{00,01,02,03,04,05,06,07,08,09,10,20,30,40,50,60,70,80,90\}$

$\therefore (X=9)\cap(Y=0)=\{09,90\}$

$$\therefore P(X=9/Y=0)=\frac{n((X=9)\cap(Y=0))}{n(Y=0)}=\frac{2}{19}$$]

(b) A box contains 5 defective and 10 non-defective lamps. 8 lamps are drawn at random in succession without replacement. What is the probability that the 8th lamp is the 5th defective ? [*Ans.* $\frac{5}{429}$]

(c) Three groups of children consists of 3 girls and 1 boy, 2 girls and 2 boys and 1 girl and 3 boys respectively. One child is selected from each group. Show that the probability that the 3 selected children consisting of 1 girl and 2 boys is 13/32. 5 + 5 + 5

12. (a) 3 identical boxes, I, II and III contain respectively 4 white and 3 red balls, 3 white and 7 red balls, 2 white and 3 red balls. A box chosen at random and a ball is drawn out of it. If the ball is found to be white, what is the probability that Box II is selected. [*Ans.* $\frac{21}{89}$]

(b) The p.d.f of a continuous distribution of a random variable is given by $f(x)=\frac{3}{4}x(2-x),\ 0<x<2$

$=0,\qquad$ otherwise

compute mean and variance of x. [*Ans.* $1, \frac{1}{5}$]

(c) It is observed that a cricket player becomes out within 10 runs in 3 out of 10 innings. If he played 4 innings, what is the probability that he will become (i) out twice

(ii) out at least once within 10 runs ? 6 + 5 + 4

[*Ans.*(i) ·2646 (ii) ·7599]

13.a) A random variable X has the following probability distribution :

$X=x:$	0	1	2	3	4	5	6	7
$p(x):$	0	k	$2k$	$2k$	$3k$	k^2	$2k^2$	$7k^2+k$

Obtain the value of k and find $P(X<6)$ and $P(0<X<5)$.

[*Ans.* $k=\frac{1}{10}; \frac{4}{5}$]

(b) A fair coin is tossed 400 times. Using normal approximation to binomial distribution find the probability of obtaining (i) exactly 200 heads (ii) between 190 and 210 heads, both inclusive. Given that the area under standard normal curve between $z=0$ and $z=0.05$ is 0.0199 and between $z=0$ and $z=1.05$ is 0.3531.

[*Ans.*(i) $0\cdot0398$ (ii) $0\cdot7062$]

(c) If X and Y are two independent Poisson variates having same mean such that $Var(X+Y)=9$. Find $P(X=3/X+Y=6)$. 5 + 5 + 5

[*Ans.* $\frac{5}{16}$]

2. If A and B are mutually independent events, prove that A^C and B^C are also mutually independent events.

3. There are 500 misprints in a book of 500 pages. What is the probability that a given page will contain at most 3 misprints ?

[*Ans.* $8/3e$]

12. (a) A car-hire firm has two cars which it hires out day by day. The number of demands for a car on each day is distributed as a Poisson distribution with mean 1.5. Calculate the proportion of days (i) on which either car is used & (ii) on which some demand is refused.

[*Ans.* ·2231, 0·19126 (Supposing $e^{-1.5}=0\cdot 2231$)]

(b) The probability density function is given by

$f(x)=kx^2, 0\le x\le 6; f(x)=k(12-x)^2, 6\le x\le 12; f(x)=0$, elsewhere.

(i) Evaluate the constant *k*. 7+4+4

(ii) Find $P(6\le x\le 9)$ [*Ans.*(i) $K=\frac{1}{144}$ (ii) $\frac{7}{16}$]

YEAR-2008

2. State and prove Bayes theorem.

5. A box contains 5 defective and 10 non-defective lamps are 8 are drawn at random in succession without replacement. What is the probability that the 8th lamp is the 5th defective ?

[*Ans.* $\frac{5}{429}$]

9. (a) If the random variable *X* has the *p.d.f.*

$f(x)=3x,\ 0<x<1$

$=0$, elsewhere

then find the *p.d.f.* of $y=4x+3$. [*Ans.* $\frac{3y-9}{10}$]

(c) 100 unbiased coins are tossed. Using normal approximation to binomial distribution calculate the probability to get

(i) exactly 40 heads

(ii) 55 heads or more.

Given $\phi(2.1)=0.9821, \phi(1.9)=0.9713, \phi(0.9)=0.8159$.

11. (a) The probability density function of a random variable *X* is given by

$$f(x)=\frac{3}{4}(2-X)$$

Compute the mean and the variance of *X*. 5

[*Ans.* $1, \frac{1}{5}$]

(b) The mean weight of 500 male students at a certain college is 150 lbs and the standard deviation is 15 lbs. Assuming that the weight are normally distributed find now many student weight

(i) between 120 and 155 lbs

(ii) more that 155 lbs

[Given $\varphi(2) = 0.9772 : \varphi(0.33) = 0.6293$]

10

. State and prove Bayes theorem.

. (b) The mean and s.d. of marks of 70 students were found to be 65 and 5.2 respectively. Later it was detected that the value 85 was recorded wrongly and therefore it was removed from the data set. Then find the mean and s.d for the remaining 69 students.

[*Ans.* $64 \cdot 71,\ 4 \cdot 64$]

. (a) If a random variable X follows normal distribution such that $P(9.6 \le X \le 13.8) = 0.7008$ and $P(X \ge 9.6) = 0.8159$, where

$$\int_{-\infty}^{0.9} \frac{1}{\sqrt{2\pi}} e^{-\frac{t^2}{2}dt} = 0.8159 \text{ and } \int_{-\infty}^{1.2} \frac{1}{\sqrt{2\pi}} e^{-\frac{t^2}{2}dt} = 0.8849.$$

find the mean and variance of X.

[*Ans.* Mean $= 11 \cdot 4$, $\text{Var}(X) = 4$]

(b) In answering a question on a multiple choice test, a student either knows the answer or he guesses. Let p be the probability that he knows the answer and $1 - p$ be the probality that he guesses. Assume that a student who guesses the answer will be correct with probability $\frac{1}{5}$. What is the conditional probability that a student knew the answer to a question given that he answered it correctly ?

(c) The overall percentage of failures in a certain examination is 40. What is the probability that out of a group of 6 candidates at least 4 passed the examination ? 5+5+5

[*Ans.* $0 \cdot 54432$]

YEAR-2009

2. If A and B are two events such that $P(A^C \cup B^C) = 5/6$, $P(A) = 1/2$ and $P(B^C) = 2/3$, show that A and B are independent.

3. A random variable X has the following probability function :

$X =$	-2	-1	0	1	2	3
$P(X) =$	0.1	k	0.2	$2k$	0.3	$3k$

(i) Calculate k

(ii) Find $P(X < 2), P(X \geq 2), P(-2 < X \leq 2)$.

4. If the chance of being killed by flood during a year is 1/3000, use Poisson distribution to calculate probability that out of 3000 persons living in a village, at least one will die in flood in a year.

6. If x follows a Normal Distribution with mean 12 and variance 16 find $P(x \geq 20)$.[Given $\int_{-\infty}^{2} 1/\sqrt{2\pi} e^{-1/2t^2} dt = 0.977725$]

8. (a) A random variable follows binomial distribution with mean 4 and standard deviation $\sqrt{2}$. Find the probability of assuming non-zero value of the variable. 7

(b) Find the mathematical expectation of the number of the points obtained in a single throw of an unbiased die. 8

9. (a) A bag contains 7 red and 5 white balls. 4 balls are drawn at random. What is the probability that (i) all of them are red (ii) two of them be white and two red? 7

(b) If a random variable follows Poisson Distribution such that $P(1) = P(2)$,

find (i) mean of the distribution

(ii) $P(4)$. 8

11. (a) A pair of dice is thrown. Find the probability of getting a sum of 7, when it is known that the digit in the first die is greater than that of the second. 5

(b) The manufacturing process of an article consists of two parts x and y. The probabilities of defect in parts x and y are 10% and 15% respectively. What is the probability that the assembled product will not have any defect? 5

(c) The probabilities of solving a problem by three students A, B and C are $\frac{2}{7}, \frac{3}{8}$ and $\frac{1}{2}$ respectively. If all of them try independently, find the probability that the problem is solved. 5

12. (a) The probability density function of a continuous distribution is given by $f(x) = \frac{3}{4}x(2-x)$. Compute mean and variance of the distribution. 5

(b) The mean weight of 500 male students at a certain college is 150 lbs and the stantdard deviation is 15 lbs. Assuming that the weight is normally distributed find how many students weight.

(i) between 120 and 155 lbs

(ii) more than 155 lbs.

[Given $\phi(2) = 0.9772$; $\phi(0.33) = 0.6293$] 10

2. Show that $f(x)$ given by

$$\begin{aligned} f(x) &= x; 0 < x < 1 \\ &= k - x; 1 < x < 2 \\ &= 0; \textit{ elsewhere} \end{aligned}$$

is a probability density function for a suitable value of *k*. Calculate the probability that the random variable lies between $\frac{1}{2}$ and $\frac{3}{2}$.

5. An urn contains 3 white and 5 black balls. One ball is drawn and its colour is noted, kept aside and then another ball is drawn. What is the probability that it is (i) black (ii) white ?

6. Find the mean and standard deviation of a bionomial distribution.

7. (a) If A and B mutually independent events, prove that A^C and B^C are also mutually independent events.

(b) There are three identical urns containing white and black balls. The first urn contains 3 white and 4 black balls, the 2nd urn contains 4 white and 5 black balls and the 3rd urn contains 2 white and 3 black balls. An urn is chosen at random and a ball is drawn from it. If the drawn ball is white, what is the probability that the 2nd urn chosen ?

(c) A random variable X has the follwoing p.d.f.

$$\begin{aligned} f(x) &= cx^2, \; 0 \le x \le 1 \\ &= 0, \text{ otherwise.} \end{aligned}$$

Find (i) c (ii) $P\left(0 \le X \le \frac{1}{2}\right)$. 5+5+5

YEAR-2010

2. Urn 1 has 2 white and 3 black balls. Urn II has 4 white and 1 black and Urn III 3 has white and 4 black balls. An Urn is selected at random and ball drawn at random is found to be white. Find the probability that Urn I was selected.

3. Show that $f(x) = x \quad 0 < x < 1$

$= k - x, \quad 1 < x < 2$

$= 0 \quad$ otherwise

is a density function for a suitable value of k. Calculate $P(1/3 < x < 1/2)$.

7. (a) Show that

$$f(x) = \begin{cases} x, & 0 \le x < 1 \\ k - x, & 1 \le x \le 2 \\ 0, & \textit{otherwise} \end{cases}$$

is a p.d.f. of a random variable x for a suitable value k. Find the value of k. Then find the distribution function of the random variable x.

Calculate the probability that the random variable lie between $\frac{1}{2}$ and $\frac{3}{2}$.

(b) A fair coin is tossed 400 times. Using normal approximation to binomial distribution. Find the probability of obtaining (i) exactly 200 heads, (ii) between 190 and 210 heads, both inclusive, given that the area under standard normal curve between $z = 0$ and $z = 5.05$ is 0.0199 and between $z = 0$ and $z = 1.05$ is 0.3531. 8+7

9. (a) Let F be a distribution function of a random variable X, then show that

(i) F is monotonically increasing on the real line.

(ii) F is continuous from the right at each point.

(iii) $\lim_{x \to \infty} F(x) = 1$ i.e., $F(\infty) = 1$ and $\lim_{x \to \infty} F(x) = 0$

i.e., $F(-\infty) = 0$.

(b) The number of students who enroll in a course conducted by an institute is a Poisson random variable with mean 100. The institute decided that if the number enrolling is 120 or more the institute will run the course in two separate section, whereas if fewer than 120 students enroll it will teach all of the students in a single section. Find the probability that the institute would make two sections. (Use central limit theorem to avoid difficult calculations).

[Given $\frac{1}{\sqrt{2\pi}}\int_{1.95}^{\infty} e^{-\frac{z^2}{2}} dz = 0.0256$] 5+10

11. (b) In a normal distribution, 41% of the items are under 55 and 7% are above 64. Find the mean and standard deviation. Given $P(-0.22 < z < 0) = 0.09$ and $P(O < Z < 1.34) = 0.43$. 8+7

10. (a) The probability density function of a random variable X is

$f(x) = K(x-1)(2-x)$, for $1 \le x \le 2$.

$= 0$, otherwise.

Determine –

(i) the value of the constant k and

(ii) $P\left(\frac{5}{4} \le X \le \frac{3}{2}\right)$.

(b) In a normal distribution, 31% of the items are under 45 and 8% are above 64. Find the mean and standard deviation. [Given that $P(0 < Z < 1\cdot405) = 0.42$ and $P(-0.496 < Z < 0) = 0.19$]

5. Prove that $\frac{d}{dx}\left\{x^{-n}J_n(x)\right\} = -x^{-n}J_{n+1}(x)$

MODULE-IV (PDE)

SHORT TYPE AND LONG TYPE QUESTIONS

YEAR-2002

2. (a) Solve the two dimensional Laplace's equation

$$\frac{\partial^2 u}{\partial x^2} + \frac{\partial^2 u}{\partial y^2} = 0, 0 \le x \le \pi,\ 0 \le y \le \pi,$$

subject to the conditions

$$u(x, 0) = x^2,\ u(x, \pi) = 0;\ \frac{\partial u}{\partial x}(0, y) = \frac{\partial u}{\partial x}(\pi, y) = 0.$$

(b) Solve the one dimensional heat flow equation

$$\frac{\partial u}{\partial t} = \frac{\partial^2 u}{\partial x^2}, x \ge 0,\ t \ge 0$$

subject to the conditions

(i) $u = 0$ when $x = 0$, $t > 0$

1 when $0 < x < 1$, $t = 0$

(ii) 0 when $x \geq 1$, $t = 0$

(iii) $u(x, t)$ is bounded.

$$[Ans.\ u(x,t) = \frac{2}{\pi}\int_0^\infty \frac{1-\cos s}{s} e^{-s^2 t} \sin sx\, ds\,]$$

6. (b) Use the method of separation of variables to obtain the general solution of the two-dimensional Laplace's equation

$$\nabla^2 u \equiv \frac{\partial^2 u}{\partial x^2} + \frac{\partial^2 u}{\partial y^2} = 0\,.$$

YEAR-2003

2. (a) Solve the equation $\frac{\partial u}{\partial t} = 2\frac{\partial^2 u}{\partial x^2}$, $x > 0$, $t > 0$ subject to the conditions $u(0,t) = 0$, $u(x,0) = e^{-x}$, $x > 0, u(x,t)$ is bounded.

$$[Ans.\ u(x,t) = \frac{2}{\pi}\int_0^\infty \frac{s e^{-2s^2 t} \sin sx}{1+s^2} ds\,]$$

(b) Solve two-dimensional Laplace's equation

$$\frac{\partial^2 u}{\partial x^2} + \frac{\partial^2 u}{\partial y^2} = 0$$

Which satisfies the following initial boundary conditions :

$$u(x,0) = x^2$$

$$u(x,\pi) = 0$$

$$\frac{\partial u(0,y)}{\partial x} = \frac{\partial u(\pi,y)}{\partial x} = 0 \text{ for } 0 \leq x \leq \pi,\ 0 \leq y \leq \pi\,.$$

6. (a) Solve the equation $\frac{\partial^2 z}{\partial x^2} - 2\frac{\partial^2 z}{\partial x \partial y} = \cos 2x \cos 3y$ 2

(b) Find a solution of the equation $\frac{\partial^2 y}{\partial t^2} = c^2 \frac{\partial^2 y}{\partial x^2}$

YEAR-2004

7. (b) Solve the partial differential equation

$$\frac{\partial u}{\partial t} = \frac{\partial^2 u}{\partial x^2}\,; \qquad x > 0,\ t > 0$$

subject to the conditions

(i) $u(0,t)=0$

(ii) $u(x,0)=1,\ 0<x<1$

$=0,\ x\geq 1$

(iii) $u(x,t)$ is bounded. [*Ans.* $u(x,t)=\frac{2}{\pi}\int_0^\infty \frac{1-\cos s}{s} e^{-s^2 t} ds$]

. (b) Solve the equation $\frac{\partial^2 z}{\partial x^2}-\frac{\partial^2 z}{\partial x \partial y}=\sin x \sin 2y$.

. (b) Use the method of separation of variables to solve the equation $\frac{\partial^2 v}{\partial x^2}=\frac{\partial v}{\partial t}$. Given that $v=0$ when $t\to\infty$, as well as $v=0$ at $x=0$ and $x=\ell$.

YEAR-2005

. Solve the diffusion equation $K\frac{\partial^2 u}{\partial x^2}=\frac{\partial u}{\partial t}$, $-\infty<x<+\infty,\ t>0$ with $u(x,t)=0$ as $x\to\pm\infty$, $\frac{\partial u}{\partial x}=0$ as $x\to\pm\infty$ and $u(x,0)=f(x), -\infty<x<+\infty$

[*Ans.* $u(x,\ t)=\frac{1}{\sqrt{4\pi kt}}\int_{-\infty}^{\infty} f(u)e^{-\frac{(x-u)^2}{4kt}} du$]

. (b) Using the method of separation of variables solve $\frac{\partial u}{\partial x}=2\frac{\partial u}{\partial t}+u$ with initial condition $u(x,0)=6e^{-3x}$.

3. (a) Find a solution of the equation $\frac{\partial^2 z}{\partial x^2}-\frac{\partial^2 z}{\partial t^2}=xt$

(b) Solve $\frac{\partial^2 u}{\partial x^2}+\frac{\partial^2 u}{\partial y^2}=0$ which satisfies the conditions

$u(0,y)=u(l,y)=u(x,0)=0$ and $u(x,a)=\sin\frac{n\pi x}{l}$.

YEAR-2006

3. (b) Solve the equation $\frac{\partial^2 u}{\partial t^2}=4\frac{\partial^2 u}{\partial x^2}$, given

$u(0,t)=u(5,t)=0, u(x,0)=0$ and $\left(\frac{\partial u}{\partial t}\right)_{t=0}=5\sin\pi x$. 8

YEAR-2007

9. (a) Solve $\frac{\partial u}{\partial t} = 2\frac{\partial^2 u}{\partial x^2}$, $x > 0, t > 0$

if $u(0,t) = 0, u(x,0) = e^{-x}, x > 0, u(x,t)$ is bounded.

[*Ans.* $u(x,t) = \frac{2}{\pi}\int_0^\infty \frac{se^{-2s^2t}\sin sx}{1+s^2}ds$

YEAR-2009

9. (a) Solve $\frac{\partial u}{\partial t} = k\frac{\partial^2 u}{\partial x^2}$, $x > 0, t > 0$

If $u(0,t) = 0, u(x,0) = e^{-x}, x > 0, u(x,t)$ is unbounded.

YEAR-2010

7. (c) Using Fourier transform slove the heat equation $\delta^2 u / \delta x^2 = (1/c^2)(\delta u / \delta x), -\infty < x < \infty, t > 0$ with boundary condition $u(x,t) \to 0, \delta u(x,t)/\delta x \to 0$ as $|x| \to \infty$ & initial condition $u(x,0) = e^{-x^3/4c^2}, -\infty < x < \infty$ 3+4+8

9. (a) Solve

$$\frac{\partial u}{\partial t} = K\frac{\partial^2 u}{\partial x^2}, x > 0, t > 0$$

if $u(0,t) = 0, u(x,0) = e^{-x}, x > 0. u(x,t)$ is unbounded.

6. Solve the wave equation $\frac{\partial^2 u}{\partial t^2} = c^2\frac{\partial^2 u}{\partial t^2}$ given that $u(0,t) = 0$

$u(x,0) = f(x)$ and $\frac{\partial u(x,0)}{\partial t} = 0$ where $0 < x < 1$.

8. (c) Use the method of separation of variables to solve $\frac{\partial u}{\partial x} = 2\frac{\partial u}{\partial t} + u$

where $u(x,0) = 6e^{-3x}$ 5+5+5

MODULE-V

MULTIPLE CHOICE QUESTIONS

YEAR-2005

1. (e) What is the Bessel's function of first kind ?

(g) Write the generating function of the Legendre polynomial.

YEAR-2006

1. (iii) $J_{\frac{1}{2}}(x)$ is given by

(a) $\sqrt{(2\pi / x)}\sin x$ (b) $\sqrt{(2\pi / x)}\cos x$

(c) $\sqrt{(\pi / 2x)}\cos x$ (d) $\sqrt{(2 / \pi x)}\sin x$

(iv) The polynominal $2x^2 + x + 3$ can be expressed in terms of Legendre's polynominals is

(a) $\frac{1}{3}(4P_2 - 3P_1 + 11P_0)$ (b) $\frac{1}{3}(4P_2 + 3P_1 - 11P_0)$

(c) $\frac{1}{3}(4P_2 + 3P_1 + 11P_0)$ (d) $\frac{1}{3}(4P_2 - 3P_1 - 11P_0)$

(x) The Rodrigues formula for $P_n(x)$ the Legendre's polynominal

of degree n, is $P_n(x) = K\frac{d^n\left[\left(x^2-1\right)^n\right]}{dx^n}$, then k is

(a) $\frac{n!}{2^n}$ (b) $\frac{2^n}{n!}$ (c) $\frac{1}{2^n n!}$ (d) $\frac{1}{2^n (n!)^2}$

(xii) Bessel's equation for order zero is

(a) $xy'' + y' + xy = 0$ (b) $xy'' + y' = 0$

(c) $xy' - y' + xy = 0$ (d) $xy'' + y' - xy = 0$

YEAR-2007

1. (v) Consider the differential equation $(x-2)\frac{d^2y}{dx^2}+x\frac{dy}{dx}+\frac{1}{x}y=0$.

Then $x=0$ is

(a) an ordinary point

(b) a singular point but not a regular singular point

(c) a regular singular point

(d) none of these.

(vi) $J_{-1/2}(x)=$

(a) $\sqrt{\frac{2}{\pi x}}\cos x$ (b) $\sqrt{\frac{2}{\pi x}}\sin x$

(c) $-\sqrt{\frac{2}{\pi x}}\left(\sin x+\frac{\cos x}{x}\right)$ (d) none of these

(vii) The value of $\int_{-1}^{1} P_n(x)dx$, where $P_n(x)$ is Legendre polynominal of degree n is

(a) 1, when $n=0$ (b) 0, when $n=0$

(c) 2, when $n=0$ (d) none of these.

YEAR-2009

1. (i) $J(x)$ is given

(a) $2\pi/x\sin x$ (b) $2\pi/x\cos x$

(c) $\sqrt{(\pi/2x)}\cos x$ (d) $\sqrt{(\pi/2x)}\sin x$

(xi) Bessel's equation for order zero is

(a) $xy''+y'+xy=0$ (b) $xy''+y'=0$

(c) $xy''-y'+xy=0$ (d) $xy''+y'-xy=0$

YEAR-2010

1. (ii) Consider the differential equation $xy''+2y'+xy=0$. Then $x=0$ is

(a) and ordinary point

(b) singular point but not a regular singular point

(c) a regular singular point

(d) none of these

SHORT TYPE AND LONG TYPE QUESTIONS

YEAR-2002

3. (a) Define ordinary point and singular point of an ordinary differential equation

$$a_0(x)\frac{d^2y}{dx^2}+a_1(x)\frac{dy}{dx}+a_2(x)y=0.$$ 3+2

Determine whether the points $x=0$ and $x=1$ are ordinary or singular points for the equation

$$(x-1)\frac{d^2y}{dx^2}+x\frac{dy}{dx}+\frac{1}{x}y=0.$$

(b) Obtain the power series solution of the initial value problem

$$(x^2-1)\frac{d^2y}{dx^2}+3x\frac{dy}{dx}+xy=0$$

$$y(0)=4,\quad y'(0)=6.$$ 5

(c) Show that the Legender Polynomial $P_n(x)$ of degree n satisfy the recursion formulae : (i) $nP_n(x)=xP_n'(x)-P_{n-1}'(x)$

(ii) $(n+1)P_{n+1}(x)-(2n+1)xP_n(x)+nP_n(x)=0$.

4. (a) Find a power series solution of the differential equation

$$\frac{d^2y}{dx^2}+x\frac{dy}{dx}+(x^2+2)y=0$$ 7

about the point $x=0$.

(b) Obtain the power series solution of the differential equation :

$$5x\frac{d^2y}{dx^2}+3\frac{dy}{dx}+2y=0.$$

YEAR-2003

3. (a) Determine the singular points of each of the following differential equations and specify whether they are regular or irregular

(i) $x^2\frac{d^2y}{dx^2}+x\frac{dy}{dx}+(x^2-n^2)y=0$

(ii) $(x-1)^4\frac{d^2y}{dx^2}+2(x-1)^3\frac{dy}{dx}+y=0$

(b) Solve in series the equation $\frac{d^2y}{dx^2}+xy=0$.

(c) Define Bessel's function of the first kind of order n, J_n.

Show that $J_{-n}(x)=(-1)^n J_n(x)$.

4. (a) Use method of Frobenius to find solution of the differentia

equation $2x^2\frac{d^2y}{dx^2}-x\frac{dy}{dx}+(x-5)y=0$ 7

(b) Show that

$$\int_{-1}^{1} P_n(x)P_m(x)dx=0, \quad \text{if } n\neq m$$

$$=\frac{2}{2n+1}, \text{ if } m=n.$$

YEAR-2004

3. (a) Express $f(x)=4x^3+6x^2+7x+2$ in terms of Legendre polynomials.

(b) Deduce Rodrigue's formula $P_n(x)=\frac{1}{2^n n!}\frac{d^n}{dx^n}(x^2-1)^n$

(c) Prove that $J_{-\frac{1}{2}}(x)=\sqrt{\frac{2}{\pi x}}\cos x$.

4. (a) Show that $x=0$ is an ordinary point of $x^2-Dy''+xy'-y=0$ but $x=1$ is a regular singular point.

(b) Show that $(2n+1)xP_n(x)=(n+1)P_{n+1}(x)+nP_{n-1}(x)$ where $P_n(x)$ denotes the Legendre's function.

(c) Use method of Frobenius to find the solution of the differential equation $x(x-1)y''+(3x-1)y'+y=0$.

YEAR-2005

5. (a) Obtain series solution of the equation

$$2x^2y''+(2x^2-x)y'+y=0.$$

(b) Express $J_4(x)$ in terms of J_0 and J_1.

6. (a) Prove that $\frac{d}{dx}\left[x^{-n}J_n(x)\right]=-x^{-n}J_{n+1}(x)$

(b) Show that $\int_{-1}^{1} P_m(x)P_n(x)dx=0$ if $m\neq n$.

YEAR-2006

3. When n is a positive integer, show that $J_{-n}(x)=(-1)^n J_n(x)$.

9. (a) Obtain series solution of the equation

$$\left(1+x^2\right)y''+xy'-y=0$$ 8

(b) Express $p(x)=x^4+2x^3+2x^2-x-3$ in terms of Legendre's polynomials.

10. (a) Prove that $P_n(x)=\frac{1}{2^n n!}\frac{d^n}{dx^n}\left(x^2-1\right)^n$ 8

(b) Prove that $\frac{d}{dx}\left(x^n J_n\right)=x^n J_{n-1}$ and if $\frac{d}{dx}\left(x^{-n}J_n/x\right)=-x^{-n}J_{n+1}(x)$

then prove that $2J_n'(x)=J_{n-1}(x)-J_{n+1}(x)$.

YEAR-2007

4. Determine the ordinary points and singular points of the

differential equation $x^2(x+1)^2\frac{d^2y}{dx^2}+\left(x^2-1\right)\frac{dy}{dx}+2xy=0$

5. Express $f(x)=2x^3+4x^2+3x+1$ in terms of Legendre polynominals.

11. (b) Show that $\int_{-1}^{1} P_m(x)P_n(x)dx=0,\ m\neq n$.

(c) Show that $J_{-n}(x)=(-1)^n . J_n(x), n \in N$ and J_n is Bessel function of first kind.

YEAR-2009

2. Express $J_4(x)$ in terms of $J_0(x)$ and $J_1(x)$.

8. (a) Prove that $\frac{d}{dx}\left[x^{-n}J_n(x)\right]=-x^{-n}J_{n+1}(x)$

9. (a) Prove that $P_n(x)=\frac{1}{2^n n! dx^n}\left(x^2-1\right)^n$

(b) Obtain the series solution of the equation

$$2x^2y''+\left(2x^2-x\right)y'+y=0$$

YEAR-2010

5. Prove that $\frac{d}{dx}\left\{x^{-n}J_n(x)\right\}=-x^{-n}J_{n+1}(x)$

7. (a) Obtain series solution of the equation

$$\left(1+x^2\right)y''+xy'-y=0$$

(b) Express $f(x)=x^4+3x^3-x^2+5x-2$ in terms of Legendre's polynomials. 8+7

8. (b) Solve : $\left(D^2-DD'+D'-1\right)z=\cos(x+2y)$

9. (b) Prove that $J_{-\frac{1}{2}}(x)=\sqrt{\frac{2}{\pi x}}\cos x$ 8+7

Questions & Solution of W. B. U. T. 3rd Semester M-302

2011

MATHEMATICS - III

GROUP - A

(Short Answer Type Questions)

1. Answer any ten of the following : $10 \times 2 = 20$

(a) Evaluate $\oint_C \frac{z}{z^2-1} dz$ where $C : |z| = 2$.

(b) Find the singularities of $f(z) = \frac{1}{\sin z - \cos z}$.

(c) Find the Fourier Sine Transform of

$$f(x) = \frac{1}{x}.$$

(d) Find the period of the function $f(x) = 2|\cos^2 x|$.

(e) Prove that every function can be represented as a sum of even and odd function.

(f) Show that $X = 0$ is a regular singular point but $X = 2$ is not a regular singular point of the equation

$$x(x-2)^3 y'' + 3(x-2)^3 y' + 4y = 0.$$

(g) Find the general value of $\log(-i)$.

(h) From the discrete distribution find k :

X	0	1	2	3	4	5	6	7
$P(X)$	0	k	$2k$	$2k$	$3k$	k^2	$2k^2$	$7k^2+k$

(i) Determine whether the function is even or odd :

$$f(x) = \log(x + \sqrt{x^2+1}).$$

(j) If $f(x) = x, \quad 0 < x < \alpha,$

$= 0, \quad x > \alpha,$

find the Fourier Sine Transform of $f(x)$.

(k) If $f(x) = e^{-x^2/2}$, then show that $F(s) \approx \sqrt{2\pi}\, e^{-s^2/2}$, where $F(s)$ is the Fourier transform of $f(x)$.

(l) If $f(x) = e^{-x}, x \geq 0$, show that Fourier sine transform of $f(x)$ is $sF'(s)$, where $F(s) = \tan^{-1} s$.

(m) A random variable X has the density function $f(x) = x$, $0 \leq x \leq 1$, $f(x) = \frac{1}{2}$, $1 < x \leq 2$. Find the mean of X.

(n) Express $x^3 + x^2$ in terms of Legendre Polynomials :

$$P_0(x), P_1(x), P_2(x), P_3(x).$$

(o) A random variable X has the following probability density function :

X	0	1	2	3	4
$P(X = x) = f(x)$:	0	$5k$	$3k$	k	k

Determine the value of k.

GROUP - B

(Long Answer Type Questions)

Answer any *five* questions from the following :

$5 \times 10 = 50$

2. (i) Let $f(x) = x, 0 < x < 2$. Find the half range coine series. Write Parseval's Identity coresponding to $f(x)$. Hence show that

$$\frac{1}{1^4} + \frac{1}{2^4} + \frac{1}{3^4} + \cdots\cdots + \frac{1}{n^4} = \frac{\pi^4}{90}.$$

(ii) For a function defined by $f(z) = \sqrt{|xy|}$; show that the Cauchy-Riemann equations are satisfied at (0, 0) but the function is not differentiable at that point. 5 + 5

3. (i) Obtain series solution of the differential equation $\frac{d^2y}{dx^2} + y = 0$ near $x = 0$ such that $y(0) = 1, y'(0) = 2$.

(ii) Expand the function $f(x) = \frac{1}{(z-1)(z-2)}$ between the annular region of $|z| = 1$ and $|z| = 2$. 5 + 5

4. (i) If $U(x, y) = 4xy - x^3 + 3xy^2$, verify that U is harmonic function and obtain its conjugate $V(x, y)$ so that $f = U + iV$ is an analytic function and also find $f(z)$, where $z = x + iy,\ x, y \in R$.

(ii) Find the Fourier sine transform of the function

$$f(x) = \begin{cases} 1 \text{ for } 0 < x \le \pi \\ 0 \text{ for } x > \pi \end{cases}$$

and hence evaluate the integral

$$\int_0^\infty \frac{1 - \cos p\pi}{p} \sin px\, dp \cdot$$

5 + 5

5. (i) A random variable X has the following :

p.d.f. $f(x) = cx^2,\ 0 \le x \le 1$

$= 0$, otherwise

Find (i) C

(ii) $P(0 \le x \le \frac{1}{2})$.

(ii) (a) Show that :

$$\int_{-1}^{1} P_m(x) P_n(x) dx = 0,\ m \ne n\ .$$

(b) Show that $J_{-n}(x) = (-1)^n \cdot J_n(x), n \in N$ and J_n is Bessel function of first kind.

6. (i) Use the method of Frobenius to solve the following differential equation :

$$x\frac{d^2y}{dx^2} + \frac{dy}{dx} + xy = 0\cdot$$

(ii) Solve the equation $\frac{\partial u}{\partial t} = \frac{\partial^2 u}{\partial x^2}$, $(x > 0, t > 0)$ subject to the conditions

(a) $u(0, t) = 0$

(b) $u(x, 0) = 1,\ 0 < x < 1$

$= 0,\ x \ge 1$

(c) $u(x, t)$ is bounded.

5 + 5

7. (i) Three bags contain respectively 4W, 7B, 9R; 5W, 8B, 7R and 6W, 9B, 5R balls. One ball is chosen at random from each bag. What is the probability that the balls are

(a) of same colour (b) 2W and 1R?

(ii) A random variable X has the following probability mass function :

x	0	1	2	3	4	5	6
$P(X=x)$	k	$3k$	$5k$	$7k$	$9k$	$11k$	$13k$

(a) Find the value of k

(b) Find $P(X<4)$, $P(X \geq 5)$, $P(3<X \leq 5)$

(c) Obtain the distribution function $F(x)$

(d) What is the smallest value of x for which $P(X \leq x) > 0.5$? 5 + 5

8. (i) The radius of a circle has distribution given by the p.d.f.

$$f(x) = 1, \; 1 < x < 2$$
$$= 0, \text{ otherwise}$$

Find the mean and variance of the area of the circle.

(ii) If the weekly wage of 10,000 workers in a family follows normal distribution with mean and standard deviation Rs. 70 and Rs. 5 respectively, find the expected number of workers whose weekly wages are

(a) between Rs. 66 and Rs. 72

(b) less than Rs. 66

(c) more than Rs. 72. 5 + 5

9. (i) Show that

$$f(z) = [x^3(1+i) - y^3(1-i)] / (x^2+y^2), \quad z \neq 0, \; f(0) = 0$$

is continuous and Cauchy's Reiman's equation are satisfied at the origin yet $f'(0)$ does not exist.

(ii) Find the first four terms of the Taylor's series expansion of $f(z) = (z+1)/(z-3)(z-4)$ about $z = 2$ and also find the region of convergence. 5 + 5

Answer

1. (a) 0.

(b) $z = n\pi + \frac{\pi}{4}$, $n = 0, 1, 2, \ldots\ldots$

(c) $\frac{\pi}{2}$.

(d) π.

(g) $(4n-1)\frac{i\pi}{2}$, $n = 0, 1, 2, \ldots\ldots.$

(h) $k = \frac{1}{10}$.

(i) odd.

(j) $\frac{1}{s^2}(\sin\alpha s - \alpha s\cos\alpha s)$.

(m) $1\frac{1}{12}$.

(n) $\frac{1}{15}(6P_3 + 16P_2 + 9P_1 + 5P_0)$.

(o) $k = \frac{1}{10}$.

3. (ii) $f(z) = -\sum_{n=0}^{\infty}\frac{z^n}{2^{n+1}} - \sum_{n=0}^{\infty}\frac{1}{z^{n+1}}$.

(i) $y = \left(1 - \frac{x^2}{2!} + \frac{x^4}{4!} \cdots\cdots\right) + 2\left(x - \frac{x^3}{3!} + \frac{x^5}{5!} \cdots\cdots\right)$

4. (i) $V = -2x^2 + 2y^2 + y^3 - 3x^2y + c$

$f(z) = -2iz^2 - z^3 + ic$

5. (i) 3 (ii) $\frac{1}{8}$.

6. (i) $y = a_1u + a_2\left[u\log x + \frac{x^2}{2^2} - \frac{x^4}{2^2 \cdot 4^2}\left(1 + \frac{1}{2}\right) + \cdots\cdots\right]$.

where $u = 1 - \frac{x^2}{2^2} - \frac{x^4}{2^2 \cdot 4^2} \cdots\cdots$

(ii) $u(x\,.\,t)=\dfrac{4}{\pi}\sum\limits_{m=1}^{\infty}\dfrac{1}{2m-1}\sin(2m-1)\pi x\, e^{-(2m-1)^2\pi^2 t}$

7. (i) (a) $\dfrac{939}{8000}, \dfrac{269}{4000}$.

(ii) (a) $\dfrac{1}{49}$.

(b) $\dfrac{16}{49}, \dfrac{45}{49}, \dfrac{20}{49}$.

(c) $F(n)=\begin{cases} 0, & -\infty<n<0 \\ \dfrac{1}{49}, & 0\le x<1 \\ \dfrac{4}{49}, & 1\le x<2 \\ \dfrac{9}{49}, & 2\le x<3 \\ \dfrac{16}{49}, & 3\le x<4 \\ \dfrac{25}{49}, & 4\le x<5 \\ \dfrac{36}{49}, & 5\le x<6 \\ 1, & 6\le x<\infty \end{cases}$

d) $x=4$.

8. (i) $\dfrac{7\pi}{3}, \dfrac{34\pi^2}{45}$.

(ii) (a) 0.4435. (b) 2119. (c) 3446.

9. (ii) $f(z)$

$$=\left(4-\frac{5}{2}\right)+\left(4-\frac{5}{2^2}\right)(z-2)+\left(4-\frac{5}{2^3}\right)(z-2)^2+\cdots\cdots\infty,$$

$|z-2|<1$ in the region of convergence

Solutions

(a) $\frac{z}{z^2-1} = \frac{z}{(z-1)(z+1)} = \frac{1}{2}\left(\frac{1}{z-1} + \frac{1}{z+1}\right)$

$$\therefore \oint_c \frac{z}{z^2-1} dz = \frac{1}{2}\oint_c \frac{dz}{z-1} + \frac{1}{2}\oint_c \frac{dz}{z+1}$$

$$= \frac{1}{2}.2\pi i + \frac{1}{2}.2\pi i = 2\pi i$$

(b) see page no. 2.61, Ex. 3 (i)

(c) see page no. 1.67, Ex. 5

(d) see page no 1.5, Ex. 2

(e) Let as consider a function $f(x)$ which can be written as

$$f(x) = \frac{1}{2}\{f(x)+f(-x)\} + \frac{1}{2}\{f(x)-f(-x)\} = \phi(x) + \psi(x).$$

where $\phi(x) = \frac{1}{2}\{f(x)+f(-x)\}$

$\therefore \phi(-x) = \phi(x)$

$\therefore \phi(x)$ is an even function

and $\psi(x) = \frac{1}{2}\{f(x)-f(-x)\}$

$\therefore \psi(-x) = -\psi(x)$

$\therefore \psi(x)$ is an odd function.

Then every function is a sum of an even function and an odd function.

(f) The given equation can be written as

$y'' + P(x)y' + Q(x)y = 0$

where $P(x) = \frac{3}{x}, Q(x) = \frac{4}{x(x-2)^3}$

$\therefore xP(x) = 3$ and $x^2 Q(x) = \frac{4x}{(x-2)^3}$ possess derivatives of all the orders is nbd of x, so x is a regular singular point.

But $(x-2)^2 Q(x) = \frac{4}{x(x-2)}$ does not possess derivatives of all orders in the nbd of $x = 2$, so $x = 2$ is a regular singular point.

(g) $-i = \cos(-\frac{\pi}{2}) + i\sin(-\frac{\pi}{2}) = e^{-i\frac{\pi}{2}}$

$$\therefore \quad \log(-i) = -i\frac{\pi}{2} + 2n\,\pi i,\ n = 0, 1, 2, ...$$

$$= (4n-1)\frac{i\pi}{2}$$

(h) see page no. 3-65, Ex (2) (i)

(i) $f(x) = \log(-x + \sqrt{x^2+1}) = \log\frac{1}{x + \sqrt{x^2+1}}$

$$= -\log(x + \sqrt{x^2+1}) = -f(x)$$

$\therefore$ $f(x)$ is odd.

(j) The Fourier sine transform of $f(x)$ is

$$F_s(s) = \int_0^{\alpha} f(x)\sin sx\,dx = \int_0^{\alpha} x\sin sx\,dx = \left[x\frac{\cos sx}{-s} + \frac{\sin sx}{s^2}\right]_0^{\alpha}$$

$$= \frac{1}{s^2}(\sin s\alpha - \alpha s\cos s\alpha)$$

(k) see page no 1-54

(l) The Fourier sine transform of $f(x) = e^{-x}$ is

$$F_s(s) = \int_0^{\infty} f(x)\sin sx\,dx = \int_0^{\infty} e^{-x}\sin sx\,dx$$

$$= \lim_{X\to\infty}\int_0^{X} e^{-x}\sin sx\,dx = \lim_{X\to\infty}\frac{s}{1+s^2}\left[1 - \frac{\cos sX}{e^X} - \frac{\sin sX}{se^X}\right]$$

$$= \frac{s}{1+s^2} = s\,F'(x) \text{ where } F(x) = \tan^{-1} s.$$

(m) mean of $x = \int_0^2 x\,f(x)\,dx = \int_0^1 x.x\,dx + \int_1^2 x.\frac{1}{2}\,dx$

$$= \left[\frac{x^3}{3}\right]_0^1 + \frac{1}{2}\left[\frac{x^2}{2}\right]_1^2 = \frac{1}{3} + \frac{3}{4} = 1\frac{1}{12}.$$

(n) We have $P_0(x)=1,\ P_1(x)=x,\ P_2(x)=\frac{1}{2}(3x^2-1)$

$$P_3(x)=\frac{1}{2}(5x^3-3x)$$

$$\therefore\ x^3=\frac{2}{5}P_3(x)+\frac{3}{5}x=\frac{2}{5}P_3(x)+\frac{3}{5}P_1(x)$$

$$x^2=\frac{2}{3}P_2(x)+\frac{1}{3}=\frac{2}{3}P_2(x)+\frac{1}{3}P_0(x)$$

$$\therefore\ x^3+x^2=\frac{2}{3}P_3(x)+\frac{3}{5}P_1(x)+\frac{2}{3}P_2(x)+\frac{1}{3}P_0(x)$$

$$=\frac{1}{15}\{6P_3(x)+10P_2(x)+9P(x)+5P_0(x)\}$$

(o) Since $\sum_x f(x)=1$, so

$$0+5k+3k+k+k=1\ \Rightarrow 10k=1$$

$$\therefore\ k=\frac{1}{10}.$$

4. i) $U=4xy-x^3+3xy^2$

$$\therefore\ \frac{\partial U}{\partial x}=4y-3x^2+3y^2,\quad \frac{\partial^2 U}{\partial x^2}=-6x$$

$$\frac{\partial U}{\partial y}=4x+6xy,\quad \frac{\partial^2 U}{\partial y^2}=6x$$

$$\therefore\ \frac{\partial^2 U}{\partial x^2}+\frac{\partial^2 U}{\partial y^2}=0$$

$\therefore$ U is a harmonic function.

Let V be the corresponding conjugate of U. Then

$$\frac{\partial U}{\partial x}=\frac{\partial V}{\partial y},\ \frac{\partial U}{\partial y}=-\frac{\partial V}{\partial x}$$

Now $$dV=\frac{\partial V}{\partial x}dx+\frac{\partial V}{\partial y}dy=-\frac{\partial U}{\partial y}dx+\frac{\partial U}{\partial x}dy$$

$$=-(4x+6xy)dx+(4y-3x^2+3y^2)dy$$

$$=-4x\,dx+4y\,dy+3y^2\,dy-(6xy\,dx+3x^2\,dy)$$

$$= -4x\,dx + 4y\,dy + 3y^2\,dy - d(3x^2y)$$

Integrating we get

$$V = -2x^2 + 2y^2 + y^3 - 3x^2y + c, \text{ where } c \text{ is a real constant.}$$

Thus $f(z) = U + iV$

$$= (4xy - x^3 + 3xy^2) + i(-2x^2 + 2y^2 + y^3 - 3x^2y + c)$$

$$= -2i(x+iy)^2 - (x+iy)^3 + ic$$

$$= -2iz^2 - z^3 + ic.$$

4. (ii) see page no 1-66, Ex-3

5. (i) Since $f(x)$ is p.d.f., so

$$\int_{-\infty}^{\infty} f(x)\,dx = 1$$

i.e., $\int_0^1 c\,x^2\,dx = 1$

i.e., $c\left[\frac{x^3}{3}\right]_0^1 = 1$

i.e., $c.\frac{1}{3} = 1.$ $\therefore\ c = 3.$

(ii) $P(0 \le x \le \frac{1}{2}) = \int_0^{\frac{1}{2}} f(x)\,dx = \int_0^{\frac{1}{2}} cx^2\,dx$

$$= 3\int_0^{\frac{1}{2}} x^2\,dx = 3\left[\frac{x^3}{3}\right]_0^{1/2} = \left(\frac{1}{8} - 0\right) = \frac{1}{8}.$$

(ii) (a) see page no. 4-98, art. 4-5.4

(b) see page no. 4-84, Ex. 1

6. (i) see page no. 4-68, Ex-1

(ii) The general solution of the equation

$$\frac{\partial u}{\partial t} = \frac{\partial^2 u}{\partial x^2}$$

is $U(x.t) = (c_1 \cos px + c_2 \sin px)e^{-p^2t}$ . . . (1)

as $U(x, t)$ is bounded

Then using the boundary condition $u(0, t) = u(1, t) = 0$,

we get, $c_1 e^{-p^2 t} = 0$

i.e., $c_1 = 0$

and $(c_1 \cos p + c_2 \sin p)e^{-p^2 t} = 0$

i.e., $c_2 \sin p e^{-p^2 t} = 0$

i.e., $\sin p = 0$

[$\because$ $c_2 \neq 0$, for $c_2 = 0 \Rightarrow u=0 \ \forall \ x, t$ which is not possible].

$\therefore$ $p = n\pi$, $n = 1, 2, 3, \ldots\ldots$

Hence (1) takes the form

$$U_n(x, t) = a_n \sin n\pi x e^{-n^2 p^2 t}, \ n = 1, 2, \ldots\ldots$$

Now we consider the most general solution as

$$U(x, t) = \sum_{n=1}^{\infty} a_n \sin n\pi x e^{-n^2 \pi^2 t} \qquad (2)$$

Then using the initial condition

$$U(x, 0) = 1, \ 0 < x < 1,$$

we get

$$1 = \sum_{n=1}^{\infty} a_n \sin n\pi x$$

which is the Fourier sine series

$$\therefore \ a_n = 2\int_0^1 \sin n\pi x \, dx, \ n = 1, 2, \ldots\ldots = 2\left[-\frac{\cos n\pi x}{n\pi}\right]_0^1$$

$$= \frac{2}{n\pi}[1-(-1)^n] = 0 \text{ when } n \text{ is even}$$

$$= \frac{4}{n\pi} \text{ when } n \text{ is odd, i.e., } n = 1, 3, 5, \ldots\ldots$$

$$\therefore \ U(x, t) = \frac{4}{\pi} \sum_{n=1, 3, 5, \ldots\ldots} \frac{1}{n} \sin n\pi x e^{-n^2\pi^2 t}$$

$$= \frac{4}{\pi} \sum_{m=1}^{\infty} \frac{1}{2m-1} \sin (2m-1)\pi x e^{-(2m-1)^2 \pi^2 t}.$$

7. (i) (a) The possible ways that the event "balls are of same colour" may occur are

1. White (W) from first bag, White (W) from second bag and White (W) from third bag in $({}^4C_1 \times {}^5C_1 \times {}^6C_1)$ ways

i.e., in $4 \times 5 \times 6$ ways

i.e, in 120 ways

2. Black (B) from first bag, Black (B) from second bag and Black (B) from third bag in $({}^7C_1 \times {}^8C_1 \times {}^9C_1)$ ways

i.e., in $7 \times 8 \times 9$ ways

i.e., in 504 ways

3. Red (R) from first bag, Red (R) from second bag and Red (R) from third bag

in $({}^9C_1 \times {}^7C_1 \times {}^5C_1)$ ways

in 315 ways.

Thus the required probability in

$$\frac{120 + 504 + 315}{{}^{20}C_1 \times {}^{20}C_1 \times {}^{20}C_1} = \frac{939}{8000}.$$

(b) Similarly, the probability that the balls are 2W and 1R is

$$\frac{{}^4C_1 \times {}^5C_1 \times {}^5C_1 + {}^4C_1 \times {}^7C_1 \times {}^6C_1 + {}^9C_1 \times {}^5C_1 \times {}^6C_1}{{}^{20}C_1 \times {}^{20}C_1 \times {}^{20}C_1} = \frac{269}{4000}.$$

(ii) (a) Since $f(x)$ is a p.m.f., $\sum_x f(x) = 1$

$\therefore \quad k + 3k + 5k + 7k + 9k + 11k + 13k = 1$

i.e., $\quad 49k = 1$

$\therefore \quad k = \dfrac{1}{49}.$

(b) $P(X < 4)$

$= P(X = 0) + P(X = 1) + P(X = 2) + P(X = 3)$

$= \dfrac{1}{49} + \dfrac{3}{49} + \dfrac{5}{49} + \dfrac{7}{49} = \dfrac{16}{49}$

$$P(X \geq 5) = 1 - P(X < 2)$$
$$= 1 - \{P(X = 0) + P(X = 1)\}$$
$$= 1 - \left\{\frac{1}{49} + \frac{3}{49}\right\} = 1 - \frac{4}{49} = \frac{45}{49}.$$

$$P(3 < X \leq 5) = P(X = 4) + P(X = 5) = \frac{9}{49} + \frac{11}{49} = \frac{20}{49}.$$

(c) The distribution function $F(x)$ is given below

$$F(x) = \begin{cases} 0, & -\infty < x < 0 \\ \frac{1}{49}, & 0 \leq x < 1 \\ \frac{1}{49} + \frac{3}{49} = \frac{4}{49}, & 1 \leq x < 2 \\ \frac{4}{49} + \frac{5}{49} = \frac{9}{49}, & 2 \leq x < 3 \\ \frac{9}{49} + \frac{7}{49} = \frac{16}{49}, & 3 \leq x < 4 \\ \frac{16}{49} + \frac{9}{49} = \frac{25}{49}, & 4 \leq x < 5 \\ \frac{25}{49} + \frac{11}{49} = \frac{36}{49}, & 5 \leq x < 6 \\ \frac{36}{49} + \frac{13}{49} = 1 & 6 \leq x < \infty \end{cases}$$

d) Now $P(X \leq 0) = \frac{1}{49} < 0.5$

$$P(X \leq 1) = \frac{1}{49} + \frac{3}{49} = \frac{4}{49} < 0.5$$

$$P(X \leq 2) = \frac{1}{49} + \frac{3}{49} + \frac{5}{49} = \frac{9}{49} < 0.5$$

$$P(X \leq 3) = \frac{1}{49} + \frac{3}{49} + \frac{5}{49} + \frac{7}{49} = \frac{16}{49} < 0.5$$

$$P(X \leq 4) = \frac{1}{49} + \frac{3}{49} + \frac{5}{49} + \frac{7}{49} + \frac{9}{49} = \frac{25}{49} > 0.5$$

Thus the minimum value of x so that $P(X \leq x) > 0.5$ is $x = 4$.

8. (i) Mean of the area of the circle

$$= E(\pi X^2) = \pi E(X^2) = \pi \int_1^2 x^2 . 1 . dx$$

$$= \pi \left[\frac{x^3}{3} \right]_1^2 = \pi \left(\frac{8}{3} - \frac{1}{3} \right) = \frac{7\pi}{3}$$

Variance of the area of the circle

$$= Var(\pi X^2) = E(\pi^2 X^4) - \{E(\pi X^2)\}^2$$

$$= \pi^2 E(X^4) - \left(\frac{7\pi}{3} \right)^2$$

Now $E(X^4) = \int_1^2 x^4 . 1 . dx = \left[\frac{x^5}{5} \right]_1^2 = \frac{32}{5} - \frac{1}{5} = \frac{31}{5}$.

$\therefore$ Variance $= \pi^2 \times \frac{31}{5} - \frac{49\pi^2}{9} = \pi^2 \left(\frac{279 - 245}{45} \right) = \frac{34\pi^2}{45}$

8. (ii) see page no. 3-138, Ex. 4

9. (i) see page no. 2-13, Ex. 4

(ii) Let $z - 2 = u$ $\therefore$ $z = u + 2$

$$\therefore \quad f(z) = \frac{z+1}{(z-3)(z-4)} = \frac{u+3}{(u-1)(u-2)} = \frac{A}{u-1} + \frac{B}{u-2} \text{ (say)}$$

$$\therefore \quad u + 3 = A(u-2) + B(u-1)$$

Putting $u = 1, 2$, we get $A = -4, B = 5$

$$\therefore \quad f(z) = -\frac{4}{u-1} + \frac{5}{u-2} = 4(1-u)^{-1} - \frac{5}{2}\left(1 - \frac{u}{2}\right)^{-1}$$

$$= 4(1 + u + u^2 + u^3 + \cdots\cdots \infty) - \frac{5}{2}\left(1 + \frac{u}{2} + \frac{u^2}{2^2} + \frac{u^3}{2^3} + \cdots\cdots \infty\right)$$

$$= \left(4 - \frac{5}{2}\right) + \left(4 - \frac{5}{2^2}\right) u + \left(4 - \frac{5}{2^3}\right) u^2 + \left(4 - \frac{5}{2^4}\right) u^3 + \cdots\cdots \infty$$

$$= \left(4 - \frac{5}{2}\right) + \left(4 - \frac{5}{2^2}\right)(z-2) + \left(4 - \frac{5}{2^3}\right)(z-2)^2$$

$$+ \left(4 - \frac{5}{2^4}\right)(z-2)^3 + \cdots\cdots \infty$$

Here the region of convergence is

$$|u| < 1 \qquad \text{i.e.,} \quad |z-2| < 1$$

2. (i) see page no. 1-17, illustration
   (ii) see page no. 2-14, Ex. 5

3. (ii) see page no. 2-51, Ex. 2
   (i) Here $x = 0$ is an ordinary point of the given equation.

So let

$$y = \sum_{n=0}^{\infty} a_n x^n \qquad \ldots (1)$$

be the series solution of the equation

$$\frac{d^2 y}{dx^2} + y = 0 \qquad \ldots (2)$$

Now, $\frac{dy}{dx} = \sum_{n=1}^{\infty} n a_n x^{n-1}$, $\frac{d^2 y}{dx^2} = \sum_{n=2}^{\infty} n(n-1) a_n x^{n-2}$

Substituting these value in (1), we get

$$\sum_{n=2}^{\infty} n(n-1) a_n x^{n-2} + \sum_{n=0}^{\infty} a_n x^n = 0$$

or, $$\sum_{n=0}^{\infty} (n+2)(n+1) a_{n+2} x^n + \sum_{n=0}^{\infty} a_n x^n = 0 \qquad \ldots (3)$$

Equating the co-efficient of x^n to zero, we have

$$(n+2)(n+1) a_{n+2} + a_n = 0$$

$$\therefore \quad a_{n+2} = -\frac{1}{(n+2)(n+1)} a_n \qquad \ldots (4)$$

Again equating to zero, the co-efficient of x^0 and x^1, we have from (3),

$$2 \cdot 1 \cdot a_2 + a_0 = 0, \; 3 \cdot 2 \cdot a_3 + a_1 = 0$$

$$\therefore \quad a_2 = -\frac{1}{2} a_0, \; a_3 = -\frac{1}{3 \cdot 2} a_1$$

Putting $n = 2, 3, 4, \ldots\ldots$ in (4) successively, we get

$$a_4 = -\frac{1}{4 \cdot 3} a_2 = \frac{1}{2 \cdot 3 \cdot 4} a_0 = \frac{1}{4!} a_0$$

$$a_5 = -\frac{1}{5 \cdot 4} a_3 = \frac{1}{2 \cdot 3 \cdot 4 \cdot 5} a_1 = \frac{1}{5!} a_1$$

$$a_6 = -\frac{1}{6!}a_0,\ a_7 = -\frac{1}{7!}a_1 \text{ and so on.}$$

Substituting there value of a_i's in (1), we get

$$y = a_0\left(1-\frac{x^2}{2!}+\frac{x^4}{4!}-\cdots\cdots\right)+a_1\left(x-\frac{x^3}{3!}+\frac{x^5}{5!}-\cdots\cdots\right)$$

$$\therefore\ y' = a_0\left(-x+\frac{x^3}{3!}-\cdots\cdots\right)+a_1\left(1-\frac{x^2}{2!}+\frac{x^4}{4!}-\cdots\cdots\right)$$

Using $y(0)=1,\ y'(0)=2$, we have

$$a_0 + a_1 \cdot 0 = 1 \Rightarrow a_0 = 1$$

$$a_0 \cdot 0 + a_1 \cdot 1 = 2 \Rightarrow a_1 = 2$$

Thus the required series solution is

$$y = \left(1-\frac{x^2}{2!}+\frac{x^4}{4!}-\cdots\cdots\right)+2\left(x-\frac{x^3}{3!}+\frac{x^5}{5!}-\cdots\cdots\right).$$

W. B. U. T. SEM-3 / M-302 / 2012-13

2012

MATHEMATICS-III

Group – A

1. Answer any *ten* from the following : 10×2=20

(i) If $f(x)=|x|,\ -2<x<2$ is a periodic of period 4, be represented in a Fourier series, then find the value of a_0.

(ii) Using Fourier transform evaluate $\int_0^\infty \frac{\cos t}{1+s^2}ds$

(iii) Find the poles of the function $f(z)=\frac{z^2}{(z-1)^2(z+2)}$

(iv) Evaluate $\int_C \frac{e^z}{z-2}dz$, where $c : |z-2|=4$

(v) Find the probability that a leap year selected at random contain 53 Tuesdays

(vi) A random variabl X has the *p.d.f.*

$f(x) = 3x,\ 0 < x < 1$
$\quad = 0,$ otherwise.

Then find $P(2x+3>2)$.

(vii) Express $J_3(x)$ in terms of $J_0(x)$ and $J_1(x)$

(viii) Obtain the partial differential equation for

$z=(x+a)(x+b)$, where a and b are arbitray constants.

(ix) Determine the singular point of the following equation :

$$x(x-1)\frac{d^2y}{dx^2}+(3x-1)\frac{dy}{dx}+y=0$$

(x) if A and B are independent events then show that A^C and B^C are independent.

(xi) If X is normally distributed with zero mean and unit variance, find $E(X^2)$.

(xii) Check whether the function

$u(x,y)=2xy+3xy^2-2y^3$ is harmonic or not.

(xiii) Write the Convolution Theorem of Fourier Transform.

(xiv) Prove that $P_n(-1)=(-1)^n$.

(xv) The mean and standard deviation of a Binomial distribution are 4 and $\sqrt{3}/3$ respectively. Find n and p.

Group – B

(Answer any *five* questions taking at least one question from each Module) 5×10=50

Module - I

2. (a) Obtain the Fourier series to represent : 6

$f(x)=x^2$ in $-\pi\le x\le\pi$

Hence show that $\frac{1}{1^2}+\frac{1}{2^2}+\frac{1}{3^2}+\cdots=\frac{\pi^2}{6}$

(b) Find the the Fourier inverse transform of the function

$F(p)=\frac{1}{p^2+4p+13}$. 4

3. (a) State Parseval's Identify on Fourier cosine transform. Find the Fourier cosine transform of e^{-ax}, $a>0$. 5

(b) Using the Parseval's identify of Fourier cosine transform, show that

$$\int_0^\infty \frac{dx}{(a^2+x^2)(b^2+x^2)}=\frac{\pi}{2ab(a+b)}$$

where $a>0,\ b>0$. 5

Module - II

4. (a) Show that by considering the function $f(z)$ defined as

$$f(z)=\frac{xy(y-ix)}{x^2+y^2}, \text{ for } z\neq 0$$

$$=0 \text{ for } z=0,$$

the C-R equations are not the sufficient conditions for a function to be analytic. 5

(b) Expand the function $f(z)=\frac{1}{(z+1)(z+3)}$ in a Laurent's series valid for $0<|z+1|<2$. 5

5. (a) Evaluate $\int_0^{2\pi}\frac{d\theta}{2+\cos\theta}$ 5

(b) Use Cauchy residue theorem to evaluate

$\int_C \frac{3z^2+z-1}{(z^2-1)(z-3)}dz$ around the circle $C:|z|=2$. 5

Module - III

6. (a) A random variable X has the following probability function :

X	0	1	2	3	4	5	6	7
$P(x)$	0	K	$2K$	$2K$	$3K$	K^2	$2K^2$	$7K^2+K$

Obtain the value of K and estimate $P(X<6)$ and $P(0<X<5)$. 5

(b) In a bolt factory, the machines M1, M2 and M3 manufacture respectively 25, 35 and 40 per cent of the total product. Of their output 5, 4 and 2 per cent respectively are defective bolts. One bolt is drawn at random from the product and is found to be defective. What is the probability that it was manufactured by machine M3? 5

7. (a) In a shooting competition the probability of a man hitting a target is $\frac{1}{5}$. If he fires 5 times, what is the probability of hitting the target at least twice? 5

(b) Assuming that the height distribution of a group is normally, find the mean and *s.d.* if 84% of the men have heights less than 65.2 inches and 68% have heights lying between 65.2 and 62.8 inches.

$$\left[\text{given} \int_{-\infty}^{0.9} \phi(t)\,dt = 0.84, \quad \int_{-\infty}^{-0.9} \phi(t)\,dt = 0.16\right.$$

Module - IV

8. (a) Use Laplace Transform to solve the one dimensional wave equation

$$\frac{\partial^2 u}{\partial t^2} = c^2 \frac{\partial^2 u}{\partial x^2} (x > 0,\ t > 0)$$

where $u(x, 0) = 0,\ \frac{\partial u}{\partial t}(x, 0) = 0,\ x > 0$

$$u(0, t) = F(t), u(\infty, t) = 0, t \geq 0$$ 10

9. (a) Solve in series the equation

$$\left(1 + x^2\right)\frac{d^2 y}{dx^2} + x\frac{dy}{dx} - y = 0.$$ 5

(b) Establish the recurrence formula for Legendre Polynomials

$$(2n+1)xP_n = (n+1)P_{n+1} + nP_{n-1}.$$ 5

SOLUTION AND ANSWER

1. (i) Here $f(x) = -x, \ -2 < x \le 0$

$$= x, \ 0 \le x \le 2$$

$$\therefore \quad a_0 = \frac{1}{2} \int_{-2}^{2} f(x)\,dx$$

$$= \frac{1}{2} \int_{-2}^{0} (-x)\,dx + \frac{1}{2} \int_{-2}^{0} x\,dx = 2$$

(ii) See Ex. 10, Page no. 1-70.

(iii) $f(z)$ has pole of order 2 at $z = 1$ and of order one at $z = -2$.

(iv) Here $z = 2$ is an interior point of c.

$\therefore$ Using Cauchy's integral formula,

$$\int_C \frac{e^{z}}{z-2}\,dz = 2\pi i \times e^2 = 2\pi i e^2$$

(v) See Ex. 1, Page no. 3-16.

(vi) $P(2x+3>2) = P(2x>-1) = P\left(x > -\frac{1}{2}\right)$

$$= \int_{-\frac{1}{2}}^{\infty} f(x)\,dx$$

$$= \int_{-\frac{1}{2}}^{0} 0.dx + \int_{0}^{1} 3x\,dx + \int_{1}^{\infty} 0.dx$$

$$= 3\left[\frac{x^2}{2}\right]_0^1 = \frac{3}{2}$$

(vii) We have the recurrence relation

$$J_{n+1}(x) = \frac{2n}{x} J_n(x) - J_{n-1}(x)$$

Putting $n=2,1$

$$J_3(x)=\frac{4}{x}J_2(x)-J_1(x)$$

$$J_2(x)=\frac{2J_1(x)}{x}-J_0(x)$$

$$\therefore \quad J_3(x)=\frac{4}{x}\left(\frac{2}{x}J_1(x)-J_0(x)\right)-J_1(x)$$

$$=\left(\frac{8}{x^2}-1\right)J_1(x)-\frac{4}{x}J_0(x)$$

(viii) $z=(x+a)(x+b)$

$$\therefore \quad \frac{\partial z}{\partial x}=1.(x+b)+(x+a).1$$

$$=2x+a+b$$

$$\therefore \quad \frac{\partial^2 z}{\partial x^2}=2$$

$$\therefore \quad r=2$$

which in the required *p.d.* equation.

(ix) $x=0$, are singular points of the equation

(x) See Ex. 11, Page no. 3-21.

(xi) See Ex. 1, Page no. 3-137.

(xii) $u=2xy+3xy^2-2y^3$

$$\therefore \quad \frac{\partial u}{\partial x}=2y+3y^2, \quad \frac{\partial^2 u}{\partial x^2}=0$$

$$\frac{\partial u}{\partial y}=2x+6xy-6y^2$$

$$\frac{\partial^2 u}{\partial y^2}=6x-12y$$

$$\therefore \quad \frac{\partial^2 u}{\partial x^2}+\frac{\partial^2 u}{\partial y^2}=6x-12y\neq 0$$

$\therefore$ u is not a harmonic function.

(xiii) See page no. 1-61.

(xiv) See Ex. 3, page no. 4-104.

(xv) See Ex. 1, page no. 3-109.

2. (a) See Ex. 18(i), page no. 1-37.

(b) See Ex. 15, page no. 1-17.

3. (a) See page no. 1-54.

4. (a) See Ex. 20(b), page no. 2-27.

(b) See Ex. 3, page no. 2-52.

5. (a) See Ex. 10(iii), page no. 2-71.

(b) See Ex. 8(f), page no. 2-71.

6. (a) See Ex. 2, page no. 3-65.

(b) See Ex. 23, page no. 3-28.

7. (a) See Ex. 12(a), page no. 3-116.

(b) See Ex. 36, page no. 3-153.

8. (a) See Ex. 1, page no. 4-30.

9. (a) See Ex. 1, page no. 4-61.

(b) See page no. 4-100.

W. B. U. T. SEM-4 / M-402 / 2012

2012

MATHEMATICS-III

Group – A

(Short Answer Type Questions)

Answer any *ten* questions : 10×2=20

1. (a) If $f(x)=x+x^2, -\pi \le x \le \pi$ be represented in a Fourier series as $a_0+\sum_{n=1}^{\infty}(a_n \cos nx + b_n \sin nx)$, then find the value of a_0.

(b) State Fourier Integral theorem.

(c) If $F(s)$ is the Fourier transform of $f(x)$, then find the Fourier transform of $f(ax)$.

(d) Find the residue of $f(z)=\dfrac{2+3\sin \pi z}{z(z-1)^2}$ at $z=1$

(e) Find the value of $\int_c \frac{z}{z-1}dz$ where C is the curve defined by $|z|=\frac{1}{2}$

(f) Find the poles of the function $f(z)=z^2/\{(z-1)^2(z+2)\}$

(g) If a Poisson variate X is such that $P(X=1)=P(X=2)$, Find $P(X=4)$.

(h) For any two events A and B (may not be mutually exclusive), prove that

$$P(A+B)=P(A)+P(B)-P(AB)$$

(i) Find the value of p so that the function $(2x-x^2+py^2)$ is harmonic.

(j) Find the value of $J_{\frac{1}{2}}(x)$

(k) Prove that $(2n+1)xP_n=(n+1)P_{n+1}+nP_{n-1}$.

(l) Find the ordinary and singular points of the differential equation

$$x^2(1+x)^2\frac{d^2y}{dx^2}+\left(x^2-1\right)\frac{dy}{dx}+2xy=0.$$

(m) Find the bilinear transformation which maps

$z=0, 1, \infty$ onto $\omega=-1, -i, 1$ respectively.

(n) Find the value of $\int_{-1}^{1} P_0(x)dx$ where $P_n(x)$ is a Legendre's polynomial of degree *n*.

(o) Write down the Bessel's equation of order 2.

Group – B

(Short Answer Type Questions)

Answer any *five* questions taking at least *one* questions from each Module 5×10=50

Module I : Fourier Series and Fourter Transform

2. (a) Expand $f(x)=x, -\pi\le x\le\pi$ in Fourier series. Hence deduce that $\frac{\pi}{4}=1-\frac{1}{3}+\frac{1}{5}-\cdots$ 4+1

(b) Find the Fourier sine transform of $f(x)$, where

$$f(x)=\begin{cases}1, & 0<x\le\pi\\ 0, & x>\pi\end{cases}$$

and hence evaluate the integral $\int_0^{\infty}\left(\frac{1-\cos p\pi}{p}\right)\sin px\, dp.$

3+2

3. (a) Find the Fourier series of $f(x)=x^2,\ -\pi \le x \le \pi$.

Hence prove that $1-\frac{1}{2^2}+\frac{1}{3^2}-\frac{1}{4^2}+\cdots=\frac{\pi^2}{12}$. 4+1

(b) Find the function whose Fourier cosine trnasform is $\frac{\sin as}{s}$. 5

Module II : Calculus of Complex variable

4. (a) Using Cauchy's Residue theorem, prove that

$$\int_0^{2\pi}\frac{d\theta}{2+\cos\theta}=\frac{2\pi}{\sqrt{3}}.$$ 5

(b) Evaluate $\oint_c \frac{e^{2z}}{(z+1)^4}dz$

where C is the circle $|z|=3$. 5

5. (a) If $u=x^2-y^2$ and $v=-\frac{y}{x^2+y^2}$, then prove that both u and v are harmonic. 5

(b) Find the zeros and their orders of the function

$$f(z)=\frac{z^5-1}{z^2+5}.$$ 5

6. (a) Expand $f(z)=\frac{z}{(z-1)(z-2)}$ in Laurent's series for $1<|z|<2$. 5

(b) Show that the function $f(z)=\sqrt{|xy|}$ is not analytic at the origin, althouth the Cauchy-Riemann conditions are satisfied at that point. 5

Module III : Probability

7. (a) Two urns contain respectively 5 white, 7 black balls and 4 white, 2 black balls. One of the urns is selected by the toss of a fair coin and then 2 balls are drawn without replacement from the selected urn. If both balls are white, what is the probability that the first urn is selected? 5

(b) X is a continuous random variable having probability density function

$$f(x)=\begin{cases} \frac{4x}{5}, & 0<x\leq 1 \\ \frac{2(3-x)}{5}, & 1<x\leq 2 \\ 0, & \text{elsewhere} \end{cases}$$

Find the mean value of x and the distribution function.

3+2

8. (a) An integer is chosen at random from the first 200 positive integers. What is the probability that the integer chosen is divisible by 6 or by 8? 5

(b) A random variable X has the following probability function :

X	0	1	2	3	4	5	6	7
$P(x)$	0	K	$2K$	$2K$	$3K$	K^2	$2K^2$	$7K^2+K$

Obtain the value of K and estimate $P(X<6)$ and $P(0<X<5)$.

Module IV : PDE and Series solution of ODE.

9. Use Laplace transform to solve the one dimensional wave equation

$$\frac{\partial^2 u}{\partial t^2}=c^2\frac{\partial^2 u}{\partial x^2}(x>0,\ t>0)$$

where $u(x,0)=0,\ \frac{\partial x}{\partial t}(x,0)=0, x>0$

and $u(0,t)=F(t),\ u(\infty,t)=0,\ t\geq 0.$ 10

10. Solve the one dimensional heat equation $\frac{\partial u}{\partial t}=c^2\frac{\partial^2 u}{\partial x^2}$ by the method of separation of variables.

11. (a) Find the series solution of *ODE*

$$\frac{d^2y}{dx^2}+x\frac{dy}{dx}+x^2y=0 \text{ about } x=0.$$ 5

(b) Show that $\int_{-1}^{1} x^2 p_{n-1}(x)\, p_{n+1}(x)dx=\frac{2n(n+1)}{(2n-1)(2n+1)(2n+3)}$

ANSWERS (2012)

1. (a) $\frac{\pi^2}{3}$ (c) $\frac{1}{a}F\left(\frac{s}{a}\right)$

(d) $-(2+3\pi)$ (e) 0

(f) $z=1$, a double pole (g) $\frac{2}{3}e^2$

$z=-2$, a simple pole

(i) $p=1$ (j) $\sqrt{\frac{2}{\pi x}}\sin x$

(l) 0, –1 are singular points. (m) $z=i\left(\frac{1+w}{1-w}\right)$

(n) 2 (o) $\frac{x^2d^2y}{dx^2}+x\frac{dy}{dx}+\left(x^2-4\right)y=0$

2. (a) $2\left[\frac{\sin x}{1}-\frac{\sin 2x}{2}+\frac{\sin 3x}{3}-\dots\right]$

(b) Fourier sine Transform is $\frac{1-\cos \pi s}{s}$

$$\int_0^\infty \frac{1-\cos p\pi}{p}\sin px\, dp=\frac{\pi}{2},\quad 0\le x<\pi$$

$$=\frac{\pi}{4},\quad x=\pi$$

$$=0,\quad x>\pi$$

3. (a) $\frac{\pi^2}{3}+4\sum_{n=1}^{\infty}\frac{(-1)^n \cos nx}{n^2}$

(b) $f(x)=1, \; x<a$

$=0, x>a$

4. (b) $\frac{8}{3}\pi i\, e^{-2}$

5. (b) $1, \cos\frac{2\pi}{5}+i\sin\frac{2\pi}{5}, \cos\frac{4\pi}{5}\pm i\sin\frac{4\pi}{5}$

order of each zeros are one

6. (a) $-\cdots\frac{1}{z^2}-\frac{1}{z}-1-\frac{z}{2}-\frac{z^2}{4}-\frac{z^2}{8}-\cdots$

7. (a) $\frac{25}{91}$

(b) $\frac{17}{15}$; $F(x)=0, x\le 0$

$=\frac{2x^2}{5}, \; 0<x\le 1$

$=\frac{6x}{5}-\frac{x^2}{5}-\frac{3}{5}, \; 1<x\le 2$

$=1, \; 2<x<\infty$

8. (a) $\frac{1}{4}$ (b) $\frac{1}{10}$; $\frac{81}{100}$; $\frac{4}{5}$

9. $u(x,t)=F\left(t-\frac{x}{a}\right)H\left(t-\frac{x}{a}\right)$

where H is the unit step Heaviside function.

11. (a) $y = a_0\left(1 - \frac{x^4}{12} + \frac{x^6}{120} - \cdots\right) + a_1\left(x - \frac{x^3}{6} - \frac{x^5}{40}\cdots\right)$

SOLUTION

1. (a) $a_0 = \frac{1}{2\pi}\int_{-\pi}^{\pi} f(x)\,dx = \frac{1}{2\pi}\int_{-\pi}^{\pi}\left(x + x^2\right)dx$

$$= \frac{1}{2\pi}\left\{\int_{-\pi}^{\pi} x\,dx + \int_{-\pi}^{\pi} x^2\,dx\right\}$$

$$= \frac{1}{2\pi} \times 2\int_0^{\pi} x^2\,dx$$

$$= \frac{1}{\pi}\cdot\left[\frac{x^3}{3}\right]_0^{\pi} = \frac{1}{\pi}\cdot\frac{\pi^3}{3} = \frac{\pi^2}{3}.$$

(b) See Art. 1.2.4, Theorem 1, page 1 – 61.

(c) See Property (2) of Art 1.2.3, Page 1–58.

(d) Here $z = 1$ is a pole of $f(z)$ of order 2

$$\therefore \quad \text{Res }(1) = \frac{1}{(2-1)!}\lim_{z\to 1}\frac{d}{dz}\left[(z-1)^2 f(z)\right]$$

$$= \lim_{z\to 1}\frac{d}{dz}\left(\frac{2 + 3\sin \pi z}{z}\right)$$

$$= \lim_{z\to 1}\frac{z.(0 + 3\pi\cos \pi z) - (2 + 3\sin \pi z).1}{z^2}$$

$$= \frac{1(3\pi\cos\pi) - (2 + 3\sin\pi)}{1^2}$$

$$= -3\pi - 2 - 3.0 = -(2 + 3\pi)$$

(e) Let $f(z)=\dfrac{z}{z-1}$

Here $z=1$ is the exterior point of $|z|=\frac{1}{2}$

So $f(z)$ is analytic within and on the circle $|z|=\frac{1}{2}$.

Hence by Cauchy's theorem,

$$\int_c \frac{z}{z-1}dz=0$$

(f) The poles of $f(z)$ are given by

$$(z-1)^2(z+2)=0$$

i.e., $z=1,\ -2$

Here $z=1$ is a double pole and $z=-2$ is a simple pole.

(g) See Ex 5, page 3-123.

(h) See Theorem 5, page 3-5.

(i) Let $f(x,y)=2x-x^2+py^2$

$$\therefore \quad \frac{\partial^2 f}{\partial x^2}=-2,\quad \frac{\partial^2 f}{\partial y^2}=2p$$

Since $f(x,y)$ is a harmonic function, so we must have

$$\frac{\partial^2 f}{\partial x^2}+\frac{\partial^2 f}{\partial y^2}=0$$

$$\therefore \quad -2+2p=0$$

$$\therefore \quad p=1$$

(j) See Ex. 3(ii), page no. 4-86.

(k) See Art 4.5.5, pag no. 4-99.

(l) Here $P_0(x) = x^2(1+x)^2$

$\therefore$ $P_0(0) = P_0(-1) = 0$.

$\therefore$ $x = 0, -1$ are singular points of the equation.

(m) Let the bilinear transform be

$$\frac{(w-w_1)(w_2-w_3)}{(w-w_3)(w_2-w_1)} = \frac{(z-z_1)(z_2-z_3)}{(z-z_3)(z_2-z_1)}$$

Here $z_1 = 0, z_2 = 1, z_3 = \infty, w_1 = -1, w_2 = -i, w_3 = 1$.

As $z_3 = \infty$, so (1) can be written as

$$\frac{(w-w_1)(w_2-w_3)}{(w-w_3)(w_2-w_1)} = \frac{(z-z_1)\left(\frac{z_2}{z_3}-1\right)}{\left(\frac{z}{z_3}-1\right)(z_2-z_1)}$$

i.e., $$\frac{(w+1)(-i-1)}{(w-1)(-i+1)} = \frac{(z-0)(0-1)}{(0-1)(1-0)}$$

$$z = \frac{w+1}{w-1} \cdot \frac{(-i-1)(-i-1)}{(-i)^2-(1)^2}$$

$$= \frac{w+1}{w-1} \cdot \frac{i^2+2i+1}{-2} = -i\frac{w+1}{w-1}$$

$\therefore$ $$z = i\left(\frac{1+w}{1-w}\right).$$

(n) we know

$P_0(x) = 1$ [For details see page no. 4-102, Ex. 1]

$\therefore$ $$\int_{-1}^{1} P_0(x)\,dx = \int_{-1}^{1} 1.dx = [x]_{-1}^{1} = 2$$

(o) The Bessel's equation of order 2 is

$$x^2\frac{d^2y}{dx^2}+x\frac{dy}{dx}+\left(x^2-2^2\right)y=0$$

i.e., $$x^2\frac{d^2y}{dx^2}+x\frac{dy}{dx}+\left(x^2-4\right)y=0$$

Module II

2. (a) Fourier series of $f(x)$ is

$$\frac{a_0}{2}+\sum_{n=1}^{\infty}(a_n\cos nx+b_n\sin nx)_i$$

where $a_0=\frac{1}{\pi}\int_{-\pi}^{\pi}f(x)\,dx=\frac{1}{\pi}\int_{-\pi}^{\pi}x\,dx$

$$=\frac{1}{\pi}\left[\frac{x^2}{2}\right]_{-\pi}^{\pi}=\frac{1}{\pi}\left(\frac{\pi^2}{2}-\frac{\pi^2}{2}\right)=0$$

$$a_n=\frac{1}{\pi}\int_{-\pi}^{\pi}f(x)\cos nx\,dx$$

$$=\frac{1}{\pi}\int_{-\pi}^{\pi}x\cos nx\,dx$$

$=0$ [$\because$ $x\cos nx$ is odd function]

$$b_n=\frac{1}{\pi}\int_{-\pi}^{\pi}f(x)\sin nx\,dx$$

$$=\frac{1}{\pi}\int_{-\pi}^{\pi}x\sin nx\,dx$$

$$= \frac{2}{\pi}\int_0^{\pi} x \sin nx\, dx \qquad [\because\ x \sin nx \text{ is even function}]$$

$$= \frac{2}{\pi}\left\{\left[x\frac{-\cos nx}{n}\right]_0^{\pi} - \int_0^{\pi} 1.\left(-\frac{\cos nx}{n}\right)dx\right\}$$

$$= \frac{2}{\pi}\left\{\frac{-\pi\cos n\pi}{n} + \frac{1}{n}\int_0^{\pi}\cos nx\, dx\right\}$$

$$= \frac{2}{\pi}\left\{\frac{-\pi(-1)^n}{n} + \frac{1}{n^2}[\sin nx]_0^{\pi}\right\}$$

$$= \frac{2(-1)^{n+1}}{n}$$

the Fourier series is

$$\sum_{n=1}^{\infty}\frac{2(-1)^{n+1}}{n}\sin nx$$

$$= 2\left[\frac{\sin x}{1} - \frac{\sin 2x}{2} + \frac{\sin 3x}{3} - \cdots\right]$$

Since $f(x) = x$ satisfies Dirichlet condition and since $f(x)$ is continuous at $x = \frac{\pi}{2}$ therefore,

$$f\left(\frac{\pi}{2}\right) = 2\left[\frac{\sin\frac{\pi}{2}}{1} - \frac{\sin 2\cdot\frac{\pi}{2}}{2} - \frac{\sin 3\cdot\frac{\pi}{2}}{3} - \cdots\right]$$

or, $$\frac{\pi}{2} = 2\left[1 - \frac{1}{3} + \frac{1}{5} - \cdots\right]$$

or, $$1 - \frac{1}{3} + \frac{1}{5} - \cdots = \frac{\pi}{4}$$

(b) See Ex. 3, page 1-66.

3. (a) Fouricr series of $f(x)=x^2,\ -\pi\le x\le\pi$ is

$$\frac{a_0}{2}+\sum_{n=1}^{\infty}(a_n\cos nx+b_n\sin nx)$$

where $a_0=\frac{1}{\pi}\int_{-\pi}^{\pi}x^2dx=\frac{2}{\pi}\int_0^{\pi}x^2dx=\frac{2}{\pi}\left[\frac{x^3}{3}\right]_0^{\pi}=\frac{2\pi^2}{3}$

$$a_0=\frac{1}{\pi}\int_{-\pi}^{\pi}x^2\cos nx\,dx=\frac{2}{\pi}\int_0^{\pi}x^2\cos nx\,dx$$

Now, $\int_{-\pi}^{\pi}x^2\cos nx\,dx$

$$=\left[x^2\frac{\sin nx}{n}\right]_0^{\pi}-\int_0^{\pi}2x\cdot\frac{\sin nx}{n}$$

$$=-\frac{2}{n}\int_0^{\pi}x\sin nx\,dx$$

$$=-\frac{2}{n}\left\{\left[x\cdot\frac{-\cos nx}{n}\right]_0^{\pi}-\int_0^{\pi}1\cdot\frac{-\cos nx}{n}dx\right\}$$

$$=-\frac{2}{n}\left\{\frac{-\pi\cos n\pi}{n}+\frac{1}{n}\int_0^{\pi}\cos nx\,dx\right\}$$

$$=-\frac{2}{n}\left\{\frac{-\pi(-1)^n}{n}+\frac{1}{n}\left[\frac{\sin nx}{n}\right]_0^{\pi}\right\}$$

$$=\frac{2.\pi(-1)^n}{n^2}$$

$\therefore\quad a_n=\frac{2}{\pi}\cdot\frac{2\pi(-1)^n}{n^2}=\frac{4(-1)^n}{n^2}$

and $\quad b_n=\frac{1}{\pi}\int_{-\pi}^{\pi}x^2\sin nx\,dx=0 \qquad \because x^2\sin xn$ is odd function.

∴ the Fourier series of $f(x)$ is

$$\frac{\pi^2}{3}+\sum_{n=1}^{\infty}\frac{4(-1)^n}{n^2}\cos nx=\frac{\pi^2}{3}+4\sum_{n=1}^{\infty}\frac{(-1)^n\cos nx}{n^2}$$

∵ $f(x)=x^2$ satisfies Dirichlet condition and continuous

$x=0$, the Fourier series is equal to $f(0)=0^2=0$.

∴ $$0=\frac{\pi^2}{3}+4\sum_{n=1}^{\infty}\frac{(-1)^n\cos 0}{n^2}$$

or, $$-\frac{\pi^2}{3}=4\sum_{n=1}^{\infty}\frac{(-1)^n}{n^2}$$

or, $$-\frac{\pi^2}{12}=\sum_{n=1}^{\infty}\frac{(-1)^n}{n^2}$$

or, $$-\frac{\pi^2}{12}=-\frac{1}{1^2}+\frac{1}{2^2}-\frac{1}{3^2}+\frac{1}{4^2}-\cdots$$

or, $$1-\frac{1}{2^2}+\frac{1}{3^2}-\frac{1}{4^2}+\cdots=\frac{\pi^2}{12}$$

(b) See Ex. 12, page 1-71.

4. (a) Let $I=\int_0^{2\pi}\frac{d\theta}{2+\cos\theta}$

$$=\int_c \frac{1}{2+\frac{1}{2}\left(z+\frac{1}{z}\right)}\frac{dz}{iz},$$ [putting $z=e^{i\theta}$ so that $d\theta=\frac{dz}{iz}$

and $\cos\theta=\frac{1}{2}\left(z+\frac{1}{z}\right)$ where c is the circle $|z|=1$.]

$$=\frac{2}{i}\int_c \frac{dz}{z^2+4z+1}$$

$$=\frac{2}{i}\int_c \frac{dz}{(z+2)^2-\left(\sqrt{3}\right)^2}$$

$$=\frac{2}{i}\int_c \frac{dz}{\left(z+2+\sqrt{3}\right)\left(z+2-\sqrt{3}\right)}$$

Let $f(z)=\dfrac{1}{\left(z+2+\sqrt{3}\right)\left(z+2-\sqrt{3}\right)}$

$=\dfrac{1}{(z+a)\left(z+\frac{1}{a}\right)}$ where $a=2+\sqrt{3}$

Then the poles of $f(z)$ are $-a, -\dfrac{1}{a}$ which are simple poles.

But $z=-\dfrac{1}{a}=-\dfrac{1}{2+\sqrt{3}}=-\left(2-\sqrt{3}\right)$

$\therefore$ $z=-\dfrac{1}{a}$ lies within the circle C where as

$z = a$ lies outside the circle.

Hence by residue theorem,

$$\int_c f(z)dz=2\pi i\times \text{Res}\left(-\frac{1}{a}\right)$$

Now, $\text{Res}\left(-\dfrac{1}{a}\right)=\lim\limits_{z\to-\frac{1}{a}}\left(z+\dfrac{1}{a}\right)f(z)$

$$=\lim_{z\to-\frac{1}{a}}\left(\frac{1}{z+a}\right)=\frac{1}{-\frac{1}{a}+a}$$

$$=\frac{1}{-\left(2-\sqrt{3}\right)+2+\sqrt{3}}=\frac{1}{2\sqrt{3}}$$

$$\therefore \quad \int_c f(z)\,dz = 2\pi i \times \frac{1}{2\sqrt{3}} = \frac{\pi i}{\sqrt{3}}$$

$$\therefore \quad I = \frac{2}{i} \times \frac{\pi i}{\sqrt{3}} = \frac{2\pi}{\sqrt{3}}$$

$$\therefore \quad \int_0^{2\pi} \frac{d\theta}{2+\cos\theta} = \frac{2\pi}{\sqrt{3}}$$

4. (b) Let $f(z) = \dfrac{e^{2z}}{(z+i)^4}$

The poles of $f(z)$ are given by $(z+1)^4 = 0$

i.e., $z = -1$ which is a pole of order 4 and lies within the circle $|z| = 3$. Hence by Cauchy's residue theorem, we have

$$\oint_c f(z)\,dz = 2\pi i \text{ Res }(-1)$$

Now Res $(-1) = \dfrac{1}{(4-1)!} \lim\limits_{z\to -1} \dfrac{d^3}{dz^3}\left[(z+1)^4 f(z)\right]$

$$= \frac{1}{6} \lim_{z\to -1} \frac{d^3}{dz^3}\left(e^{2z}\right)$$

$$= \frac{1}{6} \lim_{z\to -1} \left(8e^{2z}\right) = \frac{4}{3}e^{-2}$$

$$\therefore \quad \oint_c f(z)\,dz = 2\pi i \times \frac{4}{3}e^{-2}$$

$$\therefore \quad \oint_c \frac{e^{2z}}{(z+1)^4}\,dz = \frac{8\pi i}{3}e^{-2}$$

5. (a) Now, $u = x^2 - y^2$

$$\frac{\partial u}{\partial x} = 2x, \quad \frac{\partial^2 u}{\partial x^2} = 2$$

$$\frac{\partial u}{\partial y} = 2y, \quad \frac{\partial^2 u}{\partial y^2} = -2$$

$$\therefore \quad \frac{\partial^2 u}{\partial x^2} + \frac{\partial^2 u}{\partial y^2} = 2 - 2 = 0$$

$\therefore$ *u* is harmonic.

Again, $v = -\dfrac{y}{x^2 + y^2}$

$$\therefore \quad \frac{\partial v}{\partial x} = \frac{y.2x}{\left(x^2 + y^2\right)^2}$$

$$\frac{\partial^2 v}{\partial x^2} = 2y\left\{\frac{1}{\left(x^2 + y^2\right)^2} - \frac{4x^2}{\left(x^2 + y^2\right)^2}\right\}$$

$$\frac{\partial v}{\partial y} = -\left\{\frac{1}{x^2 + y^2} - \frac{2y^2}{\left(x^2 + y^2\right)^2}\right\}$$

$$= -\frac{x^2 - y^2}{\left(x^2 + y^2\right)^2}$$

$$\therefore \quad \frac{\partial^2 v}{\partial y^2} = -\left\{\frac{-2y}{\left(x^2 + y^2\right)^2} - \frac{2\left(x^2 - y^2\right).2y}{\left(x^2 + y^2\right)^3}\right\}$$

$$= \frac{2y}{\left(x^2 + y^2\right)^2} + \frac{4y\left(x^2 - y^2\right)}{\left(x^2 + y^2\right)^3}$$

$$\therefore \quad \frac{\partial^2 v}{\partial x^2}+\frac{\partial^2 v}{\partial y^2}$$

$$=\frac{2y}{\left(x^2+y^2\right)^2}-\frac{8x^2y}{\left(x^2+y^2\right)^3}+\frac{2y}{\left(x^2+y^2\right)^2}+\frac{4y\left(x^2-y^2\right)}{\left(x^2+y^2\right)^3}$$

$$=\frac{4y}{\left(x^2+y^2\right)^2}-\frac{4x^2y}{\left(x^2+y^2\right)^3}-\frac{4y^3}{\left(x^2+y^2\right)^3}$$

$$=\frac{4y\left(x^2+y^2\right)-4x^2y-4y^3}{\left(x^2+y^2\right)^3}=0$$

$\therefore$ **v is harmonic.**

(b) The zeros of $f(z)$ are given by equating to zero the numerator of $f(z)$ i.e. by

$$z^5-1=0$$

or, $\quad z^5=1$

or, $\quad z^5=\cos 0+i\sin 0$

$$=\cos 2n\pi+i\sin 2n\pi,\ n=0,\ \pm 1,\ \pm 2,\cdots$$

$$\therefore \quad z=\left(\cos 2n\pi+i\sin 2n\pi\right)^{\frac{1}{5}},\ n=0,1,2,3,4$$

$$=\cos\frac{2n\pi}{5}+i\sin\frac{2n\pi}{5}$$

When $n=0,\ z=\cos 0+i\sin 0=1$

$$n=1,\ z=\cos\frac{2\pi}{5}+i\sin\frac{2\pi}{5}$$

$$n=2,\ z=\cos\frac{4\pi}{5}+i\sin\frac{4\pi}{5}$$